Studies in Computational Intelligence

Volume 1126

Series Editor

Janusz Kacprzyk, Polish Academy of Sciences, Warsaw, Poland

The series "Studies in Computational Intelligence" (SCI) publishes new developments and advances in the various areas of computational intelligence—quickly and with a high quality. The intent is to cover the theory, applications, and design methods of computational intelligence, as embedded in the fields of engineering, computer science, physics and life sciences, as well as the methodologies behind them. The series contains monographs, lecture notes and edited volumes in computational intelligence spanning the areas of neural networks, connectionist systems, genetic algorithms, evolutionary computation, artificial intelligence, cellular automata, self-organizing systems, soft computing, fuzzy systems, and hybrid intelligent systems. Of particular value to both the contributors and the readership are the short publication timeframe and the world-wide distribution, which enable both wide and rapid dissemination of research output.

Indexed by SCOPUS, DBLP, WTI AG (Switzerland), zbMATH, SCImago.

All books published in the series are submitted for consideration in Web of Science.

Boris Kovalerchuk · Kawa Nazemi ·
Răzvan Andonie · Nuno Datia · Ebad Banissi
Editors

Artificial Intelligence and Visualization: Advancing Visual Knowledge Discovery

Springer

Editors
Boris Kovalerchuk
Department of Computer Science
Central Washington University
Ellensburg, WA, USA

Kawa Nazemi
Human-Computer Interaction and Visual Analytics
Darmstadt University of Applied Sciences
Darmstadt, Germany

Răzvan Andonie
Department of Computer Science
Central Washington University
Ellensburg, WA, USA

Nuno Datia
Instituto Superior de Engenharia de Lisboa (ISEL)
Polytechnic Institute of Lisbon
Lisbon, Portugal

Ebad Banissi
Department of Informatics
London South Bank University
London, UK

ISSN 1860-949X ISSN 1860-9503 (electronic)
Studies in Computational Intelligence
ISBN 978-3-031-46551-2 ISBN 978-3-031-46549-9 (eBook)
https://doi.org/10.1007/978-3-031-46549-9

This Springer imprint is published by the registered company Springer Nature Switzerland AG
The registered company address is: Gewerbestrasse 11, 6330 Cham, Switzerland

Foreword

Chess, Go, ChatGPT—what next? Lately, we have witnessed great progress in systems using Artificial Intelligence (AI). An AI-based system not only learned how to play chess better than humans—one of the dreams of early AI research, they learned how to play Go (a much more complex game) much better than humans. AI-based autopilot systems can control airplanes much better than humans, and AI-based diagnostics can compete with even the best medical doctors. It is difficult to find an area of human activity where AI applications have not made a drastic difference.

Recently, ChatGPT showed the general public that computers can write essays, answer questions, design class syllabi, and perform many other tasks requiring a highly trained professional until recently. ChatGPT and other Large Language Models have been so successful that serious people are worried that AI-based computer systems will take over the world—unless we ban them right away.

The good news is that the successes of ChatGPT are exaggerated. If you ask this system to write an essay on Shakespeare, it will do a beautiful job. However, all professionals who played with it know these systems are not technically smart. A simple question about the computational complexity of algorithms—that all junior-level Computer Science students know how to answer—can lead to a wrong answer. Even in image recognition, where, in general, AI-based systems are very good, these systems often give completely wrong answers. So no, computer-based systems are not yet ready to take over the world.

This may be good news for non-technical folks, but this is a challenge to us in computer science. AI-based systems have many practical applications, including medicine and airplane control, where mistakes can have grave consequences. How can we ensure that the AI systems do not make mistakes—at least do not make "silly" mistakes that a human would never make?

Visualization—an under-used but promising approach. Many AI researchers are working on the above-mentioned challenge. They try to develop more innovative learning algorithms; they try to make sure that the AI systems not only generate answers but also generate explanations for these answers—so that by reviewing these explanations, we can detect suspicious cases when these explanations are not convincing.

However, this mainstream activity misses an important aspect of our communication, which can be of great value to AI and other computer-related activities. This aspect is very straightforward: to a human, the most considerable portion of communication comes from the eyes, the visual channel through which we get dynamic 3-D images. In contrast, most of our communication with computer systems—including typical communications with ChatGPT—is via words, i.e., via a *1-D channel*.

When we communicate with each other, we communicate in person or—which is almost the same nowadays—via Zoom. This is how we communicate in the twenty-first century. But when we communicate with a computer, we go back, in effect, to the *1-D world of telegraph communications*, the great nineteenth-century achievement. Most of our communications with computers lack efficient, sophisticated visualization aspects. Adding this aspect will help bring our communications with computers into the twenty-first century—and hopefully lead to another boost in AI-based systems' ability to help us.

While this connection between AI and visualization is *not (yet) mainstream*, researchers worldwide are already working in this direction, and this book provides a great overview of their activity.

Visualization and AI: challenges. Visualization can help us better perceive *AI's output*. Visualization can help us better explain *to the AI system what we want*. Visualization of data can help us better understand the problem—and provide better advice to the AI system (and thus, make it better trained). Visualization of data can even allow us to discover predictive models visually in a visualization space.

The objective is clear, but there are many challenges. One of the main challenges is that while it is easy to visualize 1-D or 2-D data, most real-world *data is multidimensional*. Of course, we can restrict ourselves to only two quantities and get one or more 2-D pictures—but this is a very lossy representation of multi-D data. Even in the 3-D case, reconstructing an object from its 2-D views is often a challenging puzzle.

The good news is that visualization does not have to be lossy. Many researchers—including many authors from this volume—have come up (and are still coming up) with ways to describe multi-D data in 2-D, in the form of graphs and other structures, without losing any information. Some of these representations are more intuitive, and some are somewhat less intuitive—but we need to use them because they are the only way to grasp the full complexity of real-world data. Many of us had to learn programming languages to communicate with computers; learning different visualizations is a similar must—and just like programming languages have become increasingly convenient to use, visualization techniques become more intuitive.

There is another important aspect to it. So far, we only mentioned very lossy traditional representation and a complex lossless representation; what if we have so much data that if we represent it all, we will not be able to perceive it—an important case of what young people now call TMI, Too Much Information. In such situations, we do not want to return to simple 2-D projections; we need intermediate methods between the current lossy techniques and lossless ones. Such methods are also being actively developed, and some of the book's chapters describe them.

Most of it is already useful. The research on the relationship between AI and visualization is extremely important for the future. But many of the results are useful already, and some of these applications are described in this book. It is difficult to describe all of them; let us mention applications to predicting road traffic, predicting seabird activities, helping law officers and social workers predict and decrease crime and other types of asocial behavior—and even helping researchers collaborate.

The book is highly recommended. Read this book, study this book, and if you feel interested, join the research—let us all work together for an even brighter future!

Vladik Kreinovich
Professor of Computer Science, University of Texas at El Paso
Vice President, International Fuzzy Systems Association IFSA
Vice President, European Association for Fuzzy Logic and Technology EUSFLAT
El Paso, USA

The original version of the book has been revised: A correction to this book can be found at https://doi.org/10.1007/978-3-031-46549-9_19

Introduction

Integrating the Artificial Intelligence/Machine Learning (AI/ML) and Visual Knowledge Discovery (VKD) domains has recently become more complex. We can observe new trends and developments compared with our previously edited book "Integrating Artificial Intelligence and Visualization for Visual Knowledge Discovery" (Springer, 2022).

A critical issue for deploying ML models for high-stakes decisions is the end user's trust and understanding of the model. There is an extensive discussion on what it means to understand the model. A simple definition is informing a user how the ML algorithm produced the model or predicted a given case in terms of the algorithm. Typically, these algorithm terms are unfamiliar and outside the user's domain of expertise. A better definition is that a user is informed why the model made a given prediction or decision in the user's domain terms referring to the domain knowledge. As such, it ensures the user comprehends and verifies the explanation. Visualization is a natural way to support such a clarification of AI/ML models.

There is a tendency for dual-way deep integration between AI/ML and visualization domains rather than one-way assistance. Recent results focus on a two-way workflow between human intelligence and AI by visualization.

Another observation is that this interactive framework is a powerful instrument for building and communicating trust in ways that fit each AI/ML process stage. Extracting knowledge that a human can trust is a central AI/ML research area. The visual interaction between the human and the model facilitates trust. Through visual human/machine feedback loops at different stages of building the model, a model is created, its parameters are interactively changed, and the effects exploration and model explanations are generated in the visualization space. This process can create trust in the extracted knowledge.

A debate around ChatGPT resulted from several years of progress in Transformer models for Natural Language Processing (NLP). Trained Large Language Models (LLM) like ChatGPT learn by being exposed to vast amounts of text data. During its training process, the model tries to predict the next word in a sentence based on the context provided by the preceding words. ChatGPT doesn't have a built-in visual representation of its knowledge. However, we can use visualization techniques (word embeddings, concept maps, knowledge graphs, sentence parsing) to represent certain aspects of its knowledge. The model's knowledge is primarily embedded in its parameters and weights. Visualization of Transformers is an active area of research that aims to understand and interpret the inner workings of transformer-based models like ChatGPT.

Nevertheless, we can see from the popularity of LLM, and ChatGPT in particular, that the User Interface (UI) is important to wider adoption. A well-designed UI can simplify complex AI functionality and make it accessible to a larger audience, regardless of technical expertise. By presenting AI features clearly and understandably, users can more effectively employ AI software. The same applies to visualizations. Furthermore, users get more out of AI-supported software if they spend less effort interacting. AI models are becoming increasingly complex, as they can perform complicated undertakings, like text or image generation. Therefore, transforming their complexity into interpretable information for humans depends not solely on performance metrics. Interpretability plays an important role in establishing dependability.

Interpretability describes how humans can comprehend and analyze an AI model's internal workings and operations. It emphasizes the transparency and comprehensibility of the system's internal mechanisms. Its primary objective is to offer insights into how the AI model takes in inputs, processes them, and generates outputs. This understanding aids in assessing the reliability, robustness, and trustworthiness of the system. In this context, the term explainability is often used. Explainability and interpretability are not independent concepts; they are interconnected and often are used interchangeably along with other terms like understandability, comprehensibility, and intelligibility. Some studies limit one of these terms only to a description of the inner workings of the ML models and use others with a wider meaning covering acceptance of the explanation by the different categories of users, including developers, end users, and domain experts. To avoid confusion of terms in natural language with technical AI/ML terminology, it will be better to use some specific notation for such terms with exact definitions for emerging AI/ML domains.

Explainable AI (XAI) aims to deliver AI outcomes that are understandable to end users. XAI techniques concentrate on developing ML methods that offer explanations, trustworthiness, and comprehensible rationales for decisions made by black-box models. In contrast, black-box models are characterized by their opaque and uninterpretable internal functioning or logic.

Many ML algorithms enhance their performance by minimizing a loss function, which evaluates their ability to predict outputs for unseen inputs. Algorithms such as regression trees, decision trees, k-Nearest Neighbor, or Bayesian classifiers can be interpreted based on their model parameters, structure, and rules. To optimize the loss function, researchers have turned to more complex (in terms of parameters) and powerful models, like deep learning models. These models have demonstrated success in various domains but are often considered black-box models due to their difficulty in interpretation and explanation. Black-box models fail to provide users with a clear understanding of the internal behavior, leading to uncertainty in the output results and a lack of confidence in the outcomes.

Getting such an explanation requires the user to understand the models, their input and output attributes, and visual and analytical representations. Often, it is challenging for heavily engineered attributes. In VKD, we discover knowledge from visual data representation and original input data using computational ML algorithms. The knowledge discovered from visual data representation can be called visual knowledge, which can be challenging to translate to other forms of knowledge, like NLP or mathematical formulas.

Data can be too complex to see the pattern directly in the visual data representation without significant algorithmic work, including applying traditional ML methods or developing new methods. The good news is that VKD is not only a methodology but a growing set of actual methods, many of which are presented in this volume.

Knowledge discovery has been a major source of technological progress for centuries. Recent advances in AI/ML allow a dramatic acceleration in knowledge discovery. AI/ML achieves it by extracting knowledge from multidimensional data while leveraging the power of modern computational systems. Visualization to communicate complex knowledge to humans, like geographic maps, has been another significant way of technological progress for centuries. While visualization systems can communicate some multidimensional information, their ability to visualize abstract numeric multidimensional data without information loss, which is needed for AI/ML, is a long-standing challenge. A significant portion of this volume is devoted to resolving this challenge. It demonstrates the progress with lossless visualization of multidimensional data and merges it with knowledge discovery in AI/ML to resolve urgent issues like explaining black-box AI/ML models for high-stakes decision problems.

We are witnessing a dramatic increase in the role of (1) interactive visual methods in AI/ML and (2) AI/ML in information visualization and visual analytics. It is the emergence of a new VKD domain. In (1), it is a penetration of visual methods to the computational core of an AI/ML model discovery process. It indicates the transition of visual methods from a supporting role to the key role of model discovery and explanation. In (2), it is a penetration of AI/ML methods to automate the generation of visualization much more thoroughly. As such, intuitive and interactive visuals are needed to move AI/ML modeling to become an end user's self-service to enable data understanding.

In this way, it leads to the emergence of a new term of visual ML model to complement traditional ML model discovery with computational ML algorithms, as well as ML generative visualization process to complement the term of ML generative models. Other coined terms are AI-assisted visualization and visualization-assisted AI. These terms point to AI and visualization as respective assistants playing a supporting role. From our viewpoint, the situation is dramatically changing. AI is moving to become a core way to produce visualizations, and visualization methods are penetrating the core of machine learning model discovery. Why do we witness such dramatic shifts? What are the benefits of visual ML model discovery? What are the advantages of the ML generative visualization process?

There is a need to bring additional human domain knowledge, that is not in the training data, to the core of the ML process to discover trustable ML models. We need domain experts in the driver's seat of the model development and explanation. The model explanation is fundamentally a human-centered activity, not just a pure algorithmic one. It should not be conducted without end users. The motivation in the generative visualization process is reversed. We want to automate the visualization process. The integration of these processes is also expected in the future.

VKD focuses on combining ML models with interactive visualization. It intends to solve tasks formalized at the ML level with visual means as a core of the model-building process. Here we have the following steps: (1) posing extremely uncertain real-world task, (2) data preparation and multidimensional visual exploratory data analysis, (3) ML model discovery and explanation with visual means as a core technique like using lossless visualization space, and (4) refining (1)–(3) to the level when the model can be deployed.

VKD is connected to Visual Analytics (VA), which offers a significant advantage in facilitating human-in-the-loop decision-making. Although AI algorithms excel at processing and analyzing data, they often lack the ability to incorporate human expertise and intuition into decision-making. Through VA, human experts can interact with and guide AI algorithms, providing valuable feedback and insights that enhance the accuracy and relevance of the analysis. The human-in-the-loop approach encompasses qualitative methods and model-agnostic quantitative XAI techniques that allow for parameter refinement. These methods enable experts to conduct in-depth analysis and interact with the underlying models in real time. VA can be viewed as a visual progression in the realm of knowledge discovery. Both VA and VKD processes begin with raw data, encompass data processing steps, and culminate in knowledge generation through human involvement.

The convergence of VA and VKD is another trend demonstrated in several chapters in this volume. With any new method to deal with ML tasks that visually fit the VKD paradigm, the current focus is on promising VKD methods in the lossless visualization spaces, which preserve all high-dimensional data in the 2-D/3-D visualization space. VKD has its own ML methods based on lossless General Line Coordinates presented in Section 1 of the book.

The main idea of General Line Coordinates (GLC) is to represent *n*-D data in 2-D or 3-D without loss of information, which is a long-standing visualization problem. It ensures one can fully restore each n-D data point from its 2-D or 3-D representation. The traditional dimension reduction and visualization methods like Principal Component Analysis are lossy, converting, say, 10-D point to 2-D point. The ability to restore from 2-D points to 10-D points is limited. Such dimension reduction also leads to corruption of the *n*-D distances, as proved by the Johnson-Lindenstrauss Lemma (1984). This loss of information can prevent us from discovering n-D patterns and predictive ML models using such lossy data representation. Therefore, preserving all n-D information in the visualization space is critical for the success of VKD.

The GLC approach converts each n-D point to a directed graph in 2-D/3-D. Typically these graphs are polylines. Before the GLC representations were proposed (in 2014), only parallel and radial coordinates were known as lossless visualization methods. In contrast, GLC allows drawing coordinate axes in various ways: curved, parallel, unparalleled, collocated, disconnected, etc. Examples can be found in Section 1 of the book. Different datasets may require different visualizations to make n-D patterns visible and discoverable visually. This is an advantage of expanding the class of lossless visualization methods by GLC, which creates lossless visualization space as a core medium for discovering predictive ML models.

Lossless visualizations of multidimensional data have multiple advantages for ML, including preserving all multidimensional information for model construction and ensuring that "the baby is not thrown out with the bath water." However, a more complex visualization has an important challenge, where each n-D point is represented as a graph. In contrast, lossy methods like t-SNE represent each n-D point as a single 2-D point in the visualization. This simplicity comes with a distortion of the n-D data distribution.

A promising future research direction is a combination of lossless and lossy methods. Lossy methods can decrease, say, 500 dimensions to 30 dimensions preserving important properties of 500-D data. Then, lossless methods can visualize data in 30-D space and discover visual knowledge in that space.

This volume includes four sections: the first focuses on VKD with lossless visualization and the core technological and conceptual developments that bring visual knowledge discovery to fruition. The second section focuses on VKD to support AI/ML in a variety of ways. The third section on visual analytics applications and case studies exemplifies progress in these topics. The final section provides applications of VKD with text mining and NLP tasks. These four sections intend to deliver a roadmap to knowledge discovery with visual shine. In all, 19 chapters of this book report on the work of over 40 researchers encompassing the listed topics.

Part I: Visualizing the Unseen: Unleashing Knowledge Discovery with Lossless Visualizations

Chapter "Interactive Decision Tree Creation and Enhancement with Complete Visualization for Explainable Modeling" by Kovalerchuk et al. addresses this challenge. It explains how to visualize, create, and enhance the Decision Trees (DT) as understandable models and explanations for the black-box machine learning models. Its advantage over prior methods, illustrated in the case studies, is in (1) unique full lossless visualizations of individual n-D points in DT and (2) a deeper analysis of relations between attributes, cases, and node thresholds in the DT data flow.

A new methodology for interpretable machine learning as self-service for users is presented in Chapter "Full High-Dimensional Intelligible Learning in 2-D Lossless Visualization Space" by Kovalerchuk and Phan. It is a complete ML approach in a 2-D visualization space that does not require traditional processing of n-D data in an abstract n-D space. This study: (1) developed a subset of lossless General Line Coordinates: static and dynamic In-line-Based Coordinates, (2) classification and regression algorithms in these coordinates, and (3) confirmed their value on benchmark datasets, including multiclass imbalanced high-resolution data to produce interpretable models more general than decision trees.

A long-standing challenge of building accurate, interpretable ML models for mixed data is explored in Chapter "Explainable Machine Learning for Categorical and Mixed Data with Lossless Visualization" by Kovalerchuk and McCoy. This chapter contributes to (1) numeric coding schemes for non-numeric attributes to get explainable ML models, (2) discovering accurate explainable ML models for categorical data with a new Sequential Rule Generation (SRG) algorithm, (3) validating the SRG algorithm in multiple experiments, (4) methods for lossless visualization of n-D non-numeric categorical data in parallel coordinates, and (5) visual rule discovery in these visualizations.

Chapter "Parallel Coordinates for Discovery of Interpretable Machine Learning Models" by Hayes and Kovalerchuk advances interpretable machine learning. It proposed and evaluated a new classifier algorithm, Hyper, for numeric data combined with Parallel Coordinates (PC). The chapter offers interpretable models based on n-D rectangles (hyperblocks) that are more general than decision trees. The main contribution of this chapter is in abilities: (1) to find interpretable hyperblocks in PC, (2) to visualize n-D data with missing values, which is not supported by traditional PC, and (3) to better prevent both overgeneralization and overfitting of data than with decision trees.

Chapter "Visual Knowledge Discovery with General Line Coordinates" by Huber et al. offers several methods to build, understand, and evaluate Machine Learning models, including explaining black-box models. It is based on expansions of lossless visual representations of n-D data known as General Line Coordinates: linear, nonlinear, interactive, worst-case, and dynamic scaffolding. Its benefits, confirmed in experiments, are in (1) building accurate, interpretable models using interpretable n-D rectangles (hyperblocks), (2) the ability to analyze interpretability weaknesses of models, and (3) human-guided visual knowledge discovery with input expert knowledge.

Part II: Unveiling Insights: Empowering AI/ML Through Visual Knowledge Discovery

Chapter "Search Space Visualizations in the Context of Optimization Problems: A Literature Review" by Silva et al. presents a growing ability of interactive visualization methods to foster solutions to single and multi-objective optimization problems. Optimization is an integral part of building a machine learning model, too. Therefore, visualization of the search space and guiding search algorithms toward efficient solutions is a way to facilitate ML. It is especially critical for ML problems that rely on solving a multi-objective optimization problem, where the solution concept is very user- and task-specific.

Chapter "Relative Confusion Matrix: An Efficient Visualization for the Comparison of Classification Models" by Pommé et al. contributes to evaluating the quality of ML models. It proposes enhancing the performance metrics by introducing a Relative Confusion Matrix with visualization to compare ML models. It is a matrix-based visualization with colors and symbols highlighting class-wise performance differences between two models. The chapter shows that it is more efficient than existing approaches and applicable to pruning Deep Neural Networks.

The arrangement of points in a $\sqrt{n} \times \sqrt{n}$-sized grid offers an efficient and overlap-free way to visualize data. By making use of the Voronoi Relaxation method, Halnaut et al. provide in Chapter "Computation of Pixel-Oriented Grid Layout for 2D Datasets Using VRGrid" the Grid Layout method for dimensionality reduction within the context of Visual Knowledge Discovery. The visualization for dimension reduction eliminates visual occlusion and enhances the scalability of the visualization of high-dimensional optimization.

Kosaka and Itoh propose visualizing differences among 2-D projections of *n*-D labeled datasets produced by t-SNE in Chapter "A Method for Comparative Visualization of Labeled Multidimensional Data and Its Application to Machine Learning Data." It helps to judge how reliable different datasets are for building ML models to predict new cases by analyzing the differences in t-SNE data distributions. The t-SNE 2-D projections of images of different sizes of handwritten digits 0 and 1 from two datasets (MNIST and USPS) illustrate it.

Part III: Illuminating Insights: Visual Analytics Applications and Case Studies

Chapter "Road Traffic Flow Prediction with Visual Analytics" by Datia et al. addresses the challenge of accurate traffic flow prediction to improve traffic management, reduce congestion and pollution, and increase road safety with a real-time visual analytics solution. It proposes traffic flow prediction using a machine learning technique with the XGBoost algorithm for a short-term time delay estimation for a region of interest. It achieved a low Mean Absolute Error. The results suggest the approach could be helpful for traffic management and transportation planning.

In Chapter "Guided Visual Analytics—A Visual Analytics Guidance Approach for Systematic Reviews in Research," Sina et al. propose an AI-based visual guidance system that simplifies interaction with complex visual analytics systems through unsupervised machine learning approaches by predicting the next steps. Thereby unsupervised learning approaches are used to extract information from text. They provide a proof-of-concept through their application for generating systematic literature reviews through a visual analytics system.

Blazevic et al. propose in their Chapter "Integrating Machine Learning in Visual Analytics for Supporting Collaboration in Science" an application that showcases how AI, VA, NLP, and recommendation systems are used to support the general research process. The proposed approach addresses the need for comprehensive assistance throughout the entire research process using unsupervised learning. It includes real-time visual recommendations, similarity measurements, ideation assistance, visual exploration, and stimuli promotion allowing to mitigate the adverse effects of multitasking and multi-application for this task.

Lettieri et al., in Chapter "Visual Knowledge Discovery and Criminal Justice. Insights from a Computational Crime Analysis Research", examine the application of knowledge discovery through a visualization-developed system—CrimeMiner—in which various media—visual, audio, and textual document of different kinds. A legal and criminal case study demonstrates that such approaches easily apply to other Humanities and Social Science domains with computational heuristics.

Part IV: Illuminating Text: Unleashing Knowledge with Visual Discovery in Text Mining and Natural Language Processing

Several visual explanation methods for NLP models were recently proposed, and a major challenge is how to compare the performances of such methods. Chapter "Designing and Evaluating Context-Sensitive Visualization Models for Deep Learning Text Classifiers" by Dunn et al. investigates how to evaluate explanations resulting from visualizations of NLP models for text classification. It describes the application of several methods for evaluating text visualizations' quality, including automated techniques based on quantifiable measures and subjective techniques based on human judgments.

Pedro et al. explore, in Chapter "How NLP and Visual Analytics Can Improve Asset Management," the use of NLP to automate the analysis of maintenance work orders in asset management. Using NLP algorithms and word embedding models such as Fasttext, semantic patterns within work order descriptions can be effectively captured. The results demonstrate the superiority of Fasttext in extracting semantic information, enabling the development of machine learning algorithms for concise work order summarization using a predefined set of words.

Blazevic et al. propose in their Chapter "Similarity in Visual Analytics—A Visual Analytics Approach for Finding Similar Publications" an AI-based approach to assisting researchers in their search for relevant publications. Their approach focuses on utilizing researchers' annotation and exploratory behavior to identify their interests and, based on these interests, recommend relevant publications. VA is utilized to adequately present recommendations, while NLP approaches such as topic modeling provide the foundation for similarity analyses to identify publications of interest.

Chapter "Information Visualization with ChatGPT" investigates the capabilities of the Deep Learning chatbot "ChatGPT" for generating information visualization in comparison with OpenAI Codex and a classical natural language processing approach based on Natural Language for Data Visualization (NL4DV) toolkit. It has shown that users can create more complex visualizations with the Deep Learning approaches but sometimes need help finding the appropriate text input to solve the tasks.

Nakai and Itoh explore in Chapter "Visualization of Similarity Distribution of Lyric Collections" the use of lyric-based music search and analysis. Visualization techniques display the similarity distribution of song lyrics, aiding in exploratory searches. The study benefits music listeners, industry workers, and songwriters, enabling them to find similar songs, understand lyric trends, and use them as a reference for songwriting. The visualization utilizes scatterplots with dimensionality reduction. The chapter presents additional studies using guided dimensionality reduction and examines the differences in word selections between male and female songwriters.

Target Audience

This book addresses the interests of industry and academia researchers focusing on AI/ML and visualization that reflect various concepts, uses, and insights from the knowledge discovery and visualization fields. It demonstrates how Visual Knowledge Discovery and Artificial Intelligence have jointly evolved during the last 2 years, providing insights into some of the most exciting application areas.

Boris Kovalerchuk
Kawa Nazemi
Răzvan Andonie
Nuno Datia
Ebad Banissi

Contents

Visualizing the Unseen: Unleashing Knowledge Discovery with Lossless Visualizations

Interactive Decision Tree Creation and Enhancement with Complete Visualization for Explainable Modeling

Boris Kovalerchuk, Andrew Dunn, Alex Worland, and Sridevi Wagle

1 Introduction

Advanced methods are needed for the evaluation and improvement of Machine Learning (ML) models, including their interpretability and prediction accuracy [1, 2]. Visualization plays an important role in this [3]. Decision trees (DTs) are among a few machine learning methods, which are commonly recognized as interpretable models. Decision trees are also used to interpolate much more complex black box machine learning models. However, there are several difficulties known with the decision trees. They may not be stable and there are multiple different decision trees that have very similar accuracy and interpretation power. Therefore, efficient visualization of the decision trees can help significantly in resolving these issues, which is the focus of this work.

While decision trees are important as was outlined above, the DT topic is a part of a more general issue of visual knowledge discovery (VKD). The goal of the visual knowledge discovery is tight integration of artificial intelligence and machine learning methods with advanced visualization methods for efficient knowledge discovery [1, 4]. One of the focuses of this integration is incorporating visual methods to the core of the model discovery, enhancement, and evaluation process to make this process visual as much as possible.

B. Kovalerchuk (✉) · A. Dunn · A. Worland · S. Wagle
Department of Computer Science, Central Washington University, Ellensburg, WA, USA
e-mail: Boris.Kovalerchuk@cwu.edu

A. Dunn
e-mail: Andrew.Dunn@cwu.edu

A. Worland
e-mail: Alex.Worland@cwu.edu

S. Wagle
e-mail: Sridevi.Wagle@cwu.edu

B. Kovalerchuk et al. (eds.), *Artificial Intelligence and Visualization: Advancing Visual Knowledge Discovery*, Studies in Computational Intelligence 1126,
https://doi.org/10.1007/978-3-031-46549-9_1

In the current practice visualization often is a separate stage for the models already built by analytical methods like decision trees. Several important characteristics of the proposed approach demonstrate that it is leaning to the core of the AI/ML modeling process with visual means. First it allows building decision trees interactively visually without using analytical machine learning algorithms. In includes interactive changing of thresholds. Next, it allows losslessly visualize all training and validation data along with the DT model. This is a way to see all borderline cases to be able to adjust decision tree interactively to make it more robust. Moreover, it opens an opportunity for discovering visually models, which are more general than decision trees as it is shown in Sect. 5.

We offer new visualization methods BC-DT and SPC-DT. The concept of General Line Coordinates (GLC) [1, 5] is behind both methods. Specifically, BC-DT and SPC-DT methods rely on new Bended Coordinates (BC) and the Shifted Paired Coordinates (SPC), respectively. We also generalize these methods to several other GLCs. The Bended Coordinates are line coordinates that are bended in a specific point. In the BC-DT, this point is a threshold assigned to the given node of the DT. As the name indicates, SPC is set of shifted pairs of Cartesian coordinates. In SPC, a directed graph (digraph) visualizes an n-D point $\mathbf{x}$ in 2-D with nodes formed by consecutive pairs (x_i, x_{i+1}) of values of coordinates of $\mathbf{x}$, connected by directed edges.

These methods expand capabilities available in the traditional visualization of DT models by visualizing: (1) relations between attributes, (2) individual cases relative to the DT, (3) data flow in the DT, (4) how tight is each split threshold in the DT nodes, and (5) density of cases in parts of the n-D space. This information has a significant value for domain experts in evaluating and improving DT models.

In 2011, [6] reported on about 200 different DT representations; today, Treevis.net displays over 300 DTs. However, typically they are not detailed enough to represent the machine learning models. For example, the split thresholds in the DT machine learning models are not visible using a Treemap [7] visualization, which is particularly effective for displaying big, complicated trees. Some DT visualizations for ML present the number of cases in each node or a marginal distribution of the cases [8–10].

The decision tree's common picture, as displayed in Fig. 1, merely shows the DT's overall structure. Using the representation in Fig. 1, a new n-D point may be predicted by following a classification prediction from the root to the leaf. This does not adequately convey the accuracy of this forecast, though. The DT visualization shown in Fig. 2 from [11] is also limited not showing individual cases. We need additional knowledge to answer such questions, which is not present in such visualizations.

The fact that DT node thresholds are not unique is well known [12, 13]. We can maintain the same DT prediction on training and/or validation data by changing thresholds within certain constraints. The fluctuation of the thresholds makes prediction of new cases close to the threshold instable, which can change the predicted class dramatically. The study of these scenarios is difficult with current decision tree visualization techniques.

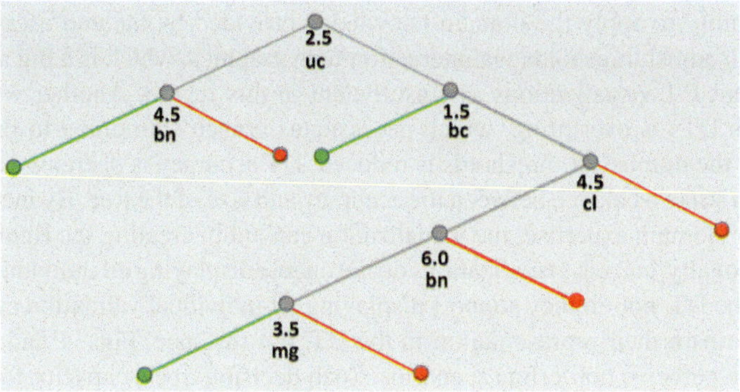

Fig. 1 Traditional decision tree representation of Wisconsin Breast Cancer (WBC) data. The benign class is denoted by green edges and nodes, while the malignant class is denoted by red edges and nodes. Cases that have not yet reached the DT terminal level are shown by gray edges and nodes

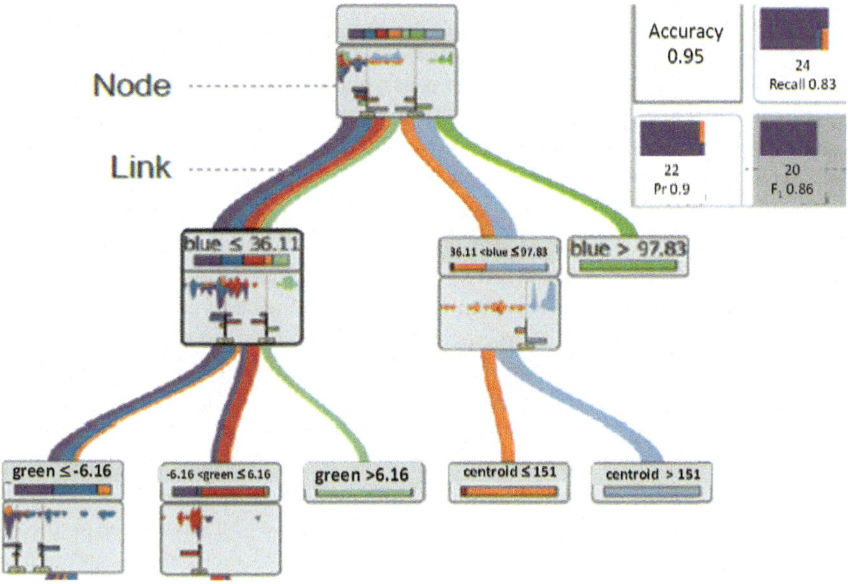

Fig. 2 DT visualization as streams [11]

Tight thresholds [12, 13] are a concern when cases from opposing classes overlap and are near to the threshold. Setting up the proper criteria automatically is challenging. As we just examined above about new borderline cases, a slight change in the threshold can result in classification changes for many training, validation, and new cases, making the prediction incorrect. Thus, for the tight borders, many training and validation cases are borderline cases that needs to be analyzed.

The ability to apply the domain knowledge provided by the end users makes it possible to adjust thresholds via interaction with graphics, which is a big advantage. The current DT visualizations are insufficient in this regard. Another well-known issue with DTs is overfitting, which necessitates branch shortening to prune DTs. Although the number of thresholds is reduced, DT accuracy is decreased. It is challenging to strike a balance between the accuracy and size of the tree. By including the end users' domain expertise, the visualization can aid in creating the right balance.

Additionally, current visualizations do not enable displaying all individual training cases in the DT, nor do they support displaying all individual validation cases in the DT to compare their representation in the DT. For instance, Figs. 1 and 2 make it difficult to perceive borderline scenarios. Such decision tree's capacity for in-depth examination and improvement is constrained by these flaws.

The suggested methods enable the visualization of all training, validation, and test data in the DT, individually or collectively. Since this visualization is lossless [1, 5], it will reveal relations concerning training, validation and test data and will assess the stability and trustworthiness of the decision tree. It also makes it possible to see data outliers.

Convincing the user that the DT prediction can be trusted is the challenge for the decision tree model. Unfortunately, the user's ability to be confident is constrained by the current visualization techniques. To solve this problem, the suggested method enables the decision tree to trace any new case in relation to the training and validation data. Also, this aids in justifying or modifying the DT thresholds for borderline cases.

Overfitting is a prominent issue with decision trees, although overgeneralization is less visible but still crucial [14]. Consider attribute x_1 in the range [0,10] and a branch of the DT, which states that if $x_1 \leq 2.5$, then the case belongs to class 1. Next, no case of class 1 has x_i that is less than 1.5 and all training cases are within that range [1.5, 2.5]. The decision three overgeneralized cases of class 1 to the interval [0, 2.5]. The suggested visualization technique makes it possible to detect overgeneralization and adjust the range to its true bounds.

This paper is organized as follows. Section 2 defines methods, Sects. 3 and 4 describe case studies on the benchmark datasets, Sect. 5 presents a discussion and generalizations of the methods and Sect. 6 summarizes the chapter.

2 Methods

2.1 Traditional Methods Versus Coordinate-Based Methods

Below the Bended Coordinates Decision Tree (BC-DT) method is explained, using the ID3 Decision Tree shown in Fig. 1 above in the traditional visualization, trained on 349 cases of for Wisconsin Breast Cancer (WBC) data [15]. These cases are about 50% of all WBC Data. Figure 3 and Table 1 describe this DT and its performance. While WBC data contain nine attributes, this decision tree uses only five attributes,

- ucellsize < 2.5
- bnuclei < 4.5 then class = **benign** (100.00 % of 200 cases)
- bnuclei ≥ 4.5 then class = **malignant** (66.67 % of 6 cases)
- ucellsize ≥ 2.5
- bchromatin < 1.5 then class = **benign** (87.50 % of 8 cases)
- bchromatin ≥ 1.5
    - clump < 4.5
    - bnuclei < 6.0
        - mgadhesion < 3.5 then class = **benign** (100.00 % of 5 cases)
        - mgadhesion ≥ 3.5 then class = **malignant** (66.67 % of 6 cases)
    - bnuclei ≥ 6.0 then class = **malignant** (100.00 % of 8 cases)
    - clump ≥ 4.5 then class = **malignant** (93.97 % of 116 cases)

Fig. 3 ID3 decision tree for WBC data

where the attribute bnuclei (bn) is used twice with two different splits. Thus, we have 5-D data in this DT that we represent as a 6-D point with bnuclei repeated.

The root of this tree is based on the attribute ucellsize (uc), the next node is based on the attribute bnuclei. The proposed methods focus on the abilities to see how the individual cases traverse through the DT, and how close they are to the thresholds associated with the DT nodes. The traditional DT visualization, in Fig. 1, has no such capability. This goal was partially reached in [16] by showing n-D data in parallel coordinates side-by-side with a traditional DT (see Fig. 4a), but not in the tree itself. It is interactive and a user can click on any node in the tree to see data reaching the node in the parallel coordinates in a separate coordinated plot.

This is an example of the use of coordinate-based methods to visualize DTs, because representation of n-D data in parallel coordinates is one of the coordinate-based methods. All these methods belong to the class of lossless General Line Coordinates [1, 5]. Another example of combining coordinate-based methods with DTs is presented in [17] using Kandogan' Star Coordinates [18] and Inselberg's Parallel Coordinates [19]. It allows a user interactively building a DT in these visualizations. To get a better data visualization a user can change Star Coordinates by moving each axis around by clicking on the endpoint of the selected axis and dragging it to the desired position.

This method produces traditional interpretable decision tree like in [16, 20] when it uses parallel coordinates. However, when it uses star coordinates the situation is very different. The result is *not a traditional decision tree.* It is a hierarchical tree

Table 1 ID3 DT performance

Error rate			0.0716			
Values prediction			Confusion matrix			
Value	Recall	1-precision		Benign	Malignant	Sum
Benign	0.9056	0.0194	Benign	202	21	223
Malignant	0.9683	0.1469	Malignant	4	122	126
			Sum	206	143	349

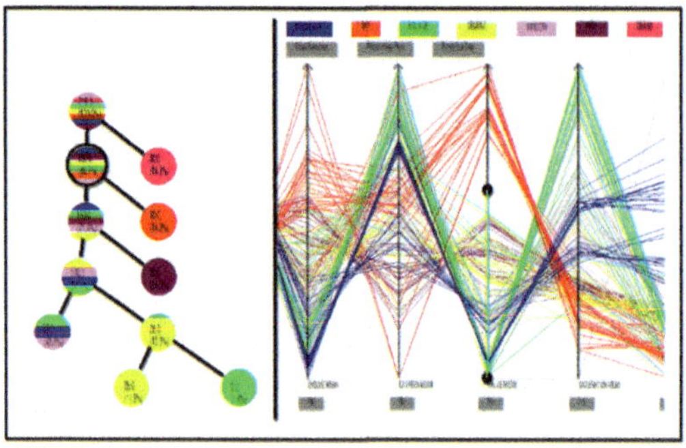

(a) n-D data in Parallel Coordinates side-by-side with a traditional DT [21].

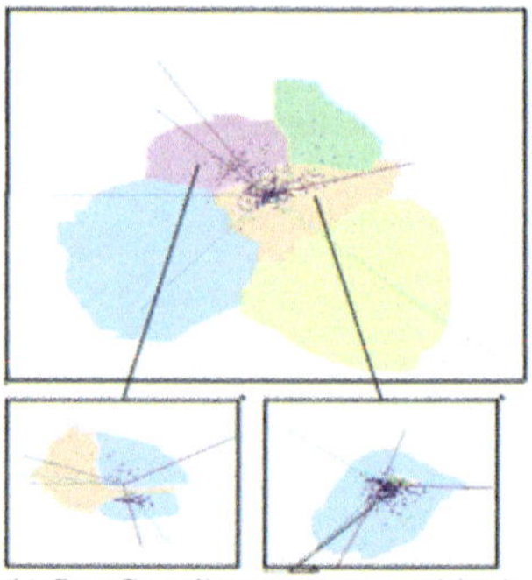

(b) Star Coordinates space partitioning to build a tree structure [17].

Fig. 4 Alternative coordinate-based methods

structure that partitions the star coordinate space. The points in this space are vector sums of the star coordinates. The application domain interpretation of these sums is not clear, in contrast with simple interpretable relations like $x_i > T_i$, where T_i is a threshold value for individual coordinate/attribute x_i captured in the traditional DTs. In [21] a method to interpret areas like found in [17] are converted to interpretable rules. Thus, an extra effort is needed to use models with Star Coordinates from [17] to derive interpretable models. In contrast with Fig. 4a, which shows a tradition DT visualization with underlying data side-by-side in parallel coordinates from [16], Fig. 4b illustrates the Star Coordinates space partitioning from [17] for a hierarchical tree structure.

In contrast Fig. 5 illustrates the method that we propose to combine DT and parallel coordinates in a single plot. It is the same DT as in Fig. 1. This visualization allows to trace individual cases in the decision three in contrast with Fig. 1. We will call this type of visualization as Parallel Coordinate Decision Tree (PC-DT). While this method allows to trace individual cases, it does not show the traditional decision tree structure. In the next section we present BC-DT method for this.

2.2 Bended Coordinates Decision Tree (BC-DT) Method

The BC-DT method augments traditional DT visualization, by using edges of DT as coordinates, which are bended in the nodes at the threshold values. Thus, this is a coordinate-based method. Figures 6 and 7 show two different ways how coordinates are bent at the threshold value.

For instance, the threshold $T = 2.5$ on coordinate uc with the interval of values [1, 10] leads to the left interval [1, 2.5) and the right interval [2.5,10]. In Fig. 6, these two unequal intervals are visualized with equal lengths, i.e., forming disproportional scale. In Fig. 7a, the lengths of edges are 1.5 to 7.5 reflecting the location of the

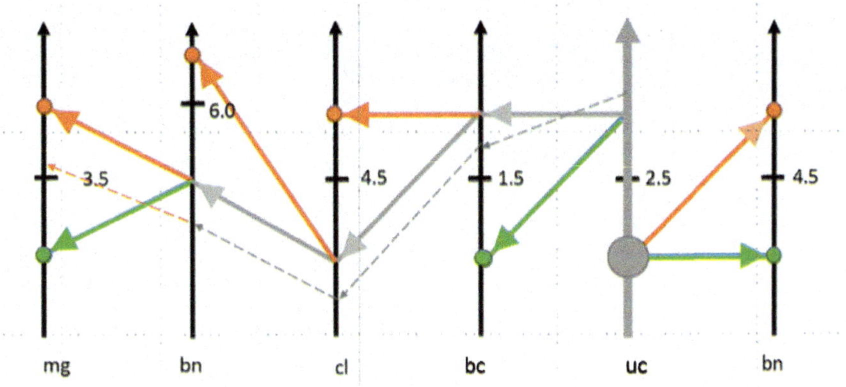

Fig. 5 DT visualized in parallel coordinates where coordinate uc is used as the root and the dotted line showing an individual case traced in the DT

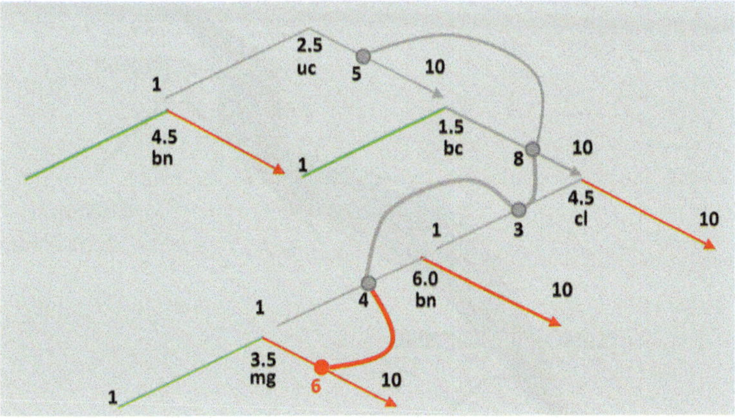

Fig. 6 Decision tree in BC-DT with edges serving as bended coordinates in disproportional scales for Wisconsin breast cancer data. The curved solid line show 5-D point (uc, bc, cl, bn, mg) = (5, 8, 3, 4, 6) that reached a point mg = 6 on this malignant edge of the DT

threshold $T = 2.5$ in the interval $[1, 10]$. In Fig. 7a, the dotted lines extend edges to make DT symmetric resembling the traditional DT visualization. Both Figs. 6 and 7 show how a 5-D case is traversed in the DT reaching a malignant node indicated by the red arrow leading to it.

Both Figs. 6 and 7a show 5-D point (uc, bc, cl, bn, mg) = (5, 8, 3, 4, 6). The distances from the thresholds to these values are different for different nodes. The shorter distances lead to less certainty in the classification of this 5-D point to the malignant class. The longer distances increase the confidence in the classification Relative to given thresholds. The unequal shoulders of the edges in Fig, 7a allow to see these differences undistorted in contrast with disproportional scaling in

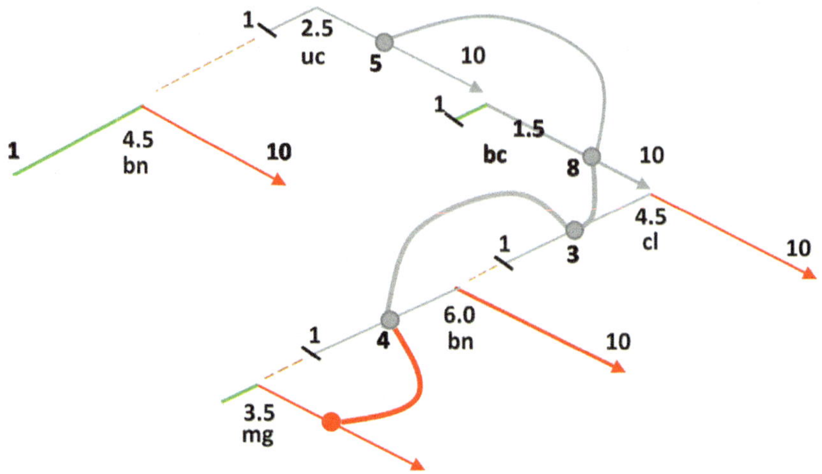

(a) 5-D point (5, 8, 3, 4, 6) in BC-DT that reaches the terminal malignant edge of the DT

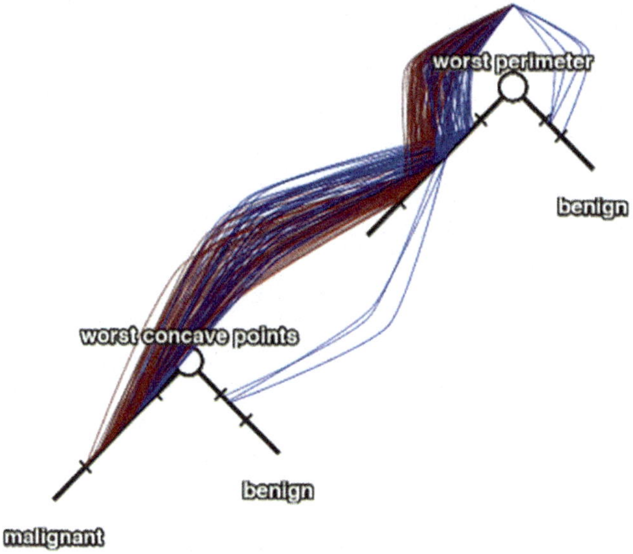

(b) Several cases in BC-DT in proportional scales

Fig. 7 BC-DT with proportional edges serving as bended coordinates

Fig. 6. Figure 7b also shows DT with unequal shoulders of edges with undistorted distribution of many cases in this DT.

The BC-DT method expands and complements capabilities available in the traditional visualization of DT models by visualizing: (1) individual cases relative to the DT, (2) detailed data flow in the DT, and (3) the tightness of each split threshold in the DT nodes. This information is of a significant value to domain experts in evaluating and improving DT models. To the best of our knowledge, none of the existing DT visualization methods support all these capabilities. Section 4 provides case studies with BC-DT visualization on several benchmark datasets.

2.3 Shifted Paired Coordinates—Decision Tree (SPC-DT) Method

The SPC-DT approach is also based on the concept of General Line Coordinates (GLC) [1]. The technique specifically uses the Shifted Paired Coordinates (SPC) representation. SPC, as its name suggests, is a collection of shifted pairs of Cartesian coordinates. An n-D point **x** is represented in 2-D by a directed graph (digraph) in SPC, where nodes are made up of consecutive pairs of **x**'s coordinate values (x_i, x_{i+1}) and edges connect them.

The SPC-DT technique enhances and completes the capabilities of the traditional DT model expanding [22] by visualizing: (1) relations between attributes, (2) individual cases relative to the DT, (3) detailed data flow in the DT, (4) the tightness of each split threshold in the DT nodes, and (5) 2-D density of cases in areas of the n-D space. Experts in the field can greatly benefit from this knowledge in the process of evaluating and enhancing DT models. To the best of our knowledge, none of the DT visualization techniques now in use provide all of these capabilities.

Below, a conceptual explanation of the Shifted Paired Coordinates-Decision Tree (SPC-DT) method is followed by examples. In order to create a DT visualization from a given decision tree model, the SPC-DT process must go through the following major steps:

(1) Parsing the DT model.
(2) Pairing attributes.
(3) Building a set of paired Cartesian coordinates.
(4) Drawing each pair of coordinates in the default location shifted one over another.
(5) Mapping a part of the decision tree to a respective pair of coordinates as rectangles based on the threshold values.
(6) Color rectangles according to classes.
(7) Drawing collection of n-D points as directed graphs in the SPC.
(8) Interactive modification of visualization. The options include:
    (a) change a set of visualized n-D points,
    (b) change the location of a coordinate pair,
    (c) change colors of the classes,
    (d) negate/flip coordinates,
    (e) swap vertical and horizontal coordinates,
    (f) condense points in a rectangle,
    (g) show overall DT structure,
    (h) adjust thresholds, and

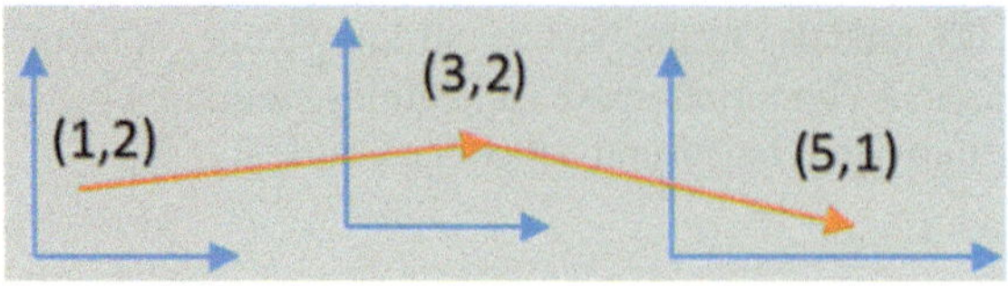

Fig. 8 6-D point $x = (x_1, x_2, \ldots, x_6) = (1,2,3,2,5,1)$ in shifted paired coordinates (SPC) visualization

(i) compute accuracy after adjusting thresholds.

The following example explains Shifted Paired Coordinates (SPC) in more detail. Take the 6-D point $\mathbf{x} = (x_1, x_2, \ldots, x_6) = (1, 2, 3, 2, 5, 1)$. In SPC representation, a point in the first pair of coordinates $(x_1, x_2) = (1, 2)$ is visualized (X_1, X_2). In the second set of coordinates (X_3, X_4), the second pair $(x_3, x_4) = (3, 2)$ is visualized, and the third pair $(x_5, x_6) = (5, 1)$ is visualized (X_5, X_6). In Fig. 8, a directed graph (1, 2) (3, 2) (5, 1) is formed by connecting these three points, which are shifted relative to one another to prevent overlap. Thus, shifted pairs of coordinates are generated.

The decision tree in SPC-DT that was designed to visualize the WBC data is depicted in Fig. 9. Performance of the tree is shown in Table 2, and initial placements for each coordinate pair are shown for all WBC data in SPC-DT in Fig. 10. The gray areas on each paired coordinate space map denote areas where a case's class cannot be determined at the current DT level and must be followed to the next DT level to be evaluated further.

To depict the various outcomes of instances in gray areas, several grayscale shades are used. There is just one destination plot for all cases that fall inside the same shade of gray. The DT has classified some cases as "malignant," while other cases have been classified as "benign," as indicated by the red and green areas. These cases terminate at the current level since a class has been determined.

The benefits of the SCP-DT visualization in Fig. 10 include:

1. relations between 2 attributes inside of each plot,
2. relations between 4 attributes using green and red arrows between two plots,
3. relations between all 6 attributes using green and red arrows between three plots and more attributes for larger dimensions,
4. the tree structure and data flow by arrows in the DT.

Many 10-D points in the SPC-DT are depicted in Fig. 10 as reaching the DT's terminal nodes. The proximity of such particular cases to each split threshold in the DT nodes can be seen on the graphs of those cases. Moreover, graphs for training instances make it possible to see how densely distributed the cases are in the n-D space. The number of instances in various areas of the space or the degree of class purity in the region can be displayed using color intensity (lighter/darker).

All these characteristics work together to boost trust in the DT model's ability to be understood and to predict accurately. Consider a DT where the actual training examples are all outside the threshold (border) area but a new case that needs to

- ucellsize < 2.5000
    - bnuclei < 4.5000
        - clump < 6.5000 then classe = **begnin** (100.00 % of 407 examples)
        - clump >= 6.5000 then classe = **malignant** (60.00 % of 5 examples)
    - bnuclei >= 4.5000 then classe = **malignant** (52.94 % of 17 examples)
- ucellsize >= 2.5000
    - ucellsize < 4.5000
        - bnuclei < 2.5000
            - normnucl < 2.5000 then classe = **begnin** (100.00 % of 19 examples)
            - normnucl >= 2.5000 then classe = **begnin** (54.55 % of 11 examples)
        - bnuclei >= 2.5000
            - clump < 6.5000
                - bchromatin < 3.5000 then classe = **begnin** (63.64 % of 11 examples)
                - bchromatin >= 3.5000 then classe = **malignant** (83.33 % of 18 examples)
            - clump >= 6.5000
                - normnucl < 7.0000 then classe = **malignant** (100.00 % of 24 examples)
                - normnucl >= 7.0000 then classe = **malignant** (88.89 % of 9 examples)
    - ucellsize >= 4.5000
        - clump < 6.5000
            - bnuclei < 8.5000
                - bchromatin < 4.5000 then classe = **begnin** (50.00 % of 8 examples)
                - bchromatin >= 4.5000
                    - mgadhesion < 8.5000 then classe = **malignant** (100.00 % of 16 examples)
                    - mgadhesion >= 8.5000 then classe = **malignant** (75.00 % of 4 examples)
            - bnuclei >= 8.5000 then classe = **malignant** (100.00 % of 44 examples)
        - clump >= 6.5000 then classe = **malignant** (100.00 % of 106 examples)

Fig. 9 Alternative ID3 decision tree for WBC data

Table 2 Performance of the decision tree from Fig. 9

Error rate			0.0401			
Values prediction			Confusion matrix			
Value	Recall	1-Precision		Benign	Malignant	Sum
Benign	0.9672	0.0285	Benign	443	15	458
Malignant	0.9461	0.0617	Malignant	13	228	241
			Sum	456	243	699

be forecasted is just on the border. Most likely, this DT model overgeneralized the training set, making the predictions excessively uncertain of such new case.

SPC-DT visualization aids domain experts in assessing the effectiveness, interpretation, overgeneralization, and overfitting of the DT model. The interactive SPC-DT can allow users to turn on and off "context" attributes, which are attributes that are not part of the tree but are a part of the input dataset. The tree root can be connected to these attributes. When such properties are disabled, they can be grayed out or made semitransparent so that the context may still be seen.

The SPC-DT design allows showing just subsets of cases when we need to visualize a large number of cases and classes. It can be "centers" of subsets, min or max cases of subsets with the remaining cases being gray or semitransparent to decrease occlusion and simplify visual patterns. The SPC-DT design also allows showing the

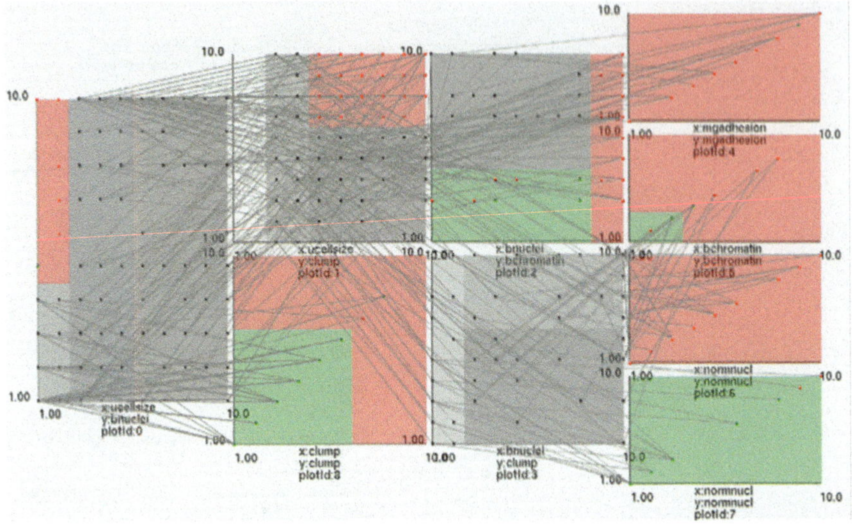

Fig. 10 WBC in default SPC-DT

error rate for each dataset, branch of the tree, and the confusion matrix to aid the user in making confident predictions.

3 BC-DT Case Studies

3.1 Case Study: Seeds Dataset

Below the Bended Coordinates DT approach (BC-DT) is presented as a case study with the seeds data. We show two versions for building the bended decision tree. The first one is based on the imported data and decision tree created outside of this program. In this case study the Tanagra system [23] was used as a source of the decision tree. In the second version, a user builds the DT interactively from scratch. A decision tree can be constructed either from scratch without any guidance or using a guidance form an existing DT, like the DT built in Tanagra [23] as a guiding prototype. A user can correct a prototype by changing thresholds and attributes or by removing some nodes or adding more nodes.

Bended DT visualization from imported DT model. Figure 11 illustrates importing data and trained DT. Next, Fig. 12. shows on the left the decision tree built using the training data with those data visualized on the DT. On the right it shows the same DTwith the testing data visualized on it. It allows comparing training and testing data side-by-side. Fig. 12 clearly shows similarity of the distribution of the training and validation data in the DT. They match each of each other quite well.

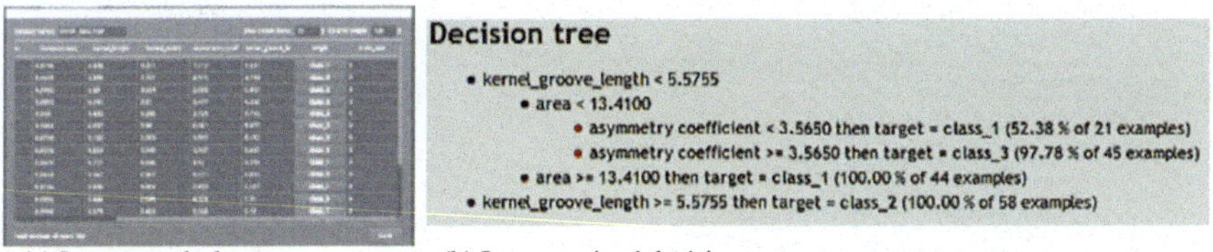

(a) Import seeds dataset. (b) Import trained decision tree.

Fig. 11 Import seeds data and trained decision tree from Tanagra system

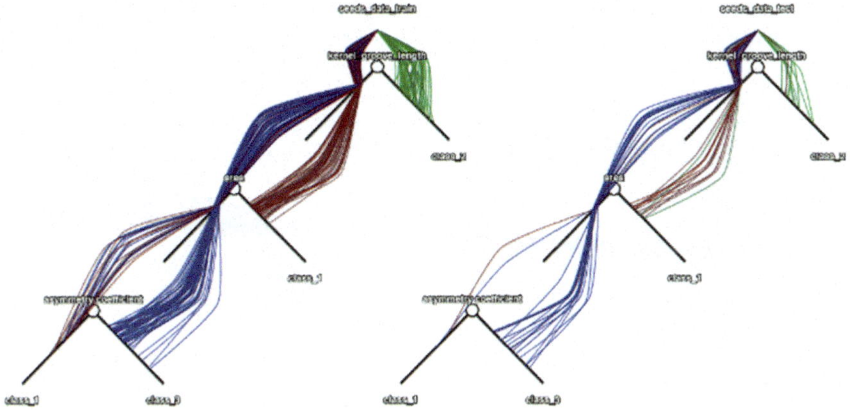

Fig. 12 DT showing both training and testing dataset for seeds dataset for the same tree

Thus, the accuracy of the DT model on both datasets should be similar. Otherwise, a user can modify DT or reject it.

Bended DT visualization from scratch. The next few figures show the process of building DT from scratch either by using existing DT as a prototype or by relying on the expert domain knowledge of the user. First the root node is identified with an attribute, a threshold *T* and an operator at the threshold. In Fig. 13, T = 5.5755 and the activated operator is "<", which means that all x < T will be on the left branch, else x will be on the on the right branch. It immediately shows data split with that threshold.

In Fig. 13, we see that the green cases are well separated from blue and brown cases with that threshold. A user can change a threshold and immediately see how the data split will be changed.

Figure 14 shows the DT growth where the left branch is added. It also shows how the data are split further. So now we can see better that blue and brown cases were not separated well at the first node, in contrast with a new split. Again, changing the threshold at this node allows to see the benefit of different thresholds. The process continues to get a full DT. Figures 15 and 16 show the adjustment of the DT thresholds by using a scroll bar.

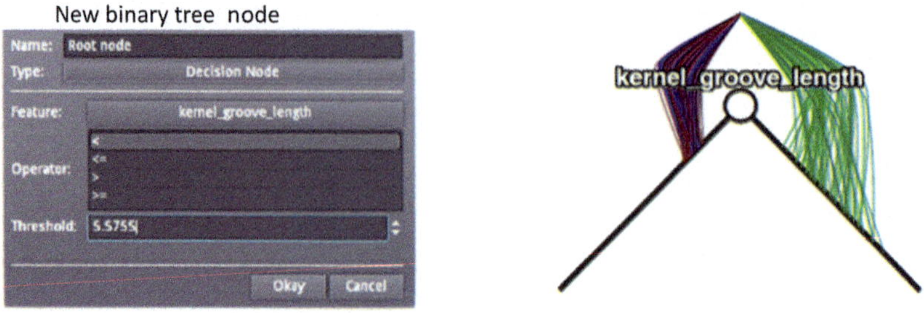

(a) Setting up the root threshold and operator, (b) Decision tree after creating root node.

Fig. 13 Creating root node in the seeds dataset decision tree

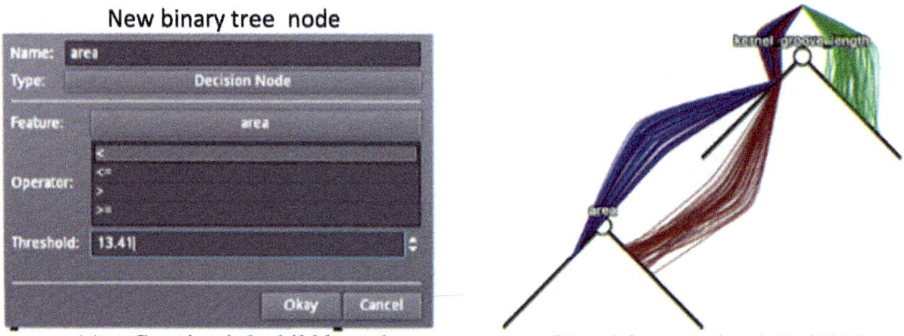

(a) Creating left child branch. (b) After creating left child branch.

Fig. 14 Creating left child branch for seeds dataset

3.2 *Case Study: Wisconsin Breast Cancer Dataset*

For the Wisconsin Breast cancer dataset, the experiment shows the abilities of SPC-DT to visualize the DT and to adjust the threshold interactively to build a better DT, which will not misclassify cancer cases as benign.

While it allowed it improvement in the number of false negatives, it's not enough to remove all false negatives in test data. Figure 17 shows a trained DT with the thresholds identified by the traditional automatic DT training algorithm with the names of attributes listed above the nodes. Figure 18 depicts the quality of this DT on training and test data, with 19 malignant cases misclassified as benign cases in the training data and 6 malignant cases misclassified as benign cases on the test data. Both failure cases are visible in Fig. 17.

We want to decrease these false negatives by building a better DT interactively in the BC-DT software tool. The first attempt was conducted by interactively adjusting the threshold at the root of the left DT that shows training data. The results are presented in Figs. 19 and 20.

The next attempt is presented in Figs. 21and 22. The DT with modified root and subtree thresholds, shows elimination of false negatives on training data more

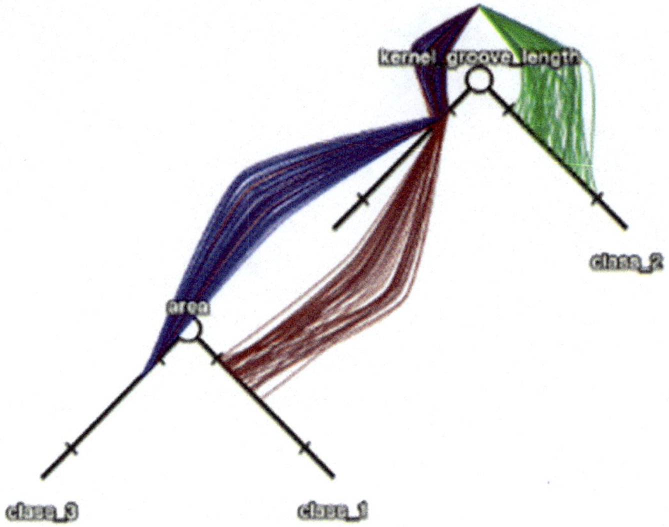

Fig. 15 Base seeds tree with trained threshold

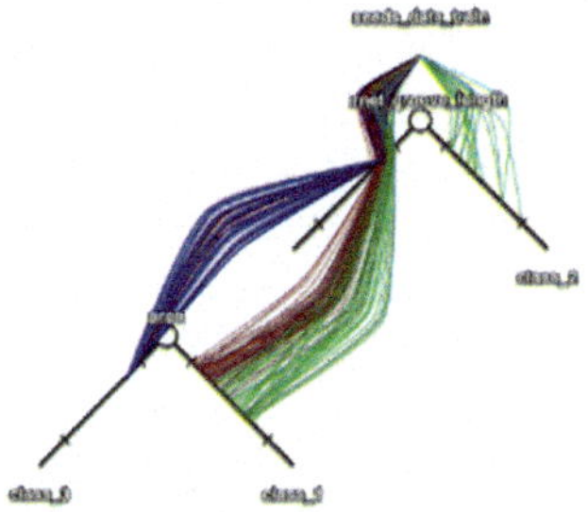

(a) Seeds tree with modified root threshold dragged to the right.

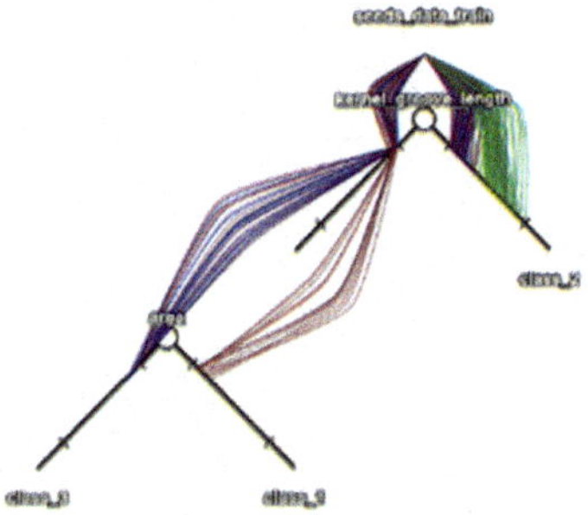

(b) Seeds tree with modified root threshold dragged to the left.

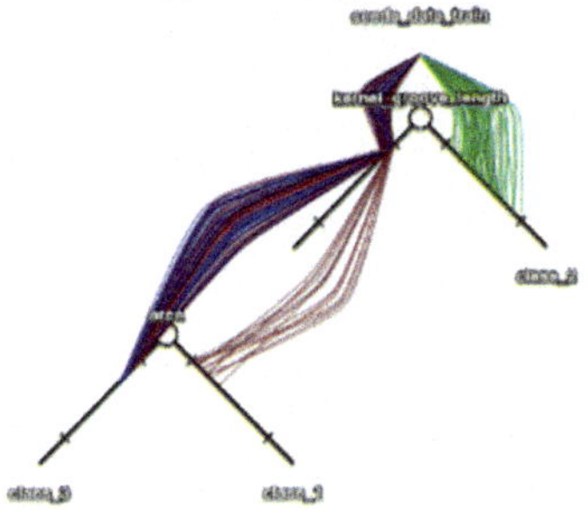

(c) Seeds tree with modified child tree threshold dragged to the right.

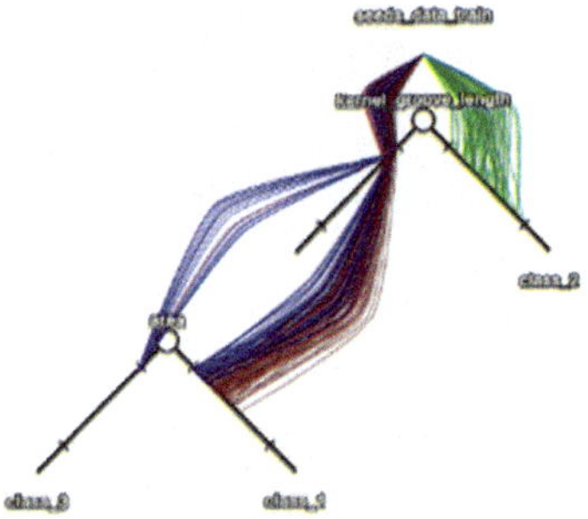

(d) Seeds tree with modified child tree threshold dragged to the left.

Fig. 16 Interactive decision tree threshold modification for seeds data

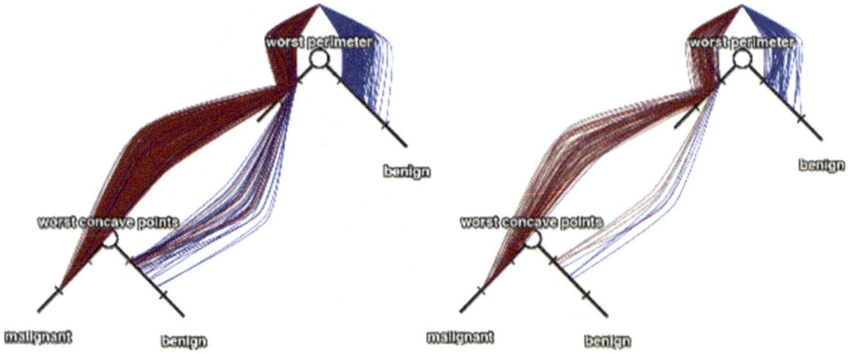

Fig. 17 Trained DT with cancer training data (left) and test data (right) visualized on the DTs

Training data			Test data		
Accuracy		93.85	Accuracy		92.11
Precision		94.44	Precision		94.00
Recall		88.95	Recall		88.68
F1 score		0.9162	F1 score		0.9126
Classified		455	Classified		114
Confusion Matrix			Confusion Matrix		
	benign	malignant		benign	malignant
benign	153	9	benign	47	3
malignant	19	274	malignant	6	58

Fig. 18 Confusion matrices for training and testing data for DT from Fig. 17

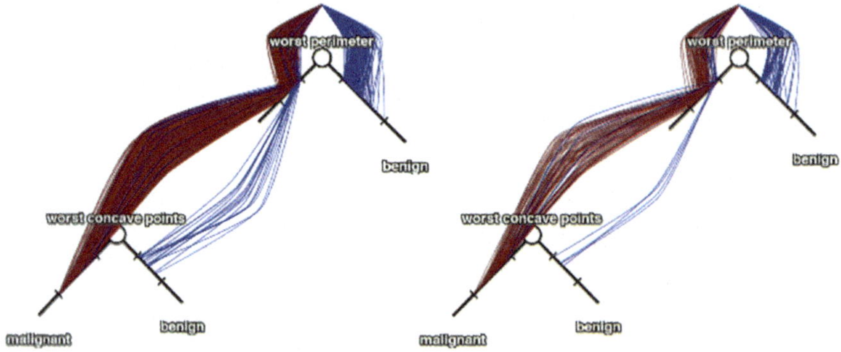

Fig. 19 DT with modified root threshold showing improvement to false negatives for cancer data

improvement to training, but there is still one false negative for the testing dataset. The DT with root threshold modified a second time to further improve testing dataset false negatives, which now both datasets have zero false negatives.

Training data			Test data		
Accuracy		92.97	Accuracy		92.98
Precision		80.25	Precision		86.00
Recall		100.00	Recall		97.72
F1 score		0.8904	F1 score		0.9149
Classified		455	Classified		114
Confusion Matrix			Confusion Matrix		
	benign	malignant		benign	malignant
benign	130	32	benign	43	7
malignant	0	293	malignant	1	63

Fig. 20 Confusion matrices for the decision tree from Fig. 19

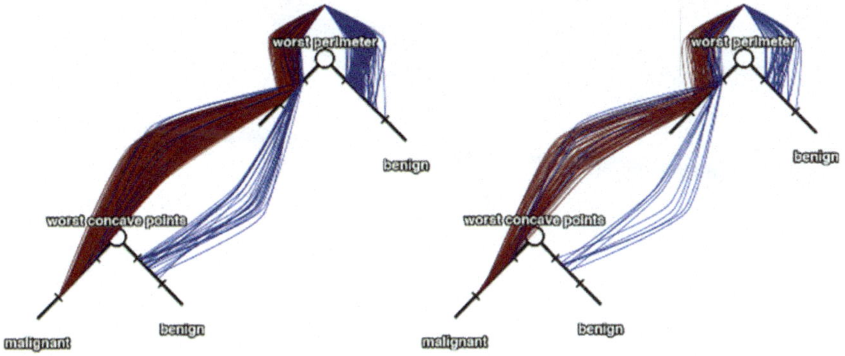

Fig. 21 DT with the fully eliminated false negatives for breast cancer data

Training data			Test data		
Accuracy		91.84	Accuracy		91.23
Precision		76.54	Precision		80.00
Recall		100.00	Recall		100.72
F1 score		0.8671	F1 score		0.8888
Classified		455	Classified		114
Confusion Matrix			Confusion Matrix		
	benign	malignant		benign	malignant
benign	124	38	benign	40	10
malignant	0	293	malignant	0	64

Fig. 22 Confusion matrices for the decision tree from Fig. 21

Figure 23 presents a DT trained on all 683 Wisconsin Breast Cancer cases. It contains edges bended in the threshold points with unequal segments allowing to see the actual distribution of attribute values. Inequality is especially visible for attribute x2. Also, it is visible that thresholds are stable because there are no points next to the thresholds.

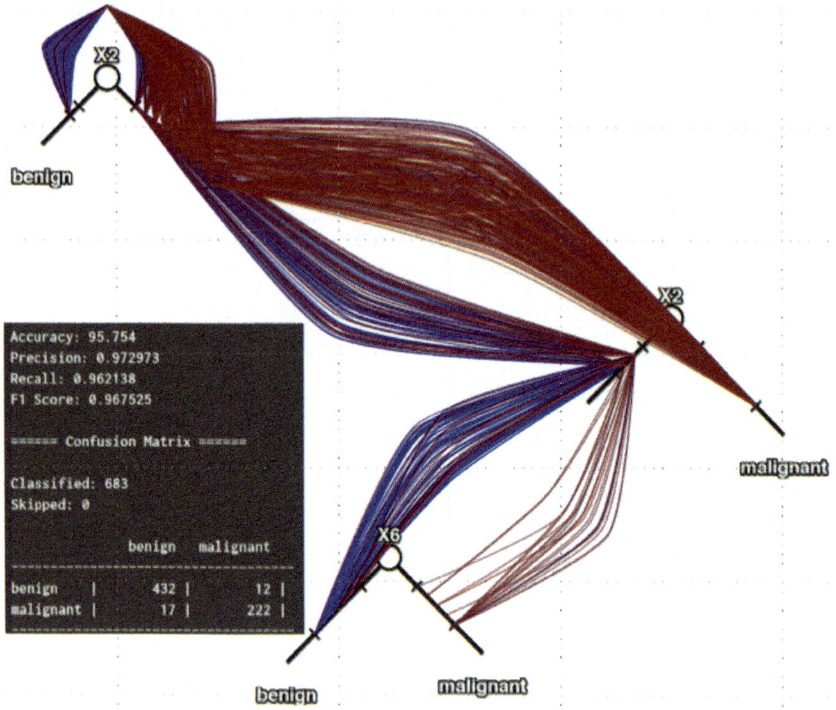

Fig. 23 DT trained on 683 Wisconsin breast cancer cases with confusion matrix

4 SPC-DT Case Studies

This section presents visualization experiments with SPC-DT on datasets from UCI ML Repository [15].

4.1 Cases Study: Wisconsin Breast Cancer

With the Wisconsin breast cancer dataset, we continue our investigation of the SPC-DT approach in this section. Our aim is to reduce the overlap of the lines to improve the visualization that was previously presented in Fig. 10 This will make it simpler to follow specific cases, particularly when there are several gray areas in the coordinate pairs. Using the interactive features of the SPC-DT software application, Fig. 24 displays the same DT as in Fig. 10 with the individual coordinate pairs moved and rearranged to improve the SPC-DT visualization.

This reorganization and placement of coordinate pair plots greatly increased the DT's analysis clarity. For this, care has been taken to minimize line overlap. Yet there's still room for improving SPC-DT visualization clarity. In several coordinate

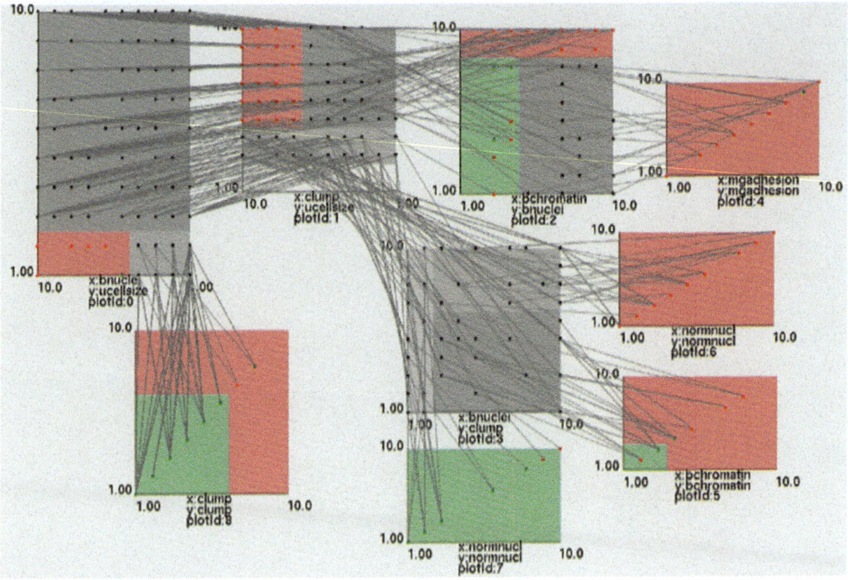

Fig. 24 WBC data in SPC-DT with rearrangement and relocation

pair graphs, lines overlap significantly, and many cases cross areas to which they do not belong.

Figure 25 shows the same data as Figs. 24 and 10, but with the condensation of points in gray zones, separated by each case's class. This allows for a significant increase of clarity, especially in coordinate pair plots that have a high number of cases passing through them, like the first plot of the left in Fig. 25, as well as plots that have a high number of gray zones like in other plots in this figure. Yet, the amount of data that may be losslessly shown is inevitably reduced when condensation is used. Condensing all points in a 2D space to a single point obscures the fact that the classification can become more uncertain the closer a point approaches the boundaries of classification zones.

4.2 Case Study: Iris Dataset

There are 150 instances of each class in this dataset. Four attributes are provided for each example. The decision tree used to visualize the Iris data in SPC-DT is shown in Fig. 26. Its performance is shown in Table 3. The DT using the SPC-DT approach is visualized in Fig. 27 with all the Iris data in the initial places for each coordinate pair.

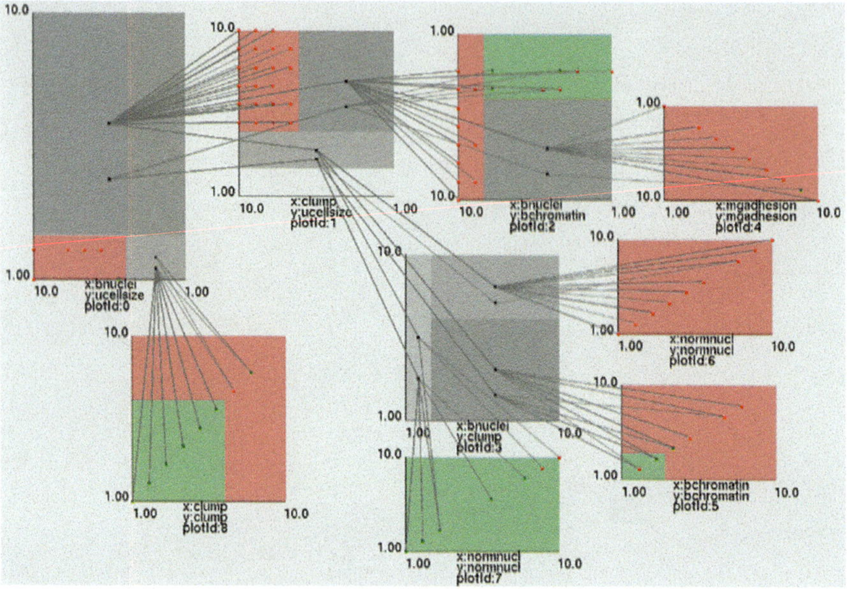

Fig. 25 WBC data in SPC-DT with rearrangement, relocation, and condensation

- petal-length < 2.4500 then class = **Iris-setosa** (100.00 % of 50 examples)
- petal-length >= 2.4500
    - petal-width < 1.7500
        - petal-length < 4.9500
            - sepal-width < 2.5500
                - sepal-width < 2.4500 then class = **Iris-versicolor** (100.00 % of 9 examples)
                - sepal-width >= 2.4500 then class = **Iris-versicolor** (80.00 % of 5 examples)
            - sepal-width >= 2.5500 then class = **Iris-versicolor** (100.00 % of 34 examples)
        - petal-length >= 4.9500 then class = **Iris-virginica** (66.67 % of 6 examples)
    - petal-width >= 1.7500
        - sepal-length < 5.9500 then class = **Iris-virginica** (85.71 % of 7 examples)
        - sepal-length >= 5.9500 then class = **Iris-virginica** (100.00 % of 39 examples)

Fig. 26 ID3 decision tree for iris data

Table 3 Performance of tree in Fig. 26

Error rate			0.0267			
Values prediction			Confusion matrix			
Value	Recall	1-Precision		Setosa	Versicolor	Virginica
Setosa	1.00	0.0000	Setosa	50	0	0
Versicolor	0.9461	0.0208	Versicolor	0	47	3
Virginica	0.9800	0.0577	Virginica	0	1	49

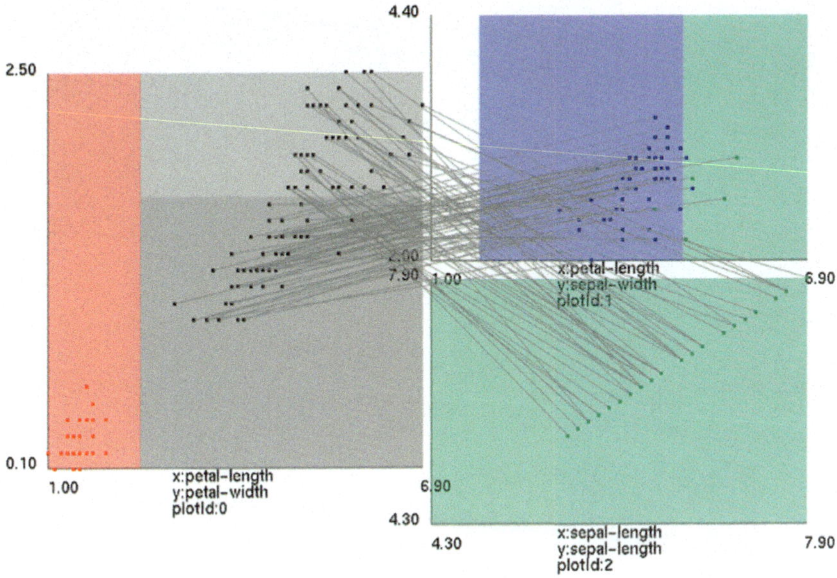

Fig. 27 Iris in default SPC-DT

The gray areas in Fig. 27 represent regions within each paired coordinate space where it is impossible to define a case's class at the current DT level and where it must be followed to the next DT level to be evaluated further.

The green areas denote cases that the DT found to be "Virginica," the blue areas denote cases that the DT classed as "Versicolor," and the red areas denote situations that the DT identified as "Setosa." As a class has already been established, these cases end at the current level. Due to the extensive line overlap, Fig. 27 does not clearly depict patterns. Although it is simpler to follow individual examples than in Fig. 24, line congestion still exists.

The repositioning of coordinate pair plots results in a clearer image, as seen in Fig. 28.

This method of moving them lessens the number of line crossings and removes the paths that go through classification zones that don't correspond to the target class. Line condensation with Iris data is shown in Fig. 29. The DT incorrectly classified the red points in plots 1 and 2. The plots were moved and adjusted to take advantage of the gray point condensation. It improved visibility of instances that were incorrectly classified and their path to their final class in the DT.

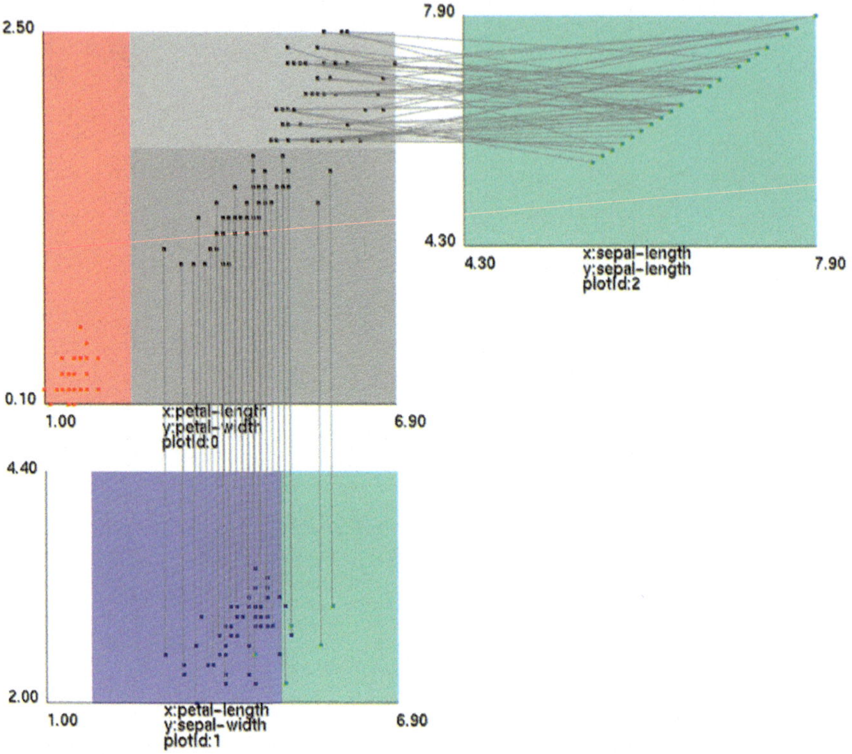

Fig. 28 Iris data in SPC-DT relocated

4.3 Case Study: Wine Dataset

The wine dataset holds 178 cases of 3 classes. Each case is presented by 13 attributes. Figure 30 and Table 4 describe the DT used.

All the data in SPC-DT are displayed in their default places in Fig. 31.

The gray region denotes areas in each paired coordinate space where it is impossible to determine a case's class at the current DT level and where it must be followed to the following DT level to be evaluated further.

After moving the plots and reordering their axes, Fig. 32 exhibits a minor gain in clarity. As a result, fewer lines cross through classification zones that do not correspond to the class of their destination, but fewer overlapping lines are created. The orange region denotes cases that the DT has classified as "class 1", the green area denotes cases that the DT has classified as "class 2", and the blue area denotes cases that the DT has classified as "class 3". As a class has already been established, these cases end at the current level. The default visualization of the Wine dataset exhibits significant line overlap across coordinate pair plots, just as the other case studies.

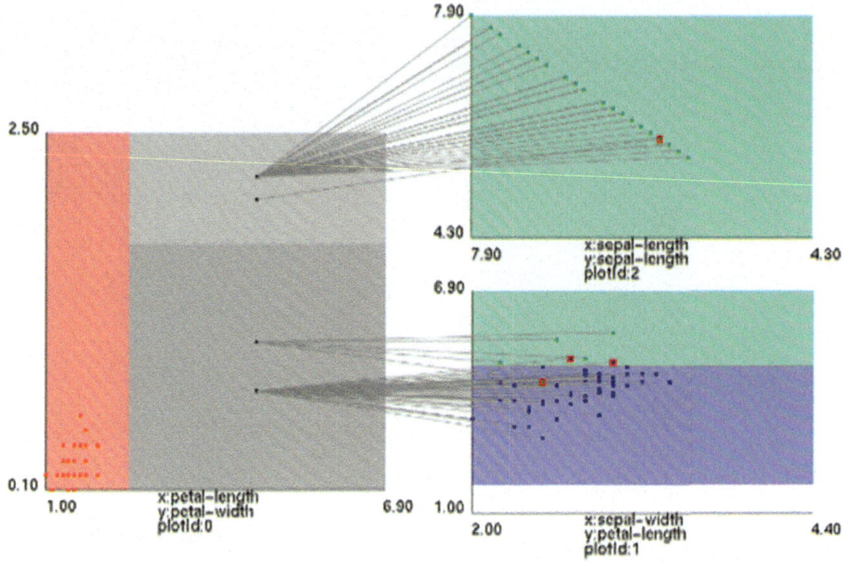

Fig. 29 Iris in SPC-DT relocated, rearranged, and condensed

- Flavanoids < 1.5750
  - Color-intensity < 3.8250 then class = class_2 (100.00 % of 13 examples)
  - Color-intensity >= 3.8250
    - Toal-Phenols < 2.0100 then class = class_3 (100.00 % of 42 examples)
    - Toal-Phenols >= 2.0100 then class = class_3 (85.71 % of 7 examples)
- Flavanoids >= 1.5750
  - Proline < 724.5000
    - Malic-Acid < 3.9250 then class = class_2 (100.00 % of 49 examples)
    - Malic-Acid >= 3.9250 then class = class_2 (80.00 % of 5 examples)
  - Proline >= 724.5000
    - Alcohol < 13.0200 then class = class_2 (66.67 % of 6 examples)
    - Alcohol >= 13.0200 then class = class_1 (100.00 % of 56 examples)

Fig. 30 ID3 decision tree for wine data

Table 4 Performance of tree in Fig. 30

Error rate			0.0225			
Values prediction			Confusion matrix			
Value	Recall	1-Precision		Class_1	Class_2	Class_3
Class_1	0.9492	0.0000	Class_1	56	3	0
Class_2	0.9859	0.0411	Class_2	0	70	1
Class_3	1.000	0.0204	Class_3	0	0	48

By combining relocation, rearrangement, and condensation, a method similar to that employed on the Iris and WBC datasets can be utilized to reduce this overlap.

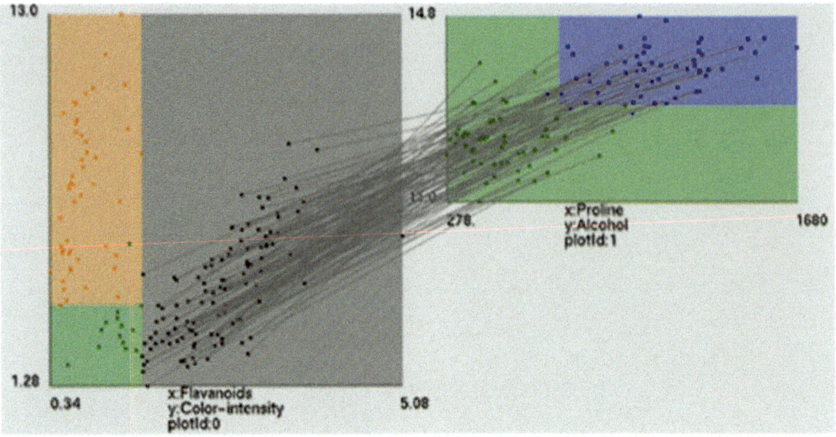

Fig. 31 Default wine dataset in SPC-DT

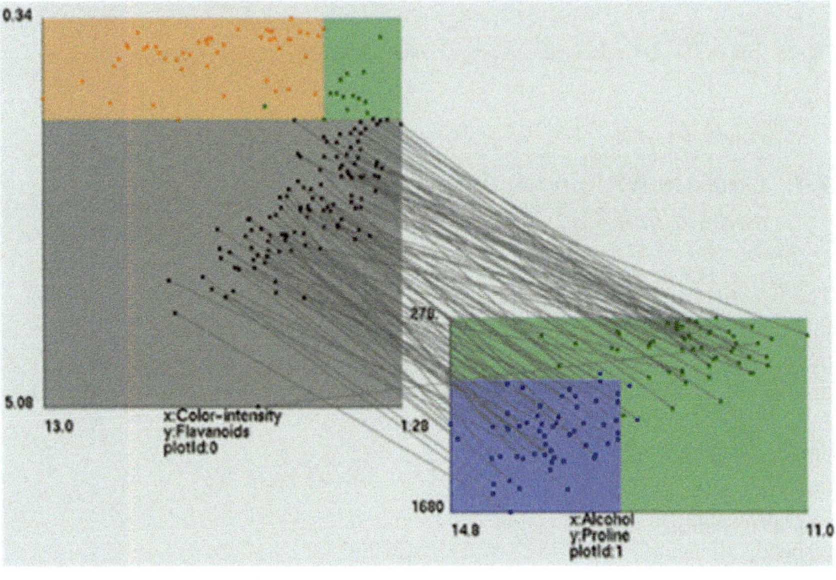

Fig. 32 Rearranged wine dataset in SPC-DT

The Wine data in SPC-DT are displayed in Fig. 33. The gray area in the plot with the caption "plotId: 0" becomes significantly clearer after applying condensation since there is less line overlap. Misclassifications are highlighted in Fig. 33 as the points with the red frames, which adds more clarity. This can make it easier to comprehend how well the decision tree is performing while being visualized.

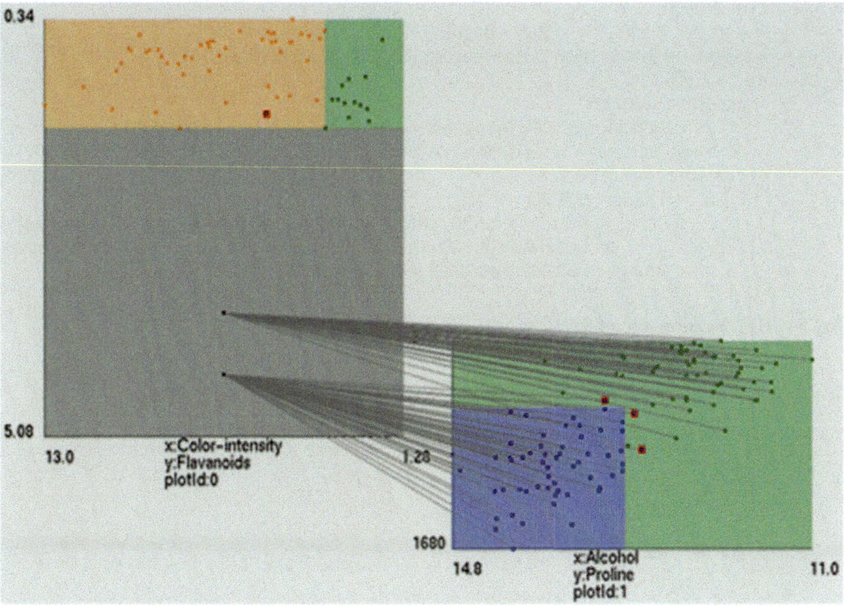

Fig. 33 Wine dataset in SPC-DT with relocation, rearrangement, and condensation

4.4 Case Study: WBC Dataset with Split Training and Validation Data

The benefits of using the SPC-DT approach to visualize datasets that are divided into training and validation data are demonstrated in this section. The split makes it possible to observe and examine relationships between training and validation data that do not pertain to training data. Separating data can improve the DT model's confidence or, alternatively, can result in updating and redesigning the decision tree, which speeds up the DT modeling process. Table 5 and Fig. 34 describe the decision tree.

Data visualizations for training and validation are shown in Figs. 35 and 36, respectively. Darker classification color shades are utilized in these representations

Table 5 Performance of tree in Fig. 34

Error rate			0.046			
Values prediction			Confusion matrix			
Value	Recall	1-Precision		Benign	Malignant	Sum
Benign	0.9709	0.0408	Benign	400	12	412
Malignant	0.9217	0.0566	Malignant	17	200	217
			Sum	417	212	629

- ucellsize < 2.5000
  - sepics < 2.5000 then class = **begnin** (99.43 % of 349 examples)
  - sepics >= 2.5000 then class = **begnin** (77.42 % of 31 examples)
- ucellsize >= 2.5000
  - ucellsize < 4.5000
    - bnuclei < 3.5000 then class = **begnin** (78.38 % of 37 examples)
    - bnuclei >= 3.5000 then class = **malignant** (85.11 % of 47 examples)
  - ucellsize >= 4.5000
    - clump < 6.5000
      - bnuclei < 8.5000 then class = **malignant** (80.00 % of 25 examples)
      - bnuclei >= 8.5000 then class = **malignant** (100.00 % of 41 examples)
    - clump >= 6.5000 then class = **malignant** (100.00 % of 99 examples)

Fig. 34 ID3 decision tree of 90/10 split WBC data

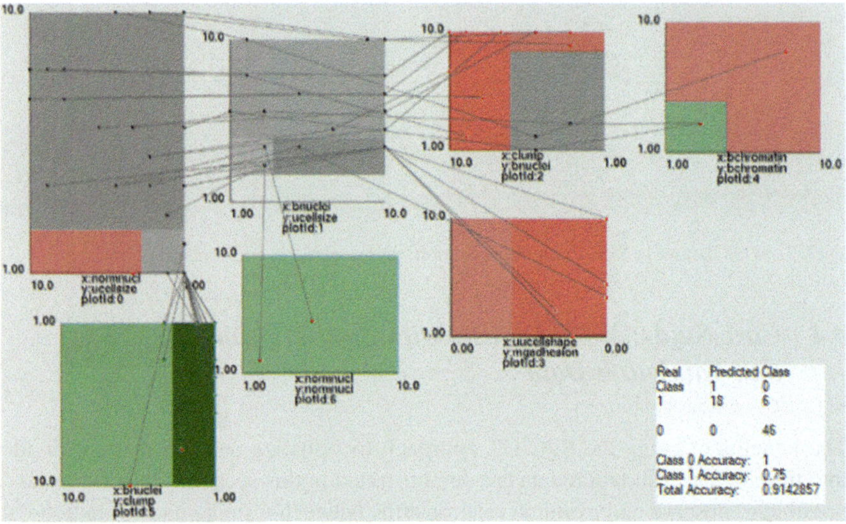

Fig. 35 Testing data in 90/10 split DT in SPC-DT

to indicate more examples that the DT rule covers. More cases are indicated by darker colors. The white borders around some cases are there to make them more visible.

To aid in the visualization of misclassified cases, overlapping classifications have been spread out. Tightly packed cases that appear in an upper left to lower right diagonal line all contain the same value, but they have been spread to better visualize misclassified cases that may be hard to see otherwise. A confusion matrix has been included in Fig. 35 because the confusion matrix in Fig. 34 is for the 90% training dataset.

Interestingly, the decision tree's structure is different from the DTs for WBC presented above. This is because earlier DTs were constructed using 100% of the WBC dataset, but the DT in Figs. 35 and 36 were constructed using just 90% of the WBC dataset. This variation shows how the DT models are unstable. Although

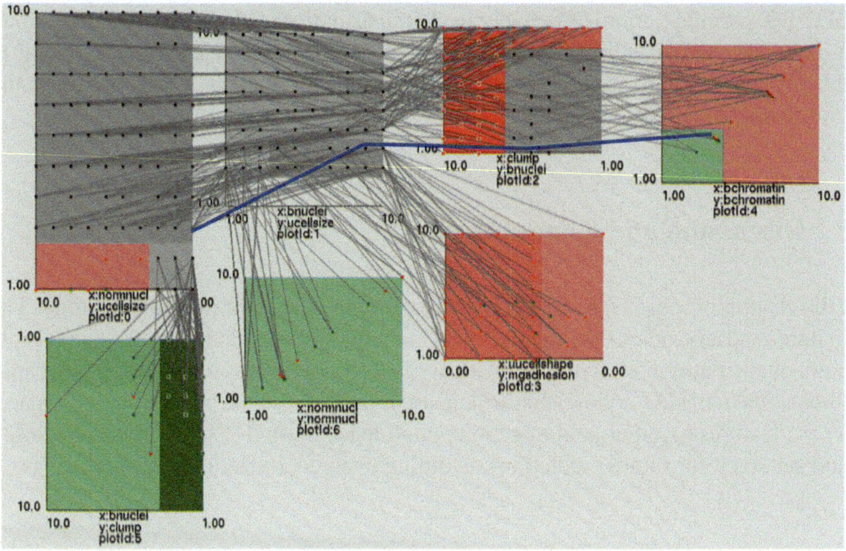

Fig. 36 Training data in 90/10 split DT in SPC-DT

though DT instability is a well-known feature, it can be challenging to distinguish the differences between the traditional DT visualizations presented above and the SPC-DT representation.

The distribution of the cases in the validation data is evidently different from that in the training data. Significant areas that were covered by the training data are not covered by it. As a result, the accuracy of the validation data cannot accurately represent the accuracy of the training data, and the validation data may not accurately represent the training data. Figures 35 and 36, which have different accuracies: 91.43% on validation data and 95.39% on training data, show this disparity.

The blue case, which begins in the gray area in the left plot very near the green area, is likewise depicted in Fig. 36. It continues to the second plot's gray region until coming to an end in the bottom plot's red area, which is once more quite near the green area. This classification is unreliable, as demonstrated by the case's visual analysis. The classification of this case will alter with a small adjustment to the borderlines. This shows the advantages of visual analysis and the SPC-DT visualization strategy.

The ability to display the actual cases and their 2-D distributions in the (X_i, X_j) coordinates given in the adjacent nodes of the DT is one of the SPC-DT method's key features. It enables for the interactive improvement of the DT model's accuracy through the tracing of incorrectly classified cases. For instance, a user can interactively alter the thresholds in these nodes in the SPC-DT visualization to change the classification of misclassified cases that are close to the threshold and see how it affects the classification of other training and validation cases.

A user can interactively tune the DT using the SPC-DT approach. While certain DT visualizations, as [8], permit interactive optimization using 1-D distributions,

SPC-DT permits it more effectively in 2-D distributions. Moreover, SPC-DT offers a more compact visualization of fewer tracing points because each tracing point in the SPC-DT visualization represents two attributes rather than just one as in the traditional DT visualization.

5 Discussion and Generalization

This chapter proposed two new DT visualization methods for machine learning. The features and advantages of these methods include showing relations between multiple pairs of attributes in each plot, individual cases, the closeness of the cases to the split thresholds in the DT nodes, and the density of the cases in the parts of the n-D space. These features together help in gaining confidence in the DT model interpretability and accuracy for future predictions, avoiding overgeneralization and overfittings.

5.1 Benefits of SPC-DT

The advantage of a combination of Shifted Paired Coordinates with a Decision Tree in SPC-DT is in allowing to discover a more efficient classification model than with DT only. Traditional decision tree algorithms search for a best spit in one attribute at the time. SPC-DT allows searching for a best spit in a pair of attributes at the time. Below we present a way to benefit from this opportunity. Consider a pair of attributes (X_i, X_j), and cases of blue and yellow classes shown as blue and yellow rectangles in Fig. 37. Every point $\mathbf{p} = (x_i, x_j)$ divides (X_i, X_j) space and produces 4 quadrants (see Fig. 37a).

We want to find point **p** to get the purest separation of blue and yellow areas, i.e., to maximize the total area all four quadrants where cases belong only to one class by selecting the location of point **p**.

In the sequential DT process with n attributes, we order all attributes by purity splits in decreasing order. If split X_i provided the highest purity, then we accept this

(a) Sequential split: black line then violet line. (b) Simultaneous split by two green lines.

Fig. 37 Sequential (**a**) versus simultaneous (**b**) splits of cases of blue and orange classes

split and repeat the same purity ordering for each obtained branch. Thus, in the DT split process, the attribute with highest purity in the branch depends on the attribute selected for the parent node. In Fig. 37a the DT algorithm first splits with a black vertical line and then with a violet horizontal line generating pure areas 1 and 2. In contrast the SPC-DT algorithm computes purity of 4 quadrants simultaneously for each pair of attributes. Figure 37b shows a better split at point **p**‘, where area 3 is greater than total areas 1 and 2 in Fig. 37a.

Computationally the search for the point **p**‘ can be done in multiple ways, including brute force grid search or optimization. Also, this SPC-DT approach can accommodate the different options of defining the purity of the quadrants. including adaptation of the Gini index to quadrants.

The deficiency of the sequential DT process was noticed long time ago and was addressed in the *random forests* algorithms by producing and combining multiple DTs. While it allowed to improve the accuracy of the resulting models it created the problem of interpretability for the random forests. In contrast the SPC-DT model is interpretable as a decision tree.

Another deficiency of the decision tree is that each attribute is split only to two parts (one-value binary split, $x < T$, $x \geq T$)). A split with more than two parts can be more efficient [16, 24]. In both [16, 24] it is conducted in parallel coordinates, using hyperblocks in [24] and intervals on individual attributes in [16]. Figure 38 from [16] shows a two-value binary split denoted as split (b).

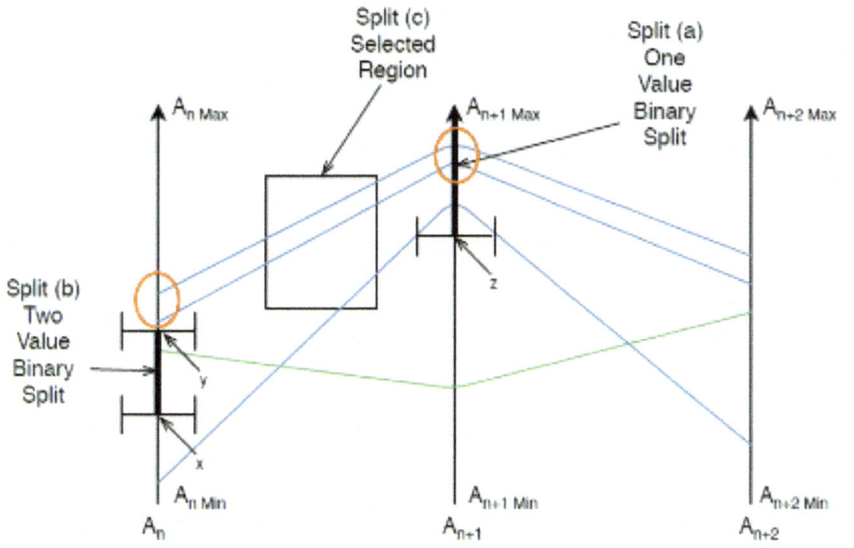

Fig. 38 Splitting techniques: **a** DT type simple binary rule x > T for the attribute; **b** a range-based rule, value in [x, y] interval; **c** region-based split for a rule, all cases that pass through the region [16]

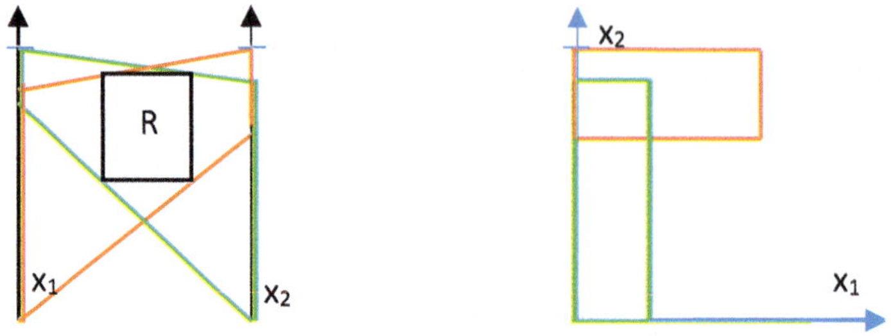

(a) Intervals for the rectangle from Fig. 38 in parallel coordinates.

(b) Red and green rectangles for the black rectangle from (a) in Cartesian Coordinates (first pair of SPC).

Fig. 39 Rectangular rule in parallel coordinates (**a**) and in cartesian coordinates (first pair of SPC) (**b**)

This figure also shows another opportunity, which is missing in the traditional decision trees. It is selecting the region (c) in the parallel coordinates that captures all instances that pass through the region between two adjacent attributes. It captures a relationship between two attributes, while each note of the DT captures a split property of a single attribute. For the region (c), the lines that go through this region can have their points in multiple intervals of these adjacent coordinates. This is how it generalizes the decision tree algorithms to richer and more complex structures, models and rules.

In Fig. 38, the red ovals show the intervals in parallel coordinates, which are mathematically equivalent to a rectangle in the Shifted Paired Coordinates. Figure 39 illustrates this equivalence, where the green quadrilateral in Fig. 39a represents green intervals in x_1 and x_2 parallel coordinates, respectively.

This green quadrilateral corresponds to the green rectangle in Fig. 39b where x_1 and x_2 form the orthogonal Cartesian Coordinates, which is a first pair of coordinates in SPC. Similarly, the red quadrilateral in Fig. 39a corresponds to a red rectangle in Fig. 39b. Also Fig. 39 illustrates how the black rectangle in Fig. 39a corresponds to the overlap area of the green and red rectangles in Fig. 39b. This single simple black rectangle R represents a more complex rule than captured by a decision tree. For example, a simple visual rule with a rectangle:

$$\text{If } \mathbf{x} \in \text{R then } \mathbf{x} \in \text{Class1}$$

is equivalent to the rule

If $(a_{G11} \le \boldsymbol{x}_1 \le a_{G12} \ \& \ a_{G21} \le \boldsymbol{x}_2 \le a_{G22}) \ \& \ (a_{R11} \le \boldsymbol{x}_1 \le a_{R12} \ \& \ a_{R21} \le \boldsymbol{x}_2 \le a_{R22}) \Rightarrow \mathbf{x} \in \text{Class1}$.

where a_{Gij} and a_{Rij} limits of the green and red intervals on attributes x_1 and x_2.

The advantage of SPC visualization in Fig. 39b relative to parallel coordinates is that it requires two times less 2-D points to visualize losslessly each n-D point [1] leading to less occlusion.

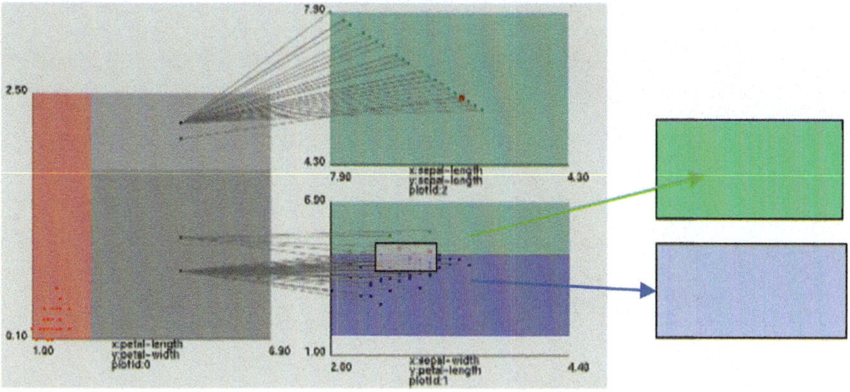

Fig. 40 Expansion of Fig. 29 for impure grey rectangle with black frame in SPC-DT

The same idea of the rectangular regions to build rules that are more general than DTs was used in [21, 25] with SPC, in [26] with Elliptic Paired Coordinates (EPC) and with In-Line Coordinates (ILC) in [27]. This approach is expandable to other general line coordinates too.

The advantage of implementing rectangles in SPC is that they are directly visible in the decision tree itself and allow interactively expanding and improving e the decision tree. Figure 40 shows this in in SPC-DT. It expands Fig. 29 with forming an impure grey rectangle in the middle of a larger area with mixed cases. This gray rectangle captures cases, which belong to different classes within a larger rectangle. It generalizes a DT by allowing rules like: If $\mathbf{x} \in$ green area then $\mathbf{x} \in$ green class, if $\mathbf{x} \in$ red area then $\mathbf{x} \in$ red class, if $\mathbf{x} \in$ grey area then $\mathbf{x}$ not classified (refused). This modification of the decision tree allows to avoid misclassification of the cases, which are in the grey rectangle.

It is also can be implemented in parallel coordinates as in (c) in Fig. 38, or in PC-DT (see Fig. 5) where DT is combined with parallel coordinates directly. It was shown in [16] that using parallel coordinates allows to build a better model than traditional decision trees due to generalization of the decision trees.

5.2 Other Related Work

Typically, alternative DT visualizations in machine learning show the number of cases in each node and/or one-dimensional (marginal) distribution of the cases. Figure 2 in Sect. 1 illustrates such alternative DT visualizations used in Machine Learning. One of the major advantages of the PC-DT and SPC-DT methods is the ability to show the actual cases. It allows tracing of the misclassified cases to improve the accuracy of the DT model interactively.

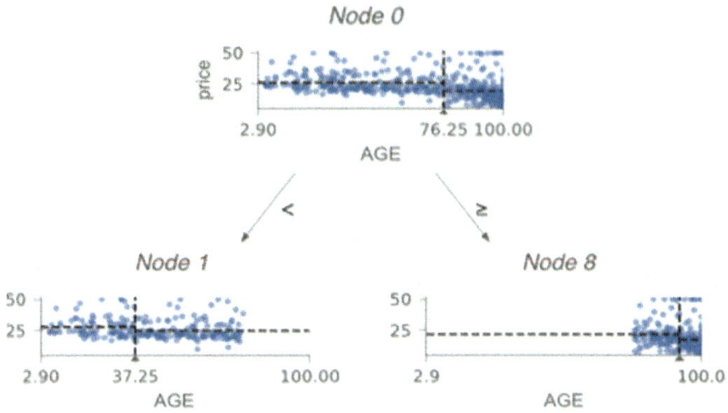

Fig. 41 Interactive splitting for DT nodes [8]

SPC-DT also shows actual 2-D distributions in the (X_i, X_j) coordinates for adjacent nodes of the DT. For instance, when misclassified cases are close to the threshold of some nodes, then a user can interactively move the thresholds in these nodes in the SPC-DT or BC-DT visualization to change classification of those cases and see how it affects the classification of other training and validation cases.

In both SPC-DT and BC-DT methods, a user can interactively optimize the DT. While some other DT visualization such as [8] allow it by using 1-D distributions (Fig. 41), SPC-DT allows it in more detail in 2-D distributions. In addition, SPC-DT provides a more compact visualization of cases with a smaller number of tracing points. The BC-DT visualization has some similarities with visualization shown in Fig. 41 [8] where the coordinates are straight lines but does not trace actual cases from the root to the terminal nodes. In BC-DT, we use edges of the DT as bended coordinate lines. This makes the BC-DT more compact. It does not need drawing edges separately outside of the DT.

Commonly methods proposed to visualize decision trees visualize an individual "best" decision tree. This tree is selected automatically by DT algorithms among the trees with user selected desired characteristics: (1) accuracy, (2) minimum number of samples in the leaf, (3) height of the tree, and (4) the features to be included to the decision tree.

Many decision trees can have similar performance, but automatic methods hide competitive alternatives from the user. A "best" DT selected by some DT algorithm may not be the best for the user. A Timbertrek visualization system [20] allows a user to select a DT interactively from a set of almost-equally-optimal DT models known as the Rashomon set [28] motivated by the Rashomon effect from [29]. This is an important conceptual advantage of this system. It allows a user to see and explore much more DTs with the same characteristics (1)–(4) listed above. The deficiency of this system is that the user will be able to see only DTs selected by properties (1)–(4). Other properties can be also important for a user and can supersede some of these properties in their importance.

For example, a user can accept DTs with a lower number of cases in a few leaves, than the threshold that was set up as the minimum number of samples for all leaves, but the system will not show these DTs to the user. The same is applicable to the height of the tree, a user can allow a few longer branches than the height threshold. Next, a user can prefer a DT with a lower total accuracy, but higher accuracy for some class, e.g., for the malignant class to minimize the number of misclassified malignant cases as benign. These are just a few examples, but there are many such scenarios where a user may want some agency in the selection of the decision tree. Often a user can come up to such decisions in the process of the detailed analysis of the decision tree without explicit requirements formalized in advance, like (1)–(4). This is a motivation to the method that we are proposing to give a user a wider capability to analyze and build a desired DT.

The operations that can be conducted in BC-DT and/or SPC-DT system include the following interactive operations: (a1) adjusting thresholds in each node, (a2) joint adjusting and optimizing a pair of attributes (nodes of the DT), (a3) expanding DT, (a4) shrinking DT, (a5) blocking classifying some areas of the feature space, (a6) observing/tracing individual cases in the DT, (a7) simplifying visual form of the DT, (a8) visualizing a DT in two times more compact form than in other systems like [20, 28] by pairing nodes. Thus, our approach allows a domain expert to adjust DT to meet expert' needs. As we pointed out above these DTs can differ from selectable in [20].

Below we discuss the ways to combine advantages of SPC-DT and Timbertrek. A user can build and store several DTs in SPC-DT and then explore them. A user also can have access to a large set of DTs from Timbertrek, select, explore and modify them using SPC-DT tools. It can be done by (a1)–(a8) listed above. For instance, it can be done by creating the areas in SPC-DT, where the DT refuses to classify cases or allows to expand the DT. SPC-DT allows to create a knowledge base of these DTs with their characteristics doing search of most appropriate DTs like in [20] but with a wider set of search criteria. A user can work with BC-DT and SPC-DT systems on the regular basis, say, after the first session the user creates ten DTs of interest, during the next session the user can modify or combine DTs.

In the combined functionalities of SPC-DT and Timbertrek a user can select: (1) a set of features to include to the decision tree, (2) branches of the DT for exploration, (3) adjust thresholds for features at the tree's nodes, (4) explore how they fit user's needs, (5) observe branches and explore specific layers of the set of DTs, (6) split layers and branches, and others.

Both SPC-DT and Timbertrek empower domain experts and data scientists to easily explore many well-performing decision trees so they can find and collect those trees that best reflect their knowledge and values. Both visually summarize DTs based on the decision paths, enabling users to seamlessly transition across different components. Repositionable DTs with details of a decision tree and multiple windows allow users to compare and modify DTs to explore, enabling users to quickly identify decision trees with desired properties within capabilities of each tool.

Shifted Paired Coordinates are used not only for decision trees but independently of the decision trees to find interpretable rules [25]. In those tasks, it is a significant challenge for Shifted Paired Coordinates to select an efficient pairing of coordinates that will allow for the discovery of patterns of cases of different classes clearly. Exploration of all possible pairings is not feasible, so a practical approach to avoid combinatorial exploration is using a decision tree to define the pairs of shifted paired coordinates as this paper illustrates with the SPC-DT approach.

The pairs are constructed using adjacent coordinates in the decision tree. In each branch of the DT, the coordinate that is at the root is paired with the coordinate in the second node of the branch. Next, the 3rd and 4th nodes form the second pair of coordinates, and so on. Some coordinates can be used in the DT several times with different thresholds. Respectively, the DT-based pairing will use them several times. This means that the decision tree approach allows us to expand the number of pairing alternatives beyond using each coordinate only in one pair.

Thus, the approach proposed in this paper allows (1) efficient visualization of the decision tree, (2) its improvement using this visualization, (3) communicating the DT to end-users, and (4) finding an efficient pairing of coordinates in SPC, which makes detecting patterns in the data much clearer. This opens an opportunity to use the SPC-DT approach beyond DTs.

5.3 Generalization of BC-DT and SPC-DT

Figure 42 shows sharp and smooth versions of the BC-DT for their comparison. The selection of a particular form of BC-DT will depends on the user preferences. This can be a visually pleasing visualization, less space used or others.

Figure 43 shows the generalization of SPC-DT with bended smooth attributes. This approach links SPC-DT with smooth BC-DT shown above. It can be beneficial because some polylines can be simpler for analysis in curvilinear coordinates than in the straight coordinates.

Figure 44 illustrates it, with a curve in the traditional straight orthogonal coordinates on the left. In contrast, on the right, this curve became a straight horizontal preattentive line, because the X coordinates became a curve.

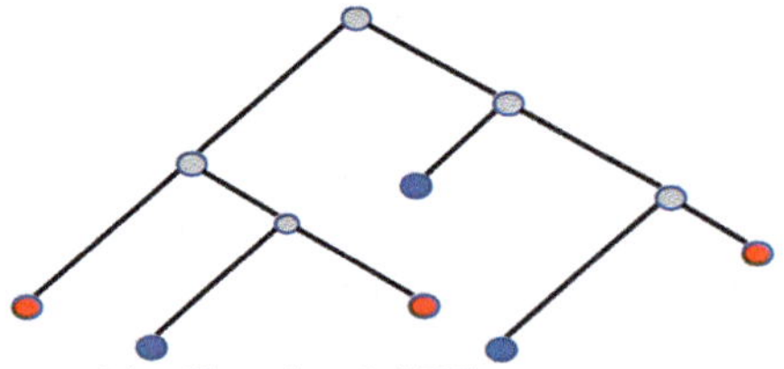

(a) Sharp bended DT.

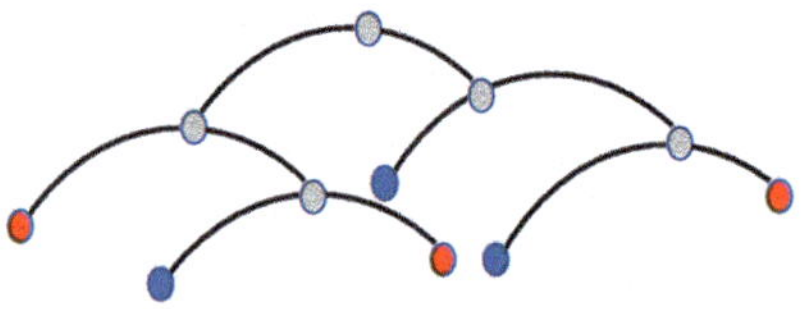

(b) Smooth bended DT.

Fig. 42 Sharp and smooth bended DTs

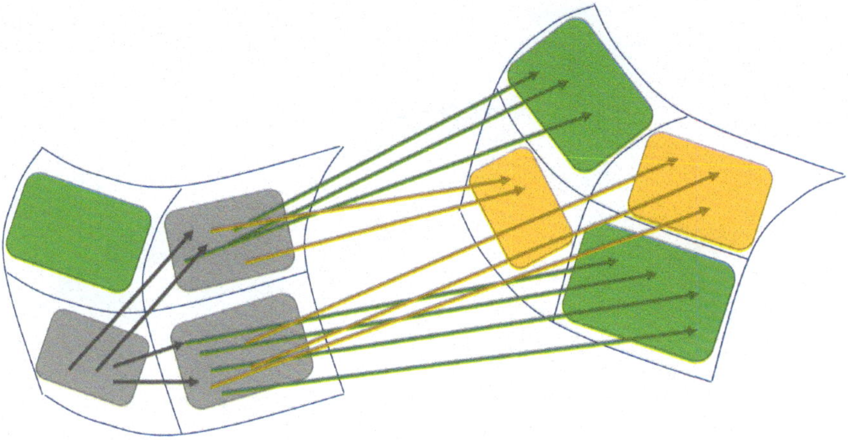

Fig. 43 SPC-DT with bended smooth attributes

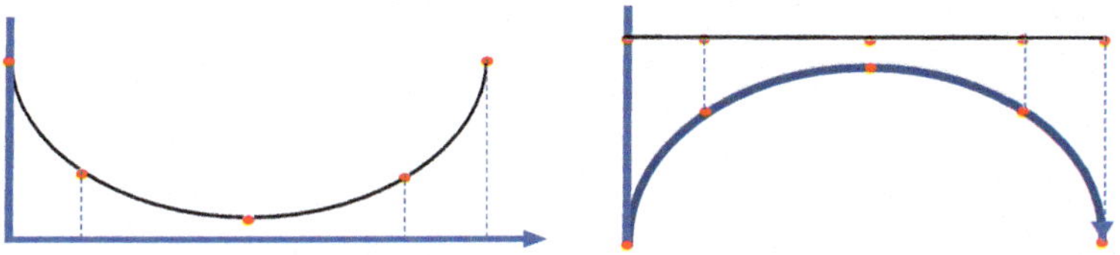

(a) Curve in Cartesian coordinates

(b) Curve transformed to a straight line in coordinates with one curvilinear coordinate.

Fig. 44 Transformation of a half circle to a straight line by transforming coordinates

Similarly to presented above generalizations we can visualize DTs in other general line coordinates. It is already demonstrated in Sect. 2.1 with DT in parallel coordinates. With this approach methods to visualize DTs for other GLCs can be developed, including inline coordinates and n-gon coordinates defined in [1].

5.4 Dealing with Larger Decision Trees

Dealing with larger decision trees requires additional features like zooming and panning and enhancement of the proposed methods. The current implementation allows to visualize DTs of the size like shown in the Fig. 45 for the Kaggle body performance dataset [30]. The problem is not only the size of the tree itself but the abilities to place the parts of the tree non-overlapping. This is a common issue for both SPC-DT and BC-DT. The current solution implemented for SPC-DT and BC-DT is interactive. A user can drag parts of the decision tree to make the visualization more efficient. A future work can be to automate this process.

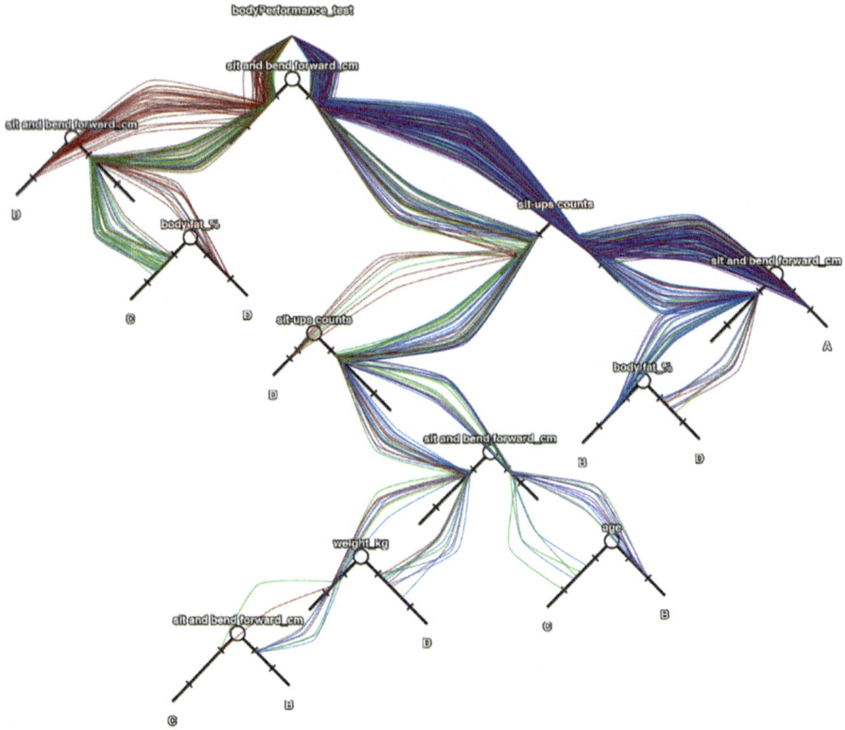

Fig. 45 Decision tree for body performance data from [30]

6 Conclusions

This paper proposed two new DT visualization methods for machine learning. The features and advantages of these methods include showing relations between multiple pairs of attributes, in each plot, individual cases, closeness of the cases to the split thresholds, in the DT nodes, and the density of the cases, in the parts of the n-D space.

These features together help in getting confidence in the DT model interpretability and accuracy for the future predictions, avoiding overgeneralization and overfittings. The future work will be directed to be able to visualize the larger decision trees with more opportunities for user interaction. The developed experimental software is available at GitHub [31].

References

1. Kovalerchuk B (2018) Visual knowledge discovery and machine learning. Springer Nature
2. Rudin C, Chen C, Chen Z, Huang H, Semenova L, Zhong C (2022) Interpretable machine learning: fundamental principles and 10 grand challenges. Stat Surv 16:1–85
3. Ming Y, Qu H, Bertini E (2018) RuleMatrix: visualizing and understanding classifiers with rules. IEEE Trans Visual Comput Graphics 25(1):342–352
4. Kovalerchuk B, Andonie R, Datia N, Nazemi K, Banissi E (2022) Visual knowledge discovery with artificial intelligence: challenges and future directions. In: Integrating artificial intelligence and visualization for visual knowledge discovery. Springer, pp 1–27
5. Kovalerchuk B, Grishin V (2019) Adjustable general line coordinates for visual knowledge discovery in n-D data. Inf Vis 18(1):3–2
6. Schulz H (2011) Treevis.net: aa tree visualization reference. IEEE Comput Graph Appl 31(6): 11–15
7. Scheibel W, Limberger D, Döllner J (2020) Survey of treemap layout algorithms. In: Proceedings of the 13th international symposium on visual information communication and interaction, pp 1–9
8. Parr T, Grover P (2019) How to visualize decision trees, https://explained.ai/decision-tree-viz/
9. R2D3 (2019) A visual introduction to machine learning. http://www.r2d3.us/visual-intro-to-machine-learning-part-1/
10. SAS: working with decision trees (2016) http://support.sas.com/documentation/cdl/en/vaug/68027/HTML/default/viewer.htm#n0q3i0zwng79kin1kb1zvpo9k312.htm
11. van den Elzen SJ, van Wijk JJ (2011) BaobabView: interactive const-ruction and analysis of decision trees. VAST 2011. IEEE, pp 151–160
12. Elkan C (2001) The foundations of cost-sensitive learning. In: International joint conference on artificial intelligence, vol 17, no 1, pp 973–978
13. Rokach L, Maimon O (2009) Classification trees. In: Data mining and knowledge discovery handbook. Springer, Boston, MA, pp 149–174
14. Kovalerchuk B, Grishin V (2018) Reversible data visualization to support machine learning. In: International conference on human interface and the management of information. Springer, pp 45–59
15. Dua D, Graff C (2019) UCI machine learning repository, http://archive.ics.uci.edu/ml. University of California, School of Information and Computer Science, Irvine, CA
16. Estivill-Castro V, Gilmore E, Hexel R (2020) Human-in-the-loop construction of decision tree classifiers with parallel coordinates. In: 2020 IEEE international conference on systems, man, and cybernetics (SMC). IEEE, pp 3852–3859
17. Teoh ST, Ma KL (2003) PaintingClass: interactive construction, visualization and exploration of decision trees. In: Proceedings of the ninth ACM SIGKDD international conference on knowledge discovery and data mining, pp 667–672
18. Kandogan E (2001) Visualizing multi-dimensional clusters, trends, and outliers using star coordinates. In: Proceedings of ACM SIGKDD '01, pp 107–116
19. Inselberg A (2009) Parallel coordinates. Springer
20. Wang ZJ, Zhong C, Xin R, Takagi T, Chen Z, Chau DH, Rudin C, Seltzer M (2022) TimberTrek: exploring and curating sparse decision trees with interactive visualization. In: 2022 IEEE visualization and visual analytics (VIS). IEEE, pp 60–64
21. Kovalerchuk B, Gharawi A (2018) Decreasing occlusion and increasing explanation in interactive visual knowledge discovery, In: Yamamoto S, Mori H (eds) Human interface and the management of information. interaction, visualization, and analytics, LNCS 10904. Springer, pp. 505–526. https://doi.org/10.1007/978-3-319-92043-6_42
22. Worland A, Wagle S, Kovalerchuk B (2022) Visualization of decision trees based on general line coordinates to support explainable models. In: 26th international conference information visualization. IEEE, pp 351–358. arXiv:2205.04035
23. Rakotomalala R, Tanagra software. http://eric.univ-lyon2.fr/~ricco/tanagra/en/tanagra.html

24. Kovalerchuk B, Hayes D (2021) Discovering interpretable machine learning models in parallel coordinates. In: 2021 25th international conference information visualisation (IV). IEEE, pp 181–188. arXiv:2106.07474
25. Wagle SN, Kovalerchuk B (2022) Self-service data classification using interactive visualization and interpretable machine learning. In: Integrating artificial intelligence and visualization for visual knowledge discovery. Springer, Cham, pp 101–139
26. McDonald R, Kovalerchuk B (2022) Non-linear visual knowledge discovery with elliptic paired coordinates. In: Integrating artificial intelligence and visualization for visual knowledge discovery. Springer, Cham, pp 141–172
27. Kovalerchuk B, Phan J (2021) Full interpretable machine learning in 2D with inline coordinates. In: 25th international conference information visualisation IV-2021, vol 1. IEEE, pp 189–196. https://doi.org/10.1109/IV53921.2021.00038. arXiv:2106.07568
28. Xin R, Zhong C, Chen Z, Takagi T, Seltzer M, Rudin C (2022) Exploring the whole Rashomon set of sparse decision trees. arXiv preprint arXiv:2209.08040
29. Breiman L (2001) Statistical modeling: the two cultures (with comments and a rejoinder by the author). Stat Sci 16(3):199–231
30. Kaggle dataset. https://www.kaggle.com/datasets/kukuroo3/body-performance-data?resource=download
31. GitHub. https://github.com/CWU-VKD-LAB. SPC-DT, Bended Attributes

Full High-Dimensional Intelligible Learning in 2-D Lossless Visualization Space

Boris Kovalerchuk and Hoang Phan

1 Introduction

Today, the Machine Learning (ML) field is heavily focused on accurate Interpretable Machine Learning (IML) [1–6]. While the term IML is popular several other terms like explainable, intelligible, comprehensible and others often are used interchangeably or with some nuanced differences. These methods cover everything from constructing explainable models from scratch to explaining black box models. Using visual methods to develop explainable machine learning models is one of the appealing possibilities. It is difficult, though, because machine learning data are multidimensional and which we cannot see by the naked eye. Therefore, new methods to make this possible are required.

Traditional visualization techniques are lossy, not retaining all multidimensional information while converting n-D data to two dimensions. General Line Coordinates (GLC) representation of n-D data, in contrast, is lossless [7]. With no information loss, this visual representation made it possible to perform full multidimensional machine learning in two dimensions. This opens the opportunity for the end user to be involved in the entire machine learning process because of its visual nature in 2D. It dramatically drops the requirement of deep mathematical knowledge of ML processed from the end users. There are two benefits to this strategy. In straightforward circumstances, a user can find the pattern visually by simply looking at the data represented in GLC. A user can find patterns in 2-D representations by applying new 2-D ML techniques in more complicated scenarios. This area is developing now into an entirely new branch of machine learning.

B. Kovalerchuk (✉) · H. Phan
Department of Computer Science, Central Washington University, Ellensburg, WA, USA
e-mail: Boris.Kovalerchuk@cwu.edu

H. Phan
e-mail: Hoang.Phan@cwu.edu

B. Kovalerchuk et al. (eds.), *Artificial Intelligence and Visualization: Advancing Visual Knowledge Discovery*, Studies in Computational Intelligence 1126,
https://doi.org/10.1007/978-3-031-46549-9_2

This chapter falls within this category and concentrates on new In-line Coordinates (ILC) expanding [8]. These coordinates are a subset of General Line Coordinates [7] and are defined in Sect. 2. The viability of full 2-D ML with various General Line Coordinates (Shifted Paired Coordinates, Elliptic Paired Coordinates, CPC-R and GLC-L) was demonstrated in a number of earlier works [8–13]. When it comes to 2-D modeling of non-image data, this field of study may be traced to [14, 15] for Parallel Coordinates and [16]. Often, 2-D studies in ML only use straightforward 2-dimensional examples to graphically demonstrate ML algorithms. The exploration of Parallel Coordinates in visual analytics studies has been quite active for clustering-related tasks [17], but much less so for supervised learning, which is the subject of our study. Among the relevant works are [18–20].

This research expands [8] with new case studies and algorithms for more challenging data, which include the heavily imbalanced high-resolution data with multiple classes like Page Block Classification data. The algorithm uses a decision tree as a model design facilitator for producing a model, which is more general than a decision tree.

We propose a full 2D ML methodology expanding [8] to consolidate prior investigations based on General Line Coordinates into a single, overarching paradigm. In the past, 2D machine learning research was viewed as merely auxiliary exploratory data/model visualization occurring primarily after or before the actual machine learning due to loss of n-D information. It was assumed that since loses of n-D information in 2-D, comprehensive interpretable n-dimensional analysis in n-D space is required to build ML models. The full 2-D ML approach demonstrates that it is not required. The main contribution is the methodology that goes *beyond a range of methods of lossless visualization* methods developed in [7] to multiple *machine learning techniques for discovering complete n-D patterns in 2-D visualization space* broadening human-assisted visual pattern discovery. Relative to the common lossy visualization of n-D data, the contribution is in abilities to actual *discovering complete n-D patterns in 2-D lossless visualization space.*

This chapter is structured as follows. In-line coordinates' primary notions are defined in Sect. 2. The Box Classification (BC), Linear Classification, and Regression algorithms are presented in Sect. 3. The case study utilizing benchmark Wisconsin Breast Cancer (WBC) data that shows the viability of the strategy is covered in Sect. 4. Section 5 presents the results with Page Block Classification (PBC) data and BC algorithm for imbalanced high-resolution data with multiple classes and compares results with other algorithms. The conclusions are presented in Sect. 6.

2 In-Line Based Coordinate Systems

With the General Line Coordinates defined in [7], n coordinate axes can be drawn in 2-D in a variety of ways, including curved, parallel, unparalleled, collocated, unconnected, etc. Testing several GLC coordinate mappings and orders are needed to discover the best visual representation of n-D data possible.

Fig. 1 Two 5-D points of two classes in In-line coordinates

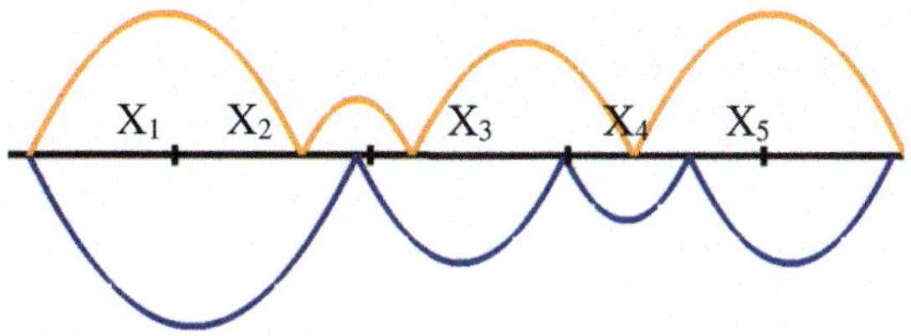

GLCs contain In-line Coordinates, which are similar to Parallel Coordinates but differ in that their axes are horizontal rather than vertical (see Fig. 1). On the same line, all coordinates $X_1, X_2, \ldots, X_n$ are collocated and may or may not overlap. The condition of lossless representation of n-D point $\mathbf{x} = (x_1, x_2, \ldots, x_n)$ in 2-D is satisfied by a sequence of directed curves or polylines, which connect adjacent points x_i, x_{i+1} in coordinates X_i, X_{i+1} as shown in Fig. 1. These curves/polylines of various heights and shapes form graphs [7]. Both Parallel Coordinates and In-line Coordinates require the same number of nodes and links and can be used similarly.

The connections between the nodes are directed edges, however when the direction matches the order of coordinates, arrowheads are not necessary. It is possible to draw n-D points from distinct classes above and below the coordinate line to more clearly illustrate the differences between them (see Fig. 1). Four ILC location modes are considered:

(L1) *Sequential* ILC with coordinates located one after another (Fig. 1).
(L2) *Collocated* ILC with coordinates drawn at the same location with full overlap (Fig. 2a).
(L3) *Generic* ILC where some coordinates can be *sequential, collocated, overlapping*, or *disjoined* (Fig. 2b).
(L4) Dynamic ILC with coordinated located *dynamically* as explained later.

By choosing a certain ILC overlap in L3, a given n-D point **c** can be collapsed to a single 2-D point while remaining fully recoverable. When n-D point **c** is the class's center point and other n-D points are concentrated around it, this is a useful visual characteristic. Another method for enhancing the visibility of the patterns of interest is to reorder the coordinates X_1-X_n.

There are numerous ways to build links connecting points x_i on coordinates X_i by giving meaning to their width, height and other charateristics, to communicate

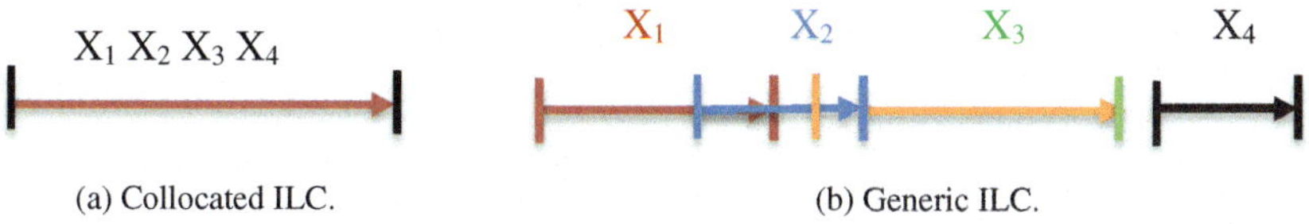

Fig. 2 Options to locate coordinates in In-line coordinates

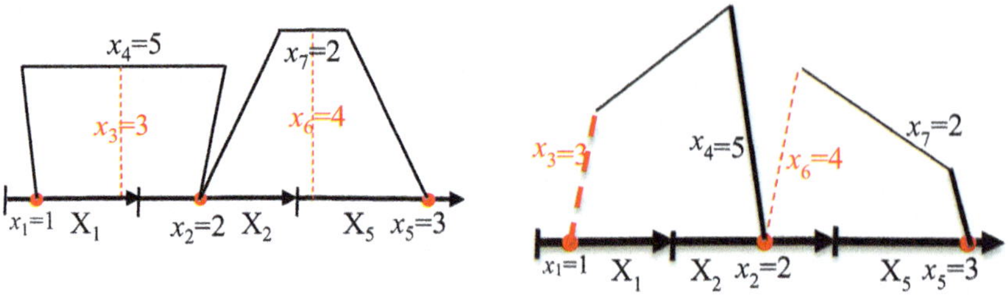

(a) x_3,x_4,x_6 and x_7 encoded by the height and width of the link lines that connect (x_1,x_2) and (x_2,x_5).

(b) x_3,x_4,x_6 and x_7 encoded by of length of sides of the link lines that connect (x_1,x_2) and (x_2,x_5).

Fig. 3 7-D point $\mathbf{x} = (x_1, x_2, x_3, x_4, x_5, x_6, x_7) = (1, 2, 3, 5, 3, 4, 2)$ in two ILBCs

more information. See an example in Fig. 3a for a 7-D point with values of x_3 and x_4 encoded as the height and the width of the line that connects (x_1,x_2), and values of x_6 and x_7 the height and width of the line that connects (x_2,x_5). This makes the ILC base line shorter because only the three coordinates x_1, x_2 and x_5 are directly encoded in it.

Figures 3b and 4 show other options. Figure 3b uses the lengths of sides of the line that connects points x_1 and x_2 to encode values of x_3 and x_4, instead of using its width and height. Similarly, lengths of sides of the line that connects points x_2 and x_5 encode values of x_6 and x_7. In-line Coordinates' primary objective is to facilitate the discovery of n-D patterns and rules with the highest levels of recall and precision.

In-Line Based Coordinates (ILBC) is the name given to these visual representations since they no longer use a single baseline as a basic ILC representation. Instead, they are based on it.

Figure 4a simplifies Fig. 3b, by making sides vertical, and Fig. 4b simplifies this Fig, 4a further by removing vertical lines, which go down to the baseline and keeping only solid lines. In this figure we use more generic simplified notation with attributes named from a to g, because any of the coordinates $\{x_i\}$ can be assigned to be on the baseline or on link lines and in any order.

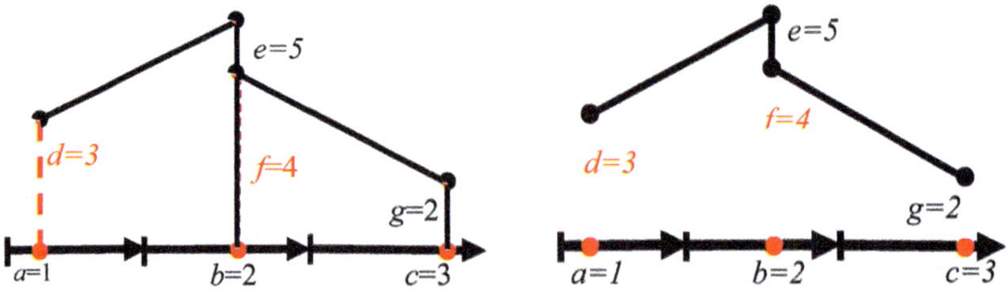

(a) Vertical simplification of Fig. 3b.

(b) Minimized representation.

Fig. 4 7-D point $\mathbf{x} = (a, b, c, d, e, f, g) = (1, 2, 3, 5, 3, 4, 2)$ in ILC2 with vertical sides

In contrast to Parallel Coordinates, which need seven nodes and six edges, this figure enables the complete restoration of all seven values and only needs four nodes and three edges. The vertical sides of the image in Fig. 4 can be understood as follows.

The ILC baseline occupies the Cartesian x-coordinate, while all vertical values *a* through *g* are situated on their corresponding coordinates *A* through *G*, which are vertically collocated on. A combination of two ILCs—horizontal and vertical—makes up ILBC in Fig. 4. Such coordinates will be denoted by ILC2.

Next, ILC2 is compared with Shifted Paired Coordinates (SPC) [7] on the same 7-D point shown in Fig. 5. In Fig. 5a SPC *d*, *e*, *f*, and *g* are also vertical, but start at the origins of individual horizontal coordinates, which are paired. SPC also requires 4 nodes and 3 edges that are longer than in ILBC in Fig. 4b. In ILC and ILBC above, the location of all coordinates on the horizontal baseline is fixed with their values located on this baseline. It is called a static mapping [7]. In contrast in the dynamic mapping of the given n-D point **x**, the location of the next value x_{i+1} in its 2-D graph x* *depends on the location and value of prior* x_i. It is a common concept for all dynamic General Line Coordinates [7], not only ILC and ILBC.

The same 7-D point is shown in Fig. 5b. Here the coordinate B starts at point $a = 2$ of coordinate A with value $b = 2$ located at the distance 2 from point $a = 1$. Respectively coordinate C starts at point $b = 2$ with point $c = 3$ located at the distance 3 from $b = 2$. The respective vertical coordinates *d*, *e*, *f*, and *g* start at the origin of the horizontal baseline. Thus, all of them are collocated and static. Here

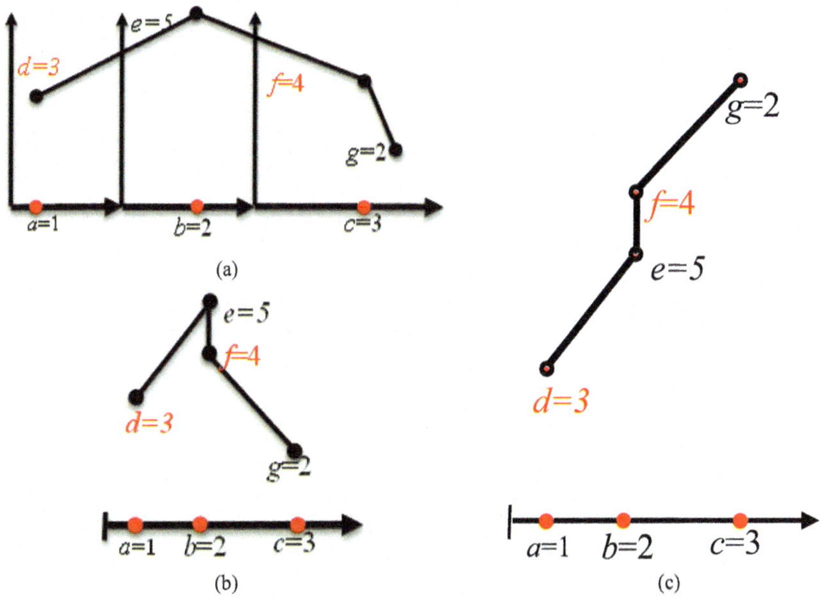

Fig. 5 **a** 7-D point **x** = (*a, b, c, d,* e, *f, g*) = (1, 2, 3, 5, 3, 4, 2) in SPC, **b** in ILBC partial dynamic, **c** in fully dynamic

horizontal coordinates are dynamic, but vertical coordinates are static, therefore, ILBC is called partial dynamic ILC2. Figure 5c shows a full dynamic ILC2 where vertical coordinates are dynamic in the same ways as horizontal coordinates, where the location of e, f and g points depends on the location of their prior points.

Figures 6 and 7 show Wisconsin Breast Cancer (WBC) data of two classes in ILC, where vertical coordinates are collocated and horizontal are static. Figures 6a and 7a are example of one case from WBC dataset for Figs. 6b and 7b. In Fig. 6a, the X-coordinate represents the values of x_1, x_4, x_7, and x_{10}. The y-coordinate represents the values of x_2, x_3, x_5, x_6, x_8, and x_9. These values are $x_1 = 5, x_2 = 1, x_3 = 1, x_4 = 1, x_5 = 2, x_6 = 1, x_7 = 3, x_8 = 1, x_9 = 1$, and $x_{10} = 1$. All patients in WBC dataset are represented in this manner.

Drawing classes "mirrored" in Fig. 6 allows to compare and see the difference and similarities of patterns of two classes without their occlusion. Figure 7b shows much better separation of WBC classes in fully dynamic ILC2.

The general formula to locate pairs (x_i, x_j) in partial dynamic IL2 is given by mapping L as follows:

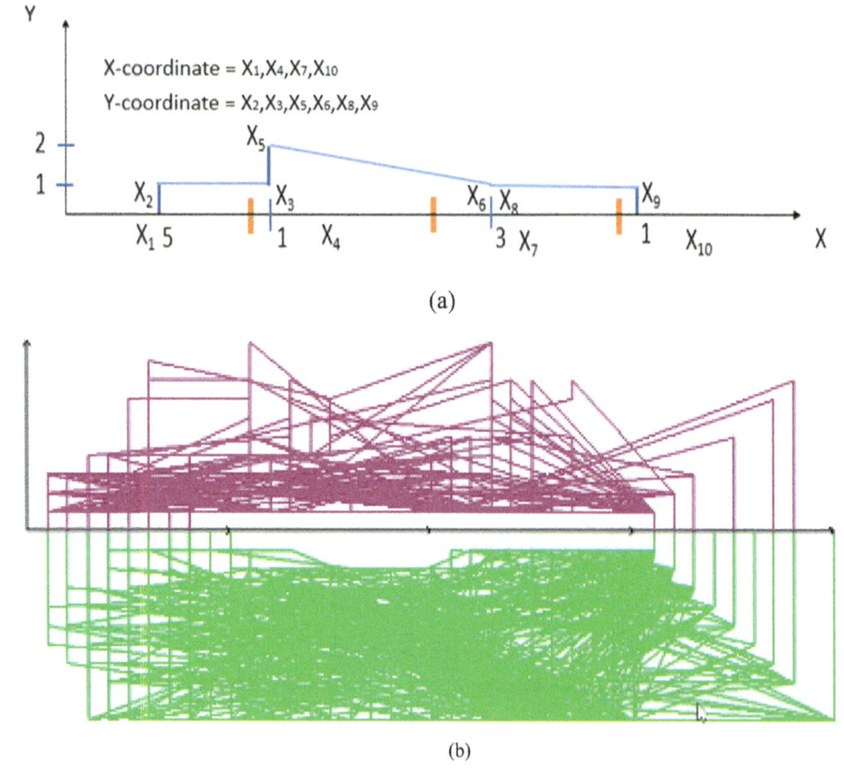

Fig. 6 WBC data in partial dynamic ILBC, **a** a single case, **b** many cases of two classes "mirrored"

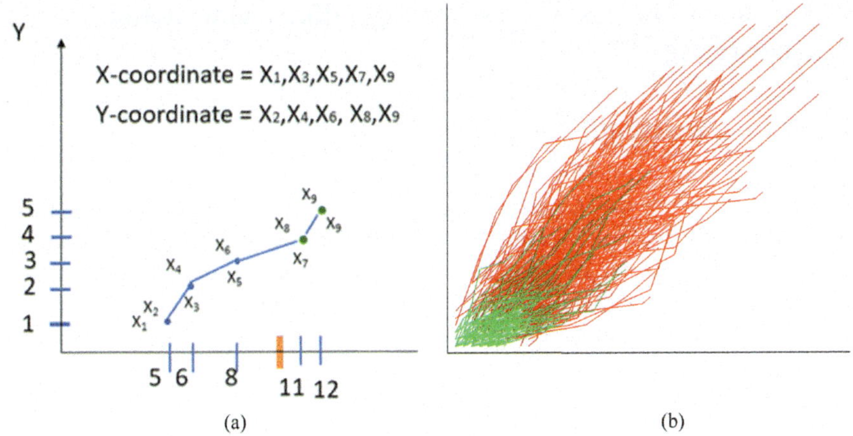

Fig. 7 **a** Example coordinates. **b** Wisconsin Breast Cancer dataset in fully dynamic ILC2

$$L(x_i, x_j) = (x_1 + x_3 + \ldots + x_i, x_j).$$

Respectively the general formula to locate pairs (x_i, x_j) in fully dynamic ILBC is given by mapping L as follows:

$$L(x_i, x_j) = (x_1 + x_3 + \ldots + x_i, x_2 + x_4 + \ldots + x_j).$$

Example: In Fig. 7a, the value (5,1) represents x_1 and x_2. The value (6,2) presents x_{3s} and x_{4s} because the position of x_{3s} depends on the location of x_1. This means that to retrieve the true value of x_3, the value of x_1 is subtracted from x_{3s}. Therefore, $x_3 = x_{3s} - x_1 = 6 - 5 = 1$. This process is repeated for all remaining points. With that, the drawn x values are $(5, 1, 6, 2, 8, 3, 11, 4, 12, 5)$, and the true x values are $(5, 1, 1, 1, 2, 1, 3, 1, 1, 1)$.

Next, weighted dynamic ILBC is introduced which is given by mapping L_w:

$$L_w(x_i, x_j) = (w_1x_1 + w_3x_3 + \ldots + w_ix_i, w_2x_2 + w_4x_4 + \ldots + w_jx_j),$$

where $W = \{w_i\}$ is a set of weights assigned to coordinates.

3 Classification and Regression Algorithms with In-Line Coordinates

3.1 Box Classification Algorithm

The main idea of the Box Classification (BC) algorithm is finding a good box with high purity and many cases, then remove all cases, which are in this box, and repeat the process of finding other good boxes in remaining cases and continue this process until all cases of all classes will be in one of the good boxes. This process is interactive and partially automated. Automation includes computing parameters of the candidates for the good boxes.

The BC algorithm operates on n-D data visualized in ILBC in the following main steps.

Step 1: Split data to training and validation data.
Step 2: Search/discover "good" boxes B_i and form rules on training data in ILBC visualizations that cover dataset cases as pure as possible. Good box' criterion is usually decided by number of cases that it covers and its purity.
Step 3: Test BC rules on independent data.
Step 4: Pruning a set of discovered rules to decrease overfitting.

Below we discuss details of these steps. For many datasets, the exhaustive grid search is possible in Step 2. For instance, it was used to discover boxes with WBC dataset. With WBC dataset, a "good" box is discovered when it covers at least more than 10% of cases of the remaining data. The search of boxes and rules is a *sequential hierarchical process for each class*. If there are multiple "good" boxes, then these "good" boxes are ranked based on the purity of the boxes. Purity of the box is the ratio of the number of cases of the *dominant class* to the number of cases of all other classes in the box. The presence of the case in the box is identified by the fact that a node of the graph of the n-D point **x** is in the box B. In the rule below, it will be denoted $\mathbf{x} \in B$ for short. This definition leads to simpler interpretation of rules based on such boxes, than an alternative definition, which requires that just an *edge* to graph x* of the n-D point **x** crosses the box B_i.

To form BC rules with the discovered boxes the algorithm starts with discovering basic rules:

$$R_i: \text{ if } \mathbf{x} \in B_i \Rightarrow \mathbf{x} \in \text{Class } C_i. \quad (1)$$

and then discovers more complex BC rules sequentially

$$R_i: \text{ if } \mathbf{x} \in B_i \;\&\; \mathbf{x} \notin (B_m \cup B_p \cup \ldots \cup B_t) \Rightarrow \mathbf{x} \in \text{Class } C_i \quad (2)$$

Here B_i is a current "good" box, and other boxes are prior "good" boxes with cases from these boxes removed before B_i is searched. For WBC dataset, the algorithm found 13 rules as presented in Table 2 in Sect. 4.

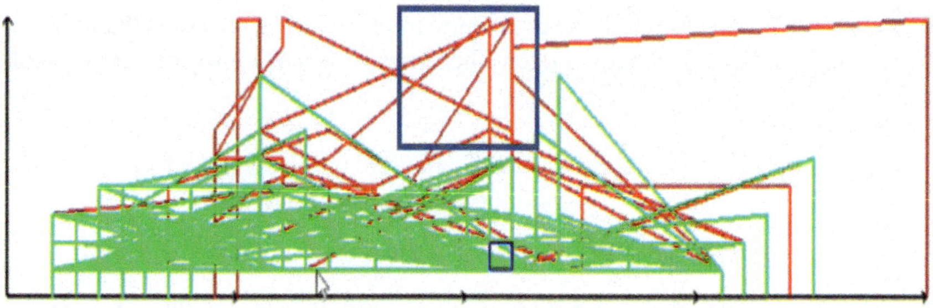

Fig. 8 Examples of boxes discovered in Wisconsin breast cancer dataset

The BC algorithm as an iterative visual logical classifier produces a series of rectangular areas (boxes). Figure 8 shows multiple WBC cases visualized in IVLC with boxes found by the BC algorithm.

The main steps of discovering boxes in Step 2 are:

Step 2.1. Create a *grid* in the ILC area. Each cell of the *grid* is a basic *box*. The box size is increased vertically and then horizontally.
Step 2.2. If the number of boxes (grid cells) and the number of n-D points is relatively small compute purity of *each* box. For the large number of boxes and n-D points use optimization and heuristic algorithms such as genetics algorithms.
Step 2.3. For each class create a list of boxes, where this class dominates. All boxes in this list must cover more cases than the desired percentage of the remaining cases. For instance, for WBC dataset it was 10% of the remaining data cases.
Step 2.4. For each class find a box of highest purity to classify new cases in this box.
Step 2.5. Create a classification rule with highest ranking box like in (1).

These steps in a pseudocode are below.

```
// Set up grid in ILC area
CreateGrid(size)

// Calculate purity of each box
For each box in grid
  CalculatePurity()

// Create list of boxes for each class
For each class
   CreateBoxList(class, coverage_threshold)

// Classify cases
For each class
   For each box in box list
       Use box with highest purity to classify cases
       UpdateClassificationRule(box, class)

// Return classification rule
Return ClassificationRule()
```

Example: for WBC dataset, there are 382 green cases and no red cases that cross box B_1 as shown in Fig. 10 in Sect. 4 later. Therefore, in this example, rule R_1 is created as follow,

$$R_1 : \mathbf{x} \in B_1 \Rightarrow \mathbf{x} \in G \text{ (Benign, 382 cases)}. \tag{3}$$

More boxes and rules for WBC data are identified later in Table 1 in Sect. 4.

Step 2.6. Exclude all cases that are in boxes used in steps 2.4 and 2.5.
Step 2.7. Conduct steps 2.2–2.6 for all remaining cases.

The pruning step is to deal with many "mini" boxes that contain less than a certain percentage of the cases of the original dataset with low level of generalization to avoid overfitting and data memorization. With WBC dataset, it was decided that if a box classified less than 17 cases of the total number of cases, which is 683, then it is very likely that this box is overfitting the data. Example: box B_5 classified only 14 red cases (about 2.05% of 683 cases). So, box B_5 is likely overfitted WBC data. This problem is also well known for decision trees. Without control the depth of decision tree and the number of cases in each terminal node, decision trees produce many terminal nodes with only few cases in each of them. The pruning of decision trees removes overfitting but decreases the accuracy of classification.

The step of the BC algorithm employs a pruning approach which

(a) associates "mini" boxes with the larger boxes interactively or
(b) refuses to predict cases that belong to "mini" boxes.

The association (a) is conducted as follows. Consider two boxes B_1 and B_2 for class C_1 the visualization allows to see their mutual location and to create a joint rule based on them. If boxes are adjacent a single bigger box $B_{1,2}$ can be produced from them. If the boxes are not adjacent than a new rule is formed:

$$\text{If } \mathbf{x} \in B_1 \text{ or } \mathbf{x} \in B_2 \text{ then } \mathbf{x} \in C_1$$

The general form for of a new rule is,

Table 1 Discovered boxes ILC2

Box	x_1, x_2, y_1, y_2	Cases	Box	x_1, x_2, y_1, y_2	Cases
B_1	15, 20.5, 1, 1.5	382	B_2	23.5, 39.5, 8.5, 10	166
B_3	1, 3. 5, 0.5, 2	28	B_4	20, 22.5, 6, 6.5	26
B_5	9.5, 10, 5, 6.5	14	B_6	16, 21, 0.5, 2	18
B_7	17.5, 18.5, 3, 3.5	23	B_8	14.5, 17, 2.5, 3	7
B_9	28.5, 29, 2.5, 3.5	4	B_{10}	17.5, 18.5, 3, 3.5	10
B_{11}	14.5, 15, 5.5, 6	4	B_{12}	26.5, 27, 7, 7.5	1
B_{13}	28, 28.5, 0.5, 9.5	10			

$$\text{If } \mathbf{x} \in B_1 \cup B_2 \text{ then } \mathbf{x} \in C_1$$

The interactivity of the BC algorithm has an advantage of allowing the end-users to observe "mini" boxes and decide to follow (a) or (b).

3.2 Linear Classification and Regression Algorithm

Design of rules based on boxes have limitations. One of them is locality of each box, i.e., the box covers only nearby cases. Typically, several boxes are needed to cover all data. In contrast, a single linear classifier can cover all data if data are linearly separable. The goal of this section is proposing an analog of a linear classifier in ILC. Figure 9 illustrates the proposed approach. First, a black line is built which is used to project all cases of both classes to this line. If the projected endpoints of cases of one class C mostly concentrate on the one side of the discrimination blue line, then a linear discrimination model M is discovered, where T is threshold on the black line shown with the blue discrimination line:

$$M(\mathbf{x}) > T \Rightarrow \mathbf{x} \in C \tag{4}$$

Another more common linear discrimination model for two classes C and Q is:

$$M(\mathbf{x}) > T \Rightarrow \mathbf{x} \in C \text{ else } \mathbf{x} \in Q \tag{5}$$

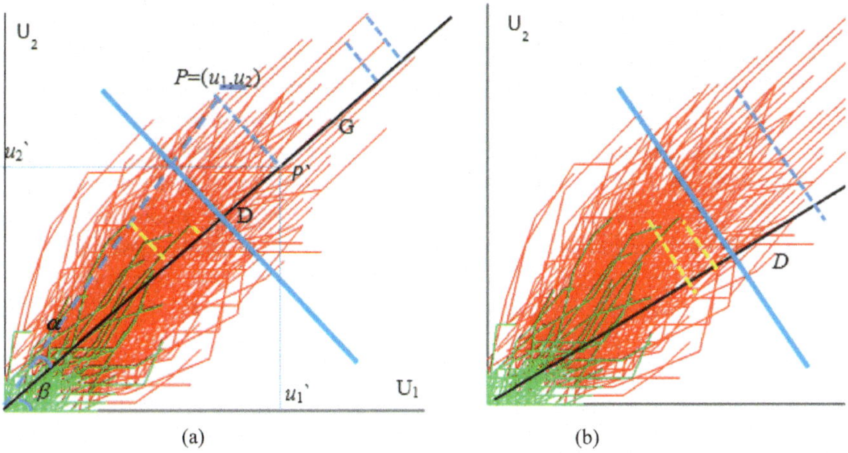

Fig. 9 Linear classification and regression algorithm with projection line at different angles in full dynamic ILP2

The model (4) only covers n-D points **x**, where $M(\mathbf{x}) > T$. This is the situation in Fig. 9 for red class above the blue line. This means that if a single model (5) cannot be built, several models like (4) need to be built that may require different black lines, where endpoints are projected as shown in Fig. 9b. Moreover, it can relax a requirement that only endpoints are projected. It can project some intermediate nodes x_k and x_u of graph of $\mathbf{x}^*$ for $k < n$ and $u < n$, where n is the dimension of n-D point **x** getting models like,

$$M(\mathbf{x}_k) > T \Rightarrow \mathbf{x} \in C.$$

$$M_1(\mathbf{x}_k) > T_k \;\&\; M_2(\mathbf{x}_u) > T_u \Rightarrow \mathbf{x} \in C.$$

Those intermediate points can be found by the BC algorithm presented above. Figure 9a and b look different because the black lines are drawn differently, which allow to optimize the prediction parameters and accuracy. The endpoints of the black lines are chosen so that a higher precision accuracy can be obtained.

4 Case Study with WBC Data in ILC2

4.1 Experiment on All WBC Data

This section presents the results of the computational experiment for discovering classification rules for WBC data encoded in partial dynamic ILC2 using the BC algorithm. The discovered 13 pure boxes are presented in Table 1. These rules can be simplified and pruned interactively. With WBC dataset, exhaustive grid search is used to discover the hyper-parameters (corners of the boxes), See column 2 and 5 in Table 1. Exhaustive grid search can be used with WBC dataset because WBC dataset has a low-resolution attributes with only ten values in each attribute.

With higher dimension and higher resolution dataset, exhaustive grid search will take much greater computational time because there will be much more cells in the grid. We avoided such computations by proposing a heuristic algorithm that is described in section V. The boxes cover all WBC cases. Hyper-parameters are values x_1, x_2, y_1, y_2, which identify left, right, bottom, and top corners of the box within ILC2 display.

Table 2 presents the rules constructed from these boxes in the hierarchical process of the BC algorithm that was described above. The benign class is denoted as B and is drawn as green with letter G used to identify this class in Table 2. Respectively, the class malignant is denoted by M and R (red) for short.

The boxes and rules in Tables 1 and 2 cover all 444 B cases and 239 M cases. Boxes B_1–B_4 and respective rules R_1-R_4) cover most of the cases (602 cases) with 100% precision without any misclassified cases. This means that 88.14% of all cases are classified by *simplest single box rules* without any other boxes involved. The

Table 2 Rules R_1–R_{13} with precision P = 100%

Benign, B (green, G) class rules
R_1: $\mathbf{x} \in B_1 \Rightarrow \mathbf{x} \in G$ (382 cases)
R_3: $\mathbf{x} \in B_3 \Rightarrow \mathbf{x} \in G$ (28 cases)
R_6: $\mathbf{x} \in B_6$ & $\mathbf{x} \notin B_2 \cup B_4 \cup B_5 \Rightarrow \mathbf{x} \in G$ (18 cases)
R_8: $\mathbf{x} \in B_8$ & $\mathbf{x} \notin B_2 \cup B_4 \cup B_5 \cup B_7 \cup B_{10} \cup B_{13} \Rightarrow \mathbf{x} \in G$ (7 cases)
R_9: $\mathbf{x} \in B_9$ & $\mathbf{x} \notin B_2 \cup B_4 \cup B_5 \cup B_7 \cup B_{10} \cup B_{13} \Rightarrow \mathbf{x} \in G$ (4 cases)
R_{11}: $\mathbf{x} \in B_{11}$ & $\mathbf{x} \notin B_2 \cup B_4 \cup B_5 \Rightarrow \mathbf{x} \in G$ (4 cases)
R_{12}: $\mathbf{x} \in B_{12}$ & $\mathbf{x} \notin B_2 \cup B_4 \cup B_5 \cup B_7 \cup B_{10} \Rightarrow \mathbf{x} \in G$ (1 case)
Malignant, M (red, R) class rules
R_2: $\mathbf{x} \in B_2 \Rightarrow \mathbf{x} \in R$ (166 cases)
R_4: $\mathbf{x} \in B_4 \Rightarrow \mathbf{x} \in R$ (26 cases)
R_5: $\mathbf{x} \in B_5$ & $\mathbf{x} \notin B_1 \cup B_3 \Rightarrow \mathbf{x} \in R$ (14 cases)
R_7: $\mathbf{x} \in B_7$ & $\mathbf{x} \notin B_1 \cup B_3 \cup B_6 \Rightarrow \mathbf{x} \in R$ (13 cases)
R_{10}: $\mathbf{x} \in B_{10}$ & $\mathbf{x} \notin B_3 \cup B_6 \cup B_8 \cup B_9 \Rightarrow \mathbf{x} \in R$ (10 cases)
R_{13}: $\mathbf{x} \in B_{13}$ & $\mathbf{x} \notin B_1 \cup B_3 \cup B_6 \cup B_8 \cup B_9 \cup B_{11} \cup B_{12} \Rightarrow \mathbf{x} \in R$ (10 cases)

other rules involve "negated" boxes requiring that the case does not belong to these boxes to satisfy the rule. Some rules in Table 2 such as rule R_8 have simplified forms too with reduced "negated" boxes, because all cases of some boxes are covered by other boxes in these rules.

Table 2 shows that class G has more rules/boxes with smaller coverage (four rules that cover from one to seven cases with total 16 cases covered by these rules). These boxes can be called "*mini*" boxes. In contrast, class R (red) has no rules and boxes with such small coverage. It means a better generalization for R class, than for G class, in these rules. When rules are analyzed with large coverage, the situation is the opposite. The first three G rules cover 428 cases (96.4% of G cases), while the first two R rules (rules R_2 and R_4) cover 192 cases, 80.3% of R cases).

Next, the rules that cover a small number of cases are more complex. This is rather a memorization of cases, than their generalization. Such complex rules are needed only for a small number of cases. This is an indication that the domain experts, who designed attributes for WBC data correctly captured/engineered critical attributes. It also can indicate superior human abilities to generate informative features manually. While, deep learning algorithms can automatically discover informative features, often it is challenging to interpret and explain them.

The tenfold cross validation of a comparable ID3 Decision Tree (DT) gave the average accuracy of 92.85%, with min of accuracy equal to 89.86% and max of accuracy equal to 94.58% [21]. The representative DT of these DTs has 30 terminal nodes that contain seven or less cases [21]. In contrast the BC algorithm produced 13 boxes/rules, and *only four* of them contain seven or less cases on all WBC data.

Figures 10, 11, 12, 13, 14, 15 and 16 illustrate all boxes showing the cases, which cross these boxes. In addition, Figs. 12, 13, 14, 15 and 16 show also the cases of other colors, which are *removed* before discovering a given box by requiring not to belong to a set of prior boxes, listed in the rules in Table 2. All these visualizations are

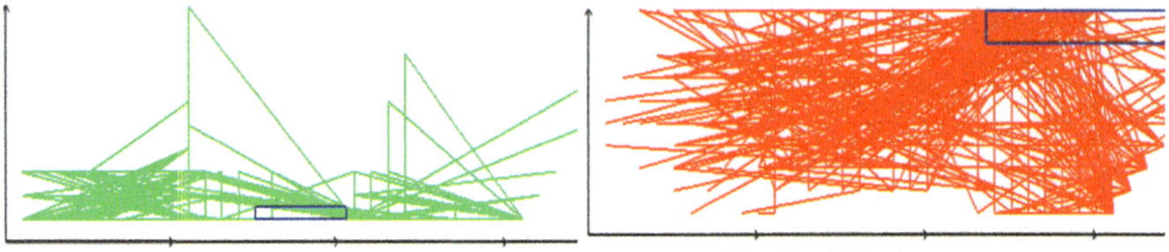

Box B_1: 382 green cases. Box B_2: 166 red cases.

Fig. 10 Boxes B_1 and B_2

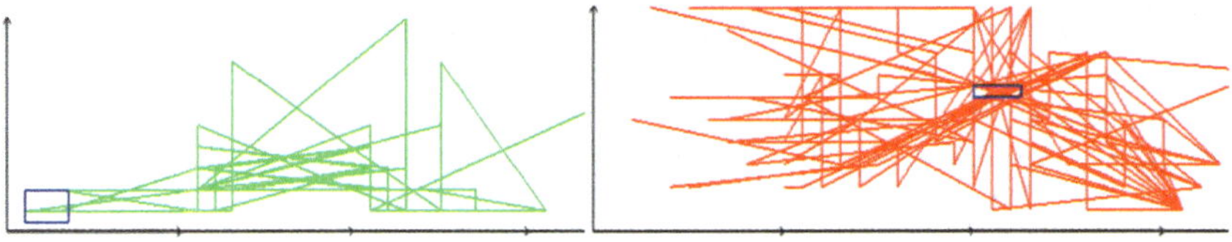

Box B_3: 28 green cases. Box B_4: 26 red cases.

Fig. 11 Boxes B_3 and B_4

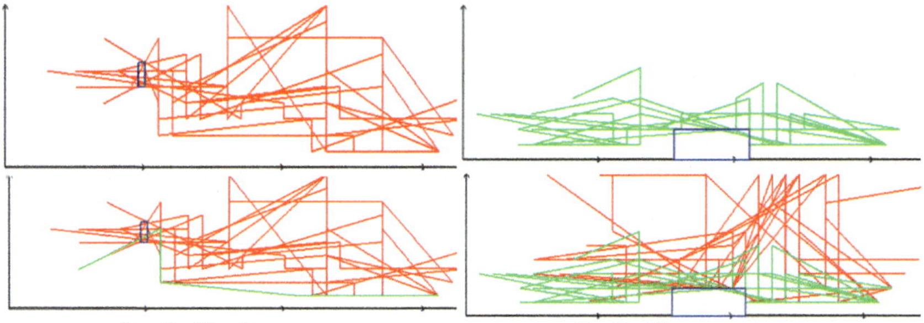

Box B_5: 14 red cases. Box B_6: 18 green cases.

Fig. 12 Boxes B_5 and B_6

done in accordance with the partial dynamic ILBC visualization method illustrated in Fig. 6a.

Example: in Fig. 10, 382 green cases are classified with box B_1. The BC algorithm removed theses 382 green cases before the next search. Because BC algorithm is sequential hierarchical process, it searches for each box in the sequential order such as B_1, then B_2 and then B_3 and so on. This means that after cases of box B_1 are removed, there will be only $444 - 382 = 62$ green cases remaining for the next search. Because of the small number of cases left, there is a good chance that a rule for them can be overfitting. Therefore, it will be beneficial to analyze all green cases without removing cases of the prior boxes to get another rule with higher coverage.

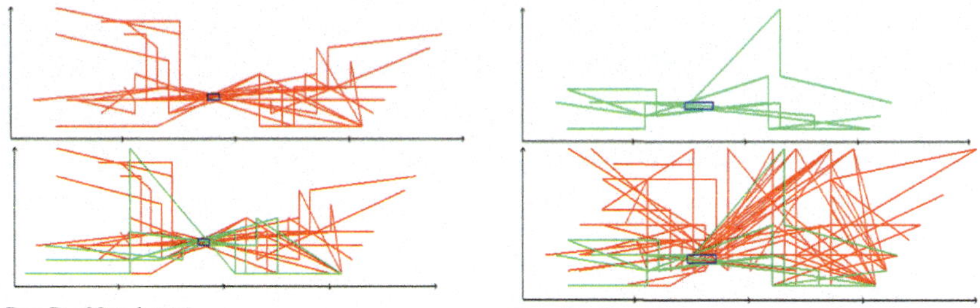

Box B_7: 23 red cases. Box B_8: 7 green cases.

Fig. 13 Boxes B_7 and B_8

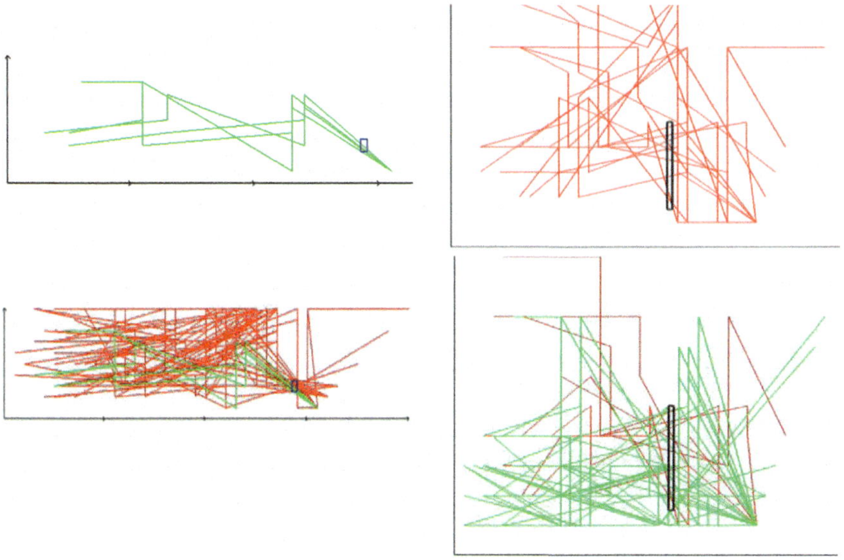

Box B_9: 4 green cases. Box B_{10}: 7 green cases.

Fig. 14 Boxes B_9 and B_{10}

Figures 10 and 11 represent situations when boxes are discovered *without preconditions* on other boxes. These pictures show boxes B_1, B_2, B_3, and B_4 with only cases that cross them. Figures 12, 13, 14, 15 and 16 represent situations with boxes discovered *with a precondition* that cases that cross prior boxes.

In these figures for each box B_i, the first picture shows only cases which cross box B_i after removing cases from B_1-B_4 and the second picture also shows cases from the opposite class that cross B_i without removing cases that cross prior boxes. For example, Fig. 12 first shows 14 red cases that cross box B_5 and then it shows both red and green cases without removing 382 cases that cross box B_1 and 28 cases that cross box B_3. For box B_6 the first picture shows 18 green cases, and the second

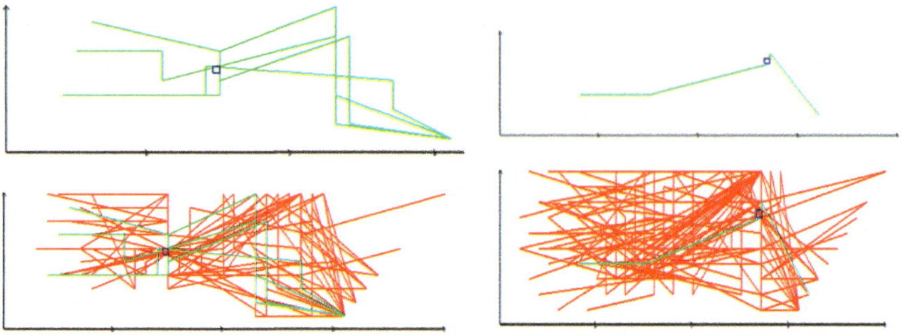

Box B_{11}: 4 green cases. Box B_{12}: 7 red cases.

Fig. 15 Boxes B_{11} and B_{12}

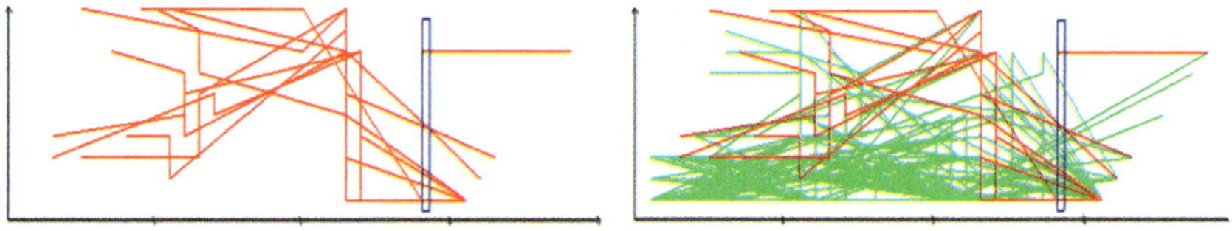

Box B_{13}: 10 red cases. Box B_{13}: 10 green cases.

Fig. 16 Box B_{13}

picture shows both green and red cases without removing red cases from boxes B_2 and B_4.

Pruning. Boxes B_9, B_{11} with four green cases each and box B_{12} with one green case can be used to prune the set of rules by creating modified rules,

$$R_{9M}: \mathbf{x} \in B_9 \Rightarrow \mathbf{x} \in R \text{ (47 red /4 green)}.$$

$$R_{11M}: \mathbf{x} \in B_{11} \Rightarrow \mathbf{x} \in R \text{ (28 red /4 green)}.$$

$$R_{12M}: \mathbf{x} \in B_{12} \Rightarrow \mathbf{x} \in R \text{ (52 red /1 green)}.$$

Here $\mathbf{x} \in B_i$ means that polyline for n-D point $\mathbf{x}$ crosses the box B_i.

As it can be seen these rules have low error rate. Currently, the pruning process is interactive, therefore the end-users can explore them and accept if the error rate is tolerable.

Joining rules. The next task is decreasing the number of rules, which is demonstrated on 13 rules shown above. The proposed approach joins rules by combining them including the use of the else condition. In contrast with the pruning, this process does not introduce any error. The result is shown in Table 3.

The steps of the Rule Joining (RJ) algorithm are:

Step 1: Combine rules with a single rectangle of a given class.
Example: $R_{1,3}$: $\mathbf{x} \in B_1 \cup B_3 \Rightarrow \mathbf{x} \in G$ (410 cases).

Table 3 Rules after joining

Expanded benign (green, G) class rules
$R_{1,3}$: $x \in B_1 \cup B_3 \Rightarrow x \in G$ (410 cases)
$R_{1,3,5}$: $\mathbf{x} \in B_1 \cup B_3 \Rightarrow \mathbf{x} \in G$ (else $\mathbf{x} \in B_5 \Rightarrow \mathbf{x} \in R$) (424 cases)
$R_{8,9}$: $\mathbf{x} \in B_8 \cup B_9$ & $\mathbf{x} \notin B_2 \cup B_4 \cup B_5 \cup B_7 \Rightarrow \mathbf{x} \in G$ (11 cases)
$R_{11,12}$: $\mathbf{x} \in B_{11} \cup B_{12}$ & $\mathbf{x} \notin B_2 \cup B_4 \cup B_5 \Rightarrow \mathbf{x} \in G$ (5 cases)
Expanded malignant (red, R) class rules
$R_{2,4}$: $\mathbf{x} \in B_2 \cup B_4 \Rightarrow \mathbf{x} \in R$ (192 cases)
$R_{2,4,6}$: $\mathbf{x} \in B_2 \cup B_4 \Rightarrow \mathbf{x} \in R$ (else $\mathbf{x} \in B_6 \Rightarrow \mathbf{x} \in G$) (210 cases)
R_7: $\mathbf{x} \in B_7$ & $\mathbf{x} \notin B_3 \cup B_6 \Rightarrow \mathbf{x} \in R$ (13 cases)
$R_{2,4,8}$: $\mathbf{x} \in B_2 \cup B_4 \Rightarrow \mathbf{x} \in R$ (else $\mathbf{x} \in B_8$ & $\mathbf{x} \notin B_5 \cup B_7 \Rightarrow \mathbf{x} \in G$) (199 cases)
$R_{2,4,6,8}$: $\mathbf{x} \in B_2 \cup B_4 \Rightarrow \mathbf{x} \in R$ (else $\mathbf{x} \in B_6 \cup B_8$ & $\mathbf{x} \notin B_5 \cup B_7 \Rightarrow \mathbf{x} \in G$) (217 cases)
R_{10}: $\mathbf{x} \in B_{10}$ & $\mathbf{x} \notin B_3 \cup B_6 \cup B_8 \cup B_9 \Rightarrow \mathbf{x} \in R$ (10 cases)
R_{13}: $\mathbf{x} \in B_{13}$ & $\mathbf{x} \notin B_1 \cup B_3 \cup B_6 \cup B_8 \cup B_9 \cup B_{11} \cup B_{12} \Rightarrow \mathbf{x} \in R$ (10 cases)

Step 2: Find the opposite class rules that are conditioned by rectangles used in Step 1.

Example: Rule 5 is created after BC algorithm has removed 602 samples (410 green cases and 192 red cases) from boxes B_1, B_2, B_3, and B_4. Without removing these 602 samples, some green samples are crossing through box B_5. Therefore, rule R_5 is conditioned by $\mathbf{x} \notin B_1 \cup B_3$.

Step 3: Combine rules from Steps 1 and 2.

Example: $R_{1,3,5}$: $\mathbf{x} \in B_1 \cup B_3 \Rightarrow \mathbf{x} \in G$ (else $\mathbf{x} \in B_5 \Rightarrow x \in R$) (428 cases). Here rule R_5 covers only 14 cases that can be viewed as a potential overfitting, while rule $R_{1,3,5}$ covers 428 cases. The else condition makes R_5 a part of the larger rule.

The analysis of rules in Table 3 shows that seven rules $R_{1,3,5}$, $R_{8,9}$, $R_{11,12}$, $R_{2,4,6,8}$, R_7, R_{10}, and R_{13} are equivalent to 13 original rules. Here rules $R_{8,9}$, $R_{11,12}$, R_{10}, and R_{13} cover 10 or less cases with total 36 cases (16 green and 20 red). Excluding these rules and, respectively refusing to classify cases that satisfy them will eliminate potential overfitting.

4.2 *Experiment with Stratified Tenfold Cross Validation*

All discovered WBC rules were presented above in Tables 2 and 3. This section analyzes rules $R_{1,3}$ and $R_{2,4}$ discovered by BC algorithm, which classify over 88% of WBC data. To get the even number of coordinates for the algorithm to run, the attribute X_9 was used twice. Table 4 presents results of BC algorithm on WBC data for these rules for *all* 444 green (benign) and 239 red (malignant) cases of WBC data without splitting data to training and validation subsets. The results are summarized in tables. Here, rule $R_{1,3}$ predicts green class and $R_{2,4}$ predicts red class based on all WBC data. Hyper-parameters of the rectangles B_1 and B_3 used in rule $R_{1,3}$ are $x_1 = 15$, $x_2 = 20.5$, $y_1 = 1$, $y_2 = 1.5$ for B_1 and $x_1 = 1$, $x_2 = 3.5$, $y_1 = 0.5$, $y_2 = 2$ for

Table 4 Precision and recall of rules $R_{1,3}$ and $R_{2,4}$ for all data

Rule	WBC dataset	Rule precision (%)	Rule recall (%)
$R_{1,3}$	683 cases	100	92.34
$R_{2,4}$	683 cases	100	80.33

Table 5 Precision and recall of rule $R_{1,3}$ and $R_{2,4}$ for stratified tenfold cross validation (average and min max interval)

Rule	Rule precision		Rule recall	
	Training (%)	Validation (%)	Training (%)	Validation (%)
$R_{1,3}$	99.67 [99.2, 99,73]	99.57 [99.65, 100]	93.38 [93.25, 99,75]	94.56 [90.91, 100]
$R_{2,4}$	95.90 [95.52, 96.5]	96.80 [95.0, 100.0]	89.21 [88.37, 89,76]	85.00 [75.0, 100.0]

B_3. Hyper-parameters of the rectangles B_1 and B_4 used in rule $R_{2,4}$ are $x_1 = 23.5$, $x_2 = 39.5$, $y_1 = 8.5$, $y_2 = 10$ for B_2 and $x_1 = 20$, $x_2 = 22.5$, $y_1 = 6$, $y_2 = 6.5$ for B_4. These rules are:

$$R_{1,3}: \mathbf{x} \in B_1 \cup B_3 \Rightarrow \mathbf{x} \in G \text{ (Benign)}.$$
$$R_{2,4}: \mathbf{x} \in B_2 \cup B_4 \Rightarrow \mathbf{x} \in R \text{ (Malignant)}.$$

The importance of the results in Table 4 is that two simple visual rules $R_{1,3}$, and $R_{2,4}$, which need only four unchanged rectangles B_1-B_4 allowed to recall of 80.3 and 92,34% cases with 100% precision on all data. Table 5 shows a similar performance in stratified tenfold cross validation of BC algorithm on WBC data for rules $R_{1,3}$ (green class) and rule $R_{2,4}$ (red class). For both rules WBC data were split to training and validation datasets in accordance with the total number of cases in the respective class.

5 Box Classification Algorithm for Multiclass Unbalanced High-Resolution Data

The goal of designing a box classification algorithm for *multiclass imbalanced high-resolution data* is motivated by the Page Block Classification (PBC) dataset, which has all these properties. PBC dataset [22] has 5473 cases with 10 attributes each. There are 4913 cases from class Text (class C_1), 329 cases from class Horizontal Line (class C_2), 28 cases from class Graphic (class C_3), 88 cases from class Vertical Line (class C_4) and 115 cases from class Picture (class C_5). This dataset is *heavily imbalanced* in the number of cases of its *five classes* that range from 28 to 4913 cases. PBC is also a high-resolution dataset where each attribute has a large number of values, where some attributes have 5 digits in each value leading to 10^5 values. Therefore, the BC algorithm is presented below in the terms of PBC data being applicable to many multiclass unbalanced high-resolution data. This expanded BC algorithm is divided into several sub-algorithms presented below.

Divide and conquer (DC) algorithm for imbalanced data. To classify imbalanced data, a common divide and conquer approach is used with three steps presented for five PBC data classes.

Step 1: *Combine* Classes C_2–C_5 into a single class C_{2345}. Classify cases between class C_1 and class $C_{2345.}$ This task is less imbalanced with 4913 cases in C_1 and 560 cases in C_{2345} than the task with 5 classes C_1–C_5.
Step 2: *Classify* cases between class C_2 and class C_{345} that combines classes C_3–C_5 which is also less imbalanced: 329 cases in C_2 versus 231 cases in C_{345}.
Step 3: *Classify* 231 cases in joint class C_{345} among three classes C_3, C_4 and C_5.

DT Guided (DTG) algorithm for high-resolution dataset PBC data. PBC is a *high-resolution dataset* with many digits on attributes. In contrast, in WBC data, each value consists of a single digit where the exhaustive grid search runs only on 1000 boxes. The exhaustive grid search in PBC data requires to run on a grid that is several orders of magnitude larger. So, the run time needs to be decreased. In [11] a random selection of grid cells (boxes) was used to decrease the search time. Here a *Decision Tree* (DT) is used as a guide for finding promising boxes. The steps of the DT Guided (DTG) algorithm are as follows.

Step 1: *Build a DT* on the same data.
Step 2: *Select high purity* DT branches, where a single class highly dominates.
Step 3: *Build boxes* based on those branches (one or more boxes from the branch).
Step 4: *Search for better boxes* in the vicinity of boxes derived from the DT step 3.

The link between branches of the DT and the boxes is shown in [21]. Each branch of DT can be a source of several boxes because boxes in static ILP are two-dimensional, while each DT branch commonly involve more attributes. The current implementation of step 3 starts from nodes of the branch that are close to the root of the DT because they typically cover more cases than the nodes, which are closer to the terminal nodes. As an illustration consider the following branch of the DT for the 4-D point $\mathbf{x} = (x_1,x_2,x_3,x_4)$ with all $x_i \in [0,10]$,

$$\text{If } (x_1 > 5 \& x_3 > 6) \& (x_2 < 3 \& x_4 < 7) \text{ then } \mathbf{x} \in \text{class } 1.$$

Here box B_1 from ($x_1 > 5$ & $x_3 > 6$) and box B_2 from ($x_2 < 3$ & $x_4 < 7$) are created. Respectively B_1 is defined by $10 \geq x_1 > 5$, $10 \geq x_3 > 6$. When the limit 10 is a known it can be simplified to ($x_1 > 5$ & $x_3 > 6$) as it is commonly written in the DTs.

BC algorithm as a generalization of Decision Tree algorithm. DT and rules based on boxes serve the same common goal of providing interpretable and easily visualizable models. The major limitation of models constructed by the DT algorithms is the need to select a start attribute (*tree root*). This *narrows the class of models* that can be discovered. It led to development of *Random Forests* (RFs) algorithms, where multiple DTs are combined by voting. RFs fundamentally expanded the class of models but with the cost of *losing interpretability*. The BC algorithm covers a *wider class of models* than DTs because they to do not require a root attribute. This is an *advantage* for the BC algorithm over the DTs.

ILC box visualization as a richer visualization of Decision Tree. The description of a DT as sequences of boxes is not only an alternative to *describe* DT but also an alternative to *visualize* it in ILC as boxes as Figs. 6, 7, 8, 9, 10, 11, 12, 13, 14, 15 and 16 show. The advantage of this visualization of DT relative to a traditional visualization of decision trees is in ability to trace each n-D point in the tree visually. The traditional DT visualization is not doing this in contrast with the box visualization in ILC. It shows all cases that go through the boxes and fully represents the DT. It makes this visualization richer and more informative. ILC allows distinguishing DT branches by using distinct colors when all branches are visualized together or by showing each branch in the separate ILC.

6 Case Study with Page Block Classification Data in ILC2

6.1 Construction of the Guiding Decision Tree

In accordance with the DT guided algorithm for PBC high-resolution data we

(a) build a DT on the PBC data,
(b) select high purity DT branches, where a single class highly dominates,
(c) build boxes based on those branches, and
(d) search for better boxes in the vicinity of boxes from step 3.

Table 6 shows the characteristics of the constructed decision tree. To compute the *total precision of k rules* for a class we use the weighted precision formula [12]

$$P = \sum_{i=1}^{k} (p_i c_i) / \sum_{i=1}^{k} c_i$$

where p_i is precision of rule R_i and c_i is the number of cases covered by R_i.

Example: Consider rule R_a that predicted 100 cases with 90 correctly classified and 10 misclassified cases ($P_a = 90\%$ precision). Another rule R_b classified 200 cases, with 160 cases correctly and 40 incorrectly classified ($P_2 = 80\%$ precision).

Here $c_a = 100$ and $c_b = 200$, $W_a = c_a/(c_a + c_b) \approx 0.33$ and $W_b = c_b/(c_a + c_b) \approx 0.67$ and the weighted precision $P = 0.33 * 90\% + 0.67 * 80\% = 83.33\%$. Table 6 uses this concept of weighted precision P that is the weighted sum of precisions of all rules R_1-R_k:

$$P = W_1 P_1 + W_2 P_2 + \cdots + W_k P_k$$

where the weight W_i of each rule R_i is computed as follows:

Table 6 Weighted precision for all classes with DT for PBC dataset

Class	Weight (%)	Precision (%)	Classified cases	Weighted precision (%)
	Training			
Class 1	89.14	98.05	3950	87.41
Class 2	7.15	74.76	317	5.35
Class 3	0.00	0.00	0	0.00
Class 4	1.62	90.28	72	1.47
Class 5	2.08	46.74	92	0.97
All	100.00		4431	95.19
	Validation			
Class 1	58.10	97.56	287	56.68
Class 2	7.89	58.97	39	4.66
Class 3	**0.00**	**0.00**	**0**	**0.00**
Class 4	31.17	5.19	154	1.62
Class 5	2.83	21.43	14	0.61
All	100.00		494	63.56
	Testing			
Class 1	87.77	98.75	481	86.68
Class 2	7.85	72.09	43	5.66
Class 3	0.00	0.00	0	0.00
Class 4	1.64	77.78	9	1.28
Class 5	2.74	60.00	15	1.64
All	100.00		548	95.26

$$W_i = \frac{c_i}{(c_1 + c_2 \ldots\ldots c_k)}.$$

where c_i is the total number of cases classified by rule R_i as above.

fThe abilities to use a decision tree as a good guidance heavily depend on the quality of the decision tree, therefore, we need to analyze this tree. An ideal DT will be highly accurate without overfitting. Much more often a DT is not perfect but with insufficient accuracy and significant overfitting.

Multiple DT branches may have terminal nodes, which include just few cases rather memorizing data than learning the patterns. Table 6 demonstrates highly insufficient precision of the constructed DT for some data subsets. *None* of the cases from class 3 is correctly classified.

This means that the guidance from the decision tree can be incomplete. The hope is that the DT guidance will decrease the number of cases and the areas to search for good boxes needs dramatically. It will allow an exhaustive search with the remaining parts of the visualization data space.

Our experiments with PBC dataset reported in the next section shows success of this strategy. An alternative approach is building a better tree or pruning the available tree. While pruning is a common way to decrease overfitting of DT it can decrease accuracy.

In the next section we show that BC algorithm allowed getting high accuracy without pruning because it is not limited by the tree structure.

6.2 Construction Rules in Stratified Tenfold Cross Validation Experiment

For stratified tenfold cross validation, PBC dataset is split into 90%:10% where 90% used for into training and validation set and 10% for independent testing. Training and validation set was then split to 90% training set and 10% validation set.

Tables 7, 8, 9, 10 and 11 present the result of BC algorithm *1st fold* in stratified tenfold cross validation. Results for folds 2–10 are summarized in Table 12. Table 7 presents hyper-parameters of rectangles *derived from the DT* for this fold and Table 8 presents rules based on these rectangles. This DT was build using respective 90% of training and validation set designated for this fold.

Table 7 Hyper-parameters of the rectangles B_1–B_{11} (BC algorithm, PBC dataset) of 1st fold in stratified tenfold cross validation

Box	Hyper-parameters	Box	Hyper-parameters
B_1	$X_6 < 0.0011$ and $0.0015 \le X_0 < 0.0214$	B_2	$0.1550 \le X_4 < 0.9394$ and $0.0065 < X_0 \le 0.107$
B_3	$0.7525 \le X_5 < 1$	B_4	$0.0750 \le X3 < 1$ and $0.0001 \le X6 < 1$
B_5	$0 \le X3 < 0.0005$	B_6	$0.3730 \le X_4 < 1$ and $0.0115 \le X_3 < 0.537$
B_7	$0 \le X_4 < 0.12$	B_8	$0 < X_3 \le 0.0005$
B_9	$0 < X_3 \le 0.0005$	B_{10}	$0 \le X_4 \le 0.2944$
B_{11}	$0.006 \le X_2 \le 0.6058$		

Table 8 Rules R_1–R_6 (BC algorithm, PBC dataset) using boxes B_1–B_{11}

Class C_1 rule	R_1: $\mathbf{x} \in B_1 \cup B_2 \cup B_3 \Rightarrow \mathbf{x} \in$ Class C_1
Class $C_{2,3,4,5}$ rule	R_2: $\mathbf{x} \in B_4 \cup B_5 \cup B_6 \cup B_7$ & $\mathbf{x} \notin B_1 \cup B_2 \cup B_3 \Rightarrow \mathbf{x} \in$ Class $C_{2,3,4,5}$
Class C_2 rule	R_3: $\mathbf{x} \in B_8 \Rightarrow \mathbf{x} \in$ Class C_2
Class C_4 rule	R_4: $\mathbf{x} \in B_9$ & $\mathbf{x} \notin B_8 \Rightarrow \mathbf{x} \in$ Class C_4
Class C_5 rule	R_5: $\mathbf{x} \in B_{10}$ & $\mathbf{x} \notin B_8 \cup B_9 \Rightarrow \mathbf{x} \in$ Class C_5
Class C_3 rule	R_6: $\mathbf{x} \in B_{11}$ & $\mathbf{x} \notin B_8 \cup B_9 \cup B_{10} \Rightarrow \mathbf{x} \in$ Class C_3

Table 9 Number of cases that satisfy rules R_1–R_6 (BC algorithm, PBC dataset) of first fold in stratified tenfold cross validation

Rule	Training		Validation		Testing	
R_1	$C_1 => C_1$	$C_{2345} => C_1$	$C_1 => C_1$	$C_{2345} => C_1$	$C_1 => C_1$	$C_{2345} => C_1$
	3782	55	420	5	445	5
R_2	$C_{2345} => C_{2345}$	$C_1 => C_{2345}$	$C_{2345} => C_{2345}$	$C_1 => C_{2345}$	$C_{2345} => C_{2345}$	$C_1 => C_{2345}$
	359	35	42	4	48	9
R_3	$C_2 => C_2$	$C_{345} => C_2$	$C_2 => C_2$	$C_{345} => C_2$	$C_2 => C_2$	$C_{345} => C_2$
	251	2	28	0	32	3
R_4	$C_4 => C_4$	$C_{235} => C_4$	$C_4 => C_4$	$C_{235} => C_4$	$C_4 => C_4$	$C_{235} => C_4$
	66	6	7	0	7	1
R_5	$C_5 => C_5$	$C_{234} => C_5$	$C_5 => C_5$	$C_{234} => C_5$	$C_5 => C_5$	$C_{234} => C_5$
	83	8	11	1	12	0
R_6	$C_3 => C_3$	$C_{245} => C_3$	$C_3 => C_3$	$C_{245} => C_3$	$C_3 => C_3$	$C_{245} => C_3$
	22	2	3	1	3	0

Table 10 Precision and recall that satisfy rules R_1–R_6 (BC algorithm, PBC dataset) of first fold in stratified tenfold cross validation

Rule	Training		Validation		Testing	
	Precision (%)	Recall (%)	Precision (%)	Recall (%)	Precision (%)	Recall (%)
R_1	98.57	95.03	98.82	95.02	98.89	90.6
R_2	91.12	79.60	91.30	80.77	84.21	84.21
R_3	99.21	94.36	100.00	93.33	91.43	96.97
R_4	91.67	92.96	100.00	87.50	87.50	77.78
R_5	91.21	90.22	91.67	100.00	100.00	100.00
R_6	91.67	100.00	75.00	100.00	100.00	100.00

The process of classification using these rules is hierarchical. Case **x** is classified between class C_1 and a joined class $C_{2,3,4,5}$ and then if **x** is in $C_{2,3,4,5}$ then **x** is classified to classes C_2-C_5. See Fig. 17.

Table 9 shows the number of cases that were predicted for training, validation, and testing sets in Table 8. The notation $C_i => C_i$ indicates situations when cases of class C_i were predicted correctly as cases of class C_i, e.g., $C_{2345} => C_{2345}$ means that cases of the joint class C_{2345} were predicted correctly as cases of this joint class. The notation $C_i => C_j$ when $i \neq j$ indicates situations when cases of class C_i were predicted *incorrectly* as cases of class C_j.

Table 10 shows precision and recall percentages of all rules R_1–R_6 of first fold in stratified tenfold cross validation.

The rules R_1–R_6 do not cover the same number of cases, therefore the balancing of their contribution to the total precision of requires using the **weighted precision**

Table 11 Weighted precision for all classes of BC algorithm for PBC dataset of 1st fold in stratified tenfold cross validation

Rule	Weight (%)	Precision (%)	Classified cases	Weighted precision (%)
	Training			
R_1	89.71	98.57	3837	88.43
R_3	5.92	99.21	253	5.87
R_4	1.68	91.67	72	1.54
R_5	2.13	91.21	91	1.94
R_6	0.56	91.67	24	0.51
All	100.00		4277	98.29
	Validation			
R_1	89.29	98.82	425	88.24
R_3	5.88	100.00	28	5.88
R_4	1.47	100.00	7	1.47
R_5	2.52	91.67	12	2.31
R_6	0.84	75.00	4	0.63
All	100.00		476	98.53
	Testing			
R_1	88.58	98.89	450	87.60
R_3	6.89	91.43	35	6.30
R_4	1.57	87.50	8	1.38
R_5	2.36	100.00	12	2.36
R_6	0.59	100.00	3	0.59
All	100.00		508	98.23

Table 12 Precision for stratified tenfold cross validation of BC algorithm for PBC dataset

Fold's number	Weighted precision (%) for testing data
1st fold	98.23
2nd fold	96.65
3rd fold	98.62
4th fold	97.16
5th fold	97.84
6th fold	97.64
7th fold	96.36
8th fold	96.18
9th fold	97.16
10th fold	97.11
Average	97.30

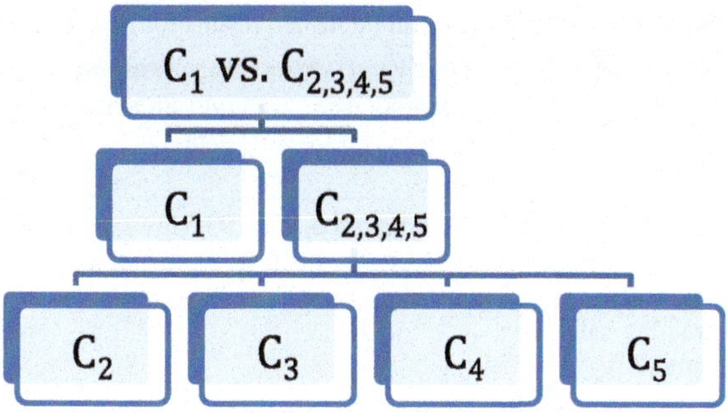

Fig. 17 Classification process

that will account this difference. The formula for the weighted precision is given in the previous section.

The rule R_2 is excluded from computing the weighted precision because it is intermediate rule, which predicts a joined class $C_{2,3,4,5}$ is and only the rules, which predict actual classes C_1-C_5 called terminal level rules are included. The main steps to calculate weighted precision for PBC dataset are below.

Step 1: Calculate weighted precision of the terminal level rules R_1, R_3, R_4, R_5, and R_6 that predict actual classes C_1, C_2, C_4, C_5, and C_3, respectively.
Step 2: Calculate weighted precision of R_1, R_3, R_4, R_5, and R_6.
Step 3: Sum up the weighted precisions from Step 2.

The results for 1st fold in stratified tenfold cross validation are presented in Table 11, where the weighted precision of all rules is 98.29% for the training set, 98.53% for the validation set, and 98.23% for the testing set. The summary of other folds is in Table 12. It shows that average precision of BC algorithm tenfold cross validation for PBC dataset is 97.30%.

6.3 Comparison of Box Classification with Decision Tree and Published Results

The exact comparison of our results with published results is difficult because the authors use different splits of data. Table 13 presents results of comparisons with clearly stated what is the difference in conducted experiments. These differences are discussed in detail below.

The KNN result is obtained with 80:20 split. With the KNN method, the precision is 93.51%, which is lower than our BC algorithm at 97.30%. However, this lower precision is to be expected because typically precision with 90:10 split is greater than with 80:20 split. The C4.5 DT precision is obtained with 90:10 split in tenfold cross validation.

Therefore, C4.5 DT result would give us a better comparison with BC algorithm because PBC dataset is split in the same ratio. With BC algorithm precision is slightly

Table 13 Comparisons with published results for PBC dataset

Algorithm	Precision (%) on training data	Precision (%) on validation data	Precision (%) on test data	Classes with 0% precision (completely misclassified)
K-nearest neighbor with a single 80%:20% training-validation split [23]	Not reported	93.51	Precision % on validati-on data	Not reported
C4.5 decision tree with tenfold cross validation 90%:10% training-validation split [24]	Not reported	96.95[b]	Not reported	Not reported
C4.5 decision tree with 100% training data	96.02%	N/A	N/A	Class 3
ID3 decision tree with a single 81%:9%:10% split	95.19%	63.56%	95.26%	Class 3
Block classification with tenfold cross validation 81%:9%:10%	**98.26%**	96.34%[a]	**97.30%**[a]	No such classes

[a] Average; [b] presumed average

higher at 97.30% compared to C4.5 DT precision at 96.95%. Table 13 also presents C4.5 DT precision with all 100% of PBC data used as training data, ID3 DT with a single 81%:9%:10% split of PBC data, and Block Classification with tenfold cross validation 81%:9%:10% of PBC data.

Both C4.5 DT and ID3 DT did not correctly classify any cases from class 3, indicating insufficient accuracy and precision. The BC algorithm classified all classes with higher precision on independent test data. Furthermore, BC algorithm with In-line coordinate can visualization of PBC dataset allowing the end-user easier and faster understanding data.

Table 14 is produced as follows. Out of all data, 90% were selected randomly for training and validation and 10% for testing. Then the first 90% of data was split in 90:10 ratio for training and validation data. The precision of all classes in training is 95.19 and 95.26% in testing. Table 14 shows that precision of DT for PBC dataset is 95.26% which is slightly lower BC algorithm. In Table 14, it can also be seen that the disadvantage of using DT for PBC dataset such that DT cannot classify any cases from class 3. With BC algorithm for PBC dataset, it was classified that all 5 classes with precision higher than 90% compared to DT algorithm where class 2,3,4, and 5 have precision lower than 80%. BC algorithm with In-line Coordinates also allows

Table 14 Weighted precision for all classes with Decision Tree ID3 for PBC dataset for a single 81%:9%:10% training: validation: test split

Class	Weight (%)	Precision (%)	Classified cases	Weighted precision (%)
	Training			
Class 1	89.14	98.05	3950	87.41
Class 2	7.15	74.76	317	5.35
Class 3	**0.00**	**0.00**	**0**	0.00
Class 4	1.62	90.28	72	1.47
Class 5	2.08	**46.74**	92	0.97
All	100.00		4431	95.19
	Validation			
Class 1	58.10	97.56	287	56.68
Class 2	7.89	58.97	39	4.66
Class 3	**0.00**	**0.00**	0	0.00
Class 4	31.17%	5.19	154	1.62
Class 5	2.83	21.43	14	0.61
All	100.00		494	63.56
	Testing			
Class 1	87.77	98.75	481	86.68
Class 2	7.85	72.09	43	5.66
Class 3	**0.00**	**0.00**	0	0.00
Class 4	1.64	77.78	9	1.28
Class 5	2.74	60.00	15	1.64
All	100.00		548	95.26

us to show how high dimensional dataset like PBC can be visualized on Cartesian coordinates compared to non-visualization method of DT.

7 Interpretability

A major advantage of the Box algorithm on the static ILC visualization is in direct interpretability of boxes and rules it produces when the input attributes are interpretable themselves. The limitation of this Box algorithm is in capturing only relatively simple interval rules.

For data with more complex patterns, we proposed the dynamic ILC Box algorithms, where boxes depend on the value of several attributes. Boxes at the very beginning of the 2-D visualization space capture dependences of just a few attributes but the boxes at the end capture dependences of all attributes. It allows capturing a variety of simple and complex relations. These relations are varied depending on the sequences and locations of attributes in the visualization space. It leads to combinatorial exploration of different sequences and locations of attributes in the visualization space to capture complex relations that are specific for a given data. Therefore, these dynamic algorithms are flexible but limited computationally.

Each box found by ILC-bases algorithms can be identified by the values of two functions: f_{hor} and f_{ver}, which identify its low left corner. For the fully dynamic ILC in Fig. 5c, the horizontal function $f_{\text{hor}} = a + b + c$, is the sum of attributes/coordinates, which are located horizontally, and the vertical function $f_{\text{ver}} = d + e + f + g$ is the sum of values of the attributes/coordinates located vertically. In general, for other dynamic ILC-based algorithms the exact mathematical class of rules is a subject of further exploration.

To justify that the boxes identified by linear functions are interpretable we use the method presented in [25] in this volume. The goal is to explain why a new n-D point **c** in the box B of a given class L belongs to that class L. We do this by finding two n-D points **a** and **b** in this box B, such that **c** is between them, $\mathbf{a} \le \mathbf{c} \le \mathbf{b}$, i.e., for each $i = 1{:}n\ a_i \le c_i \le b_i$. The points **a** and **b** can be actual training points, or artificial ones. If actual such points exist then we output them in addition to artificially created points that can cover wider data intervals. All these n-D points are visualized in ILC-based space along with the box B to provide easy to capture visual explanation. When the rule contains several boxes the same is done for all these boxes.

A common external method today to explain a non-interpretable model $F(\mathbf{x})$ is using an explainable model $G(\mathbf{x})$ like a decision three to approximate $F(\mathbf{x})$ **locally** around n-D point **x**. We expand the pool of such external explainable models $G(\mathbf{x})$ with rules generated by the Box algorithm in the ILC-based lossless visualization spaces.

The approximation process with the Box algorithm is as follows. Consider a new n-D point **c** and prediction of its class by computing $F(\mathbf{c})$. This prediction by non-interpretable F needs to be explained.

Step 1: Set up an initial grid resolution r for the selected ILC-based lossless visualization space S.
Step 2: Plot all training data in S.
Step 3: Find all boxes $\{B\}$, which **c** belongs, i.e., boxes where the polyline **c*** of n-D point **c** crosses or ends.
Step 4: Find all $\{B\}$ with a dominant class purity greater or equal to the threshold T.
Step 5: If no such B exist decrees the grid r by value Δ.
Step 6: Repeat steps 4–5 until a box with purity(B) $\ge T$ will be found. If no box found stop.
Step 7: If required box B exists find two *training* n-D points **a** and **b** in B from class L, such that **c** is between them, $\mathbf{a} \le \mathbf{c} \le \mathbf{b}$, i.e., $\forall i = 1{:}n\ a_i \le c_i \le b_i$.
Step 8: If required box B exists generate two *artificial* n-D points **d** and **e** in B, such that **c** is between them, $\mathbf{d} \le \mathbf{c} \le \mathbf{e}$, i.e., $\forall i = 1{:}n\ d_i \le c_i \le e_i$.
Step 9: Visualize **a,b,c, d,e** in ILC-based space along with the box B.

If several such boxes exist then we will have multiple explanations. If none of such boxes exist then a more complex areas than simple boxes like defined by B_1&(notB_2) can be explored to be pure enough.

Can we get a more interesting or richer explanation with the Box algorithm than with a decision tree? A local DT explanation for **c** means tracing a DT branch that contains **c**. Each DT node conducts a simple test like $x_i \le T_i$. The Box algorithm tests the box, i.e., two intervals like, $T_{1x} \le x \le T_{2x}$ and $T_{1y} \le y \le T_{2y}$ for x, y coordinates of the visualization plot grid. These two intervals are equivalent to a DT with a single branch with 4 nodes:

$$T_{1x} \le x \ \& \ x \le T_{2x} \ \& \ T_{1y} \le y \ \& \ y \ \le T_{2y}$$

For the static ILC x and y are original data attributes x_i. The rules, which include more than one box, like $R_{1,3}$: $\mathbf{x} \in B_1 \cup B_3 \Rightarrow \mathbf{x} \in G$ from Table 3, will require several such small DTs with a single branch each. These DTs can be completely independent in contrast with branches of the DT which have a common root. It means that they can capture properties that an individual decision tree cannot capture. In the decision tree theory this limitation was lifted by building a set of decision trees in random forests algorithms.

Another fundamental distinct property of boxes is that the polyline (graph) $\mathbf{x}^*$ of the n-D point **x** can go through that box, not having any node of $\mathbf{x}^*$ inside of the box B. The rules based on such boxes express the properties **two nodes** of $\mathbf{x}^*$ that form a segment of the polyline that go through the box B. Each of this node require more than one attribute. This means that this box B captures relations between two or more attributes. Such properties are not captured by decision trees at least directly.

The dynamic ILC-based visualizations allow to express more complex models than static ILCs, because here coordinates of the visualization space x and y are functions of original attributes $x_1, \underline{x_2}, \ldots, x_n$ not these attributes themselves.

In contrast with the random forest (RF) models the rules produced by the Box algorithm do not require voting, which create interpretability problem for RF. Thus, rules produced by the Box algorithm have a form of general logical rules on intervals of values without requiring a root attribute as it is for each DT. The abilities to produce all general logical rules on intervals is a subject of the further research. Major current limitations of the Box algorithm are computational to deal with the data of large dimensions, high resolution, and a large number of cases.

In summary the box algorithm can be used like a decision tree: (a) to build interpretable models from scratch and (b) to explain locally any black box model with advantages that it can capture more general and complex relations than Decision trees.

8 Conclusions

In this research, a novel 2-dimensional machine learning (2-D ML) methodology in the context of inline coordinates was developed. This is a full machine learning approach, which does not necessitate dealing with n-dimensional data in n-dimensional abstract space. It functions without any information loss with n-dimensional data in a 2-dimensional visualization space. Several static and dynamic alternative in-line-based coordinates have been developed for this. Following that, techniques for classification and regression based on these inline coordinates were introduced. A successful case study using WBC data showed that the strategy is feasible.

With a new methodology, this work advances full 2-D machine learning to become a new research area. It benefits from the ability to actively involve end users/domain exerts in the machine learning discovery process and model justification. It enables the delivery of ML models that are comprehensible.

With In-line Coordinates data visualization, this study has shown the power of interpretable data classification techniques that are implemented in automatic and interactive modes, which have been tested with WBC and PBC datasets. The proposed BC algorithm allowed to successfully classify WBC and PBC datasets.

It was observed that BC algorithm worked well with lower feature resolution lesser for a higher resolution and larger dataset. Higher dimension of features makes finding the best order of coordinates difficult because this process require a great amount of time. This led to further development of BC algorithm with using a Decision Tree for guidance. It allowed finding an efficient order of coordinates and a set of boxes in a practical run time.

With In-line Based Coordinates, it was demonstrated that the power of visualization can reduce the overgeneralization of hyper-parameters produced by the guiding DT algorithm. The BC algorithm showed that it is possible to achieve better results compared to both published results of KNN and C4.5 DT algorithms. In comparison with the DT, the BC algorithm did not miss any class on highly imbalanced PBC dataset.

Further work will focus on improving algorithms for greater accuracy and expanding the range of potential patterns that can be found in in-line coordinates. Boxes can be substituted by a more complex interpretable structure. Multithreading with GPU will be developed to deal with larger datasets.

References

1. Lipton Z (2018) The mythos of model interpretability. Commun ACM 61:36–43
2. Caruana R, Lou Y, Gehrke J, Koch P, Sturm M, Elhadad N (2015) Intelligible models for healthcare: predicting pneumonia risk and hospital 30-day readmission. In: Proceedings of the 21st ACM SIGKDD. ACM, pp 1721–1730

3. Kovalerchuk B, Ahmad MA, Teredesai A (2021) Survey of explainable machine learning with visual and granular methods beyond quasi-explanations. In: Pedrycz W, Chen SM (eds) Interpretable artificial intelligence: a perspective of granular computing. Springer, pp 217–267. https://arxiv.org/abs/2009.1022
4. Kovalerchuk B (2020) Enhancement of cross validation using hybrid visual and analytical means with shannon function. In: Beyond traditional probabilistic data processing techniques: interval, fuzzy etc. methods and their applications. Springer, pp 517–554
5. Molnar C (2020) Interpretable machine learning. https://christophm.github.io/-interpretable-ml-book/
6. Rudin C (2019) Stop explaining black box machine learning models for high stakes decisions and use interpretable models instead. Nat Mach Intell 1(5):206–215
7. Kovalerchuk B (2018) Visual knowledge discovery and machine learning. Springer
8. Kovalerchuk B, Phan H (2021) Full interpretable machine learning in 2D with inline coordinates. In: 2021 25th international conference information visualisation. IEEE, pp 189–196
9. Dovhalets D, Kovalerchuk B, Vajda S, Andonie R (2018) Deep learning of 2-D images representing n-D data in general line coordinates. In: International symposium on affective science and engineering, pp 1–6. https://doi.org/10.5057/isase.2018-C000025
10. Kovalerchuk B, Kalla DC, Agarwal B (2022) Deep learning image recognition for non-images. In: Integrating artificial intelligence and visualization for visual knowledge discovery. Springer, pp 63–100
11. Kovalerchuk B, Gharawi A (2018) Decreasing occlusion and increasing explanation in interactive visual knowledge discovery, In: Human interface and the management of information. Interaction, visualization, and analytics. Springer, pp 505–526
12. McDonald R, Kovalerchuk B (2022) Non-linear Visual Knowledge Discovery With Elliptic Paired Coordinates. In: Integrating artificial intelligence and visualization for visual knowledge discovery. Springer, pp 141–172
13. Wagle SN, Kovalerchuk B (2020) Interactive visual self-service data classification approach to democratize machine learning. In: 2020 24th international conference information visualization (IV). IEEE, pp 280–285. https://doi.org/10.1109/IV51561.2020.00052
14. Inselberg A (1998) Visual data mining with parallel coordinates. Comput Stat 13(1)
15. Inselberg A (2009) Parallel coordinates: visual multidimensional geometry and its applications. Springer
16. Sharma A, Vans E, Shigemizu D, Boroevich KA, Tsunoda T (2019) Deep insight: a methodology to transform a non-image data to an image for convolution neural network architecture. Nat: Sci Rep 9(1):1–7
17. Sansen J, Richer G, Jourde T, Lalanne F, Auber D, Bourqui R (2017) Visual exploration of large multidimensional data using parallel coordinates on big data infrastructure. Informatics, vol 4, no 3, p 21. Multidisciplinary Digital Publishing Institute
18. Estivill-Castro V, Gilmore E, Hexel R (2020) Constructing interpretable decision trees using parallel coordinates. In: International conference on artificial intelligence and soft computing. Springer, pp 152–164
19. Tam GK, Kothari V, Chen M (2016) An analysis of machine and human analytics in classification. IEEE Trans Visual Comput Graphics 23(1):71–80
20. Xu Y, Hong W, Chen N, Li X, Liu W, Zhang T (2007) Parallel filter: a visual classifier based on parallel coordinates and multivariate data analysis. In: International conference on intelligent computing. Springer, pp 1172–1183
21. Kovalerchuk B, Hayes D, Discovering interpretable models in parallel coordinates, international information visualization conference. https://doi.org/10.1109/IV53921.2021.00037
22. Dua D, Graff C (2019) UCI machine learning repository, University of California, School of Information and Computer Science, Irvine, CA. https://archive.ics.uci.edu/ml
23. Zaky D, Gunawan PH (2020) Computational parallel of K-nearest neighbor on page blocks classification dataset. In: 2020 8th international conference on information and communication technology (ICoICT). IEEE, pp 1–4

24. Eschrich S, Chawla NV, Hall LO (2002) Generalization methods in bioinformatics. In: BIOKDD, vol 2, pp 25–32
25. Huber L, Kovalerchuk B, Recaido C (2023) Visual knowledge discovery with general line coordinates. https://arxiv.org/pdf/2305.18429

Explainable Machine Learning for Categorical and Mixed Data with Lossless Visualization

Boris Kovalerchuk and Elijah McCoy

1 Introduction

Many Machine Learning (ML) datasets contain mixed/heterogeneous non-numeric data, but multiple ML algorithms cannot discover models on such diverse data [1]. Mixed data include text, graphics, numeric and non-numeric data. Non-numeric data frequently take the form of ordinal (ordered) data, such as large, medium, and tiny, or nominal data with values like red, green, and blue.

The success of explainable ML algorithms for mixed data heavily depends on abilities of lossless visualization of multidimensional data and ML models. The latter enables end-users, who can contribute valuable domain knowledge that is missing from the training data, to create explainable ML models using visual knowledge discovery [2, 3].

Additionally, the interpretability and appeal of visual machine learning models are frequently higher for the end user than those of analytical ML models. This creates a new window of opportunity for developing and advancing the field of explainable ML. It requires to visualize n-D data *without loss of n-D information*, because it is not known in advance which of this n-D information will be critical for discovering ML model in a visual form.

However, creating methods for lossless visualization for multiple data types is a long-standing challenge [4–6], because it is more difficult than producing lossy visualizations of high dimensional data, which are less demanding on data preservation.

B. Kovalerchuk (✉) · E. McCoy
Department of Computer Science, Central Washington University, Ellensburg, WA, USA
e-mail: Boris.Kovalerchuk@cwu.edu

E. McCoy
e-mail: Elijah.McCoy@cwu.edu

B. Kovalerchuk et al. (eds.), *Artificial Intelligence and Visualization: Advancing Visual Knowledge Discovery*, Studies in Computational Intelligence 1126,
https://doi.org/10.1007/978-3-031-46549-9_3

For instance, [7] focuses on existing lossy methods like 2-D projections and limited Star coordinates with visualizing only end points of the graphs making this visualization lossy relative to n-D data. In general, research on the lossless visualization of mixed nominal, ordinal, and numerical attributes and its contribution to the explainable machine learning is in the nascent stage.

Numeric coding is a popular method for adapting existing ML algorithms to non-numeric data [8–10]. If the features of these non-numeric data, such as value similarities, are corrupted by the coding, it can result in non-explainable ML models. For distance-based algorithms like kNN and others it happened with a common integer coding. Unfortunately, there is little focus on interpretability in the coding literature, even though it is a significant barrier to expanding the use of ML with mixed data in the domains with high error costs, like medicine.

Building Explainable Machine Learning (EML) models is a significant challenge for all types of data [11–17]. It requires understanding the *nature of the information* contained in the input data relative to the domain theory, and the nature of the knowledge that learning algorithms discover [11, 18]. It is also a *considerable computational challenge* especially for rule lists/sets [14], which is very evident for categorical data for large datasets with multiple value of each attribute. A *brute force* approach for categorical data is a computationally intractable problem with the exponential number of computations. A feasible approach to deal with this challenge is a sequential hierarchical approach that is presented in this based on the theory of monotone Boolean functions.

Often EML models are *unintelligible* being large and complex leading to the need to build smaller models [17, 18]. An extensive review of current approaches to build smaller explainable models as sets of rules is presented in [17]. Our approach differs from [17] in treating combinatorics of all possible rules. We use the computational advantages derived from the theory of monotone Boolean functions [19, 20] with controlled levels of model precision and coverage to build an EML model as a set of explainable rules. We also do not require that a very complex explainable model is already available to be simplified as it is assumed in [17].

Most of the typical ML algorithms, like neural networks, SVM, kNN, Linear Discriminant Functions (LDF), CNN, and others are designed for quantitative numeric attributes such as pixel intensities in images. However, many machine learning data have a qualitative nature measured either in nominal or order scale, like good, bad, very good. Those attributes typically are encoded using one-hot key method which produces binary attributes [8].

The problem with one hot key is that it dramatically increases the sizes of the space. For instance, if an attribute has ten nominal values, one hot key will produce 10 binary attributes. Moreover, the interpretability of ML models such as kNN is questionable with one hot key. It counts the number of binary values equal or not equal mixed with differences between values of numeric attributes. It creates a *scaling problem* to make the total distances meaningful.

Moreover, many ML algorithms require distance metrics, which makes questionable the interpretability of those models on qualitative data. In contrast, we create the logical rules for qualitative data, which resolve such interpretability issue.

This approach generates fully explainable and visualizable rules presented in this chapter expanding [21]. The algorithm to generate such rules is denoted as Sequential Rule Generation (SRG) algorithm. This chapter presents several versions of this algorithm starting from the initial SRG0 algorithm.

Respectively, the main contribution of this chapter is in

(1) Integrating numeric coding schemes for non-numeric attributes for ML algorithms to support accurate and explainable ML models.
(2) Developing accurate and explainable ML models for categorical and mixed data.
(3) Developing methods for lossless visualization and visual knowledge discovery of n-D non-numeric categorical data and visual rule discovery in these visualizations.

The main innovation of this chapter in (1) is in the *tight match and control of knowledge* of input data types and operations of ML algorithms that can be allowed. The main innovation in (2) is in the *algorithms that cut out computations* using the theory of Monotone Boolean Functions. The main innovation in (3) is in *frequency-based visualization* and *visual knowledge discovery* of qualitative data in parallel coordinates. In contrast with [8] it is *lossless* without merging different values with equal frequencies including frequencies referencing the target attribute. This lossless visualization is a step toward the solution of the long-standing challenge of lossless visualization for multiple data types outlined above.

This chapter is organized as follows. Section 2 proposes a classification of mixed data types and analyzes their important role in Machine Learning. Section 3 describes the proposed approach and the methodology for enforcing interpretability of internal operations of ML algorithms on mixed data. Section 4 presents different version of the Sequential Rule Generation (SRG) algorithm with a summary of verification experiments. Section 5 describes the proposed visual exploration approach and mixed data visualization methods. Section 6 summarizes the results and outlines the future work.

Appendixes "Experiment 1 with Algorithm SRG0 for Discovering Rules on Mushroom with Sequential Triples"–"Experiment 10 with SPG 5 and Comparison of Rule Complexity" detail experiments conducted with SRG algorithms on real-world data and Appendix "Toolkit" describes the toolkit developed.

2 Heterogeneous Data Types and Their Role in Machine Learning

In this section, mixed data types are categorized along with analyses of how important they are for machine learning. The discussion of various coding schemas for measurement data types is included.

2.1 Classification of Mixed Data Types for Machine Learning

Heterogeneity/mixture of data is not limited by numeric, nominal, and ordinal data types that are measurement data types [22]. Numeric data are not homogeneous, but heterogeneous too. They have several measurement types such as absolute, ratio, interval, and cyclical. The strongest one (*absolute*) has the largest set of meaningful arithmetic relations and operations, which ML algorithms can conduct with them to produce explainable models.

This data type includes attributes that represent the *counting* the number of elements, which has unique not arbitrary 0 and unit 1. The next type is *ratio data type*, which has a unique zero, but not a unique unit, like Kelvin temperature with unique physical zero, but the unit is defined by convention. Other data types include *interval data type* (e.g., Celsius and Fahrenheit temperature where both zero and the unit are defined by convention) and values can be negative, cyclical (e.g., azimuth) [23] and others.

These data types can provide useful *arithmetic operations, distances, frequency distributions, distinct averages, standard deviation*, and other quantities that ML algorithms can use to create explainable models. However, the meaningful difference between two values in the cyclical azimuth attribute between 1° and 359° is 2° not 358 as it is for the attribute like weight. The *ordinal data*, in contrast, *forbid arithmetic operations* and only allow testing relations like $a \leq b$. Therefore, ML algorithms that perform these operations on ordinal data will result in models that cannot be explained.

Finally, only testing $a \neq b$ is meaningful for ML algorithms with *nominal data*. Other actions will result in unexplainable ML models. Graphs and texts are other examples of data heterogeneity, which are not given by a set of well-defined attributes. The attributes need to be made from them. We will call such data as non-attribute-based data types.

Real-world physical modality data types. Both height and weight are ratio attributes with physical zeros of weight and height, but arbitrary units and very different real-world physical meaning, modalities. Their sum and other mathematical operations, such as 3kg + 5m, are meaningless. This makes popular linear machine learning models non-explainable for data of different modalities (*heterogeneous*) and with mutual dependencies [12, 18]. In contrast, ML models with data of the same modality type (homogeneous) as temperature in time series can be explainable.

Thus, each dataset is characterized by the several categories of data types: (1) *type of measurement*, (2) *attribute/non-attribute*, and (3) *type of modality*. A dataset is *heterogeneous* or *mixed* if it contains data of various such types. Differences in data types have a significant impact on the precision and interpretability of ML models. The ML algorithm likely will produce incorrect and unintelligible models when used with data types for which it was not intended.

While the effects of measurement and attribute/non-attribute data types [8–10] are widely acknowledged, the effects of physical modality are less recognized in the

ML literature. This has led to several unsupported claims like that linear models are always comprehensible/interpretable, which was critically analyzed in [18].

The process of embedding, which focuses on converting non-numeric data like graphs and texts [24] into quantitative data (strings of numbers, vectors), is the subject of numerous recent studies. Although this technique is widely used and aims to maintain the structure of the original data, it has a serious flaw. The interpretability of such vectors is unclear or nonexistent, while these models are frequently quite accurate.

Encoding individual attributes based on their measurement data types [8–10] allows building explainable models simpler than with embedding from mixed data. To reduce the attribute space and to make attributes more relevant to the ML task, the data preparation stage frequently encodes the original attributes to other attributes by many-to-one mappings. It results in loss of some information and can corrupt the data type.

Example. The absolute numeric data in [0,100] interval can be converted to ordinal data, where code 1 is for interval [0,50), code 2 for [50,70) and code 3 for [70,100]. The resulting codes will be cd(a) = 1, cd(b), and cd(c), respectively, for the values a = 10, b = 60, and c = 100. In the original absolute data, c = 10a and b = 6a, but in codes cd(b) = 2cd(a) and cd(c) = 3cd(a), if we treat codes as numeric data of the same measurement type as original data.

This example demonstrates that the data type cannot be automatically expanded to a modified attribute. The validity of the code ratios cd(b) = 2cd(a) and cd(c) = 3cd(a) in the domain should be confirmed by a domain expert. Otherwise, the ML model that uses these codes is not explainable in that domain.

We cannot consider codes as a numeric data type without such confirmation from a domain expert; instead, we must treat them as ordinal data, where computing ratios is both forbidden and meaningless. The disparities cd(b)-cd(a) = 2–1 = 1 and cd(c)-cd(a) = 3–1 = 2 also have little significance in the absence of equivalent confirmation. Therefore, distance-based ML algorithms like kNN will result in unexplainable models in this situation. Unfortunately, the focus of the kNN literature reviewed in [1] is out of the interpretability issues.

Calculating the dissimilarity between each pair of nominal attribute values in relation to the target attribute for the ML classification problem is one method for enriching nominal attributes. This can be achieved by constructing the adaptive dissimilarity matrices and minimizing an error function on the training samples while accounting for attribute correlation [25]. This method called ADM generalizes the Value Difference Metric (VDM) [26]. A similar idea is presented in [9] for SVM. Such approach improved the ML model accuracy and interpretations of the relationships among the different values of nominal attributes [25] for some datasets from UCI ML repository. However, for the mushroom data [27], which we consider, the explainable C4.5 Decision Tree (DT) algorithm produced the lowest error of 0.4% reported in [25] without using ADM or VDM. Because DTs can overlook some superior possibilities when examining each attribute only one at a time during node splitting, this lowest error is still not zero [25].

The main challenge with ADM is that the error function is computed using a machine learning algorithm on training data. Another ML algorithm will lead to a different result. Next, if the ML algorithm is unexplainable, then the ADM outcomes will likewise be unexplainable too.

The radial basis function (RDF) classifier was used in [25] as the ML algorithm. It sums up the exp functions of the distances. The interpretability of it for the mushroom task is not clear, in contrast with decision trees or logical rules. The result with RDF needs to be independently confirmed by the domain knowledge, which was done in [25] for the odor mushroom attribute. The coding that is explainable in the domain and established without using the target attribute does not require such domain confirmation afterward.

2.2 Coding of Measurement Data Types

The major coding techniques for nominal, ordinal, and other attributes used in machine learning [8–10] are briefly discussed below, along with their relations to interpretability. The development of these techniques to accommodate knowledge discovery in multidimensional data utilizing visualization will be presented in Sect. 5.

One Hot Encoding maps a nominal attribute to a binary vector, where 1 denotes the existence of the provided attribute value and 0 denotes the absence of that value of the attribute. It greatly lengthens the run time and space dimension for the ML algorithms. This coding is meaningful for nominal attributes because it preserves the equal Hamming distances between any two values of the nominal attribute.

The *label encoding* simply assigns numbers from 1 to k to k values of the nominal attribute. It does not preserve the equal distance for values of the nominal attribute. Thus, it is not explainable, which is especially applicable to the ML algorithm that exploit the differences of distances for classification like kNN. Such coding can be used for other techniques, such as decision trees (DTs), which do not use n-D distances, but it may result in less effective DTs than with alternative encodings including another numbering of values of the attribute.

Hashing maps categorical attributes to vectors in n-D space, where the distance between two vectors is roughly preserved. Compared to One Hot Encoding, the final dimension is significantly lower [8]. This is a subcategory of a more comprehensive *embedding* notion for various kinds of heterogeneous data. Not every ML algorithm can benefit from hashing. Preserving the structure for nominal attributes means all equal distances between values. If the ML algorithm exploits the differences of the distances, then it cannot benefit from such hashing, where all distances are the same. Establishing meaningful distances for ordinal data require additional information, which may not be available. It limits the use of the hashing method for these data. A hashing space can be not explainable, which is critical.

Ordinal encoding is applicable to the ordinal data like very short, short, medium, tall, very tall. It differs from the label coding by the fact that now numbering cannot be arbitrary, but it must follow the order of the values, e.g., 1 for very short, 2 for short, 3, for medium, 4 for tall, and 5 for very tall. The explainable operations with ordinal data are limited by less or equal relation and should not include arithmetic operations. This limits application of many existing ML algorithms that use arithmetic operations.

Below we summarize statistics-based coding methods [8]. These methods are applicable to many data types but can assign the same code to different values of the attribute making this coding lossy. We resolve this issue for visualization in Sect. 5.

Frequency Encoding assigns codes to values of the attribute by their frequency. It allows exploring the link between attribute and the values of the target variable too.

Mean Encoding or *Target Encoding* computes the mean of the number of times the value of the attribute appears in the target class in the two-class classification task.

Probability Ratio Encoding for each value uses ratio P(1)/P(0) of the frequency of this value of the attribute for class 1 to the frequency for class 0 as codes.

James–Stein estimator assigns as a code a weighted average of the mean target value for the observed feature value and the mean target value (regardless of the feature value).

3 Approach and Methodology

This section describes the proposed approach and the methodology for enforcing *interpretability of all internal operations of ML algorithms on mixed data*. The Toolkit that implements this approach is outlined in the Appendix "Toolkit".

Enforcing interpretability of all internal operations of ML algorithms. The main idea behind the proposed methodology is *to produce explainable models, where the internal operations of the ML algorithm must be comprehensible/explainable.* It is a significant restriction for models that deal with mixed data. For instance, the algorithm cannot perform any action for nominal data other than verifying whether two values are equal or not because all other operations cannot be justified for nominal data.

This leads to algorithms, which discover logical rules like (1–3) or similar:

$$\text{If } [(x_1 = \text{a} \,\&\, x_2 = \text{b}) \vee (x_3 = \text{c})] \,\&\, x_4 \neq \text{d} \,\&\, (\bar{x}_5 \neq \text{e}) \text{ Then } \mathbf{x} \in \text{class } \text{C}_1 \tag{1}$$

$$\begin{aligned} &\text{If } [(x_1 = \text{a} \,\&\, x_2 = \text{b}) \vee (x_3 = \text{c})] \,\&\, x_4 \neq \text{d} \,\&\, (\bar{x}_5 \neq \text{e}) \text{ Then } \mathbf{x} \in \text{class } \text{C}_1 \\ &\text{else } \mathbf{x} \in \text{class } \text{C}_2 \end{aligned} \tag{2}$$

$$\begin{aligned} &\text{If } [(x_1 = \text{y}_1 \,\&\, x_2 = \text{y}_2) \vee (x_3 = \text{c})] \,\&\, x_4 \neq \text{y}_4 \,\&\, (\bar{x}_5 \neq \text{e}) \,\&\, \mathbf{x} \in \text{class } \text{C}_1 \\ &\text{Then } \mathbf{y} \in \text{class } \text{C}_1 \end{aligned} \tag{3}$$

for nominal n-D points $\mathbf{x} = (x_1, x_2, \ldots x_n)$, $\mathbf{y} = (y_1, y_2, \ldots y_n)$. Thus, the rules can include only logical operations &, $\vee$, $\neg$ and tests if values are equal or not for one or more n-D points.

Even decision trees which are traditionally considered as explainable ML algorithms are not formally explainable for the nominal data because they check $\leq$ relation, which is prohibited for nominal data. Coding of nominal values by integers (known as label encoding) allows technically to apply decision trees to nominal data. However, randomness of this coding leads to different decision trees and does not guarantee finding the most accurate model. Next, the produced DT must be converted to logical rules with elimination of all $\leq$ relations and their thresholds to make it fully explainable.

The advantage of ML algorithms, which build logical rules is that they are data type universal, i.e., can produce models that include *heterogenous* data of all types that are expressed in the propositional or the First order Logic (FoL) [28]. Therefore, this work focuses on logical rules. Above rules (1) and (2) are propositional and rule (3) is an example of FoL rules, because it includes the FoL clause $x_4 \neq y_4$.

Addressing all heterogeneous data types. The actual number of data types in ML tasks with mixed data is much greater than what is usually listed. Consider a nominal data type for which we cannot say that **a** is closer to **b** than to **c**, like occupations with categories: doctors, engineers, and teachers. Next, we can add a nurse to this list and can continue to call it the nominal data type. Alternatively, we can say that doctors and nurses are closer to each other than to engineers and teachers. Then, occupations are *not nominal anymore* because more relations have meaning. Similarly, we can add technicians and teaching assistants. This creates more relations, where engineers are closer to technicians and teachers are closer to teaching assistants than to other occupations. To treat these occupations as a nominal data type we can create groups: (1) doctors and nurses, (2) engineers and technicians, and (3) teachers and teaching assistants and assign codes to these groups.

While grouping allows us to go back to the nominal data, we lose important similarity information about occupations. The resulting ML model on groups can be less accurate than without grouping. Instead of grouping we can assign the following codes: nurse (1), doctor (2), technician (5), engineer (6), teaching assistant (10), teacher (11) with limiting the set of operations and relations that are considered as explainable and allowable. For instance, relation,

$$\mathrm{c(doctor)} - \mathrm{c(nurse)} < \mathrm{c(teacher)} - \mathrm{c(engineer)} \quad (4)$$

is allowed. Here, c(.) is an integer code for the occupation, e.g., c(nurse) = 1. Thus, the ML algorithm can use relation (4), but not (5):

$$\mathrm{c(doctor)} - \mathrm{c(engineer)} < \mathrm{c(teacher)} - \mathrm{c(engineer)} \quad (5)$$

This example shows that to get an explainable ML model with non-numeric data like occupations we can use a numeric coding of them, but we need to limit operations and relations, which can be conducted with these numbers by ML algorithms to ensure that algorithms will produce explainable models. In summary, this example shows that the design of ML algorithms for mixed data needs to address a wide variety of data types. The relational ML algorithms [28] fit well this task for mixed data.

Building explainable models based on the explainable atoms (hyperblocks). The next important concept in our methodology for explainable ML on mixed data is the concept of the n-D data hyperblocks (HBs). Some HBs serve as *explainable data atoms.*

A *numeric hyperblock* (hyperrectangle, n-orthotope) is a set of numeric n-D points $\{\mathbf{x} = (x_1, x_2,..., x_n)\}$ with center in n-D point $\mathbf{c} = (c_1, c_2,..., c_n)$ and *lengths* $\boldsymbol{L} = (L_1, L_2,..., L_n)$ such that

$$\forall i \in I_u ||x_i - c_i|| \leq L_i/2 \tag{6}$$

This definition assumes that attributes are numeric and the difference between values is meaningful. This means that data are of the interval data type at least. For heterogenous data we need another definition which will include different data types. We first define the hyperblock for ordinal data and then for nominal data.

An *ordinal hyperblock* (hyperrectangle, n-orthotope) is a set of ordinal n-D points $\{\mathbf{x} = (x_1, x_2,..., x_n)\}$ with *edge ordinal* n-D points $\mathbf{s} = (s_1, s_2,..., s_n)$ $\mathbf{e} = (e_1, e_2,..., e_n)$ such that

$$\forall i \in I_o \; s_i \leq x_i \leq e_i \tag{7}$$

For instance, if X_i has values,1,2,3,4,5 and $s_i = 2$, and $e_i = 4$ then n-D points with values 2,3, or 4 of X_i will be in this hyperblock

A *nominal hyperblock* is a set of nominal n-D points $\{\mathbf{x} = (x_1, x_2,..., x_n)\}$ such that

$$\forall i \in I_n \; \ldots x_i \in Q_i \tag{8}$$

where Q_i is a subset of values of attribute X_i. For instance,

X_i = {doctor, teacher, engineer} and Q_i = {doctor, teacher}.

A *heterogeneous hyperblock* is a set of n-D points $\{\mathbf{x} = (x_1, x_2,..., x_n)\}$ such that

$$\forall i \in \mathrm{I}_u \; ||x_i - c_i|| \leq L_i/2 \quad \text{for all numeric attributes}$$

$$\forall i \in \mathrm{I}_o \; s_i \leq x_i \leq e_i \; \text{ for all ordinal attributes and}$$

$$\forall i \in \mathrm{I}_n \quad x_i \in \mathrm{Q}_i \text{ for all nominal attributes}$$

All numeric HBs are explainable for data with numeric attributes of different modality types like temperature and blood pressure. HBs are explainable because: (i) the distance in (6) is defined within each numeric attribute, where it is meaningful and (ii) do not include arithmetic operations between heterogenous attributes, which are not meaningful, but only combine them with the meaningful logic operation. Thus, each such HB is a logical model, e.g., a HB contains all cases with temperature in [35,37] interval and blood pressure within [100,120] interval. Similarly, the hyperblocks for ordinal and nominal data are explainable. As a result, this hyperblock for heterogeneous data is also explainable.

Next larger explainable heterogeneous datasets can be formed from smaller HBs. These small HBs serve as explainable atoms for such larger datasets. We can formulate a stronger conjecture that all explainable heterogeneous datasets with numeric, ordinal, and nominal attributes can be formed from HBs.

A hyperblocks is called a *pure HB* if it contains only cases of a single class. For every n-D point **x** it is possible to find a single class HB if there is no n-D point **y** = **x** that belongs for another class. Another advantage of numeric HBs is that they can be visualized losslessly, which is shown in [29].

The practice of coding the values of a nominal attribute by consecutive integers 1,2,…,*n* known as label encoding [8] is *contradictory* from stating that (1) it can always be used to encode nominal attributes without any limitation to stating that (2) this coding should never be used because it is not explainable. In fact, it should not be used in ML algorithms, which conduct arithmetic operations with these codes like subtraction, squaring and so on. It is common in ML algorithms that use distances between n-D points like kNN. The use of this coding with the ML algorithms that do not make arithmetic operations with values of nominal attributes does not create interpretability problems. The logical algorithms, which we outlined above belong to this category. Another advantage of creating and visualizing logical rules is that often they are used as an efficient tool to explain deep learning and other black box models by mimicking behavior of these models.

4 Sequential Rule Generation (SRG) Algorithms

In this section we present the Sequential Rule Generation (SRG) algorithm for finding rules in categorical data analytically. After such rules will be discovered they can be visualized in parallel coordinates similarly to shown in Sect. 5. This approach generates fully explainable and visualizable rules.

4.1 Rule Generation Process

We consider qualitative data, which are represented by nominal and order attributes without obvious numeric values. Below first we focus on nominal attributes. Discovering rules with these data is challenging computationally. A brute force algorithm to discover rules on the available data will need to test all rules like:

$$\text{R}_1\text{: If } x_1 = 3 \;\&\; x_4 = 1 \;\&\; x_{10} = 4 \text{ then } \mathbf{x} \text{ belongs to class } C \tag{9}$$

for all possible subsets of *n* attributes, all possible values of these attributes and on all n-D points (cases/samples) of available data. The number of these subsets is 2^n

and is growing exponentially with n. For each subset of attributes, we need to test several rules. Assume that in rule (1) attribute x_1 has three values (1,2,3), x_4 has two values (1,2) and x_{10} has 4 values (1,2,3,4), then $k_1k_2k_3 = 3{\cdot}2{\cdot}4 = 24$ such rules must be tested. If each attribute has only 2 values, then each subset with q attributes will have 2^q rules. Next each rule must be tested on all m given n-D points. This shows the exponential complexity of the brute force algorithm. We cannot generate and test all rules in this way for large n, m, and q. Therefore, Sequential Rule Generation algorithm generates rules sequentially and filters out not promising rules without full testing them.

The filtering methods in the Sequential Rule Generation algorithm is based on the principle of monotone Boolean functions [19, 20] as follows. Consider rule R_1 from (1) and rule R_2, which contains rule R_1 as a subrule, where R_2 is generated by adding an extra clause $x_6 = 3$:

$$R_2\text{: If } x_1 = 3 \;\&\; x_4 = 1 \;\&\; x_{10} = 4 \;\&\; x_6 = 3 \text{ then } \mathbf{x} \text{ belongs to class } C \quad (10)$$

If R_1 has low coverage of cases, then R_2 will also have low coverage because an additional requirement $x_6 = 3$ will at most keep the number of cases that satisfy R_2 the same as for R_1 but most typically it will be less than R_1. If we already *filtered out* R_1 as covering too few cases, then we can filter out R_2 without testing it on the data. Similarly, if R_1 has greater precision than R_2, then R_2 will also have the same or greater precision. If we already satisfied with R_1 precision (e.g., 100%), then we can omit testing R_2 because it is more complex and keeping R_1 is sufficient for the selected level of precision. Thus, SRG algorithm allows cutting out brute force generation of rules by using monotonicity properties, which filter out not promising rules. It is done with the techniques of monotone Boolean functions [19, 20].

To cut out the brute force computations further the SRG algorithms splits the attributes to groups and for each group the search for rules is conducted based on monotonicity approach described above. In our experiments, groups contained 3–5 attributes. For instance, in one of the experiments with the mushroom data, which contain 22 attributes, the groups were formed as follows: Group 1: x_1, x_2, x_3, Group 2: $x_4, x_5, x_6, \ldots$, Group 7: $x_{19}, x_{20}, x_{21}, x_{22}$.

According to the design of SRG algorithm it runs at the several levels of thresholds that the user sets up. We experimented with 3 level of precision: 75%. 85%, 95% and with different levels of coverage. These thresholds limit only the low margin of the rule quality, but do not limit their upper level. Therefore, we also computed the actual precision and coverage reached for the poisonous class at each level.

The monotone Boolean process for $n = 3$ is presented below. We build three Hansel chains [19], which cover all 8 triples that represent subsets of attributes to be used to generate rules. For instance, the triple (110) means that only attributes x_1 and x_2 will be used to build rules, but x_3 is excluded.

Example Figures 1 and 2 show the following chains:
Chain 1: (010), (110); Chain 2: (100), (101); Chain 3: (000), (001), (011), (111).

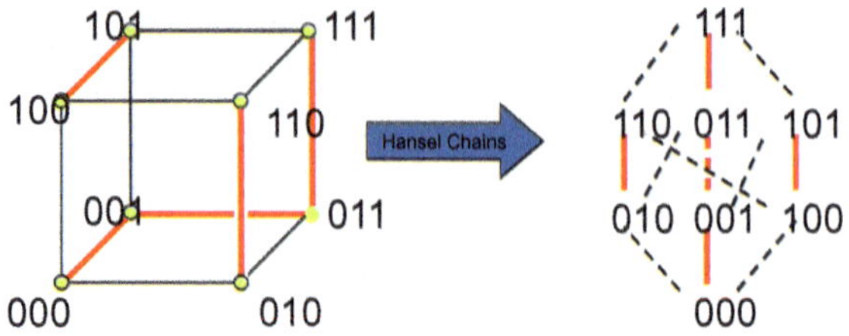

Fig. 1 3-D Hansel chains unlabeled

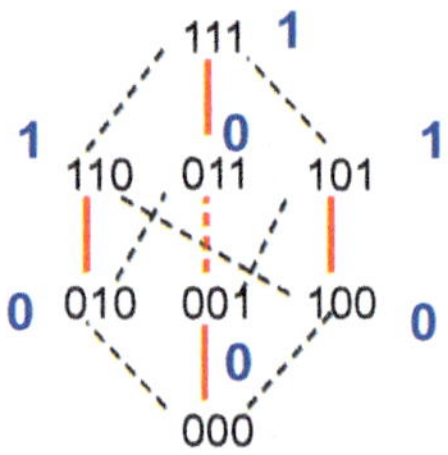

Fig. 2 Labeled Hansel chains

First, we check triple (010) on Chain 1 if any rule for this triple of attributes satisfies the requirement for 75% precision and 0.5% coverage. This triple means that only attribute x_2 is used, because only the second bit in (010) is 1. If such rule would be found then no other triples on chain 1 is tested. The illustrative example in Fig. 2 shows that no such rule is found for attribute x_2. Therefore, we need to test the next triple (110) on chain 1, i.e., rules with attributes x_1 and x_2.

Figure 2 shows that such rule was found, and the triple (110) is labeled by 1. It immediately implies that (111) on chain 3 is excluded from testing because it is greater than (110). The process moves to test the first triple (100) on chain 2 and fails to find a rule that satisfies the requirements of 75% precision and 0.5% coverage in x_1 only, which corresponds to (100) triple. Then we successfully check the next triple (101) on chain 2. Then the process moves to chain 3 until all remaining triples on this chain are tested starting from (001). In Fig. 2. it will require to test (001) and (011).

Here we illustrated the process starting from the bottom of each chain, Alternative processes that start from the top of each chain or from the middle of each chain are also implemented and depending on the data set the cutting of computations can be more efficient with one of these options. Starting from the longest chain also can change the number of tests required. The domain knowledge can help selecting one of these processes. If the domain expert expects that 2 or more of the attributes are required for rules to be successful out of three attributes then starting from the top will be more beneficial to cut computations than starting from the bottom of chains.

4.2 General Algorithm Design and SRG0 Algorithm

The SRG algorithm attempts to discover a set of rules for class C such that each rule R captures some properties of a single class:

$$\text{if } R(\text{x}) = 1 \text{ then case } \mathbf{x} \text{ belongs to class } C$$

with the following desirable properties: (1) a high number of cases that R predicts/covers, $N_{covered}$, (2) a high number of cases that R predicts correctly, $N_{correct}$., (3) a small set of rules, and (4) simple rules.

Below we will use the following concepts for every rule R, which predicts class C:

Recall is $N_{correct}$/N,, where $N_{correct}$ is the number of cases **x** that satisfy rule R and *belong* to class C, N is the *total number* of cases of class C.

Precision is $N_{correct}/(N_{correct} + N_{incorrect})$, where $N_{incorrect}$ is the number of cases **x** that satisfy rule R, but do *not* belong to class C.

Coverage is $(N_{correct} + N_{incorrect})$.

Examples Let for rule R_2: If $x_1 = 3$ & $x_4 = 1$ & $x_{10} = 4$ & $x_6 = 3$ Then $\mathbf{x} \in C$ the value $N = 100$ (the number of cases from class C which satisfy the if-part of this rule). If $N_{correct} = 80$, $N_{incorrect} = 10$ then we get recall = 80/100, precision = 80/(80 + 10), and coverage = (80 + 10)/100. Note that coverage can be greater than 1 when $N_{incorrect}$ is large, e.g., if 50 cases of another class would satisfy the rule in this example. If the rule predicts cases of two classes, then N will be the number of cases from these two classes. The values of $N_{correct}$ and $N_{incorrect}$ also will be adjusted for two classes. The advantage of discovering rules that are limited to a subset of cases of a single class is that in many tasks we do not have enough knowledge to discover rules that will classify every possible case of all classes correctly and with high confidence. So, it is better to leave those cases unclassified, than to predict them incorrectly or with a low confidence.

While the SRG algorithm discovers only single class rules, datasets contain cases of several classes. In the two-class classification, the rule R' for the opposite class C' can be generated from R by reversing it, $R'(\mathbf{x}) = 1 \Leftrightarrow R(\mathbf{x}) = 0$, i.e., If $R'(\mathbf{x}) = 1$ then **x** belongs to class C'. Respectively, the coverage and precision of rule R' heavily depend on these properties of rule R. This shows the importance of maximizing the coverage and precision of all cases of class C by rule R to be able to predict accurately cases of the opposite class. In many situations, precision, recall, and coverage contradict each other and the tradeoff between them is needed. This chapter presents different versions of the Sequential Rule Generation algorithm with different tradeoffs and different options to form groups of attributes (sequentially, randomly, by an expert, and by guidance of existing ML model) to decrease the search time.

The complexity of a rule and set of rules are defined as the number of the base clauses in the rules relative to the number of cases covered. A base clause is a base predicate in the rule like $x_i = 5$, or $x_i \neq 5$.

Complexity of rule $R =$ (Number of base clauses in R)/(Number of cases covered by rule R).

Complexity of a set of rules= (Number of clauses in the set)/ (Number of cases covered by the set of rules).

The number of cases covered by a set of rules can be computed as: (1) a sum of cases covered by each rule independently or (2) dependently counting each case only once, while it can be covered by multiples rules. The selection of this method depends on the goal of the complexity analysis. To measure the rule confidence (1) can be more informative, but to measure coverage (2) can be more informative.

Example. Rule R1 has a complexity of 1/1000 and rule R2 has a complexity of 100/1000 then it is intuitively clear that R1 is less complex than R2 because the last one requires 100 times more base clauses to cover the same number of cases as the first rule R1.

The base algorithm (SRG0) consists of the following steps:

(1) Form groups of attributes sequentially, e.g., group 1: (x_1,x_2,x_3), group 2: (x_4,x_5,x_6,x_7) and so on. In the current experiments each groups contained 3–5 attributes per group typically.
(2) Generate a set of rules in each group by using Monotone Boolean Functions (MBFs) with computing their precision and coverage
(3) Sort rules by the number of cases they cover in decreasing order, e.g., R_5:1000 cases, R_3: 900 cases, R_2: 800 cases, R_7: 700.
(4) Search for rules in the sorted list with 100% precision.
(5) If such rules found (e.g., R_3 with 900 cases and R_7 with 800 cases) record. Thus, SRG0 algorithm is for the "ideal" data, where rules with 100% precision exist.

4.3 Algorithm SRG1 with Rule Overlap Analysis

Algorithm SRG1 improves rule discovery by analyzing the overlap of the rules with each other on underlying data. If rules do not overlap, then they are fully complementary. If rules overlap, then we have more options to modify and combine them. In addition, rule overlap is useful for increasing the confidence in prediction. For example, consider case **x** covered by three rules. These rules use different attributes and different their values, but still predict that case **x** belongs to the same class *C*. This gives a *higher confidence* to the prediction that **x** belongs to class *C* like poisonous mushroom than if we would have only one rule. This is in line with the voting approach that is actively and successfully used in machine learning, especially in the random forest algorithm and other ensemble-based methods.

SRG1 algorithm specifies how rules are selected from generated rules for the selected class *C*, which we call a target class. This process orders generated rules by the number of cases in class *C* that are covered by each individual rule but are not covered by already selected rules. It uses the advantage of selecting rules with the highest number of class *C* cases that are not covered by already selected rules to cover class *C*.

Now consider a situation when one *rule dominates and covers majority* of the cases of the given class with high precision. This rule must be highlighted for the deep analysis by the domain expert/analyst. The reason is that it is unlikely that this rule is spurious and overfitting data, especially if it is short containing only few attributes and few values of those attributes. This rule indicates that attributes and their values are most informative ones. So, it allows to focus attention on these attributes to limit later data collection by those attributes and measuring only their values.

If other rules cover only a few cases, then they have a low generalizability and low trust/confidence. The modification of them by increasing their coverage will increase the confidence in rules and will avoid data overfitting and memorization. This is another way to look at the confidence of the rules. Thus, all rules that cover many cases increase the confidence in the set of rules and those rules should be preferred rules to be included instead of rules that cover only a few remaining cases not covered by the dominant rule.

The SRG1 algorithm consists of the following steps:

(1) Form groups of attributes sequentially, e.g., group 1: (x_1,x_2,x_3), group 2: (x_4,x_5,x_6,x_7) and so on.
(2) Generate a set of rules in each group by using Monotone Boolean Functions (MBF) with analysis of the *overlap* of rules in the process of rule generation.
(3) Sort rules by the number of cases they cover in decreasing order, e.g., R_5:1000 cases, R_3: 900 cases, R_2:800 cases, R_7:700.
(4) Determine the first candidate rule $R_{candidate}$ and record the value of $N_{notCovered}$ for it.
(5) Select all other rules, denoted as *OtherRules,* with the same value of $N_{notCovered}$.
(6) If any rule in *OtherRules* has greater $N_{covered}$ and greater precision *P,* select it and repeat the process for finding better rules.

Thus, SRG1 algorithm selects rules with less overlap or with more cases that are not covered by prior rules. It differs from SRG0 in steps (4)–(6).

4.4 Algorithm SRG2 with Complementary Rules for Precision Improvement

As shown above, the mushroom dataset allowed to achieve full coverage of the target value "poisonous" with 100% precision by requiring 100% precision in the rule discovery. Above the SRG0 algorithm required all rules to have 100% precision. This however is not feasible for datasets that have no rules with 100% precision. Therefore, in this section we present a rule selection process that first allows selecting less precise rules and then update the rules to increase precision. It is done by discovering rules that predict an *opposite class,* which is eatable for the mushroom data. We will also call it non-target class. The algorithm records all incorrectly predicted cases by initially selected rules, and then generates new rules to predict the opposite class on these data. For example, if 900 cases are incorrectly predicted as class 1 that are in fact in class 2, then the algorithm would generate rules to predict **class 2** using those 900 cases.

A produced algorithm is denoted as SRG2 algorithm. It assumes a prior run of algorithms SRG0 and SRG1 on the same data with grouping attributes. After that the steps of the SRG2 algorithm are as follows:

(1) Choose most frequent attributes to be used to generate rules in Step (2). The most frequent attributes are derived from a prior run of algorithms SRG0 and SRG1 on the same data with grouping attributes. The attribute is considered as a most frequent if it appeared in 50% of the rules that each group generated.
(2) Sort chosen attributes by the percentage of appearance per attribute grouping and form groups based on this sorting with the first group with the most frequent attributes.
(3) Generate all rules for the target class (e.g., poisonous) with chosen attributes and select the rules that lead to the highest coverage and precision while recording incorrectly predicted cases.
(4) Generate all rules for the non-target class (e.g., eatable) using the incorrectly predicted cases. These rules are called complementary rules.
(5) Add rules that predict the non-target class cases to the final selected rules if the rules increase overall precision and meet the following requirements,

    Requirement 1: Rule must cover previous incorrectly predicted cases.
    Requirement 2: Rule must not incorrectly predict target class cases.

(6) Combining rules for two classes (see the next section for details).

4.5 Algorithm SRG3 Based on 30 Randomly Generated Triples of Attributes

Version 3 of Sequential Rule Generation algorithm, SRG3, differs from versions SRG0-SRG2 in the selection attributes to the groups. In the previous versions groups are created sequentially, like (x_1, x_2, x_3), (x_4, x_5, x_6) and so on. In SRG3 groups of three attributes are formed randomly and 30 random such triples are created.

Accordingly, we have versions of SRG3 based on the SRG1 and SRG2 as follows. SRG3 based on SRG1 uses 30 randomly created triples of attributes. SRG3 based on SRG2 uses most frequent attributes from 30 randomly created triples in the run of SRG3 based on SRG1.

The exploration of algorithm SRG3 based on algorithm SRG1 allows to see if it could achieve 100% coverage and 100% precision of class C_1 with a small set of rules without relying on the original attribute order or any other specific attribute groupings. This is done in the experiment 4 on Mushroom data.

The exploration based on the SRG2 allows to see how the frequency of attributes in the rules impacts the quality of the rules. Those most frequent attributes are selected from the results of running SRG3 based on SRG1. This is done in the experiment 5 on the same Mushroom data.

4.6 Algorithm SRG4 Based on Expert Selected Groups

The next version of the sequential rule generation algorithm, denoted as SRG4 differs from the prior versions in that the groups of the attributes are generated by the domain expert. For the mushroom data it was a biologist. This was done to see if the mushroom data attributes could be grouped using expert knowledge to produce high coverage and high precision rules.

The main difficulty when creating attribute groupings is that there is usually little knowledge of how the attributes relate to each other. By allowing the expert biologist to create attribute groupings, the hope is that this difficulty will decrease.

4.7 Algorithm SRG5 Based on Prior Suceesful Attributes

The idea of the next version of the sequential rule generation algorithm, denoted as SRG5 is selecting attributes that have been successful in the other models on the same data. This is a form on the knowledge transfer.

The sources of these attributes can be attributes that are revealed in previous high precision rules on these data. The expectation is that it will allow the rule generation process to generate and select less rules while keeping a high precision and total coverage of the target class. To shorten computation time this to the different groups.

4.8 Summary of 10 Experiments with SRG Algorithms

The detailed results of 10 experiments with SRG algorithm on the mushroom data from [27] are presented in appendices "Experiment 1 with Algorithm SRG0 for Discovering Rules on Mushroom with Sequential Triples"–"Experiment 10 with SPG 5 and Comparison of Rule Complexity". Table 1 provides a summary of these results. The target class of rules in all experiments is poisonous. This table allows us to conclude that SRG algorithm is capable to discover useful and compact interpretable rules from the mushroom data with high precision and coverage reaching 100% in both. This indicates the feasibility to discover such rules on other data sets with qualitative attributes, which needs additional exploration.

Table 1 Summary of ten experiments with SRG algorithm on mushroom data

#	Algorithm	Description	Rules	Precision %	Coverage %	Highlight
1	SRG0	Sequential attribute triples	7	100	100	96.94% coverage of the best rule R_1
2	SRG1	Rule overlap analysis	7	100	100	The non-overlap of rules R_1 and R_6, increases coverage
3	SRG2	Complimentary rules, 95% precision threefold	13	[95.2,100]	100	Rule with largest coverage R_1, 98.47% coverage and 95.2% precision
4	SRG3	30 random attribute groups	7	100	100	96.94% coverage of the best rule R_1
5	SRG3	13 most frequent attributes	7	100	100	96.94% coverage of the best rule R1. Did not reduce the number of rules below 7
6	SRG3	Sequential attribute triples, tenfold cross validation	4	100	100	All tenfold cross validation folds acquire 100% accuracy and precision
7	SRG3	30 random attribute triples (4 times), tenfold cross validation	4–5	100	100	All tenfold cross validation folds acquire 100% accuracy and precision. Complexity of rule sets [13/3525, 16/3520]
8	SRG4	Expert selected groups. 95% precision threshold	3	100	99.81	Smallest number of rules with a good coverage
9	SRG5	7 successful attributes and groups. 95% precision threshold	7	100	100	Did not reduce the number of rules below 7
10	SRG5	Attributes and groups from [1, 30]. 95% precision threshold	4	100	100	Lower complexity of rules. Complexity 0.002 vs. 0.0027 for [1, 30] with 11 clauses versus 13 clauses for [1, 30]

Table 2 Situations and goals appropriate for versions of SRG algorithm

Algorithm	Situation	Goal
SRG0	Given order of attributes is good for attribute grouping	100% precision and coverage
SRG1	Given order of attributes is good for attribute grouping	Short set of rules
SRG2	Given order of attributes is good for attribute grouping	Precision and coverage above thresholds
SRG3	No information on good attribute grouping	100% precision and coverage or short set of rules
SRG 4	Expert selected attribute groups	Precision and coverage above thresholds
SRG 5	Prior successful attributes in rules are given	Precision and coverage above thresholds

Analysis of these results shows that selecting a specific SRG algorithm for the task at hand depends on a combination of several properties of the task and data. If the goal is to get 100% precision and coverage then the SRD0 can be applied. While it happened with the mushroom data, we cannot expect it for all possible data.

For such data SRG will be more appropriate with abilities to set up a desired level of precision and coverage, like 75, 85, or 95%. If additional knowledge is available like expert grouping of attributes, or attributes that have been successful in prior models on the same data then SRG45 are preferable.

Table 2 summarizes the appropriate situations and goals for SRG versions.

5 Mixed Multidimensional Data Visualization and Visual Rule Discovery

In the previous sections we demonstrated abilities to discover rules analytically with an appropriate coding of categorical data. This section presents methods to visualize mixed multidimensional data and rules along with rule discovery in parallel coordinates.

The rules produced in Experiment 10 are presented in Appendix "Experiment 10 with SPG 5 and Comparison of Rule Complexity" in detail. These rules are less complex than CR rules from [30, 31]. Our rules use 11 clauses covering all poisonous cases with duplication (5428 cases, complexity 11/5428 = 0.002). CR rules use 13 clauses to cover all poisonous cases with less duplication (4780 cases, complexity 13/4780 = 0.0027).

R_1: $[(x_5 = 3) \vee (x_5 = 4) \vee (x_5 = 5) \vee (x_5 = 6) \vee (x_5 = 8) \vee (x_5 = 9)] \Rightarrow \mathbf{x} \in C_1$
R_2: $[(x_{20} = 5)] \Rightarrow x \in C_1$
R_3: $[(x_{12} = 3)\ \&\ (x_{21} = 5)] \Rightarrow x \in C_1$

R_4: & $(x_{21} = 2)] \Rightarrow \mathrm{x} \in C_1$

CR Rules [30, 31]

CR_1: $[(x_5 = 3) \vee (x_5 = 4) \vee (x_5 = 5) \vee (x_5 = 6) \vee (x_5 = 8) \vee (x_5 = 9)] \Rightarrow \mathbf{x} \in C_1$,
CR_2: $[(x_{20} = 5)] \Rightarrow \mathbf{x} \in C_1$
CR_{31}: $(x_8 = 2)$ & $(x_{12} = 3) \Rightarrow \mathbf{x} \in C_1$
CR_{32}: $(x_8 = 2)$ & $(x_{12} = 2) \Rightarrow \mathbf{x} \in C_1$
CR_{33}: $(x_8 = 2)$ & $(x_{21} = 2) \Rightarrow \mathbf{x} \in C_1$

Our rules have been discovered using extensive analytical computations by machine learning SRG5 algorithm described in Sect. 4. Similarly, CR rules also have been discovered using extensive computations by other machine learning methods. These computations are hidden from the user to be able to trust the produced rules. A user should be able to get some kind of justification of that the rules correct. Visualization is an efficient method to help users to understand and trust rules.

These rules are visualized in Fig. 3 in parallel coordinates. They contain only 5 attributes out of 22 attributes providing 100 precision and coverage of all mushroom poisonous cases. The visualization of rules in Fig. 3 is quite simple and can be understood by the end users easily.

Rules R_1 and R_2 are the same as rules CR_1 and CR_2. In Fig. 3, visually rule R_3 is simpler than a joint rule of rules CR_{31} and CR_{32} shown in blue. In contrast the rule R_3 visually looks a more complex than rule CR_{33} in gray, while they have the same computational complexity, R_4 checks $x_8 \neq 1$ and CR_{33} checks $x_8 = 2$. This shows the deficiency of parallel coordinate visualization, which does not allow compactly visualize the negation property $x_8 \neq 1$. It requires showing all values which are not equal to 1 on X_8.

It is relatively easy to provide to the user a visual confirmation of these rules by plotting all poisonous cases in these rules in Fig. 3. For rule R_1 it will locate each case that satisfy R_1 as a point in the red area on attribute X_5. For R_2 it will be points in the in the green area on attribute X_{20} with value 5. For R_3 it will be all lines connecting value 3 in X_{12} with value 5 in X_{21} and for rule R_4 it will be all lines connecting value 2 of X_{21} with all other values in X_8 but 1.

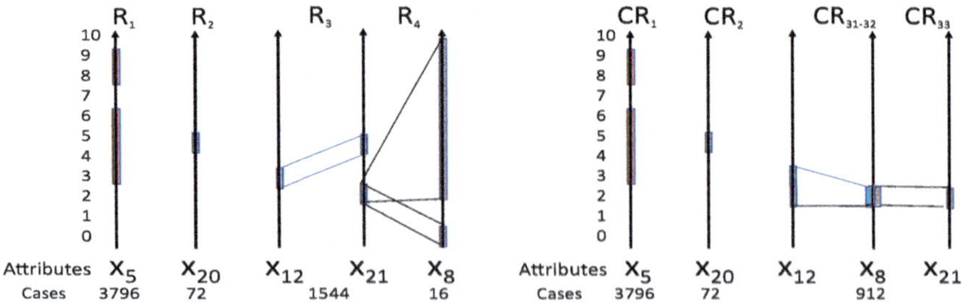

(a) Visualization of rules from Experiment 10.

(b) Visualization of rules from [29,32].

Fig. 3 Visualization of rules generated in Experiment 10 and from [30, 31] for poisonous mushroom

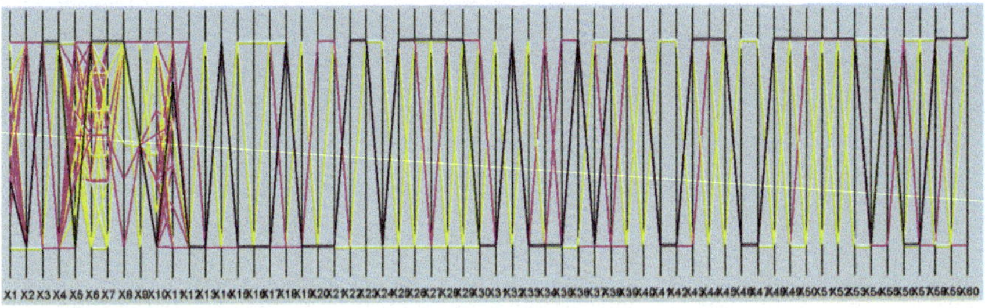

Fig. 4 Binary encoded mushroom data in parallel coordinates

However, drawing in Fig. 3 all mushroom cases fully showing their values not only in 1–2 attributes but 5 or all 22 attributes will produce significant overlap and occlusion of cases. It creates difficulties to analyze cases fully, to observe discovered patterns that rules present and to discover new patterns visually. Solving this task requires visual knowledge discovery (VKD) methods that we present in this section.

Figure 4 shows all mushroom data visualized in parallel coordinates in VisCanvas [29, 32], where nominal attributes are encoded by binary attributes using the Data Type Editor of our Toolkit described in Appendix "Toolkit" resulted in over 70 attributes. The uninformative attributes with the same values for all cases are omitted leaving 60 attributes shown in Fig. 4.

These 60 attributes from 22 original attributes demonstrate deficiency of such coding for visualization. The data are quite hard to look at since there are so many attributes. Moreover, in general, binary attributes do not fit well to be visualized in parallel coordinates because binary attributes do not have much variability and do not fill the area between 0 and 1 in each coordinate. As a result, many lines cover each other making different classes practically undistinguishable as was shown before in [5].

Figure 5 shows visualization of the Census income dataset from the UCI ML repository [27] in parallel coordinates. It was selected for its size of 48,842 cases of mixture of 14 integer and categorical attributes. Conversion several nominal attributes to binary attributes produced total 26 attributes. Binary coding of nominal attributes requires domain knowledge to make it meaningful, which is easier when attributes are a part of the common knowledge like in the Census data.

Figures 4, 5 and 6 show the deficiency of visualization of data with multiple nominal attributes by converting to binary attributes. Figure 6 visualizes the Teaching Assistant (TA) evaluation dataset from the UCI ML repository [27], which is a mix of categorical and integer data. In these data values of some attributes are grouped and converted to binary attributes using the Data Type Editor. The cases are colored according to the values of the attribute X9, which has over 20 values. It shows high overlap of cases making visual discovering of patterns difficult. Therefore, below we present alternative methods.

Frequency based visualization of nominal attributes. Figure 7 shows the visualization of the same TA dataset in parallel coordinates where bars are sized by

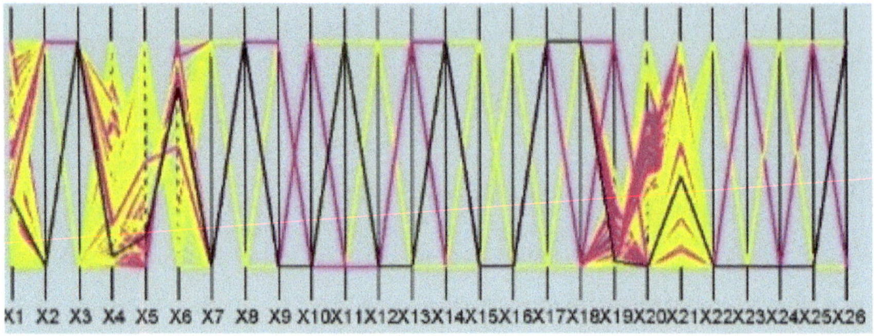

Fig. 5 Mixed census income data in parallel coordinates after meaningful coding

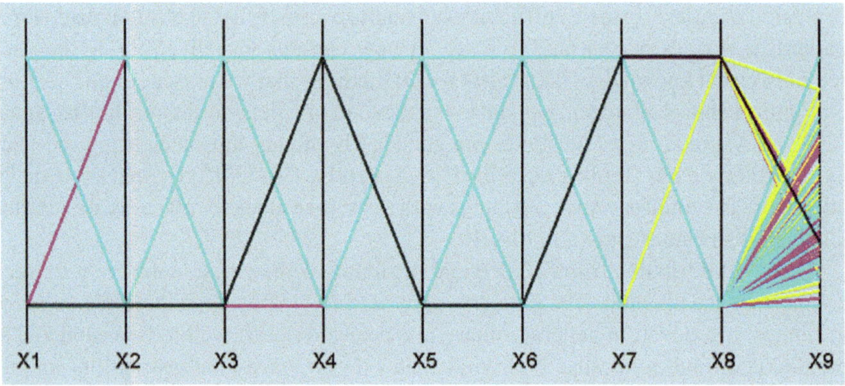

Fig. 6 TA data in parallel coordinates

the frequency of occurrence of the respective nominal values. The user interface in VisCanvas allows a user to generate this frequency-based visualization. This visualization is similar to the parallel sets visualization [33].

The important difference is that the lines going from one attribute to another do not have the same width as the bar on the attribute. Use thin lines making visualization less occluded. The frequencies of values of X9 also have been computed and visualized by the height of the bars.

The bars are ordered according to their frequencies in the descending order with bars of the greatest frequency at the bottom of X9. This visualization is more informative than shown in Fig. 6, which helps to solve the occlusion problem that we see in Fig. 6. Also, a user can select a specific color scheme for the nominal blocks.

Reference frequency visualization. The frequency visualizations shown in Fig. 7 adapted from parallel sets [33] have an important deficiency for machine learning tasks. They are *not related to the classes,* the height of the bar for a given value of the attribute is based on the *frequency* of this value *itself* in the dataset, which contain cases of several classes. Below we present visualizations that are more relevant to

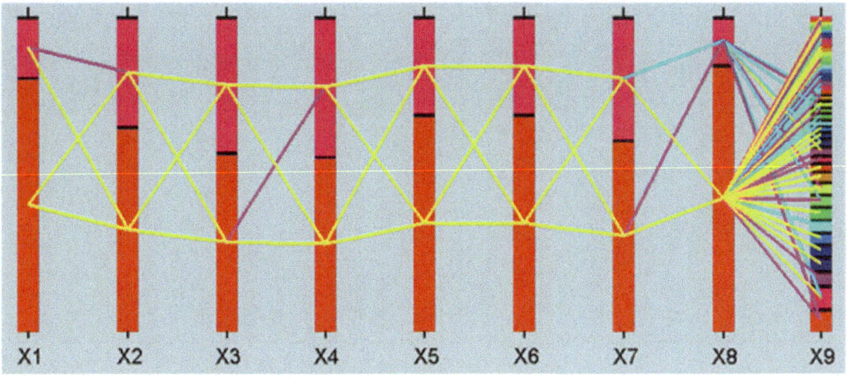

Fig. 7 TA data in adapted parallel sets

machine learning. This method computes the number of bars, and their heights in each attribute X_i based on the relations of these values with values of another reference attribute X_t (e.g., target attribute).

The coloring of lines and bars is based on the values of the reference attribute/classes. For instance, reference attribute can have 106 cases with $x_i = \text{a}$, which split to 10 cases with $x_t = 0$, 70 cases with $x_t = 1$, 12 cases with $x_t = 2$, and 8 cases with $x_t = 3$. Here $x_t = 1$ is a dominant value of X_i.

The bars for all non-dominant values can be joined to a single bar (see grey bars in Fig. 8). It visually emphasizes the dominant value of the target attribute/class. In Fig. 8 the portion of each bar is colored by the dominant class (magenta or blue) and the non-dominant part is grey, the black horizontal lines separate bars.

Figures 9 and 10 illustrate advantages of reference frequency-based visualization on the mushroom classification dataset with class colors. Figure 9 shows dominant class frequency with all bars, while Fig. 10 shows only bars of high purity ($\geq$80%).

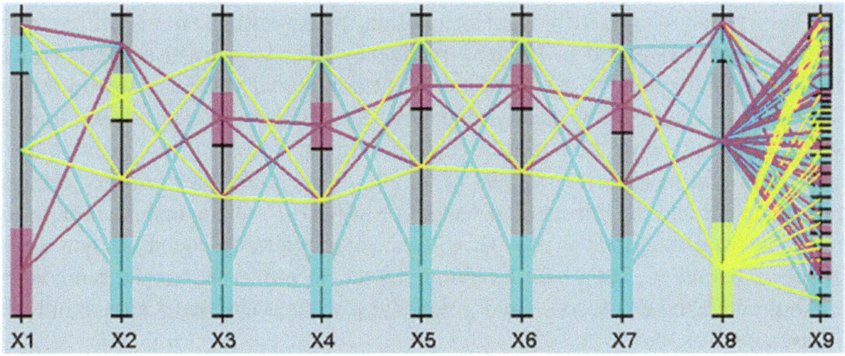

Fig. 8 Visualization with class colors and dominant frequency for TA dataset with weighted lines

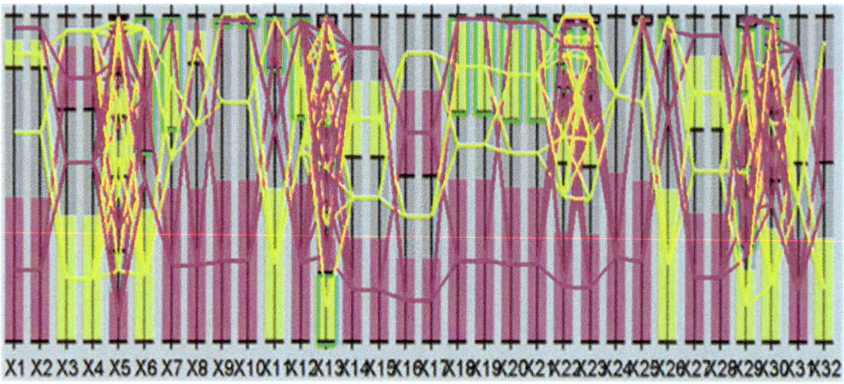

Fig. 9 Mushroom dataset with class colors and dominant frequency with all bars

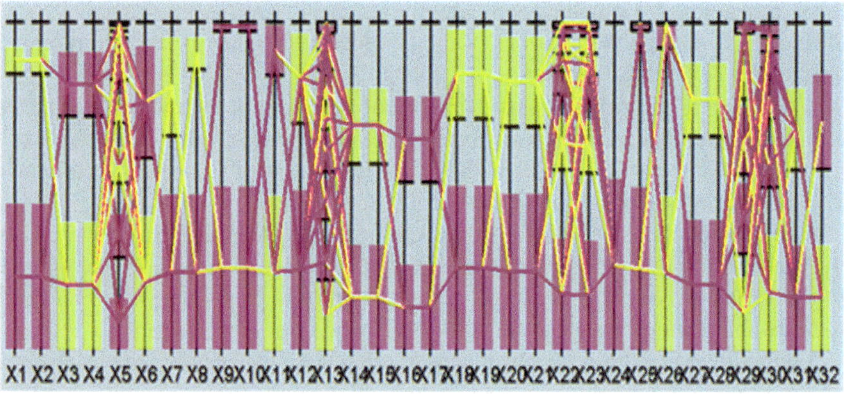

Fig. 10 Mushroom dataset with class colors with bars above 80% purity

The frequency approach implemented in these visualizations differs from frequency methods listed in [8]. The Frequency Encoding in [8] converts textual data into numeric data by assigning the frequency of that value as its code. In this coding if two values get the same frequency, say, 0.3, then 0.3 will be used as code for both values making them indistinguishable. In our frequency visualization these two values will have their own bars of length 0.3 each. So, they will not collapse to a single bar and the information will be preserved. To distinguish our frequency encoding from described in [8] we will call our frequency encoding as a *visual frequency encoding*. The Mean Encoding/Target Encoding [8] has the same issue. It can produce the equal codes and glue values, while our visualization avoids the information loss.

To help a user to get more information from the visualization their linguistic descriptions are automatically generated (Fig. 11). It describes blocks/bars with purity of 80% or higher and attributes with small frequency blocks/bars.

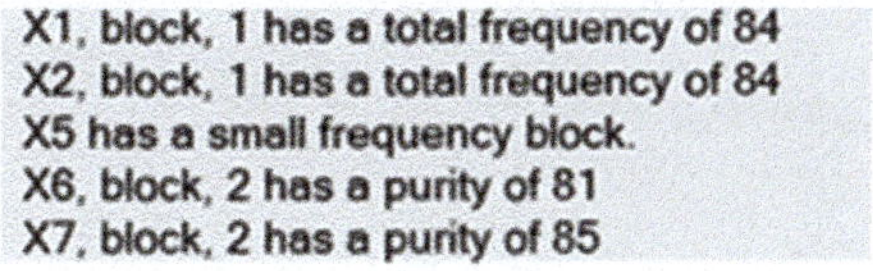

Fig. 11 Dominant blocks/ bars on mushroom dataset

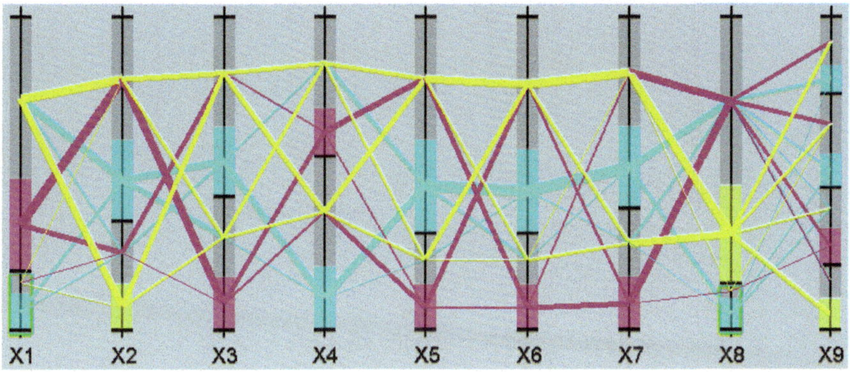

Fig. 12 TA dataset with blocks sorted by purity and wider frequent cases

Figure 11 describes a few attributes of the mushroom dataset visualized in Fig. 10. On this dataset many attributes have small blocks with purity above 80% purity. To emphasize larger blocks the limit on block size of 10% was introduced in addition to 80% of purity.

In Fig. 8, attribute X_9 was not very useful in discovering rules. In contrast, Fig. 12 shows the TA dataset with splitting values of attribute X_9 (class size) into 4 groups. Now it shows the pattern that TAs rated the best (blue blocks and lines) were dominant in the first and the fourth groups. The worst rated TAs were dominant in the second group and the average TAs were dominant in the third group.

Flipping attributes allows making visual patterns clearer and VisCanvas supports it. Figure 13 shows flipping (negating) some attributes in the TA dataset. All attributes are normalized to [0,1] and flipping creates 1-x for the attribute x. As a result, blue cases concentrate at the bottom, magenta cases are in the middle and yellow cases split between the top and the middle.

Reordering attributes is another option in VisCanvas to make patters clearer. Figure 14 shows the mushroom data with reordered attributes, where the attributes with the puree blocks are on the left and the less pure blocks are on the right. It makes visual patters clearer. A user can change the purity threshold for more distilled visual patterns.

Grouping, relocating blocks, and sorting by color is another way to make patterns simpler and clearer. Some datasets have attributes with a few large blocks and many small blocks. Often these small blocks are of high purity and are next to the large blocks, which making them hard to see as in Fig. 14 on the right. Therefore, the *smaller blocks* (under 20%) are moved to the top of those attributes (see Fig. 15).

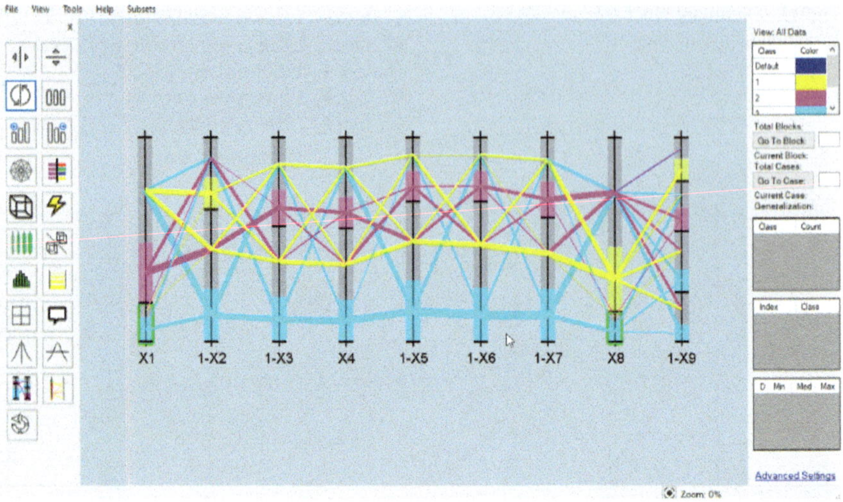

Fig. 13 TA dataset with flipped attributes to simplify visuals

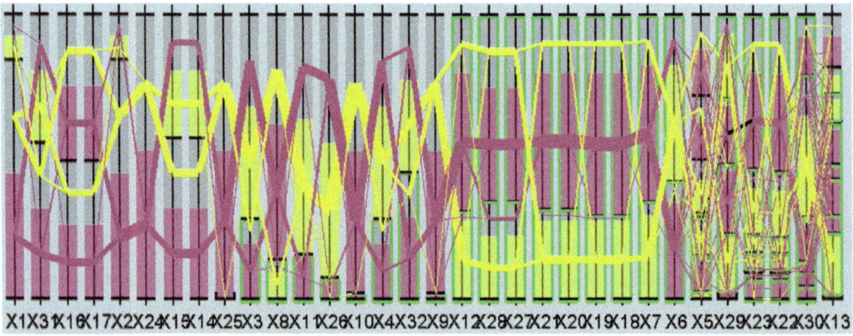

Fig. 14 Mushroom data with attributes ordered by purity of blocks decreasing from left to right and wider frequent cases

Figure 15 also shows the results of *sorting* the mushroom data by *color* and putting the yellow blocks on the top. Here the attributes are sorted by the number of purity blocks.

Visual rule generation. Such strong visual patterns in the data allow to make classification rules like

$$\text{If case a is in magenta block in X11 \& is NOT in yellow block X12 then a is in magenta class} \quad (12)$$

These blocks are outlined by the black ovals in Fig. 15. These magenta and yellow boxes have green frames indicating their high purity. This rule is an example

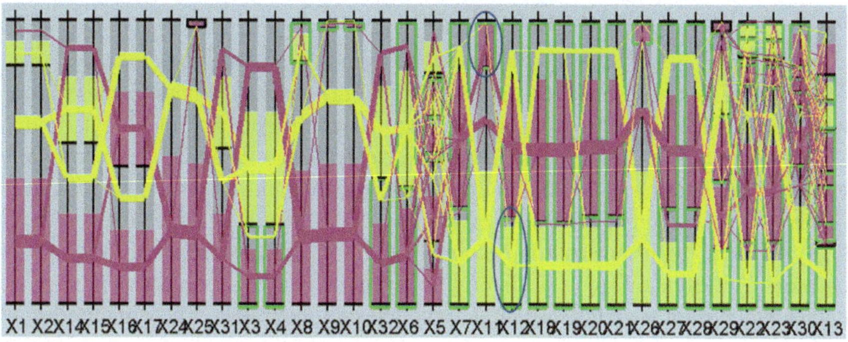

Fig. 15 Mushroom dataset sorted by color

of explainable rules that end users can find themselves by such visual knowledge discovery process. The comparison of Figs. 4 and 15 with the same data shows the benefit of the toolkit. Extracting rules like (7) is impossible in Fig. 4.

The benefits of visual knowledge discovery derived from visualization in Fig. 13 are described below for the TA dataset. This dataset is a collection of attributes of students who were rated 1 to 3 on their effectiveness of being a teaching assistant. The value 3 is a ‘great’ rating, 2 is an ‘average’ rating, and 1 is a ‘bad’ rating. In Figs. 3 and 13 is blue, 2 is yellow, and 1 is magenta. After flipping X2-X7, and X9 we can see a clear pattern in the bottom of Fig. 13 that leads to a teaching assistant being great.

The attribute X1 shows a purity of over 60% dominant class 3 (great) in the bottom block and a dominant class of class 1(bad) in the top block. This attribute represents a Boolean value of whether the teaching assistant was a native English speaker or not where the top block is false, and the bottom block is true. This means that a teaching assistant is more likely to be rated as ‘great’ if they are native English speakers.

The first block in X2 is dominantly blue with most “great” TAs going there. It represents the value 0 meaning that a majority of student’s label “great” had instructors from either group 1, 2, or 3. The upper block in the ovals with magenta dominant block is much less pure. Similarly, in X3 most of the “great” teaching assistants either had instructor in group 1 or group 4. Next, value 1 of the blue block in X4 means that the “great” teaching assistants came from group 1 or 3.

The analysis of attributes X5, X6, and X7 shows that “worst” rated TAs were assisting in classes from group 4 since the first block value is one. The attribute X8 represents whether the semester was a normal semester or a summer semester. The bottom block of this coordinate is dominantly blue. Its value 2 means that most “great” teaching assistants were helping in classes during summer semesters. Figure 13 shows that the bottom two blocks are dominantly blue in attribute X9 (class size). This means that “great” teaching assistants dominantly helped in classes with 3 to 19 students or 37 to 66 students.

6 Conclusion and Future Work

The focus of this study has been on developing: (1) numeric coding schemes for non-numeric attributes for ML algorithms to support accurate and explainable ML models, (2) accurate and explainable ML models for categorical data, and (3) methods for lossless visualization of n-D non-numeric categorical data with visual rule discovery in these visualizations. For this we proposed a classification of mixed data types and analyzes their important role in Machine Learning, an approach, and a toolkit for enforcing interpretability of all internal operations of ML algorithms on heterogeneous data with a visual exploration approach on mixed data. These methods have been demonstrated on Mushroom, Census income, and Teaching Assistants evaluation datasets. For (2) we proposed different versions of the Sequential Rule Generation (SRG) algorithm for explainable ML with categorical data. These versions offer options to group attributes (sequential, random, expert based, and by knowledge transfer from successful models).

These version of SRG algorithms have been successfully evaluated in multiple experiments on the benchmark Mushroom dataset. In experiments the different versions of SRG algorithm have been able to produce rules that accurately classify all mushroom data. The comparison with the rules generated in [30] had shown advantages of SRG generated rules relative to rules produced in [30]. SRG rules reached the same accuracy with less complexity. The versions of the SRG algorithm varied in the ways how the groups of attributes are formed, types of rules generated and how the selection of priority rules. The major issue, which we discovered in this study is computational limitations to explore the groups with the larger number of attributes. Further experiments are needed to test SRG algorithms, which will allow to discover other limitation of current versions of SRG and to develop new versions of SRG algorithm which will fit different datasets.

This study provides a user with the adaptation of mixed data types and their coding schemas for lossless visualization of multidimensional mixed data in parallel coordinates. The developed experimental Toolkit combines the Data Types Editor and VisCanvas system for mixed multidimensional data visualization and explainable rule discovery. It is available on GitHub [32]. It supports (a) numeric *coding schemes* for non-numeric attributes for *explainable ML* with mixed data, (b) *lossless visualization* of n-D *non-numeric* data, and (c) *visual rule discovery* in these visualizations, and (d) analytical rule discovery with Sequential Rule Generation algorithms. The future work is developing new *full scope ML algorithms* for mixed data integrated that will generalize SRG algorithms with lossless visualization of n-D heterogeneous data and other types of General Line Coordinates beyond Parallel Coordinates.

Table 3 Characteristics of discovered rules R1–R7

Characteristic	R1	R2	R3	R4	R5	R6	R7
Precision, %	100	100	100	100	100	100	100
Coverage, %	96.94	44.74	49.34	12.56	9.91	1.84	1.33
Total cases predicted, $N_{covered}$	3796	1752	1184	492	388	72	52
Correctly predicted cases, $N_{correct}$	3796	1752	1184	492	388	72	52
Misclassified cases	0	0	0	0	0	0	0

Appendix

Experiment 1 with Algorithm SRG0 for Discovering Rules on Mushroom with Sequential Triples

In this experiment we tested SRG0 algorithm for discovering rules on Mushroom data [27] to solve a two-class classification problem (poisonous - eatable). We found rules with 100% precision for the *target class* C_1 *– poisonous mushrooms.* One of these rules is a rule reported in the literature [30]. Below we present discovered rules R1-R7 using notation of the case $\mathbf{x} = (x_1,x_2,\ldots,x_n)$. If any of these rules is true, then $\mathbf{x}$ belongs to class C_1.

Rules R1–R7
R_1: $[(x_5 = 3) \vee (x_5 = 4) \vee (x_5 = 5) \vee (x_5 = 6) \vee (x_5 = 8) \vee (x_5 = 9)] \Rightarrow \mathbf{x} \in C_1$
R_2: $[(x_9 = 6) \vee (x_9 = 3)] \Rightarrow x \in C_1$
R_3: $[(x_{19} = 2)\ \&\ (x_{20} = 8)\ \&\ (x_{21} \neq 2)\ \&\ (x_{22} \neq 2)] \Rightarrow \mathbf{x} \in C_1$
R_4: $[(x_{15} = 3) \vee (x_{15} = 2) \vee (x_{15} = 9)] \Rightarrow \mathbf{x} \in C_1$
R_5: $[(x_{19} \neq 2)\ \&\ (x_{20} \neq 6)\ \&\ (x_{21} = 5)\ \&\ (x_{22} = 1)] \Rightarrow \mathbf{x} \in C_1$
R_6: $[(x_{19} = 6)\ \&\ (x_{20} = 5)\ \&\ (x_{21} \neq 1)\ \&\ (x_{22} = !6)] \Rightarrow \mathbf{x} \in C_1$
R_7: $[(x_{20} = 8)\ \&\ (x_{21} = 2)\ \&\ (x_{22} = !6)] \Rightarrow \mathbf{x} \in C_1$

Table 3 presents characteristics of discovered 7 rules, R1-R7, which cover 100% of the mushroom data for the "poisonous" class with 100% precision. The total number of cases in the poisonous N_{total} class is 3916 cases. The rules were with attributes broken up into six sequential triples and one group with 4 attributes: x_{19}, x_{20}, x_{21}, x_{22}. These 6 groups are (x_1, x_2, x_3), (x_4, x_5, x_6), (x_7, x_8, x_9), (x_{10}, x_{11}, x_{12}), (x_{13}, x_{14}, x_{35}), (x_{16}, x_{17}, x_{18}).

Experiment 2 with SRG1 Algorithm for Discovering Rules on Mushroom Data with Rule Overlap Minimization

The analysis of rules presented in the previous section shows that, Rule R_1 covers 96.94% of the cases with target value "poisonous", which means that the other 6 rules add to the coverage only 3.04% of these cases (118 cases). For this reason, to get a better understanding of these rules we calculated the overlap between cases classified by different rules. See Tables 4 and 5. In these tables, the overlap, OL, is the total number of cases that are in the intersection of rules R_i and R_j, $OL(R_i,R_j) = |Cases(R_i) \cap Cases(R_j)|$, the overlap percentage is $OL(R_i,R_j)$ divided by the total number of cases in the union of cases covered by both rules, $OL(R_i,R_j)/|cases(R_i) \cup cases(R_j)|$. Table 4 shows the relations of rule R_1 with other rules and Table 5 of remaining rules with each other.

For instance, the analysis of the relations between rules R_1 and R_2 show that together they cover 3820 cases of poisonous mushrooms and overlap in 1728 cases (45.24% of 3820 cases). Rule R_1 covers 3796 cases, thus rule R_2 adds only 3820–3796 = 24 new cases to R_1, which is 20.34% of 120 cases not covered by rule R_1. R_6 does not overlap with rule R_1 it adds 72 new cases to R_1, which is 60% of 120 cases not covered by rule R_1.

Thus, rules R_1&R_2, R_1&R_6 differ in an important characteristic. Heavy overlap of R_1 and R_2 (45.24%) *increases the confidence* in classification of those overlapped

Table 4 Overlap between dominant rule R_1 and other rules

		R_2	R_3	R_4	R_5	R_6	R_7
	Cases covered	1752	1184	492	388	72	52
R_1 3796 cases	Overlap, %	45.24%	30.09	12.25%	9.19%	0%	0.94%
	Total cases	3820	3828	3820	3832	3868	3812
	Overlap cases	1728	1152	468	352	0	36
	Added cases	24	32	24	36	72	16
	Added cases, %	20.00%	26.67%	20.00%	30.00%	60.00%	13.33%

Table 5 Overlap percentage and number of cases for rules R_2–R_7

	R_3 1184 cases	R_4 492 cases	R_5 388 cases	R_6 72 cases	R_7 52 cases
R_2 1752 cases	64.00% Total: 1784 Overlap: 1152	0% Total: 2244 Overlap: 0	0.56% Total: 2128 Overlap: 12	1.33% Total: 1800 Overlap: 24	0% Total: 1804 Overlap: 0
R_3 1184 cases		0.96% Total: 1660 Overlap: 16	0% Total: 1572 Overlap: 0	0% Total: 1256 Overlap: 0	0% Total: 1236 Overlap: 0
R_4 492 cases			8.01% Total: 808 Overlap: 72	0% Total: 564 Overlap: 0	8.8% Total: 500 Overlap: 44
R_5 388 cases				8.49% Total: 424 Overlap: 36	0% Total: 440 Overlap: 0
R_6 72 cases					0% Total: 124 Overlap: 0

Table 6 Relations between rules CR_1–CR_3

		CR_2 72 cases	CR_3 912 cases
CR_1 3796 cases	Overlap, %	0%	22.48%
	Total cases	3868	3844
	Overlap cases	0	864
	Added cases	72	48
	Added cases, %	60%	40%
$CR_2$72 cases	Overlap, %	100%	0%
	Total cases	72	984
	Overlap cases	72	0
	Added cases	0	912
	Added cases, %	0%	23.73%

cases. The non-overlap of R_1 and R_6, *increases coverage* expanding the number of cases that the rule predicts two more than with adding R_2 to R_1.

For rules R_2-R_7, Table 5 shows overlap percentage for each pair of rules, the total number of distinct cases covered by two rules and the number of cases in their overlap. Only rules R_2 and R_3 heavily overlap. All other rules either do not overlap (7 pairs) or overlap in no more than 72 cases, which less than 10% of the total number of cases in each pair. As expected, many of the rules overlap with each other. The heaviest overlapping rules are R_2 and R_3, where the overlap is 64% together of the total cases predicted. This means that these two rules are closely related.

Table 6 shows the relations between rules CR_1-CR_3 from [30]. Rules CR1-CR3 use 5 attributes (x_5, x_8, $\times_{12}$, $\times_{20}$, $\times_{21}$), while our rules R1-R7 use 7 attributes (x_5, x_9, $\times_{15}$, $\times_{19}$, $\times_{20}$, $\times_{21}$, $\times_{22}$).

Rule R_1 is the same as rule CR_1 from [30]. Tables 4 and 5 show that rules R_2-R_7 are more general than rules CR_2 and CR_3 discovered in [30]. Rule CR_2 covers 72 cases and rule CR_3 covers 912 cases, while our rule R_7 with the smallest coverage cover 52 cases. Rules R_2-R_7 cover in total 2188 cases, while rules CR_2 and CR_3 cover only 984 cases together.

Experiment 3 with SRG2 Algorithm on Mushroom Data with Complimentary Rules Generation

Rule Generation

Below we present rules generated at Step 3 first for the poisonous class and then complementary rules for the eatable class on Mushroom data with most frequent attributes generated from all 22 mushroom attributes. In this experiment the following attribute groups are used: Group A1: x_9, x_5, x_7, x_{11}; Group A2: x_{13}, x_{14}, x_{15}, x_6; Group A3: x_1, x_2, x_4, x_{21}, x_{22}.

According to the design of SRG algorithm it runs at the several levels of thresholds that can be set up by a user. We experimented with 3 level of precision: 75%. 85% and 95% with fixed level of coverage of 0.5%. This means that rules that have lower precision coverage are filtered out and not selected. For coverage with 3916 cases in the poisonous class it means that rules that cover less than 20 cases are filtered out considered as overfitting rules.

The thresholds 75%, 85%, and 95% limit only the low margin of the rule quality, but do not limit their upper level. Therefore, we also computed the actual precision and coverage reached for the poisonous class at each level.

Table 7 shows that rules at all levels missed only 4 cases from the poisonously class, giving 99.89% coverage and none of the poisonous cases was misclassified as eatable, giving 100% precision. All classification errors came from classifying some eatable cases as poisonous, that ranged from 800 cases for 75% threshold to 192 cases for 95% threshold.

We conducted the further analysis for 13 rules selected at level 3 with 95% threshold for rules. See Table 8. This table shows that all 192 misclassified cases belong to the rule R_1 that has 98.47% coverage and 95.2% precision. All other rules have 100% precision and coverage that smaller than for dominant rule R_1.

Rules R_1–R_{13} (Poisonous)

R_1: $[(x_5 \neq 7)\ \&\ (x_7 \neq 2)\ \&\ (x_9 \neq 7)\ \&\ (x_{11} \neq 2)] \Rightarrow \mathbf{x} \in C_1$

R_2: $[(x_5 = 3) \vee (x_5 = 4) \vee (x_5 = 5) \vee (x_5 = 6) \vee (x_5 = 8)\ \mathrm{V}\ (x_5 = 9)] \Rightarrow \mathbf{x} \in C_1$

R_3: $[(x_9 = 3) \vee (x_9 = 6)] \Rightarrow \mathrm{x} \in C_1$

R_4: $[(x_6 \neq 1)\ \&\ (x_{13} \neq 4)\ \&\ (x_{14} \neq 8)\ \&\ (x_{15} \neq 1)] \Rightarrow \mathbf{x} \in C_1$

R_5: $[(x_1 \neq 6)\ \&\ (x_2 \neq 3)\ \&\ (x_4 \neq 1)\ \&\ (x_{21} \neq 6)\ \&\ (x_{22} = 7)] \Rightarrow \mathbf{x} \in C_1$

R_6: $[(x_9 = 5)\ \&\ (x_{11} = 1)] \Rightarrow \mathrm{x} \in C_1$

R_7: $[(x_{15} = 3) \vee (x_{15} = 3) \vee (x_{15} = 9)] \Rightarrow \mathbf{x} \in C_1$

R_8: $[(x_1 = 5)\ \&\ (x_4 = 2)\ \&\ (x_{21} = 5)\ \&\ (x_{22} \neq 2)] \Rightarrow \mathbf{x} \in C_1$

Table 7 Results of sequential rule generation for poisonous class in Experiment 3: Rules R_1–R_{13}

Characteristic	*Level 1*	*Level 2*	*Level 3*
Low rule precision threshold. %	75	85	95
Low rule coverage threshold, %	0.5	0.5	0.5
Actual number of cases of poisonous class	3916	3916	3916
Number of rules selected	14	14	13
Number of cases covered by all rules	4712	4520	4104
Number of cases correctly classified by all rules	3912	3912	3912
Number of unclassified cases of the poisonous class	4	4	4
Number of misclassified cases by all rules	800	608	192
Actual coverage of the poisonous class, %	99.89	99.89	99.89
Actual precision, %	100	100	100

Table 8 Characteristics of discovered rules R_1–R_{13} for poisonous class in Experiment 3

Characteristic	R_1	R_2	R_3	R_4	R_5	R_6	R_7
Precision, %	95.02	100	100	100	100	100	100
Coverage, %	98.47	96.94	44.7	33.76	15.42	12.84	12.56
Total cases predicted	3856	3796	1752	1322	604	504	492
Correct cases	3664	3796	1752	1322	604	504	492
Misclassified cases	192	0	0	0	0	0	0
	R_8	R_9	R_{10}	R_{11}	R_{12}	R_{13}	
Precision, %	100	100	100	100	100	100	
Coverage, %	10.01	9.91	1.48	1.23	1.12	0.92	
Total cases predicted	392	388	58	48	44	36	
Correct cases	392	388	58	48	44	36	
Misclassified cases	0	0	0	0	0	0	

R_9: $[(x_{21} = 5) \& (x_{22} = 1)] \Rightarrow x \in C_1$
R_{10}: $[(x_6 = 3) \& (x_{13} = 2) \& (x_{14} \neq 1) \& (x_{15} \neq 8)] \Rightarrow \mathbf{x} \in C_1$
R_{11}: $[(x_2 = 3) \& (x_{21} = 2) \& (x_{22} \neq 6)] \Rightarrow \mathbf{x} \in C_1$
R_{12}: $[(x_1 = 1) \& (x_{21} = 5) \& (x_{22} \neq 2)] \Rightarrow \mathbf{x} \in C_1$
R_{13}: $[(x_1 \neq 6) \& (x_2 \neq 1) \& (x_4 \neq 2) \& (x_{21} = 5) \& (x_{22} = 3)] \Rightarrow \mathbf{x} \in C_1$

The rules R14 and R15 generated for the eatable class C_2 are presented below:

R_{14}: $[(x_5 = 1) \& (x_9 \neq 1)] => x\ C_2$
R_{15}: $[(x_5 = 2) \& (x_9 \neq 1)] => x\ C_2$

Table 9 shows the analysis of both R14 and R15 for class eatable class C_2. Rules R_{14} and R_{15} together cover and correctly predict all 192 cases misclassed by Rule R_1. While both R_{14} and R_{15} cover 336 cases each, these cases are different. In fact, rules R_{14} and R_{15} have 0% overlap and combined cover 672 cases of eatable class C_2.

Table 9 Characteristics of discovered rules R_{14}, R_{15} for eatable class for Experiment 3

	R_{14}	R_{15}
Precision, %	100	100
Coverage, %	7.98	7.98
Total cases predicted	336	336
Correctly predicted cases	366	336
Misclassified cases	0	0

Combining Rules for Two Classes

Now we have rules R_1-R_{13} for the target class and rules R_{14}-R_{15} for the non-target class and can accomplish step 3 of combining them. The only rule for the target class that misclassified some cases is R_1. So, we need to improve only this rule. It is done by creating a new rule R_N that combines rule R_1 with R_{14} and R_{15} as follows.

$$R_N(x) = R_1(x) \& \neg(R_{14}(x) \vee R_{15}(x))$$

resulted in

$$R_N: [(x_5 \neq 7) \& (x_7 \neq 2) \& (x_9 \neq 7) \& (x_{11} \neq 2)] \&$$

$$\neg([(x_5 = 1) \& (x_9 \neq 1)] \vee [(x_5 = 2) \& (x_9 \neq 1)]) => x \in C_1$$

after putting actual rules R_1, R_{14} and R_{15} to the formula. Rule R_N is false, $R_N(\mathbf{x}) = 0$, for all 192 cases $\mathbf{x}$ misclassified by rule R_1 as poisonous, for which $R_1(\mathbf{x}) = 1$, because for those $\mathbf{x}$ rule R_{14} or R_{15} is true. If each rule R_{14} and R_{15} would independently cover all 192 cases misclassed by R_1 then R_N can be defined simpler in two ways by using any of these rules:

$$R_N(x) = R_1(x) \& \neg R_{14}(x), \; R_N(x) = R_1(x) \& \neg R_{15}(x)$$

Experiment 4 with SRG3 on Mushroom Data with All 30 Random Groups

To evaluate this algorithm, we performed the test with 30 groups of 3 attributes were my generated from the total 22 mushroom attributes, as shown in Table 10. The algorithm generated all possible rules for these groups on the mushroom data. Then, with all these rules the rule combination process and selection process were

Table 10 30 Randomly generated groups using 22 attributes

16, 19, 5	16, 21, 11	1, 9, 10	19, 14, 3	3, 22, 1	14, 3, 11
22, 16, 14	22, 17, 20	1, 7, 19	13, 11, 5	5, 18, 14	16, 9, 18
7, 4, 22	6, 13, 17	12, 1, 14	4, 21, 14	2, 22, 1	6, 2, 8
13, 18, 19	6, 18, 8	5, 9, 8	16, 14, 21	1, 11, 3	9, 15, 6
6, 8, 9	7, 20, 19	22, 7, 5	9, 17, 3	12, 20, 7	4, 12, 15

ran and the final rules were selected, as shown in both Table 11 and 12. Below are the randomly generated groups and the result of this test.

Rules R1–R7 (Poisonous)
R_1: $[(x_5 = 3) \vee (x_5 = 4) \vee (x_5 = 5) \vee (x_5 = 6) \vee (x_5 = 8) \vee (x_5 = 9)] \Rightarrow \mathbf{x} \in C_1$
R_2: $[(x_{12} = 3) \& (x_{20} \neq 6) \& (x_7 \neq 2)] \Rightarrow \mathbf{x} \in C_1$
R_3: $[(x_9 = 3) \vee (x_9 = 6)] \Rightarrow x \in C_1$
R_4: $[(x_{19} \neq 6) \& (x_3 = 10)] \Rightarrow x \in C_1$
R_5: $[(x_3 = 2) \& (x_{11} = 1)] \Rightarrow x \in C_1$
R_6: $[(x_{12} \neq 1) \& (x_{20} = 5) \& (x_7 = 1)] \Rightarrow \mathbf{x} \in C_1$
R_7: $[(x_{16} = 1) \& (x_{21} = 2) \& (x_{11} \neq 7)] \Rightarrow \mathbf{x} \in C_1$

At all levels tests there is no unclassified cases of the poisonous class, and misclassified cases by all rules, with 100% coverage of the poisonous class and 100% accuracy.

The analysis of the tables above shows that with 30 random groups the SRG algorithm was able to achieve 100% coverage and 100% precision requiring only 7 rules in the level 3 test. In addition, this result is also positive because it was able to pick up a better rule with small coverage (52 cases vs. 42 cases in the alternative rule).

Table 11 Sequential rule generation with 30 randomly generated attribute groupings of 3

Characteristic	Level 1	Level 2	Level 3
Low rule precision threshold. %	75	85	95
Low rule coverage threshold, %	0.5	0.5	0.5
Actual number of cases of poisonous class	3916	3916	3916
Number of rules selected	11	9	7
Number of cases covered by all rules	4716	4524	3916
Number of cases correctly classified by all rules	3916	3916	3916

Table 12 Characteristics of discovered rules R_1-R_7 for poisonous class in Experiment 4

	R_1	R_2	R_3	R_4	R_5	R_6	R_7
Precision, %	100	100	100	100	100	100	100
Coverage, %	96.94	56.89	44.74	17.16	3.06	1.84	1.33
Total cases predicted	3796	2228	1752	672	120	72	52
Correctly predicted cases	3796	2228	1752	672	120	72	52
Misclassified cases	0	0	0	0	0	0	0

Experiment 5 with SRG3 on Mushroom Data with 13 Most Frequent Attributes from 30 Groups

Here we tested the sequential rule generation algorithm SRG3 with the 13 most frequent attributes used in the 30 random triples test as seen above, where 7 rules reached 100% precision and 100% coverage. The 13 most frequent attributes were broken up into three different groups to ensure that the test would finish in reasonable time (Tables 13 and 14).

Rules R_1–R_7 (Poisonous):
R_1: $[(x_5 = 3) \vee (x_5 = 4) \vee (x_5 = 5) \vee (x_5 = 6) \vee (x_5 = 8) \vee (x_5 = 9)] \Rightarrow \mathbf{x} \in C_1$
R_2: $[(x_9 = 3) \vee (x_9 = 6)] \Rightarrow x \in C_1$
R_3: $[(x_9 = 5) \& (x_{11} = 1)] \Rightarrow x \in C_1$
R_4: $[(x_{16} = 1) \& (x_{19} \neq 2) \& (x_{20} = 5) \& (x_{21} = 5)] \Rightarrow \mathbf{x} \in C_1$
R_5: $[(x_{11} = 2) \& (x_{12} \neq 4)] \Rightarrow x \in C_1$
R_6: $[(x_{16} = 1) \& (x_{19} \neq 2) \& (x_{20} = 8) \& (x_{21} = 2)] \Rightarrow \mathbf{x} \in C_1$
R_7: $[(x_3 = 10) \& (x_5 = 7)] \Rightarrow x \in C_1$

Here the SRG3 algorithm did not reduce the number of selected rules below 7.

Table 13 13 used attributes in 30 random triple test results

Characteristic	Level 1	Level 2	Level 3
Low rule precision threshold. %	75	85	95
Low rule coverage threshold, %	0.5	0.5	0.5
Actual number of cases of poisonous class	3916	3916	3916
Number of rules selected	8	9	7
Number of cases covered by all rules	4700	4396	3900
Number of cases correctly classified by all rules	3900	3900	3900
Number of unclassified cases of the poisonous class	16	16	16
Number of misclassified cases by all rules	0	0	0
Actual coverage of the poisonous class, %	99.59	99.59	99.59
Actual precision, %	100	100	100

Table 14 Characteristics of discovered rules R_1-R_7 for poisonous class in Experiment 5

	R_1	R_2	R_3	R_4	R_5	R_6	R_7
Precision, %	100	100	100	100	100	100	100
Coverage, %	96.94	44.74	12.87	1.84	1.12	1.12	0.61
Total cases predicted	3796	1752	504	72	44	44	24
Correctly predicted cases	3796	1752	504	72	44	44	24
Misclassified cases	0	0	0	0	0	0	0

Table 15 Tenfold cross validation results for sequential triple attribute groups

	Test 1	Test 2	Test 3	Test 4	Test 5	Test 6	Test 7	Test 8	Test 9	Test 10
Correctly predicted	381	397	397	398	398	373	395	388	392	397
Misclassified	0	0	0	0	0	0	0	0	0	0
Total classified	381	397	397	398	398	373	395	388	392	397
Validation cases	812	812	812	812	812	812	813	813	813	813
Rules for class C_1	4	4	4	4	4	4	4	4	4	4
Rules for class C_2	0	0	0	0	0	0	0	0	0	0
C_1 validation cases	381	397	397	398	398	373	395	388	392	397

Experiment 6 with SRG3 Algorithm on Mushroom Data and Tenfold Cross Validation with Generated Rules

Here we tested the abilities of SRG3 algorithm to generate beneficial rules in the tenfold cross validation with the sequential triples attribute groups. These groups are G1: $\{x_1, x_2, x_3\}$; G2: $\{x_4, x_5, x_6\}$, ..., G7: $\{x_{19}, x_{20}, x_{21}, x_{22}\}$. The tenfold cross validation test was run with these groups to ensure that performance on the training data can be confirmed in the validation data. Table 15 shows the result of this test with 95% precision threshold for rule generation. In all 10 tests all rules provided 100% precision, 100% coverage of the target class.

The Table 15 shows that the tenfold cross validation achieved **100% accuracy** in every test/fold. It generated and selected the **four rules** that were previously generated using all the data and the given attribute groups. This confirms the efficiency of the SRG algorithm to train and generate rules for newly added cases.

Experiment 7 with SRG3 Algorithm on Mushroom Data and Tenfold Cross Validation with 30 Random Randomly Generated Triples

Here we tested the abilities of SRG3 algorithm to generate beneficial rules using the tenfold cross validation algorithm with 30 random triples of attribute groups as defined below. We ran this test four times to validate the accuracy of results. The result validation is necessary due to the variability of the randomly generated triples.

Run 1

See Tables 16 and 17.

In all 10 tests all rules provided 100% precision, 100% coverage of the target class and 100% accuracy.

Rules generated using 10-Fold Cross Validation:
$CR_1 = R_1$: $[(x_5 = 3) \vee (x_5 = 4) \vee (x_5 = 5) \vee (x_5 = 6) \vee (x_5 = 8) \vee (x_5 = 9)] \Rightarrow \mathbf{x} \in C_1$.
R_2: $[(x_4 \neq 2) \& (x_{20} = 5)] \Rightarrow x \in C_1$
R_3: $[(x_{12} = 3) \& (x_{16} = 1) \& (x_{19} \neq 6)] \Rightarrow \mathbf{x} \in C_1$
R_4: $[(x_4 \neq 2) \& (x_7 = 2) \& (x_{10} = 1)] \Rightarrow \mathbf{x} \in C_1$
R_5: $[(x_{19} \neq 6) \& (x_3 = 10)] \Rightarrow x \in C_1$
Complexity 16/3520 = 0.004545.

Table 16 30 Randomly generated triples of attributes for Run 1

12, 15, 19	1, 4, 7	11, 15, 18	22, 3, 7	11, 14, 17	21, 3, 6
19, 22, 3	7, 11, 14	18, 21, 3	7, 10, 13	17, 21, 2	6, 9, 12
3, 6, 10	14, 17, 20	3, 6, 9	13, 16, 20	2, 5, 8	12, 16, 19
10, 13, 16	20, 2, 5	9, 12, 16	20, 1, 4	8, 12, 15	19, 22, 4
16, 20, 1	5, 8, 11	16, 19, 22	4, 8, 11	15, 18, 21	4, 7, 10

Table 17 Tenfold cross validation results for sequential triple attribute groups (95% precision threshold) for run 1

	Test 1	Test 2	Test 3	Test 4	Test 5	Test 6	Test 7	Test 8	Test 9	Test 10
Correctly predicted	393	394	409	386	385	394	383	391	385	396
Misclassified	0	0	0	0	0	0	0	0	0	0
Total predicted	393	394	409	386	385	394	383	391	385	396
Rules selected for C1	5	5	5	5	5	5	5	5	5	5
Rules selected for C2	0	0	0	0	0	0	0	0	0	0
Target validation cases	393	394	409	386	385	394	383	391	385	396

Run 2

See Tables 18 and 19.

At all 10 tests precision of all rules is 100% and target class coverage by rules is 100%.

Rules generated using 10-Fold Cross Validation:
$CR_1 = R_1$: $[(x_5 = 3) \vee (x_5 = 4) \vee (x_5 = 5) \vee (x_5 = 6) \vee (x_5 = 8) \vee (x_5 = 9)] \Rightarrow \mathbf{x} \in C_1$.
R_2: $[(x_4 \neq 2)\ \&\ (x_{20} = 5)] \Rightarrow x \in C_1$
R_3: $[(x_{12} = 3)\ \&\ (x_{18} \neq 3)] \Rightarrow x \in C_1$
R_4: $[(x_8 \neq 1)\ \&\ (x_{14} \neq 8)] \Rightarrow x \in C_1$
R_5: $[(x_{22} = 2)\ \&\ (x_4 \neq 2)] \Rightarrow x \in C_1$
Complexity = 14/3533 = 0.00396.

Table 18 30 randomly generated triples for run 2

11, 14, 17	21, 3, 6	10, 13, 16	21, 2, 5	9, 13, 16	20, 1, 4
17, 20, 2	6, 9, 12	16, 20, 1	5, 8, 12	16, 19, 22	4, 8, 11
8, 12, 15	12, 16, 19	1, 4, 8	12, 15, 18	22, 4, 7	11, 14, 18
10, 13, 16	19, 22, 3	1, 4, 8	18, 21, 3	7, 10, 13	18, 21, 2
15, 18, 21	3, 7, 10	14, 17, 21	3, 6, 9	13, 17, 20	2, 5, 9

Table 19 Tenfold cross validation results for sequential triple attribute groups (95% precision threshold) for run 2

Characteristic	Test 1	Test 2	Test 3	Test 4	Test 5	Test 6	Test 7	Test 8	Test 9	Test 10
Correctly predicted	395	405	380	388	400	396	375	396	398	383
Incorrectly predicted	0	0	0	0	0	0	0	0	0	0
Total predicted	395	405	380	388	400	396	375	396	398	383
Rules selected for C1	5	5	5	5	5	5	5	5	5	5
Rules selected for C2	0	0	0	0	0	0	0	0	0	0
Validation case of class 1	395	405	380	388	400	396	375	396	398	383

Table 20 30 randomly generated triples for run 3

17, 21, 2	6, 9, 13	17, 20, 1	5, 9, 12	16, 19, 1	5, 8, 11
2, 5, 8	13, 16, 19	1, 5, 8	12, 15, 18	1, 4, 7	11, 14, 18
8, 12, 15	19, 22, 4	8, 11, 14	18, 22, 3	7, 10, 14	18, 21, 2
15, 18, 22	4, 7, 10	14, 18, 21	3, 6, 9	14, 17, 20	2, 6, 9
22, 3, 6	10, 13, 17	21, 2, 5	9, 13, 16	20, 1, 5	9, 12, 15

Table 21 Tenfold cross validation results for sequential triple attribute groups (95% precision threshold) for run 3

Characteristic	Test 1	Test 2	Test 3	Test 4	Test 5	Test 6	Test 7	Test 8	Test 9	Test 10
Correctly predicted	378	398	394	406	397	397	383	382	390	391
Misclassified	0	0	0	0	0	0	0	0	0	0
Total predicted	378	398	394	406	397	397	383	382	390	391
Rules selected for C1	4	4	4	4	4	4	4	4	4	4
Rules selected for C2	0	0	0	0	0	0	0	0	0	0
Target validation cases	378	398	394	406	397	397	383	382	390	391

Run 3

See Tables 20 and 21.

In all 10 tests all rules provided 100% precision, 100% coverage of the target class.

Rules generated using 10-Fold Cross Validation:
$CR_1 = R_1$: $[(x_5 = 3) \vee (x_5 = 4) \vee (x_5 = 5) \vee (x_5 = 6) \vee (x_5 = 8) \vee (x_5 = 9)] \Rightarrow \mathbf{x} \in C_1$.
R_2: $[(x_5 \neq 2) \,\&\, (x_{20} = 5)] \Rightarrow \mathrm{x} \in C_1$
R_3: $[(x_{13} = 2) \,\&\, (x_{16} = 1) \,\&\, (x_{19} \neq 6)] \Rightarrow \mathbf{x} \in C_1$
R_4: $[(x_{15} = 8) \,\&\, (x_{22} = 2)] \Rightarrow \mathrm{x} \in C_1$
Complexity = 13/3525 = 0.003688.

Run 4

See Tables 22 and 23.

In all 10 tests all rules provided 100% precision and 100% coverage of the target class.

Rules generated using 10-Fold Cross Validation:
$CR_1 = R_1$: $[(x_5 = 3) \vee (x_5 = 4) \vee (x_5 = 5) \vee (x_5 = 6) \vee (x_5 = 8) \vee (x_5 = 9)] \Rightarrow \mathbf{x} \in C_1$.
R_2: $[(x_1 \neq 6) \ \&\ (x_{16} = 1) \ \&\ (x_{20} = 5)] \Rightarrow \mathbf{x} \in C_1$
R_3: $[(x_{13} = 2) \ \&\ (x_{16} = 1) \ \&\ (x_{19} \neq 6)] \Rightarrow \mathbf{x} \in C_1$
R_4: $[(x_{15} = 8) \ \&\ (x_{22} = 2)] \Rightarrow \mathrm{x} \in C_1$
Complexity = 14/3531 = 0.00396 (Table 24).

In all runs all rules provided 100% precision, 100% coverage of the target class and 100% accuracy. While the tests are accurate and precise, the generated rules have a moderate amount of variation in complexity and attributes used. The variation is due to the random group generation and the tenfold data partition.

Table 22 30 randomly generated triples for run 4

3, 6, 9	14, 17, 20	2, 5, 9	13, 16, 19	2, 5, 8	12, 15, 19
9, 13, 16	20, 1, 5	9, 12, 15	19, 1, 4	8, 11, 15	19, 22, 3
16, 19, 22	5, 8, 11	15, 19, 22	4, 7, 10	15, 18, 21	3, 6, 10
22, 4, 7	11, 14, 18	22, 3, 6	10, 14, 17	21, 2, 6	10, 13, 16
7, 10, 14	18, 21, 2	6, 10, 13	17, 20, 2	6, 9, 12	16, 20, 1

Table 23 Tenfold cross validation results for sequential triple attribute groups (95% precision threshold) for run 4

Characteristic	Test 1	Test 2	Test 3	Test 4	Test 5	Test 6	Test 7	Test 8	Test 9	Test 10
Correctly predicted	398	398	374	379	401	388	401	408	384	385
Misclassified	0	0	0	0	0	0	0	0	0	0
Total predicted	398	398	374	379	401	388	401	408	384	385
Rules selected for Class 1	4	4	4	4	4	4	4	4	4	4
Rules selected for Class 2	0	0	0	0	0	0	0	0	0	0
Validation cases of class 1	398	398	374	379	401	388	401	408	384	385

Table 24 Summary of runs

Characteristic	Run 1	Run 2	Run 3	Run 4
Number of Rules	4	5	4	4
Number of clauses used in rules	16	14	13	14
Cases covered by rules in tenfold CV	3520	3533	3525	3531
Incorrectly predicted cases	0	0	0	0
Complexity of rules	16/3520 = 0.0045	14/3533 = 0.0037	**13/3525 = 0.0037**	14/3531 = 0.00396

Rule Generation

Below we present rules generated at Step 3 first for the poisonous class and then complementary rules for the eatable class on Mushroom data with most frequent attributes generated from all 22 mushroom attributes. In this experiment the following attribute groups are used: Group A1: x_9, x_5, x_7, x_{11}; Group A2: x_{13}, x_{14}, x_{15}, x_6; Group A3: x_1, x_2, x_4, x_{21}, x_{22}.

According to the design of SRG algorithm it runs at the several levels of thresholds that can be set up by a user. We experimented with 3 level of precision: 75%. 85% and 95% with fixed level of coverage of 0.5%. This means that rules that have lower precision coverage are filtered out and not selected. For coverage with 3916 cases in the poisonous class it means that rules that cover less than 20 cases are filtered out considered as overfitting rules.

The thresholds 75%, 85%, and 95% limit the low margin of the rule quality, but not their upper level. Therefore, we computed the actual precision and coverage reached for the poisonous class at each level. Table 25 shows that rules at all levels missed only 4 cases from the poisonously class, giving 99.89% coverage and none of the poisonous cases was misclassified, giving 100% precision. All misclassified are eatable cases that ranged from 800 cases for 75% threshold to 192 cases for 95% threshold.

We conducted the further analysis for 13 rules selected at level 3 with 95% threshold for rules. See Table 26. This table shows that all 192 misclassified cases belong to the rule R_1 that has 98.47% coverage and 95.2% precision. All other rules have 100% precision and coverage that smaller than for dominant rule R_1.

Rules R_1–R_{13} (Poisonous)
R_1: $[(x_5 \neq 7) \& (x_7 \neq 2) \& (x_9 \neq 7) \& (x_{11} \neq 2)] \Rightarrow \mathbf{x} \in C_1$
R_2: $[(x_5 = 3) \vee (x_5 = 4) \vee (x_5 = 5) \vee (x_5 = 6) \vee (x_5 = 8) \vee (x_5 = 9)] \Rightarrow \mathbf{x} \in C_1$
R_3: $[(x_9 = 3) \vee (x_9 = 6)] \Rightarrow x \in C_1$
R_4: $[(x_6 \neq 1) \& (x_{13} \neq 4) \& (x_{14} \neq 8) \& (x_{15} \neq 1)] \Rightarrow \mathbf{x} \in C_1$
R_5: $[(x_1 \neq 6) \& (x_2 \neq 3) \& (x_4 \neq 1) \& (x_{21} \neq 6) \& (x_{22} = 7)] \Rightarrow \mathbf{x} \in C_1$
R_6: $[(x_9 = 5) \& (x_{11} = 1)] \Rightarrow x \in C_1$

Table 25 Results of sequential rule generation for poisonous class in Experiment 7: Rules R_1–R_{13}

Characteristic	Level 1	Level 2	Level 3
Low rule precision threshold. %	75	85	95
Low rule coverage threshold, %	0.5	0.5	0.5
Actual number of cases of poisonous class	3916	3916	3916
Number of rules selected	14	14	13
Number of cases covered by all rules	4712	4520	4104
Number of cases correctly classified by all rules	3912	3912	3912
Number of unclassified cases of the poisonous class	4	4	4
Number of misclassified cases by all rules	800	608	192
Actual coverage of the poisonous class, %	99.89	99.89	99.89
Actual precision, %	100	100	100

Table 26 Characteristics of discovered rules R_1–R_{13} for poisonous class in Experiment 7

Characteristic	**R_1**	**R_2**	**R_3**	**R_4**	**R_5**	**R_6**	**R_7**
Precision, %	95.02	100	100	100	100	100	100
Coverage, %	98.47	96.94	44.7	33.76	15.42	12.84	12.56
Total cases predicted	3856	3796	1752	1322	604	504	492
Correct cases	3664	3796	1752	1322	604	504	492
Misclassified cases	192	0	0	0	0	0	0
	R_8	**R_9**	**R_{10}**		**R_{11}**	**R_{12}**	**R_{13}**
Precision, %	100	100	100		100	100	100
Coverage, %	10.01	9.91	1.48		1.23	1.12	0.92
Total cases predicted	392	388	58		48	44	36
Correct cases	392	388	58		48	44	36
Misclassified cases	0	0	0		0	0	0

R_7: $[(x_{15} = 3) \vee (x_{15} = 3) \vee (x_{15} = 9)] \Rightarrow \mathbf{x} \in C_1$
R_8: $[(x_1 = 5) \& (x_4 = 2) \& (x_{21} = 5) \& (x_{22} \neq 2)] \Rightarrow \mathbf{x} \in C_1$
R_9: $[(x_{21} = 5) \& (x_{22} = 1)] \Rightarrow x \in C_1$
R_{10}: $[(x_6 = 3) \& (x_{13} = 2) \& (x_{14} \neq 1) \& (x_{15} \neq 8)] \Rightarrow \mathbf{x} \in C_1$
R_{11}: $[(x_2 = 3) \& (x_{21} = 2) \& (x_{22} \neq 6)] \Rightarrow x \in C_1$
R_{12}: $[(x_1 = 1) \& (x_{21} = 5) \& (x_{22} \neq 2)] \Rightarrow x \in C_1$
R_{13}: $[(x_1 \neq 6) \& (x_2 \neq 1) \& (x_4 \neq 2) \& (x_{21} = 5) \& (x_{22} = 3)] \Rightarrow \mathbf{x} \in C_1$

The rules R14 and R15 generated for the eatable class C_2 are presented below:

R_{14}: $[(x_5 = 1) \& (x_9 \neq 1)] = > x\ C_2$
R_{15}: $[(x_5 = 2) \& (x_9 \neq 1)] = > x\ C_2$

Table 27 shows the analysis of both R14 and R15 for class eatable class C_2. Rules R_{14} and R_{15} together cover and correctly predict all 192 cases misclassed by Rule

Table 27 Characteristics of discovered rules R_{14}, R_{15} for eatable class for Experiment 7

	R_{14}	R_{15}
Precision, %	100	100
Coverage, %	7.98	7.98
Total cases predicted	336	336
Correctly predicted cases	366	336
Misclassified cases	0	0

R_1. While both R_{14} and R_{15} cover 336 cases each, these cases are different. In fact, rules R_{14} and R_{15} have 0% overlap and combined cover 672 cases of eatable class C_2.

Combining Rules for Two Classes

Now we have rules R_1–R_{13} for the target class and rules R_{14}–R_{15} for the non-target class and can accomplish step 3 of combining them. The only rule for the target class that misclassified some cases is R_1. So, we need to improve only this rule. It is done by creating a new rule R_N that combines rule R_1 with R_{14} and R_{15} as follows.

$$R_N(\mathbf{x}) = R_1(\mathbf{x}) \& \neg(R_{14}(\mathbf{x}) \vee R_{15}(\mathbf{x}))$$

resulted in

$$R_N{:}[(x_5 \neq 7)\&(x_7 \neq 2)\&(x_9 \neq 7)\&(x_{11} \neq 2)]\&$$

$$\neg([(x_5 = 1)\&(x_9 \neq 1)] \vee [(x_5 = 2)\&(x_9 \neq 1)]) => \mathbf{x} \in C_1$$

after putting actual rules R_1, R_{14} and R_{15} to the formula.

Rule R_N is false, $R_N(\mathbf{x}) = 0$, for all 192 cases $\mathbf{x}$ misclassified by rule R_1 as poisonous, for which $R_1(\mathbf{x}) = 1$, because for those $\mathbf{x}$ rule R_{14} or R_{15} is true. If each rule R_{14} and R_{15} would independently cover all 192 cases misclassed by R_1 then R_N can be defined simpler in two ways by using any of these rules:

$$R_N(\mathbf{x}) = R_1(\mathbf{x}) \& \neg R_{14}(\mathbf{x}),\ R_N(\mathbf{x}) = R_1(\mathbf{x}) \& \neg R_{15}(\mathbf{x})$$

Experiment 8 with Algorithm SRG4 Based on Expert Selected Groups

The results of this test are shown in Tables 28 and 29. The groups used are: Group 1 (Cap): $\{x_1, x_2, x_3\}$; Group 2 (Odor): $\{x_5\}$; Group 3 (Gill): $\{x_6, x_7, x_8, x_9\}$; Group 4 (Stalk): $\{x_{10}, x_{11}, x_{13}, x_{15}\}$; Group 5 (Veil): $\{x_{17}, x_{18}, x_{19}\}$; Group 6 (Spore): $\{x_{20}\}$; Group 7 (Distribution): $\{x_{21}, x_{22}\}$. The analysis of Tables 28 and 29 shows that using the groups generated by the biologist the SRG algorithm was able to get a good result of 99.81% coverage of the target class with 100% precision using just **3 rules**. Although this is considered to be a good result, it did not cover the whole of class C_1. One way this can be achieved is by altering the coverage threshold to a value less than 0.5%. Making this adjustment would allow more small coverage rules to be generated an in turn would allow the combination phase to combine these smaller coverage rules with larger coverage rules to produce high coverage and general rules that cover more class C_1 cases.

R_1: $[(x_5 = 3) \vee (x_5 = 4) \vee (x_5 = 5) \vee (x_5 = 6) \vee (x_5 = 8) \vee (x_5 = 9)] \Rightarrow \mathbf{x} \in C_1$
R_2: $[(x_{20} = 5)] \Rightarrow \mathrm{x} \in C_1$
R_3: $[(x_{11} \neq 1) \,\&\, (x_{13} \neq 4) \,\&\, (x_{15} \neq 8)] \Rightarrow \mathbf{x} \in C_1$

Table 28 Sequential rule generation test using expert groups from Biologist (95% precision threshold) for Experiment 8

	Level 1	Level 2	Level 3
Low rule precision threshold. %	75	85	95
Low rule coverage threshold, %	0.5	0.5	0.5
Actual number of cases of poisonous class	3916	3916	3916
Number of rules selected	3	4	3
Number of cases covered by all rules	3908	3908	3908
Number of cases correctly classified by all rules	3908	3908	3908
Number of unclassified cases of the poisonous class	8	8	8
Number of misclassified cases by all rules	99.81	99.81	99.81
Actual coverage of the poisonous class, %	100	100	100
Actual precision, %	100	100	100

Table 29 Characteristics of discovered rules R_1–R_3 for poisonous class (95% precision threshold) for Experiment 8

Characteristic	R_1	R_2	R_3
Precision, %	100	100	100
Coverage, %	96.94	1.84	12.97
Total cases predicted	3796	72	508
Correctly predicted cases	3796	72	508
Misclassified cases	0	0	0

Experiment 9 with SRG5 and 7 Successful Attributes

Here we tested SRG algorithm with the 7 used attributes in our previous 100% precision test. This test was run with these attributes in hopes that only using the 7 used attributes will allow the rule generation process to generate and select less rules while keeping the 100% precision and total coverage of the target class.

The 7 attributes were broken up into two different groups to be able to finish the test in reasonable time, where group 1: x_5, x_9, x_{15}; and Group 2: x_{19}, x_{20}, x_{21}, x_{22}. The results are shown in Tables 30 and 31.

Rules R_1–R_7 (Poisonous):
R_1: $[(x_5 = 3) \vee (x_5 = 4) \vee (x_5 = 5) \vee (x_5 = 6) \vee (x_5 = 8) \vee (x_5 = 9)] \Rightarrow \mathbf{x} \in C_1$
R_2: $[(x_9 = 3) \vee (x_9 = 6)] \Rightarrow \mathrm{x} \in C_1$
R_3: $[(x_{19} = 2)\ \&\ (x_{20} = 8)\ \&\ (x_{21} \neq !2)\ \&\ (x_{22} \neq !2)] \Rightarrow \mathbf{x} \in C_1$
R_4: $[(x_{15} = 3) \vee (x_{15} = 2) \vee (x_{15} = 9)] \Rightarrow \mathbf{x} \in C_1$
R_5: $[(x_{19} \neq 2)\ \&\ (x_{20} \neq 6)\ \&\ (x_{21} = 5)\ \&\ (x_{22} = 1)] \Rightarrow \mathbf{x} \in C_1$
R_6: $[(x_{19} = 6)\ \&\ (x_{20} = 5)\ \&\ (x_{21} \neq 1)\ \&\ (x_{22} \neq 6)] \Rightarrow \mathbf{x} \in C_1$
R_7: $[(x_{20} = 8)\ \&\ (x_{21} = 2)\ \&\ (x_{22} \neq 6)] \Rightarrow \mathbf{x} \in C_1$

At all levels all rules provided 100% precision, 100% coverage of the target class.

These tables allow to conclude that the sequential rule generation algorithm using the attribute groups created by using the 7 attributes was unable to reduce the number of rules selected to cover all of class C_1. This result is likely due to the SRG2 process used. Other existing or new versions SRG algorithm can be more successful in future.

Table 30 Sequential rule generation test using successful attributes for Experiment 9

	Level 1	Level 2	Level 3
Low rule precision threshold. %	75	85	95
Low rule coverage threshold, %	0.5	0.5	0.5
Actual number of cases of poisonous class	3916	3916	3916
Number of rules selected	10	11	7
Number of cases covered by all rules	4716	4492	3916
Number of cases correctly classified by all rules	3916	3916	3916
Number of unclassified cases of the poisonous class	0	0	0

Table 31 Characteristics of discovered rules R_1-R_7 for poisonous class for Experiment 9

	R_1	R_2	R_3	R_4	R_5	R_6	R_7
Precision, %	100	100	100	100	100	100	100
Coverage, %	96.94	44.74	30.23	12.56	9.91	1.84	1.33
Total cases predicted	3796	1752	1184	492	388	72	52
Correctly predicted cases	3796	1752	1184	492	388	72	52
Misclassified cases	0	0	0	0	0	0	0

Experiment 10 with SPG 5 and Comparison of Rule Complexity

In this experiment we used SRG5 algorithm based on SRG1 algorithm, with attributes that have been successful for mushroom data in [1, 30]. We compared the complexity of rules generated in this process with rules from [1, 30], which we denote as CR rules.

The formulas for computing complexity of rules and sets of rules are given in Sect. 4.2. Several sets of rules were generated in [1, 30]. We use only the final rules from [1, 30] listed below in our notation.

CR Rules [1, 30]
CR_1: $[(x_5 = 3) \vee (x_5 = 4) \vee (x_5 = 5) \vee (x_5 = 6) \vee (x_5 = 8) \vee (x_5 = 9)] \Rightarrow \mathbf{x} \in C_1$, Complexity 6/3796 = 0.0016.
CR_2: $[(x_{20} = 5)] \Rightarrow \mathbf{x} \in C_1$, Complexity 1/72 = 0.014.
CR_3: $[(x_8 = 2) \& (x_{12} = 3)]$ v $[(x_8 = 2) \& (x_{12} = 2)] \vee [(x_8 = 2) \& (x_{21} = 2)] \Rightarrow \mathbf{x} \in C_1$.
Complexity 6/912 = 0.0066.
Complexity of the set of rules: (6 + 1 + 6)/(3796 + 72 + 912) = 13/ 4780 = **0.0027**.

Attribute Groups that we used in SRG algorithm are $G1 = \{x_5\}$; $G2 = \{x_{20}\}$; $G3 = \{x_8, x_{12}, x_{21}\}$, which directly correspond to CR rules above.

Our Rules

R_1: $[(x_5 = 3) \vee (x_5 = 4) \vee (x_5 = 5) \vee (x_5 = 6) \vee (x_5 = 8) \vee (x_5 = 9)] \Rightarrow \mathbf{x} \in C_1$ Complexity 6/3796 = 0.0016.
R_2: $[(x_{20} = 5)] \Rightarrow \mathbf{x} \in C_1$, Complexity 1/72 = 0.014
R_3: $[(x_{12} = 3) \& (x_{21} = 5)] \Rightarrow \mathbf{x} \in C_1$, Complexity 2/1544 = 0.0013
R_4: $[(x_8 \neq 1) \& (x_{21} = 2)] \Rightarrow \mathbf{x} \in C_1$, Complexity 2/16 = 0.125
Complexity of a set of rules = (6 + 1 + 2 + 2)/(3796 + 72 + 1544 + 16) = 11/ 5428 = **0.002**.

The algorithm SRG1 generated simpler rules (11 clauses vs. 13 clauses) with the same precision as in in [1, 30] using the attribute groups derived from CR rules. This result shows that the algorithm SRG1 can generate rules that are less complex than CR rules and suggests that better rules are possible with more testing and preprocessing of attribute groups.

Toolkit

The toolkit includes the Data Type Editor integrated with visualization system VisCanvas 2.0 [29, 32] for multidimensional data visualization based on the adjustable parallel coordinates. The data type editor supports saving data in the

Original data

ID	X_1	X_2	X_3	X_4
3	A	A	A	B
1	A	B	C	F
4	C	D	A	C
2	B	D	F	I

Coding scheme

X_i	code
A	4
B	3
C	2
D	1
F	0
I	-1

Coded data

ID	X_1	X_2	X_3	X_4
	4	4	4	3
1	4	3	2	0
4	2	1	4	2
2	3	1	0	-1

Fig. 16 Example of applying coding schema for class grades

explainable measurement coding format for pattern discovery and data visualization. Figure 16 illustrates setting up and applying a coding scheme for data that converts letter grades for 4 classes X_1-X_4 as follows: A to 4, B to 3, C to 2, and so on.

The toolkit supports Nominal, Ordinal, Interval, and Ratio data measurement types. The attributes of the absolute measurement data type are encoded as the ratio data type because it differs from it only by presence of the fixed measurement unit.

The data type editor provides helpful descriptions and examples in case the user is not familiar with these data types. The user interface allows a user: to assign data type for each attribute, and to group values of each attribute. The scheme loader allows a user to assign data measurement type: nominal, ordinal, interval, and ratio.

A typical example of mixed data are the mushroom data [27], which contain 8124 instances and 22 attributes. These attributes include nominal data such as habitat (grass, leaves, meadows, paths, urban, waste, woods), ordinal data, such as gill size (broad, narrow) and gill spacing (crowded, close, distant), absolute data such as the number of rings (0,1,2), which the scheme loader treats as ratio data.

The colors of different parts of the mushroom such as cap color represent an interesting data type. It can be treated as: (1) nominal (red, blue, green and so on), (2) three numeric attributes like R,G,B, (3) some scalar function from R,G,B like R + G + B, and (4) a single numeric ratio data type that uses wavelength. The first one does not capture the similarity between colors, the second one is expanding the number of attributes, the third one corrupts similarity relations between some colors, and the last one is the most physically meaningful.

Since, each color covers a wavelength interval, grouping wavelength values according to colors is a natural way to encode the colors. These groups are ordered and can be encoded by integers starting from 0. In general, grouping attribute values decreases the space size and run time of the algorithms.

Manual coding is time consuming and tedious work for the tasks with many attributes and multiple values of each attribute. The toolkit allows to speed up this process. The editor has the "All Ordinal" and "All Nominal" options that allow to assign initially Nominal or Ordinal type and to assign integer code values from 1 to n to all attributes with abilities to edit this assignment later.

Grouping. Figure 17 illustrates grouping and binary coding keys for a nominal attribute. Figure 18 illustrates setting up groups for the numeric interval and ratio attributes by creating intervals where user sets on the starting value for the group and the length of its interval. Original values of the attributes may not correctly represent its data type for the task. The user can select keeping the existing values or generate new ones.

Hierarchy of attribute groups. When we have hundreds of attributes, a hierarchy of attributes allows to deal with them efficiently. The system supports a user in constructing a hierarchy and picking up a level at which attributes will be visualized.

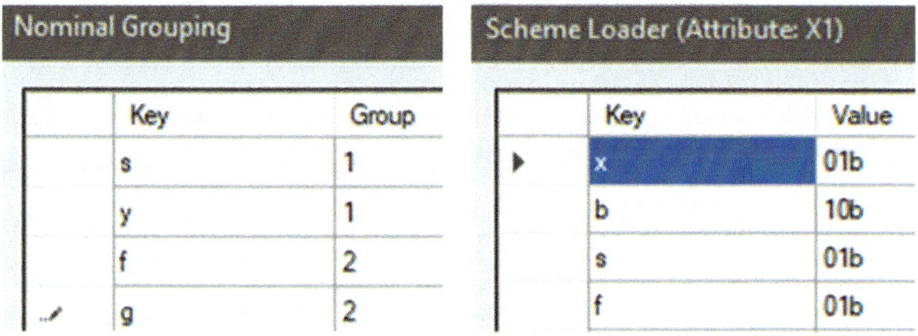

Fig. 17 Grouping keys and assigning binary codes for a nominal attribute

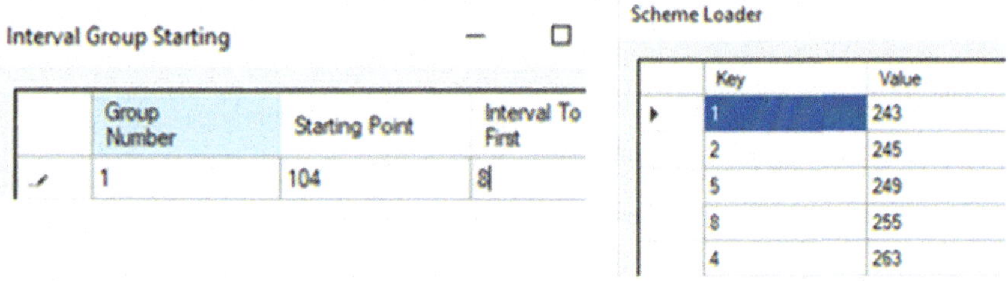

Fig. 18 Setting up groups for the numeric interval and ratio attributes

References

1. Ali N, Neagu D, Trundle P (2019) Evaluation of k-nearest neighbour classifier performance for heterogeneous data sets. SN Appl Sci 1:1–5
2. Kovalerchuk B, Grishin V (2019) Adjustable general line coordinates for visual knowledge discovery in n-D data. Inform Visualiz 18(1):3–32
3. Kovalerchuk B (2018) Visual knowldege discovery and machine learning. Springer
4. Rosario GE, Rundensteiner EA, Brown DC, Ward MO, Huang S (2004) Mapping nominal values to numbers for effective visualization. Inf Vis 3(2):80–95
5. Kovalerchuk B, Delizy F (2004) Visual data mining using monotone Boolean functions. In: Visual and spatial analysis. Springer, pp 387–406
6. Friendly M (2000) Visualizing categorical data: data, stories, and pictures. In: Proceedings of the 25th annual SAS users group intern. Conference. https://www.datavis.ca/papers/sugi/vcdstory/vcdstory.pdf
7. Shahid ML, Molchanov V, Mir J, Shaukat F, Linsen L (2020) Interactive visual analytics tool for multidimensional quantitative and categorical data analysis. Inf Vis 19(3):234–246
8. Roy B (2020) All about categorical variable encoding. https://towardsdatascience.com/all-about-categorical-variable-encoding-305f3361fd02
9. Peng S, Hu Q, Chen Y, Dang J (2015) Improved support vector machine algorithm for heterogeneous data. Pattern Recogn 48(6):2072–83
10. Potdar K, Pardawala TS, Pai CD (2017) A comparative study of categorical variable encoding techniques for neural network classifiers. Int J Comp Appl 175(4):7–9
11. Vityaev EE, Kovalerchuk BY (2008) Relational methodology for data mining and knowledge discovery. Intell Data Anal 12(2):189–210
12. Lipton Z (2018) The mythos of model interpretability. Commun ACM 61:36–43
13. Letham B, Rudin C, McCormick TH, Madigan D (2015) Interpretable classifiers using rules and bayesian analysis: building a better stroke prediction model. Ann Appl Stat 9(3):1350–1371
14. Fryer D, Strümke I, Nguyen H (2021) Shapley values for feature selection: the good, the bad, and the axioms. IEEE Access. 8(9):144352–144360
15. Watson DS (2022) Conceptual challenges for interpretable machine learning. Synthese 200:65
16. Watson DS (2021) Rational shapley values. arXiv preprint arXiv:2106.10191
17. Adilova L, Kamp M, Andrienko G, Andrienko N (2023) Re-interpreting rules interpretability. J Data Sci Anal 5:1–21. https://www.researchsquare.com/article/rs-1525944/latest.pdf
18. Kovalerchuk B, Ahmad MA, Teredesai A (2021) Survey of explainable machine learning with visual and granular methods beyond quasi-explanations. In: Pedrycz W, Chen SM (eds) Interpretable artificial intelligence: a perspective of granular computing. Springer, pp 217–267
19. Kovalerchuk B, Triantaphyllou E, Deshpande AS, Vityaev E (1996) Interactive learning of monotone Boolean functions. Inf Sci 94(1–4):87–118
20. Kovalerchuk B, Triantaphyllou E, Ruiz J (1996) Monotonicity and logical analysis of data: a mechanism for evaluation of mammographic and clinical data. In: Computer applications to assist radiology. Carlsbad, CA, Symposia Foundation, pp 191–196
21. Kovalerchuk B, McCoy E (2022) Explainable mixed data representation and lossless visualization toolkit for knowledge discovery. In: 26th International conference information visualization. IEEE, pp 314–321. arXiv:2206.06476
22. Krantz DH, Luce RD, Suppes P, Tversky A, et al (1971) Foundations of measurement, vol. 1. Acad. Press
23. Kovalerchuk B (1975) On cyclical scales. Comput Syst 61:51–59
24. Ji S, Pan S, Cambria E, Marttinen P, Philip SY (2021) A survey on knowledge graphs: representation, acquisition, and applications. IEEE Trans Neural Netw Learn Syst 33(2):494–514
25. Cheng V, Li CH, Kwok JT, Li CK (2004) Dissimilarity learning for nominal data. Pattern Recogn 37(7):1471–1477
26. Stanfill C, Waltz D (1986) Toward memory-based reasoning. Comm ACM 29(12):1213–1228
27. Dua D, Graff C (2019) Machine learning repository. University of California, Irvine, CA. https://archive.ics.uci.edu/ml/datasets/Mushroom

28. Kovalerchuk B, Vityaev E (2000) Data mining in finance: advances in relational and hybrid methods. Kluwer
29. Kovalerchuk B, Hayes D (2021) Discovering explainable machine learning models in parallel coordinates. In: 2021 25th International conference information visualisation (IV). IEEE, pp 181–188
30. Duch W, Setiono R, Zurada JM (2004) Computational intelligence methods for rule-based data understanding. Proc IEEE 92(5):771–805
31. Duch W, Adamczak R, Grabczewski K (2001) A new methodology of extraction, optimization and application of crisp and fuzzy logical rules. IEEE Trans Neural Networks 12(2):277–306
32. GitHub. https://github.com/CWU-VKD-LAB. VisCanvas 2.0
33. Bendix F, Kosara R, Hauser H (2005) Parallel sets: visual analysis of categorical data. In: Symposium on information visualization. IEEE, pp 133–140

Parallel Coordinates for Discovery of Interpretable Machine Learning Models

Dustin Hayes and Boris Kovalerchuk

1 Introduction

Acceptance, interpretability, and comprehensibility of different classifiers are crucial for future Machine Learning (ML) advancements. For many machine learning models this is a very significant challenge due to their black box specifics making these models incomprehensible. Users are reluctant to deploy such models for high-risk, high-stakes decisions. A promising solution to this problem is (VKD) [1–5]. We outline a parallel coordinates-based visual knowledge discovery approach below that includes supervised learning, data and model visualization, dimensionality reduction, and model simplification.

The supervised classification models are the main topic of this work. Often, developing a *reliable interpretable, explainable, and comprehensible ML model* necessitates placing the end-user in control of the creation of a model. The end users are frequently *subject matter experts* rather than machine learning professionals. Often for them formal ML models are opaque black boxes. The visual knowledge discovery methodology makes it possible to identify ML models, explain them for the end users, and put them in charge of model development. This work extends in: (1) visualization method for data with empty values and more elaborated examples, (2) dealing with the large data, (3) in presenting the features of software tool developed to support visual knowledge discovery in parallel coordinates denoted as VisCanvas 2.0, and (4) generalization of hyperblock approach to other general line coordinates.

The suitability of parallel coordinates for visual knowledge discovery is demonstrated in many prior works [6–12]. Parallel coordinates accomplish so without losing any of the multidimensional information, and they support interpretability and comprehensibility by using the original attributes, which have clear domain

D. Hayes · B. Kovalerchuk (✉)
Department of Computer Science, Central Washington University, Ellensburg, WA, USA
e-mail: BorisK@cwu.edu

B. Kovalerchuk et al. (eds.), *Artificial Intelligence and Visualization: Advancing Visual Knowledge Discovery*, Studies in Computational Intelligence 1126,
https://doi.org/10.1007/978-3-031-46549-9_4

meaning for the domain's end users. However, the use of parallel coordinates for supervised learning as a primary space for actual solving supervised learning tasks still is in the nascent stage.

The suggested approach is based on the ideas of hyperblocks and hypercubes, which generalize 2-D squares and rectangles to n-D space. They are naturally represented in parallel coordinates and have a distinct interpretation. It gives the end users the chance to actively participate in developing and enhancing these models in addition to joust using models. It has been suggested to develop decision tree models with parallel coordinates in [7, 13]. We demonstrate that the hyperblock technique produces *more general models* than decision trees (DTs). It allows us to deal better with both overfitting and overgeneralization of data.

The *major contributions* of this work are in:

- employing *hypercubes* and *hyperblocks* to assist users discover patterns in parallel coordinates for the supervised learning tasks,
- combining *hyperblocks, visualizations,* and *language descriptions* to discover and present patterns,
- allowing *users to lead* the processes of development, enhancement, and usage of the models interactively,
- handling *missing values* in large datasets, which overcomes a weakness of traditional parallel coordinates.

The chapter is organized as follows. Section 2 defines (1) hypercubes and hyperblocks, together with their parallel coordinate visualization, and (2) algorithms to find them automatically and interactively. We first concentrate on individual HCs and HBs, then on multiple HBs. This includes finding and visualizing pairs of non-overlapping HBs and HB combinations along with linguistic descriptions of visual patterns.

Supervised learning in parallel coordinates is presented in Sect. 3. We first discuss its difficulties before introducing the Hyper classification technique for pure and mixed/impure hyperblocks. Following the comparison with decision trees, hyper models are presented as generalized decision trees. The case study uses benchmark Wisconsin Breast Cancer (WBC) data from UCI ML repository [14] to test the methodology.

To accomplish this, we first provide learning of hyperblocks using all WBC data, followed by tenfold cross validation using training and validation data. The relationship between hyperblocks, dimension reduction, and the visualization of lower dimensional hyperblocks has been established.

Section 4 presents the features of VisCanvas 2.0 software [15], which implements Hyper methodology and is based on the prior VisCanvas 1.0 [13]. Both are available on GitHub [13, 16]. This description includes the method of visualizing cases with missing values. Section 5 covers discussion, generalization, conclusion, and future work.

2 Interactive and Automatic Visual Discovery of Hyperblocks

2.1 Hypercubes and Hyperblocks

Interpretable concepts that appear naturally in parallel coordinates are hypercubes (HCs) and hyperblocks (HBs).

A *hypercube* (*n-cube*) is a set of n-D points $\{\mathbf{x} = (x_1,x_2,\ldots,x_n)\}$ with a *center* in n-D point $\mathbf{c} = (c_1,c_2,\ldots,c_n)$ and *side length L* such that

$$\forall i ||x_i - c_i|| \leq L/2 \tag{1}$$

A hypercube is an n-dimensional generalization of ordinary square ($n = 2$) and cube ($n = 3$). A *unit hypercube* has $L = 1$. A *binary hypercube* is an n-cube in the binary n-D space E^n.

A *hyperblock* (*hyperrectangle, n-orthotope*) is a set of n-D points $\{\mathbf{x} = (x_1,x_2,\ldots,x_n)\}$ with a center in n-D point $\mathbf{c} = (c_1,c_2,\ldots,c_n)$ and side lengths $\boldsymbol{L} = (L_1, L_2,\ldots, L_n)$ such that

$$\forall i ||x_i - c_i|| \leq L_i/2 \tag{2}$$

As a result, the *n*-cube is a unique instance of a hyperblock. Figures 1 and 2 depict the idea of a hypercube. The hypercubes in these figures are sized in relation to one another in 20-D space based on their length. Compared to Fig. 1b, c, the HCs in Fig. 1a are bigger. The central 20-D points of these HCs are shown by the middle lines.

Figure 2 shows hypercube patterns in *adjustable parallel coordinates* (APC), where coordinates can be shifted up and down, to make them easier to perceive [2]. Figure 3 shows all Wisconsin Breast Cancer (WBC) data in parallel coordinates. In this study we explore WBC data and build multiple hyperblock on them to discover an interpretable supervised model classifying benign and malignant cases.

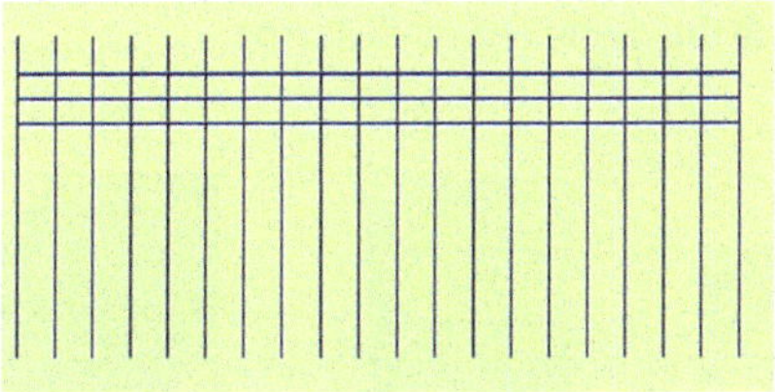

(a) hypercube A with equal coordinates

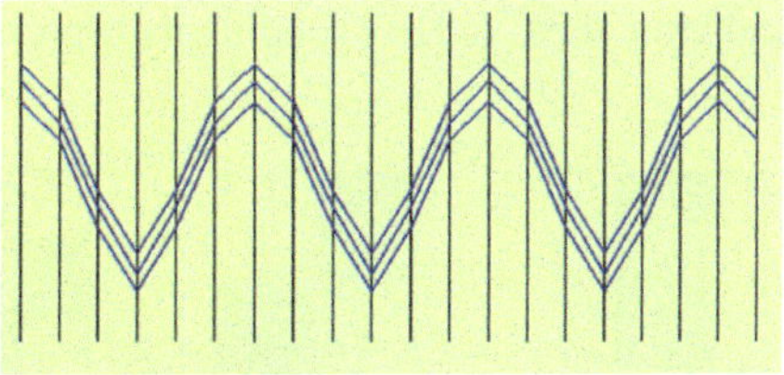

(b) zigzag hypercube B

Fig. 1 Examples of hypercubes

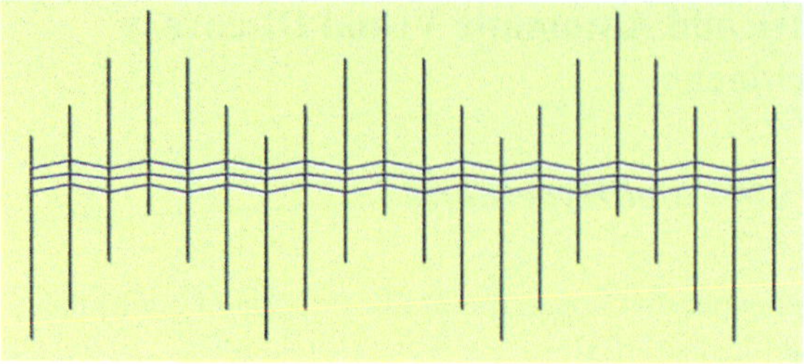

Fig. 2 Hypercubes in adjustable parallel coordinates shifted up and down to simplify hypercube patterns

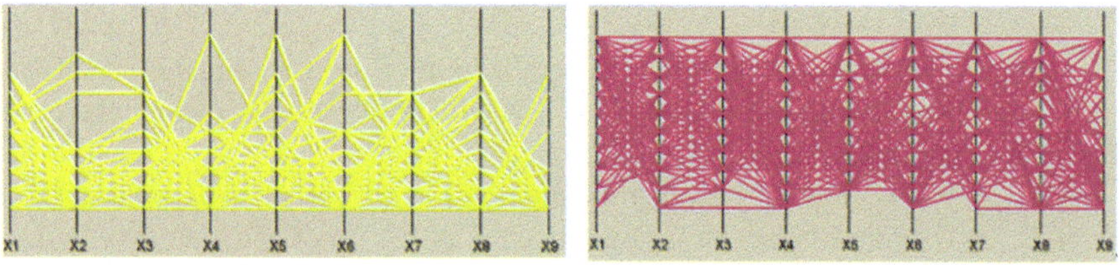

Fig. 3 All cases of WBC data (yellow–benign, magenta- malignant) in parallel coordinates

2.2 *Algorithm to Generate Hyperblocks*

Hyperblocks can be made up of hypercubes whose centers are close to one another and share some dimensions. Not all HC pairings, nevertheless, will result in hyperblocks.

Figure 4 shows different alternatives how HCs can be combined. Figure 4a shows two adjacent "2-D hypercubes" (squares) that share *n*-1 dimensions (1-D line for the squares) that can form a single hyperblock. Figure 4b shows two adjacent "2-D hypercubes" (squares) that share *n*-2 dimensions (a single point for the squares) that do not form a single hyperblock. Figures 4c, d show more n-1 and n-2 adjacent HCs for the square in the center, while Fig. 4e shows overlapping HCs that do not from a single hyperblock. Due to these complex relations between HCs and HBs, in this work, we start with small pure HBs seeded in the individual n-D points as centers and grow them to HBs keeping their purity as the steps below describe.

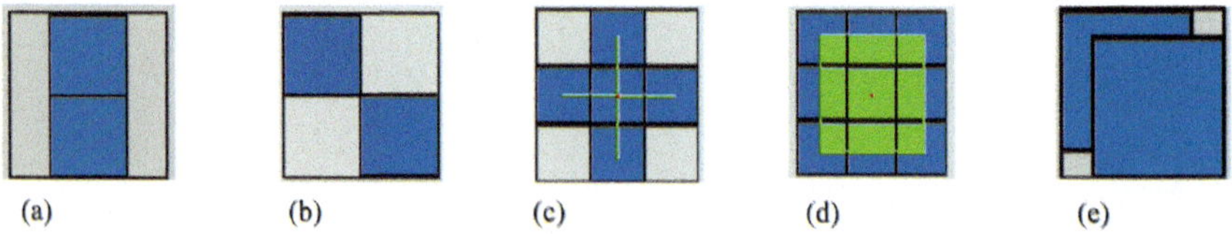

Fig. 4 Adjacency options for 2-D hypercubes

Steps of the Merger Hyperblocks (MHyper) algorithm for pure and dominant hyperblocks.

Step M1: Seed an initial set of **pure** HBs with a single n-D point in each of them (HBs with length equal to 0).
Step M2: Select a HB **x** from the set of all HBs.
Step M3: Start iterating over the remaining HBs. If HB_i has the same class as HB **x** then attempt to combine HB_i with HB **x** to get a pure HB.

Step M3.1: Create a joint HB from HB_i and **x** that is an envelope around HB_i and HB **x** using the minimum and maximum of each dimension for HB_i and HB **x**.
Step M3.2: Check if any other n-D point **y** belongs to the envelop of HB_i and HB **x**. If **y** belongs to this envelope add **y** to the joint HB.
Step M3.3: If all points **y** in the joint HB are of the same class then remove HB **x** and HB_i from the set of HBs that need to be changed.

Step M4: Repeat step 3 for all remaining HBs that need to be changed. The result is a *full pure HB* that cannot be extended with other n-D points and continue to be pure,
Step M5: Repeat step 2 for n-D points do not belong to already constructed full pure HBs.
Step M6: Define an *impurity threshold* that limits the percentage of n-D points from opposite classes allowed in a **dominant** HB.
Step M7: Select a HB **x** from the set of all HBs.
Step M8: Attempt to combine **x** with remaining HBs.

Step M8.1. Create a joint HB from HB_i and **x** that is an envelope of HB_i and **x**.
Step M8.2: Check if any other n-D point **y** belongs to the envelop of HB_i and **x**. If **y** belongs to this envelope add **y** to the joint HB.
Step M8.3: Compute *impurity* of the HB_i (the percentage of n-D points from opposite classes introduced by the combination of **x** with HB_i.)
Step M8.3 Find HB_i with lowest impurity. If this lowest impurity is below predefined impurity *threshold* create a joint HB.

Step M9: Repeat step 7 until all combinations are made.

2.3 Individual Hyperblocks

The overlapping of lines (polylines) that represent individual cases (n-D points) is a significant problem for finding patterns, as was seen above. Even when the values are different, hundreds of lines can follow the same course and be displayed on top of one another in a given visualization resolution. Those lines are indistinguishable to the user. The user is therefore unaware of the number of lines in the hypercube and which line segments correspond to which n-D locations.

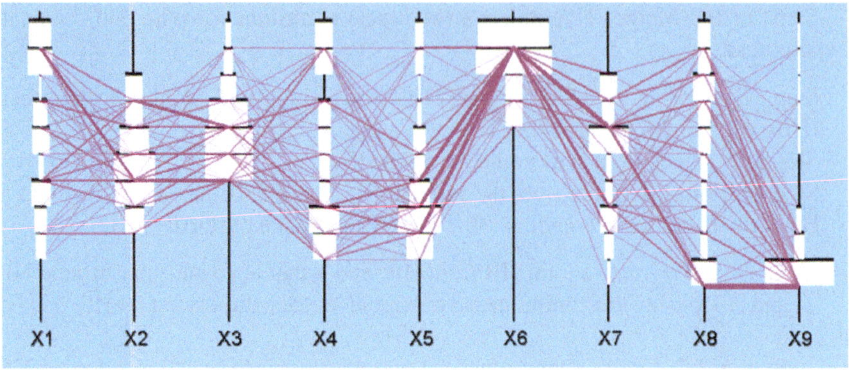

Fig. 5 Distribution of 9-D points with white rectangles to show quantiles for two WBC HBs

Histograms for individual hyperblocks. This problem is addressed in part by histograms. A *distribution* of n-D points (pink lines) within a hyperblock is shown in Fig. 5 as white rectangles and identical-length black lines. A *data quantile* is represented by the *height* of each white rectangle and black separation lines. The pink lines' *widths* represent their *frequencies.* This image demonstrates the distribution of these 9-D points in the 9-D WBC space.

The visual representation of the distribution of polylines in each hyperblock presented side-by-side with the display of quartiles in a separate window is shown in Figs. 6 and 7. The distribution of 9-D points in HBs A and B is shown in Figs. 6 and 7. It also make it possible to distinguish between them clearly, particularly in coordinate X6 where their distributions do not overlap while both HBs A and B from Figs. 6 and 7 belong to the same class.

In Fig. 6 the white HB A contains pink and yellow lines from opposing classes with other HBs visible only partially on the background. Several yellow lines and a single pink line from the opposite class are used to depict another HB B in Fig. 7.

In Fig. 8a, an illustration of well-separated HBs from opposing pink and yellow classes is shown. Their frequencies are presented by the width of the lines. As a result, we have embedded frequencies in these parallel coordinates.

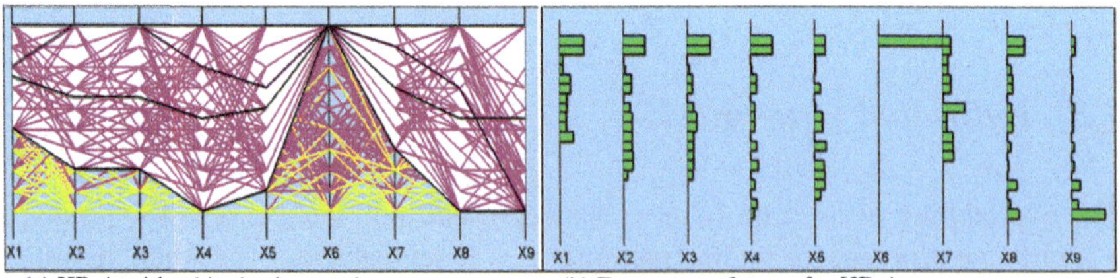

(a) HB A with white background (b) Frequency of cases for HB A.

Fig. 6 Hyperblock A and its distribution using quantiles

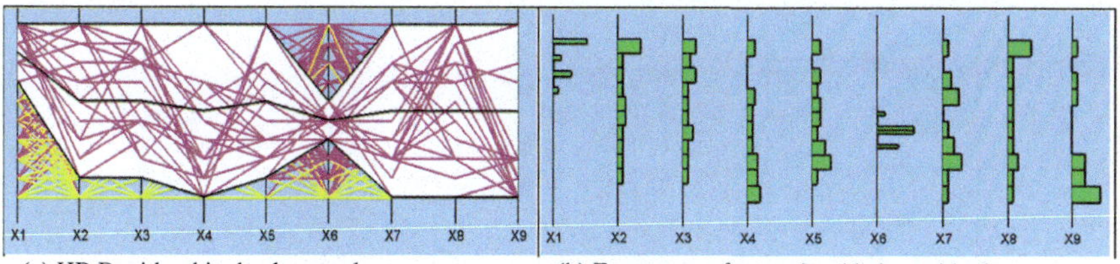

(a) HB B with white background (b) Frequency of cases for this hyperblock

Fig. 7 Hyperblock B and its distribution using quantiles

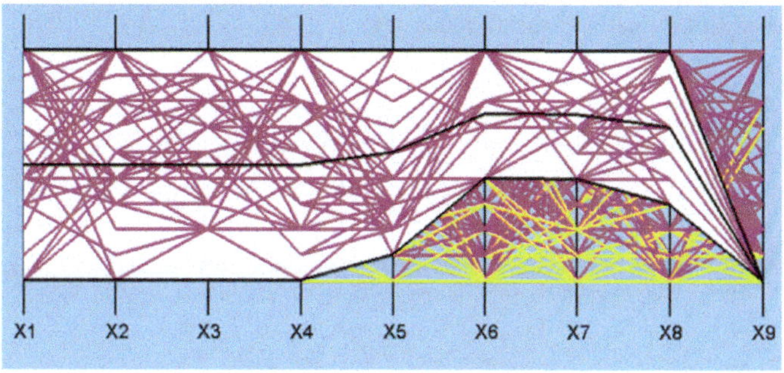

Fig. 8 Normal visualization of a hyperblock

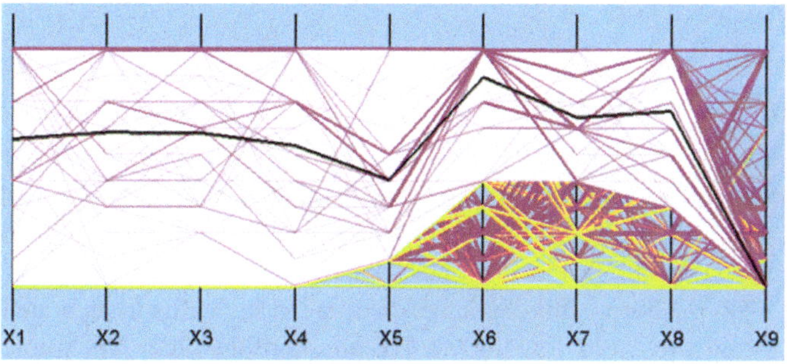

Fig. 9 A hyperblock with line frequencies embeded

These illustrations demonstrate that if we can find such different HBs, we can develop understandable categorization rules. Further visualization examples in parallel coordinates with embedded frequency by line width are shown in Figs. 8, 9 and 10.

The selected hyperblock is displayed with a white background and a black line indicates the hyperblock's mean. A pure HB example is displayed in Fig. 10, whereas mixed hyperblock examples are shown in Fig. 11, where cases from various classes are depicted in various colors.

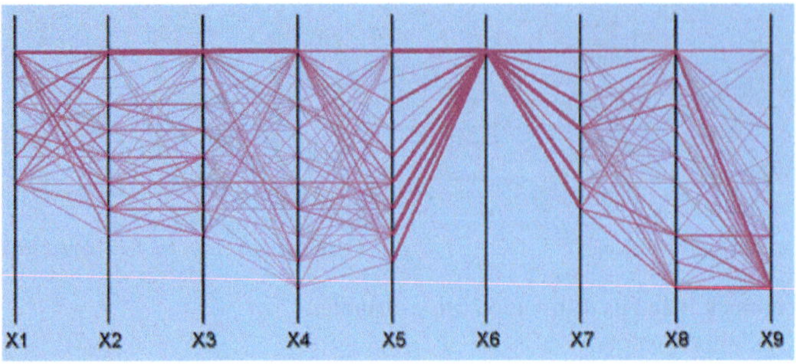

Fig. 10 Hyperblocks' lines with their frequencies

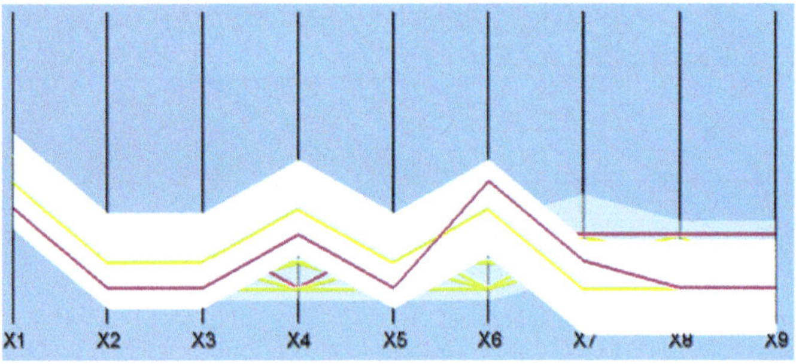

Fig. 11 Mixed hyperblocks

2.4 *Multiple Hyperblocks*

Hyperblocks viewed side by side. It is difficult to visualize hyperblocks that do not overlap in n-D space also not overlapping in 2-D space of parallel coordinates. Figures 6 and 7 depict HBs that do not overlap in 9-D space but do so in parallel coordinates in 2-D, illustrating this challenge. We placed such HBs in separate Figs. 6 and 7, or side by side, to avoid drawing one HB on top of the other. The implementation of this side-by-side option in the VisCanvas 2.0 system is shown in Fig. 12.

Another strategy is to display the data hyperblocks in three dimensions. First, the camera is set up to only to view two dimensions. The user can then tilt the camera to display the three-dimensional space, where each hyperblock is discernible in a different plane in the z-coordinate. HBs would be situated on the same plane if they overlap.

Areas without overlap. The following strategy involves displaying the variations in HB distribution in parallel coordinates in n-D space. First, all distinct hyperblock pairings are created. Then, these pairings are iterated over, and a rectangle is drawn on the dimensions where a given pair **do not** overlap.

This was implemented in several different ways as Fig. 13 shows. The more pairs that do not overlap in a coordinate, the darker the rectangle is. It is clear from this visualization in Fig. 13 that HBs less overlap in X6 coordinate.

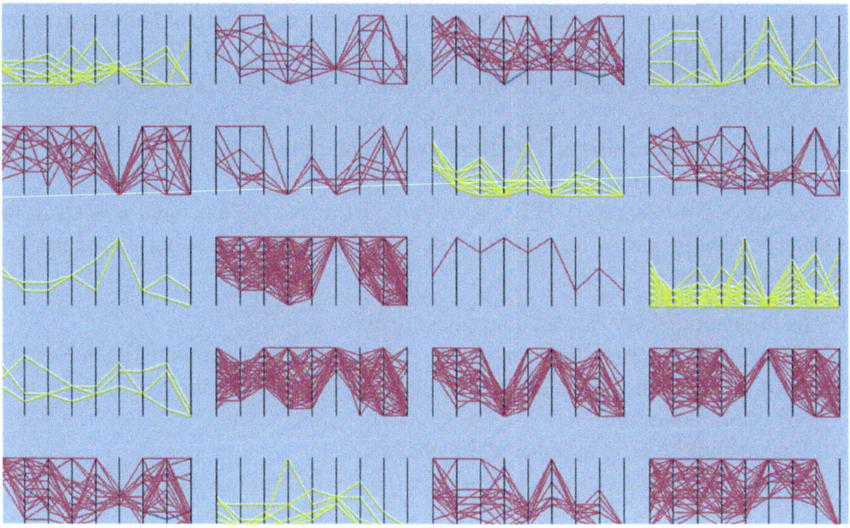

Fig. 12 All WBC discovered pure HBs of two classes side by side

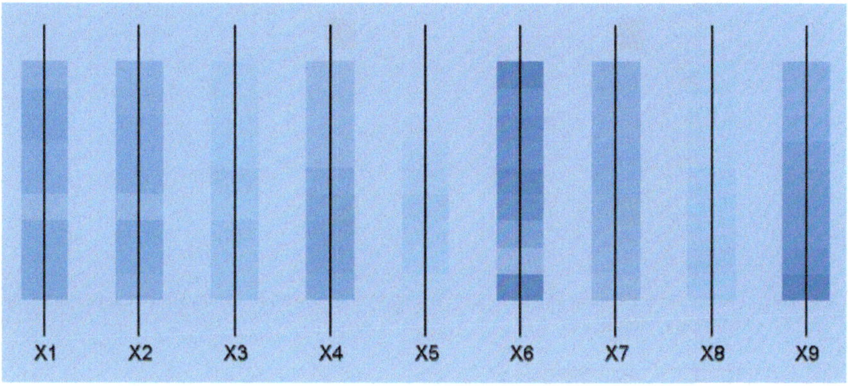

Fig. 13 Locations where pairs of hyperblocks do not overlap

2.5 Discovering and Visualizing Pairs of Non-overlapping Hyperblocks

When polylines of opposing classes widely overlap and occlude one another, it becomes much more difficult to visually detect patterns of various classes in parallel coordinates. To improve the distinction computational and interactive techniques are required. One of the approaches is finding subsets of coordinates where the difference is quite clear and visualizing only data in those coordinates.

This search can be successful only if such subsets exist. The challenge is in multiplicity of HBs that represent each class. An individual pair of HBs from opposing classes can be quite distinct in some coordinates. However, as the number of HBs that represent classes increases, the likelihood of a subset of coordinates where all HBs not overlap goes down. Therefore, we search only for distinctive pairs of HBs and visualize each such pairs individually.

The successful search in WBC data that VisCanvas 2.0 executed is seen in the results below. To get to this point, VisCanvas takes the following automatic and interactive steps:

1. Auto-generate hyperblocks for opposing classes,
2. Chose hyperblocks which contain the most of data n-D points for each class,
3. Auto-reduce dimensions of these hyperblocks to dimensions where there these HBs do not overlap,
4. Visualize these HBs in reduced dimensions,
5. Ascertain a classification rule based on these HBs and evaluate its accuracy.

These steps resulted in 93.7% accuracy for these WBC data. Figure 14 shows two such discovered HBs.

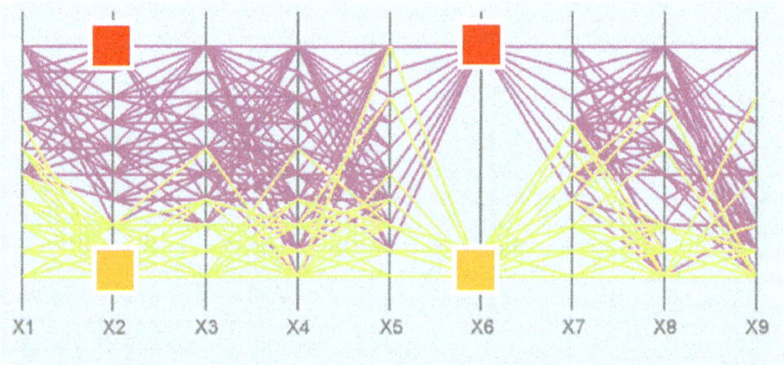

(a) Two non-overlapping hyperblocks(pink and yellow) with red and yellow squares marking separating coordinates.

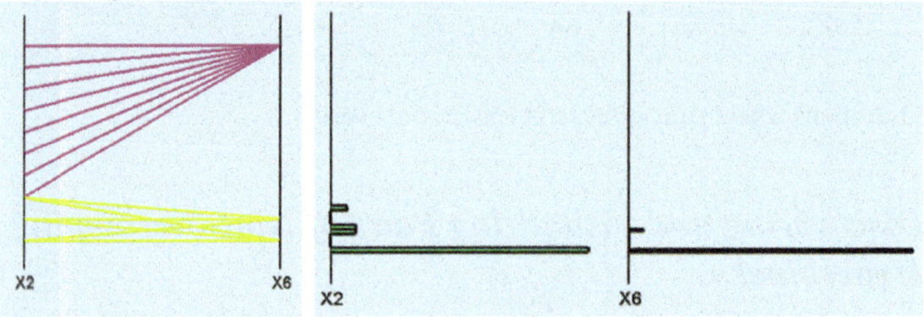

(b) Frequency visualization of yellow HB in most informative coordinates X2 and X6.

Fig. 14 Two hyperblocks of two classes in parallel coordinates

2.6 Exploration and Combination of Hypelocks

Exploration of hyperblocks. With VisCanvas 2.0, exploring hyperblocks begins by choosing a line that corresponds to an n-D point of interest. Then we build HB with this line as its center line and at a certain distance from it, such as 0.2. Hyperblocks with and without their lines are depicted in Figs. 15 and 16.

Combination Modes. We investigated three alternative ways to put hyperblocks together. The first mode (M1) examines the overlap of these HBs, or if all dimensions of each center are within range of one another. The second mode (M2) determines whether both HBs contain at least one point. Further verification of the presence of cases from other classes that may be in the joint hyperblock is necessary for both approaches to succeed. The third mode (M3) determines if one of each center's n-1 dimensions is inside one of the other's ranges and whether both HB n-1 dimensions are identical. This situation can arise rarely. It is shown in Fig. 16.

In addition, setting a threshold allows showing hyperblocks that have a certain number of elements within them. Some hyperblocks overlap. Finding and removing

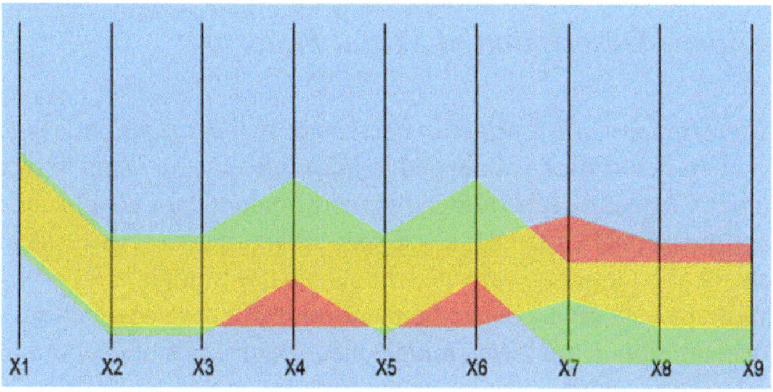

Fig. 15 Hyperblocks without lines

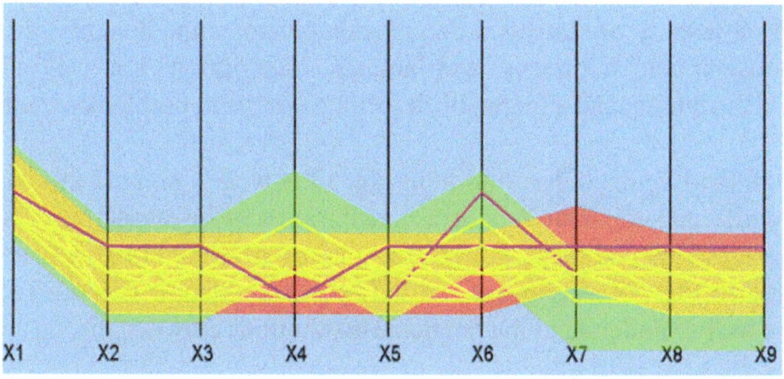

Fig. 16 Hyperblocks with lines

overlaps and defining joined new HBs will minimize the number of them. HBs that have a low purity ratio are not good candidates to combine with other hyperblocks.

Therefore, it is better to hold them in their own. These blocks can then be drawn in a different color, toggled on/off, etc. Below we present algorithms and results of experiments for combining HBs and VisCanvas 2.0 features that support it.

VisCanvas 2.0 checks whether several n-D points produce the same HBs. It creates a single set of HBs from adjacent HBs. This is accomplished by comparing each prospective HB's dimension to each other HB's dimension.

They are considered as adjacent if the sum of all differences divided by 2 is less than or equal to the threshold value. Finally, VisCanvas 2.0 counts and displays the total number of unique HBs and the number of them with cases of multiple classes. It can be used to check for overfitting.

A user can view the ratio of HBs that have cases of multiple classes compared to the total number of unique HBs. For instance, it can show a HB with 93 cases from class 1, and 2 cases from class 2 in different colors. Multiple algorithms can be designed to combine HBs. One of them is presented below.

2.7 Linguistic Description of Visual Patterns

Linguistic descriptions of hyperblocks combined with distribution visualization. A linguistic pattern description is intended to make it easier for users to recognize the visual patterns. Frequently it is shorter and more "natural" for people and represents the entire n-dimensional space, not a sub-space that a specific HB resides in. With a straightforward UI, a user can activate a linguistic description.

While more precise divisions are being developed, these descriptions now point out data concentration in the lower, middle, and upper thirds of the coordinates. The current implementation allows merging descriptions of dimensions that are concentrated in the same thirds of them into a compact single description as Fig. 17 shows.

The distribution of n-D points might become distorted due to occlusion, as shown in Fig. 17b. A linguistic description prevents this contamination because it is based on a study of statistical data. It demonstrates the benefits of integrating visual representation, frequency, and language description. The n-D points, their frequency distribution, and a linguistic description of the hyperblock are all displayed in Fig. 17a.

Although the linguistic description in Fig. 17 is well organized in relation to the bottom, middle, and upper areas of data concentration in the respective coordinates, it is not a conventional natural language (NL) phrase. Modified linguistic descriptions are depicted in Fig. 18 as conventional NL phrases. Also, it is more succinct and enables the user to understand the language distinctions between the various classes.

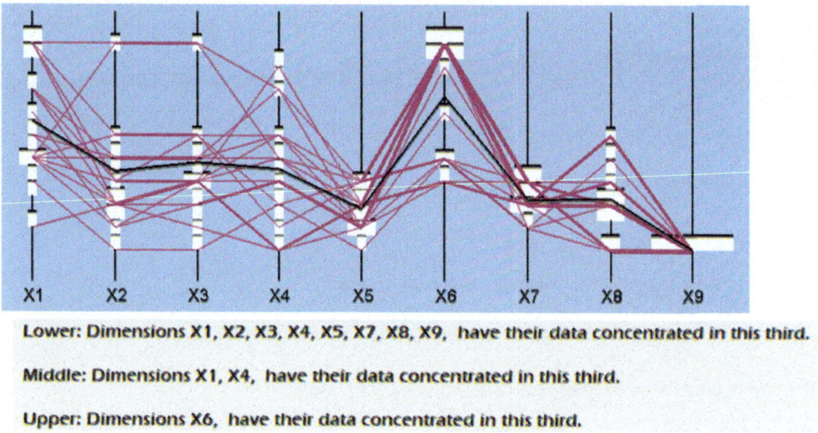

(a) A hyperblock with frequency pattern with its linguistic description.

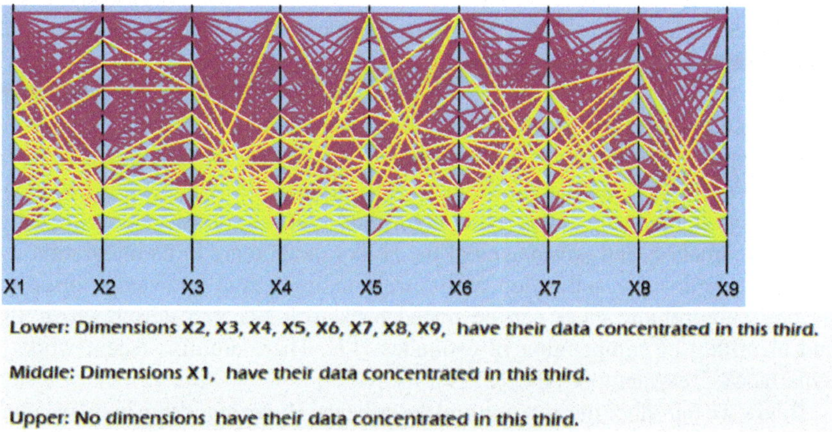

(b) All WBC data of both classes with linguistic description.

Fig. 17 WBC data visualized with linguistically annotated patterns

3 Supervised Learning in Parallel Coordinates

3.1 Hyperblock-Based Classification Algorithm (HYPER)

3.1.1 Mixed and Pure Hyperblocks

A supervised learning system called a *hyperblock-based classification* (*Hyper*) algorithm learns and displays interpretable rules in parallel coordinates. VisCanvas 2.0 software uses the Hyper algorithm. The core concept behind the Hyper algorithm is the classification of n-D points using n-D hyperblocks. First, it looks for dominant

Class malignant

The data in dimensions X1, X2, X3, X4, X5, X6, X7, X8, and X9 are concentrated in the lower third.

No dimensions have their data concentrated in the middle third.

No dimensions have their data concentrated in the upper third.

Class benign

The data in dimension X9 is concentrated in the lower third.

The data in dimensions X2, X3, X4, X5, X7, and X8 are concentrated in the middle third.

The data in dimensions X1, and X6 are concentrated in the upper third.

Fig. 18 Linguistic description of all WBC data

hyperblocks, which are regions where the majority of n-D points fall under one class. Let's say that HB1 dominates in class 1, HB2 dominates in class 2, and so on.

If a new n-D point **x** *belongs* to a respective HB where class C_i *dominates* then **x** is classified to class C_i. If **x** does not belong to any dominant HB, then **x** is classified to the *nearest HB*. If there are several competing HBs nearby, then the those nearest k HBs *vote* and **x** is classified to the class with *majority votes*. If there is not enough HBs to vote in the vicinity, the Hyper algorithm *refuses* to classify this n-D point **x**.

The uniqueness and advantages of the Hyper method are in its integration with parallel coordinate visualization and corresponding visual knowledge discovery. Its basic algorithmic ideas can be traced back to k-NN first developed in 1951 and algorithm of computation of estimates [17]. There are also recent works on hyperblocks relevant to k-NN [18, 19].

Below we introduce the notation and describe the Hyper algorithm in more details. Let HB_i be a hyperblock where n-D points of *class* C_i *dominate*. Also let $\{HB\}_k$ be a set of k *nearest HBs* for n-D point **x**, which *vote for class* C_i, i.e., *most* of these hyperblocks are class C_i dominant HBs.

The Hyper algorithm learns rules in the following forms:

Rule 1: If n-D point **x** belongs to any dominant hyperblock HB_i, for class C_i, then **x** belongs to class C_i.
Rule 2: If HB_i is a nearest hyperblock for n-D point **x,** then **x** belongs to class C_i.
Rule 3: If $\{HB\}_k$ is a set of k nearest hyperblocks for n-D point **x** and the majority of $\{HB\}_k$ vote for class C_i, then **x** belongs to class C_i.

More formally these rules are:

$$R_1: \exists HB_i\ (\mathbf{x} \in HB_i) \Rightarrow \mathbf{x} \in C_i \tag{3}$$

$$R_2: (HB_i \text{ is a nearest HB for } \mathbf{x}) \Rightarrow \mathbf{x} \in C_i \tag{4}$$

$$R_3: (\{HB\}_k \text{ is } k - NN \text{ HBs for } \mathbf{x}) \;\&\; (\text{Vote}(\{HB\}_k) = C_i) \Rightarrow \mathbf{x} \in C_i \quad (5)$$

The major steps of Hyper algorithm to learn R_1-R_3 are:

Step H1: Split data to *training data* T_r and *validation data* V_d (e.g., by tenfold cross validation) and split *Tr* to T_{rh} for learning HBs and T_{rk} learning the number of nearest neighbors HBs.
Step H2. Learn *dominant HBs*.
Step H3. Learn the number k of nearest neighbors HBs.
Step H4. Generate rules R_1–R_3.
Step H5. Validate rules R_1–R_3 using validation data V_d.

Below we present Steps H2 and H3 in more details.

Step H2. Learn *dominant HBs*.

Step H2.1. Construct all pure HB on training data T_r by the algorithm presented in Section II.B.
Step H2.2. Collect all n-D points $\{\mathbf{b}_j\}$ that belong to *single-point HBs* (HBs without other n-D points).
Step H2.3. For each $\mathbf{b}_j$ find a nearest HB. If the class of $\mathbf{b}_j$ and the nearest pure HB is the same, record it as a positive classification, else record it as negative classification.
Step H2.4. Attempt to combine HBs to larger dominant HBs by adding small HBs of other classes to adjacent large HBs keeping HB purity above in a given threshold T.

Step H3. Learn the number of nearest neighbors HBs k.

Step H3.1. Set up an interval of possible number k of nearest HBs.
Step H3.2. Loop through k is this interval to find k with most accurate voting classification for all n-D points in T_{rk}. If predefined accuracy threshold Q is reached, finish learning process, else leave n-D points in T_{rk} unclassified and exit.

3.1.2 Sets of Hyperblocks as Generalized Decision Trees

Below we discuss learning simple dominant HBs such as common in decision trees. According to the HB definition it satisfies inequalities $|| x_i\text{-}c_i || \le L_i/2$ for a respective *center* n-D point $\mathbf{c} = (c_1, c_2,\ldots, c_n)$ and *lengths* $\boldsymbol{L} = (L_1, L_2,\ldots, L_n)$.

Consider an example of a hyperblock $|| x_i\text{-}c_i || \le c_i$ in parallel coordinates that are in [0,10] interval each. This inequality is true for all non-negative $x_i \le 2c_i$, in other words, $x_i \in [0,2c_i]$. Such simple HB needs only c_i values that identify its center. Similarly, consider a complimentary HB where $x_i \in (2c_i,10]$, i.e., $x_i > 2c_i$. Its center $\boldsymbol{h} = (h_1, h_2,\ldots, h_n)$ is defined by $h_i = (10\text{-}2c_i)/2$. As we see these HBs are defined by a simple set of inequalities.

The Hyper algorithm allows discovering and visualizing hyperblocks of the type of branches of decision trees (DTs). It is based on a direct link between DTs and hyperblocks (HBs). Consider, a branch of DT: $x_1 > 5$ & $x_2 < 6$ & $x_3 > 2$ that is labeled, say, by class 1. Assume that all coordinates X_i are in [0.10] interval, then HB for this branch is defined by three intervals:

$$x_1 \in (5, 10],\ x_2 \in [0, 6)\ \textit{and}\ x_3 \in (2, 10]$$

Alternatively, instead of starting from a DT, we can start from a hyperblock. Let HB be given by three intervals $x_1 \in (5,7)$, $x_2 \in (3,6)$, and $x_3 \in (2,4)$ within interval [0,10]. It is equivalent to the DT branch:

$$x_1 > 5\ \&\ x_1 < 7\ \&\ x_2 < 6\ \&\ x_2 > 3\ \&\ x_3 > 2\ \&\ x_3 < 4 \quad (6)$$

These examples show a 1:1 mapping between a DT branch and a hyperblock without loss of generality.

A user can create a DT outside of VisCanvas 2.0 and then convert each DT branch to a respective HB to be visualized in VisCanvas 2.0. Alternatively, a set of HBs produced in VisCanvas 2.0 can be viewed as a set of DTs as follows. Assume that each discovered HB is represented by a DT branch like (4). Can we combine these branches to a single tree? It is possible only in a special situation when branches have a common root, like in the following example. Say, branch *A* contains inequality $x_i \le$ T, and branch *B* contains the opposite inequality $x_i > T$, then coordinate X_i can serve as a common root for *A* and *B*. Thus, outside of special cases, a set of HBs (HB "forest") is a more general model than a DT model. A set of HBs removes a limitation of a single DT requiring a root.

3.2 WBC Case Study

3.2.1 Learning of Hyperblocks with All Data

WBC data have been used to compare the Hyper method to the ID3 decision tree algorithm used in VisCanvas 2 and Scikit-learns, respectively. First, we try to discover a hyperblock HB_i that is dominant for class C_i and then to build a rule on all WBC data:

$$\text{If } \mathbf{x} \in HB_i \Rightarrow \mathbf{x} \in \text{class } C_i \text{ dominant in } HB_i \quad (7)$$

Investigating the rule's accuracy is part of the study. The best-case accuracy estimate on this WBC data with HBs can be achieved by using all WBC data for training (discovering HBs) without reserving a subset for validation.

Table 1 shows the HBs generated by the Hyper algorithm when using a 0.2 distance from the center of HB in WBC data normalized to [0,1]. Total 22 HBs have been produced: 20 pure and 2 mixed. Out of 20 pure HBs 7 belong to the class B (benign)

Table 1 Hyperblocks with using a 0.2 center distance with the WBC dataset

Type of HCs	Number of HCs	Number of n-D points in HCs
Pure	20	588
Mixed	2	95
Total	22	683

Table 2 Number of n-D points in pure overlapped HBs for WBC dataset

Class	Number of n-D points in hyperblock											
B	404	392	92	34	16	5	4	1				
M	76	44	42	39	34	25	18	18	12	9	7	1

and 13 to class M (malignant). Two mixed HBs are very different. One is almost pure class M dominant, 92/1, but the second HB is not with 2 cases and ratio 1/1. In this situation when HB_i is not dominant for any class, all n-D points of this HB are classified to class M (malignant) to avoid more dangerous misclassification to benign class.

The number of n-D points in each HB also varies very significantly from 404 n-D points to a single n-D point for class B, and from 76 n-D points to single n-D point for class M (see Table 2). The produced HBs overlap due to the version of the algorithm used. A more elaborated version allows removing overlaps. Having only two mixed HBs with ratios 92/1 and 1/1 the Hyper algorithm misclassified only two benign cases with total accuracy 681/683, i.e., 99.70%.

Note, this rule relies on several very small HBs, e.g., 7 HBs include 9 or less n-D points, which can be interpreted as overfitting. To avoid it we can refuse to classify n-D points that are in these HBs, i.e., remove these HBs from the rule. These HBs contain 29 cases. It will decrease the number of cases covered by the rule to 683–29 = 654 (recall 95, 75%) with precision 99.85% (653/654) that is slightly greater than the accuracy 99.7%. The increase the threshold to 25 cases in a HB will remove 11 HBs with total 91 n-D points, resulted in recall of 86.68% and precision of 99.83. These precisions set up two other best-case benchmarks for HB-based algorithms on WBC data.

3.2.2 Supervised Learning with Training and Validation Data

The WBC dataset was subjected to tenfold cross validation using VisCanvas 2.0 and Scikit-learns ID3 decision tree implementation. For each tenfold split, the Hyper algorithm in VisCanvas 2.0 automatically built hyperblocks, and rules (1) through (3) derived from these HBs were evaluated on the validation folds. Three versions of k-NN were explored for each n-D point $\mathbf{x}$ with $k = 1,3, 5$:

(N1) the distance from **x** to the *center n-D points* of discovered hyperblocks,
(N2) the distance from **x** to the *mean n-D point* of discovered hyperblocks,
(N3) the distance from **x** to the *nearest n-D point*.

For N1-N3 with pure HBs, Table 3 presents the results that are similar to each other with average accuracy above 95% and reaching 97.61%. The version N3 produced the lowest results. It means that this method is more sensitive to the distribution of closest points, while methods N1 and N3 based on means and centers are more robust.

The average accuracy for the ID3 decision tree is 92.85%, with min equal to 89.86% and max equal to 94.58%. All of them are below the averages for N1-N3 of the *k*-NN hyperblock algorithm Hyper, which shows the advantages of hyperblocks for this dataset. The decision tree has size depth 8, it contains 23 branches which are larger than the average number of HBs which is 18.1 in the worst case.

Likewise, for N1-N3 with mixed HBs, Table 3 presents the results with similar accuracy scores. This could be because the number of HBs has been reduced to

Table 3 Summary of experiments with WBC data using pure and mixed HBs

Model type	Tenfold accuracy			Average # of HB
	Average	Min	Max	
$k = 1$				
N1 pure	**95.52**	91.04	100	16.9
N2 pure	95.22	89.55	100	16.9
N3 pure	93.58	86.57	100	16.9
N1 mix	93.43	86.57	97.01	5.4
N2 mix	**93.58**	86.57	97.01	5.4
N3 mix	93.28	86.57	98.51	5.4
$k = 3$				
N1 pure	96.57	92.54	98.51	17.3
N2 pure	**97.61**	94.03	100	17.3
N3 pure	94.78	91.04	97.01	17.3
N1 mix	**93.88**	88.06	98.51	7.1
N2 mix	**93.88**	88.06	98.51	7.1
N3 mix	93.73	88.06	97.01	7.1
$k = 5$				
N1 pure	**96.42**	92.54	100	18.1
N2 pure	**96.42**	92.54	100	18.1
N3 pure	94.78	89.55	100	18.1
N1 mix	**96.12**	91.04	98.51	6.4
N2 mix	**96.12**	91.04	98.51	6.4
N3 mix	95.82	91.04	98.51	6.4

Table 4 Confusion matrixes for the fold that is closest to the average accuracy

	Pure hyperblocks								
k=1	N1	B	M	N2	B	M	N3	B	M
	B	44	0	B	44	0	B	44	0
	M	3	20	M	3	20	M	4	19
k=3	N1	B	M	N2	B	M	N3	B	M
	B	43	1	B	43	1	B	44	0
	M	1	22	M	1	22	M	3	20
k=5	N1	B	M	N2	B	M	N3	B	M
	B	42	2	B	43	1	B	44	0
	M	0	23	M	2	21	M	3	20
	Mixed dominant hyperblocks								
k=1	N1	B	M	N2	B	M	N3	B	M
	B	44	0	B	42	2	B	42	2
	M	3	20	M	2	21	M	3	20
k=3	N1	B	M	N2	B	M	N3	B	M
	B	43	1	B	43	1	B	44	0
	M	1	22	M	1	22	M	3	20
k=5	N1	B	M	N2	B	M	N3	B	M
	B	41	3	B	43	1	B	43	1
	M	0	23	M	2	21	M	3	21

approximately 6 in comparison with the pure HBs average of 18. These HBs cover more space and it is easier to place n-D points within them for classification. As a result, the accuracy would be consistent and there would be no requirement for N1, N2, or N3 in the k-NN version of the algorithms. Moreover, it consistently outperformed decision tree averages, with the exception of minimal accuracies, where it lagged behind the decision tree.

For pure and mixed hyperblocks applied in the Hyper method, Table 4 displays confusion matrices for the fold that comes the closest to the average accuracy in tenfold cross validation for $k = 1,3,5$.

Comparable ID3 Decision tree on training data is shown in Table 5 and Fig. 33 in the appendix. The confusion matrix for these ID3 decision trees on the representative validation fold is shown in Table 6, where the number of errors for the cancer class is higher than that obtained by the Hyper approach.

The complexity of HBs vs. DT can be measured by comparing the number of values needed to store HBs and DT. Our estimates had shown that this DT will require significantly more values with ratio above 1.5.

3.2.3 Merger

The steps of the Merger Hyperblocks (MHyper) algorithm (see Sect. 2) were performed on the WBC dataset in conjunction with tenfold cross validation. The average number of HBs created for each fold was 17.3. Four of the folds contain HBs that include only a single n-D point. Three folds had one HB with only a single n-D point and another fold had two HBs with only a single n-D point. To help this approach gain more generality it was adapted to also consider dominant HBs.

Table 5 Number of n-D points in nodes of ID3 for WBC dataset

Cases	Number of n-D points of two classes in the node											
Node	*1*	*2*	*3*	*4*	*5*	*6*	*7*	*8*	*9*	*10*	*11*	*12*	*13*
Class B	397	364	33	363	1	15	18	363	0	1	0	15	0
Class M	217	9	208	2	7	3	205	1	1	0	7	0	3
Node	*14*	*15*	*16*	*17*	*18*	*19*	*20*	*21*	*22*	*23*	*24*	*25*	*26*
Class B	15	3	362	1	9	6	2	1	1	0	8	1	5
Class M	149	156	0	1	4	45	9	147	0	1	0	4	16
Node	*27*	*28*	*29*	*30*	*31*	*32*	*33*	*34*	*35*	*36*	*37*	*38*	*39*
Class B	1	2	0	1	0	0	1	4	1	0	1	2	0
Class M	29	2	7	24	123	4	0	16	0	22	7	0	2
Node	*40*	*41*	*42*	*43*	*44*	*45*	*46*	*47*	*48*	*49*	*50*	*51*	*52*
Class B	0	1	3	1	1	0	1	0	3	0	1	0	0
Class M	20	4	4	12	1	6	0	4	1	3	1	11	1
Node	*53*	*54*	*55*	*56*	*57*								
Class B	1	0	3	0	1								
Class M	0	1	0	1	0								

Table 6 ID3 confusion matrix

ID3	2	4
B	45	2
M	5	17

Next, we reduced the number of single points HBs including impure combinations. All steps of MHyper algorithm were performed with a threshold of 10% impurity. It produced 5 HBs in average for each fold without any single n-D point HB. This approach of defining a threshold for the percentage of n-D points from an opposite class allowed a balance between generalization and accuracy to be made. Table 7 shows the resulting 5 HBs in a representative fold. The merger algorithm allows overlap of merged HBs, which is seen in Table 7. The average number of HBs decreases about 3 times from the initial number of pure HBs. These HBs have impurity less than 10% (see Table 7).

Table 7 Results of MHyper for impure HBs for a representative fold

	Number of n-D points of two classes in HBs				
Hyperblock	**1**	**2**	**3**	**4**	**5**
Cases of class B	25	0	439	432	12
Cases of class M	226	53	47	30	137
Impurity	9.9%	0	9%	6.5%	8%

We also compared complexity of HBs vs. the decision tree by using the number of cases in the smallest merged HB. This merged HB contains 53 cases, which is 7.76% of the overall data. The DT's minimum number of n-D points within its nodes is 1. This is 0.15% of the overall data. It has 10 of these nodes with a single element within it. It also has 46 nodes that contain less than 53 cases. So, 80.7% of the DT's nodes are less generalized than the HB's least generalized block. The merger of HBs and pruning of DTs leads to both higher generalization and higher error rate that is controlled by the impurity threshold for HBs.

3.2.4 Simple Hyper-Boxes and Dimension Reduction

Below we present a rule discovery by the Hyper algorithm with simple HBs of type of decision tree branches. The simplest discovered rule on WBC data is a one-dimensional rule

$$\text{if } x_6 < 3 \text{ then class 2 (benign) else class 4 (malignant)} \quad (8)$$

This rule has accuracy of 91.22% with correctly classified 623 cases out of 683 total cases.

A two-dimensional rule is

$$x_6 < 3 \ \& \ x_8 < 4 \text{ then class B (benign) else class M} \quad (9)$$

This rule has accuracy of 93.85% with correctly classified 641 cases out of 683 total cases.

A three-dimensional rule is

$$\text{if } x_6 < 3 \ \& \ x_8 < 4 \ \& \ x_5 < 6 \text{ then class 2 (benign) else class M} \quad (10)$$

This rule has accuracy of 94.58% with correctly classified 646 cases out of 683 total cases. These rules have been discovered in a dimension reduction process in VisCanvas 2.0 where a user can toggle coordinates on and off in the "Dimension" tab of the settings to reduce data dimension. This is resulted in reduction of WBC dataset from 9-D down to the 3-D where patterns were *perceived previously visually*, the resulting patterns become even more obvious.

Two clearly separated hyperblocks of opposite classes are displayed in Fig. 14. It shows the pink hyperblock that contains 95 malignant cases, and the yellow hyperblock that contains 404 benign cases. It is a large majority of all data and the mirror like difference of these two HBs of two classes makes the pattern easier to distinguish. This is in line with rule (10) that involves the same 3 coordinates. The general form of the rules above is.

$$\text{if } x_i < T_i \ \& \ x_j < T_j \ldots x_k < T_k \text{ then class B (benign) else class M}$$

4 VisCanvas 2.0 Software Features

4.1 Interaction Features

Observing and analyzing n-D data, hypercubes, and hyperblocks in parallel coordinates is the major goal of VisCanvas 2.0. Figure 19 shows two hyperblocks generated in VisCanvas 2.0 along with the frequency distribution of values.

The data table menu includes options for analyzing hyperblocks. On opening a file, a hyperblock is created around every single data point for the use in this table. The table displays the class structure within the hyperblock. A user can select/deselect HBs for viewing. It allows a user to view the space around n-D points of interest. Figure 20 shows an example of the data table and respective user interface.

Creating subsets of n-D points. VisCanvas 2.0 has a feature that allows creating subsets of n-D points. It includes a tool strip menu (see Fig. 20) and can be accessed through the subsets drop down. It has a column of checkboxes to make n-D points visible or not. Another column indicates the number of n-D points of other classes this n-D point crosses. The remaining columns show the values of coordinates of n-D points.

This table is populated by choosing one of the classes from the dropdown menu above it. The "All" button selects/de-selects all points in the given class. The "Apply" button sets the visibility according to checked checkboxes.

The hyperblock selection utilities have their own tab in the settings page and buttons with icons. The GUI options panel contains the data for the currently selected

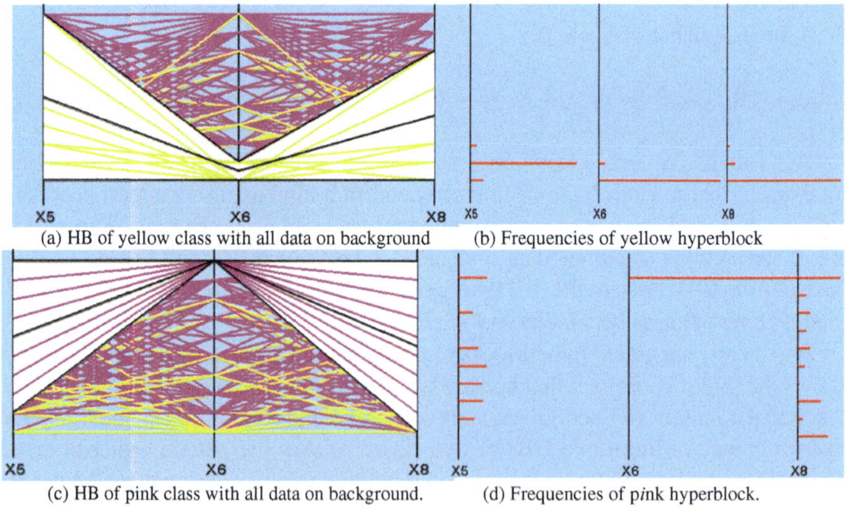

Fig. 19 Two distinct HBs of two classes with frequency distributions of n-D points in these HBs in respective coordinates

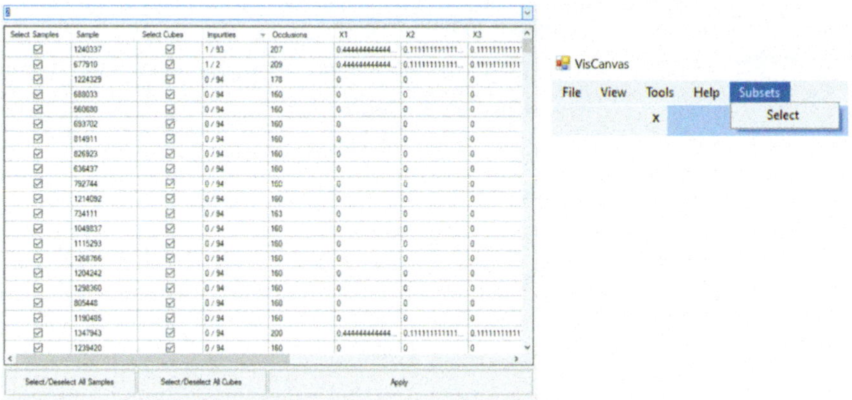

Fig. 20 Accessing the subsets feature via the toolstrip menu dropdown and *data table* example

hyperblock and updates when the currently selected hyperblock changes. It also allows the user to go directly to a point or hyperblock by using its index.

The min, center, and max lines of a HB can also be toggled on and off. Figure 21 shows a selected hyperblock in main VisCanvas GUI window. The black line is its center **c**. the panel on right shows the number of n-D points on this hyperblock.

Toggled drawings. The user can independently toggle different lines on/off (see Fig. 22). This allows the user to control which visual data to observe instead of showing all or none of it.

The minimum/maximum lines are black and represent the lower and upper borders of the HB. The center line is also black and represents the median of the HB, its center.

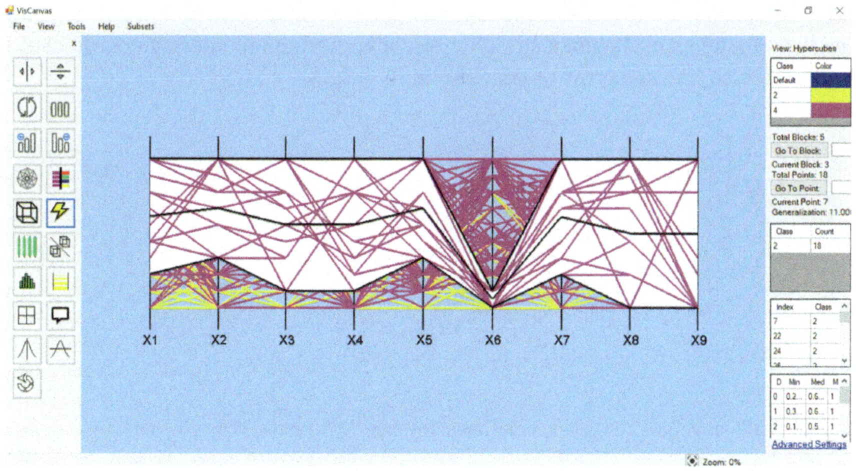

Fig. 21 A hyperblock in main VisCanvas GUI window

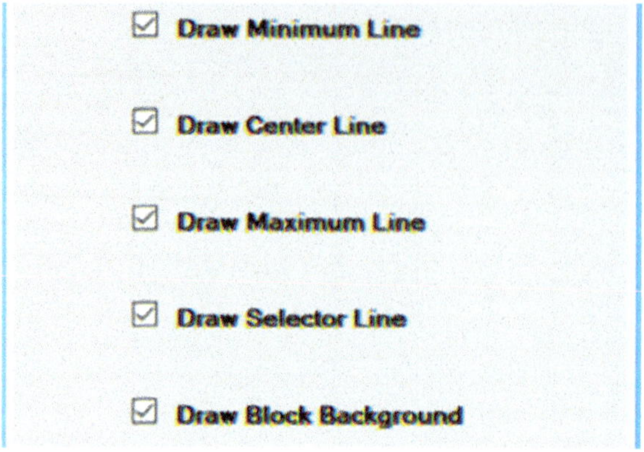

Fig. 22 Hyperblock selection utilities

The selector line represents the currently selected line and is black by default, but the color can be changed by the user. See Fig. 21 with white background.

Coloring lines in different colors. Coloring each line in a different color allows distinguishing individual lines on all dataset and distinguishing different lines in subsets such as classes, hypercubes, and hyperblocks. Figure 23 shows coloring every line a different color by using two coloring algorithms C2 and C3.

Other features that have been demonstrated above include showing a *histogram* of each coordinate to see where points overlap, *merging hyperblocks* to increase visual discovery generalization and automatically generating a *linguistic description* of a hyperblock and display it to the user.

The listed features implemented in VisCanvas 2.0 complement and expand features that were already implemented in VisCanvas 1.0 [20] making parallel coordinates adjustable. It includes abilities such as shifting coordinates up and down making the line of a given n-D point a straight horizontal line and reordering coordinates so that a selected line can grow or decrease in its values.

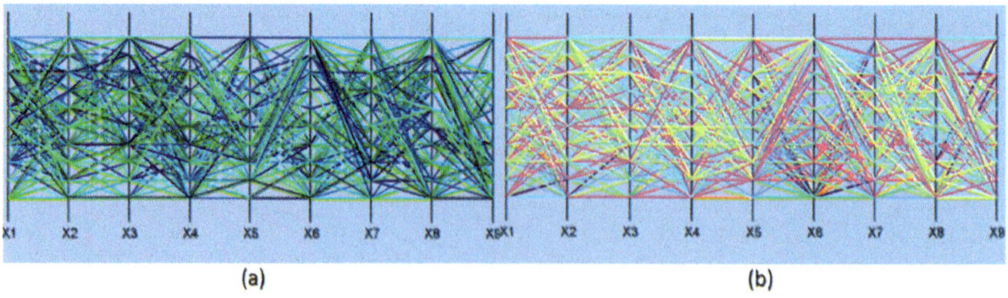

Fig. 23 Individual lines colored by coloring algorithm C2 (a) and C3 (b)

4.2 *Missing Value Visualization*

Commonly many datasets have a huge number of missing values (empty cells). Traditional approaches range from removing data with empty cells, to filling the missing values with some approximations computed automatically or conducted manually. Removing empty cells dramatically decreases the available data. In many datasets it can completely wipe out all data because by excluding the whole row or column with a single empty cell we are not only removing that single cell but all data in the respective row or column in the data table.

The missing values approximation is often risky because it can be incorrect. It also can be very time consuming if conducted manually. Therefore, it is necessary to develop a method which will overcome these difficulties, From the visualization viewpoint the major difficulty is that we cannot visualize attributes with missing values in parallel coordinates and in other general line coordinates too.

Below we present the *Empty Cell Visualization method* denoted as **ECV** method. Figure 24 shows a typical dataset with missing values. The VisCanvas program analyzes such data files. If the value in the cell is not a number, but a text, then it is extracted and put as a label below the parallel coordinates plot as shown in Fig. 25.

If the cell is empty the word "Empty" is used as a label. See Fig. 25. In Figs. 24 and 25 the labels include: n/c (not collected), did not record, Empty, ?, in other place, n /a (not applicable). Respectively, Fig. 25 visualizes that three blue cases have first two of these missing values in SL attribute. Other cases have different missing values in other attributes. Other datasets can have different terms and those terms are automatically extracted and put to the visualization. Thus, ECV method

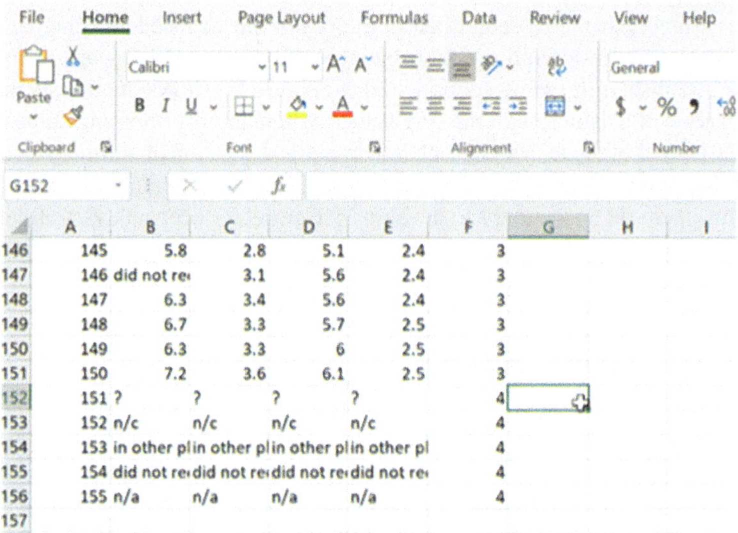

	A	B	C	D	E	F	G
146	145	5.8	2.8	5.1	2.4	3	
147	146 did not re		3.1	5.6	2.4	3	
148	147	6.3	3.4	5.6	2.4	3	
149	148	6.7	3.3	5.7	2.5	3	
150	149	6.3	3.3	6	2.5	3	
151	150	7.2	3.6	6.1	2.5	3	
152	151 ?	?	?	?		4	
153	152 n/c	n/c	n/c	n/c		4	
154	153 in other pl	in other pl	in other pl	in other pl		4	
155	154 did not re	did not re	did not re	did not re		4	
156	155 n/a	n/a	n/a	n/a		4	
157							

Fig. 24 Example of the dataset with missing values of different types

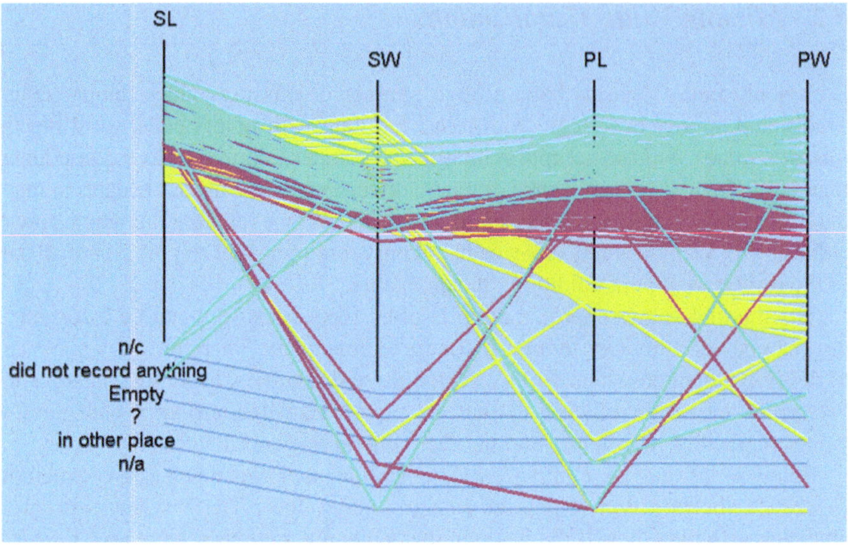

Fig. 25 Missing values visualized in adjustable parallel coordinates

allows "lossless" visual representation of missing values preserving types of missing values.

In addition, Fig. 25 illustrates abilities of VisCanvas 2.0 to support *Adjustable Parallel Coordinates* (APC), where each attribute can be drugged up and down to make any individual line straight.

This simplifies the comparison of that line with all other. See also Figs. 1, 2 in Sect. 1. This line can be a center line of the HB or one of its edges. The perceptual simplification follows from the fact that straight horizontal lines are preattentive as it is established in the perception studies. Figure 26 shows another example of visualization of a dataset with missing values with markers of missing values on the top. The benefit of the ECV approach is a dramatic expansion of abilities to visualize data with missing values, which was impossible before implementing in VisCanvas 2.0. This opens an opportunity to develop VisCanvas for deep analysis of data with missing values.

4.3 Dealing with Large Datasets

One of the significant issues is abilities interactively operate with the large data sets in parallel coordinates in VisCanvas 2.0. Specifically, we tested abilities of doing this in CWU testbed. We generated simulated 250,000 4-dimensional records. At the beginning rendering these data was very slow. Interactive manipulation like moving

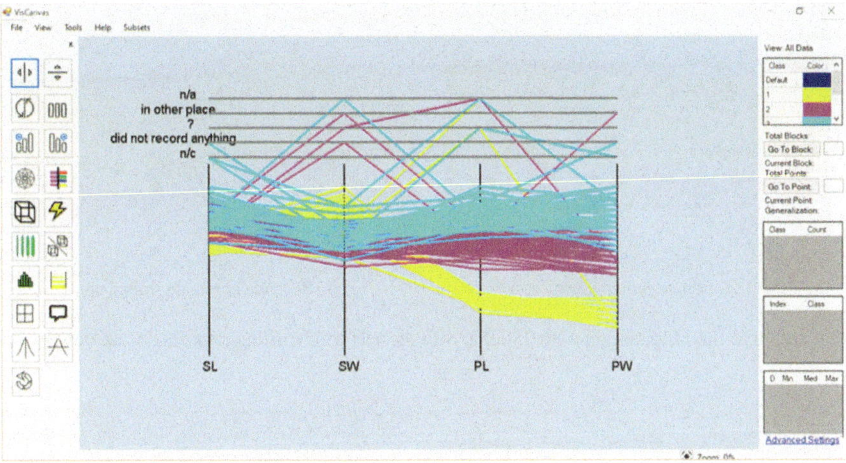

Fig. 26 Visualization of 4-dimentisonal data of 3 classes with markers of unknow values on the top

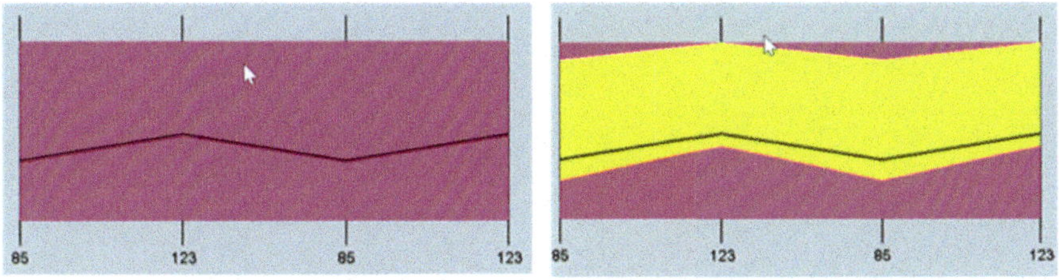

(a) Magenta class on the top.

(b) yellow class on the top.

Fig. 27 250,000 4-D records with a selected case (black)

coordinates up/down was also very slow. Later, with modification of the code with a more efficient looping sequence we were able to do this instantly. See Figs. 27, 28 show the result of manipulation with 250,000 records.

5 Discussion, Generalization, Conclusion and Future Work

Hyperblocks in parallel coordinates. This chapter contributes to interpretable machine learning via visual knowledge discovery in parallel coordinates by visual pattern discovery, data and model classification visualization, dimension reduction and model simplification. This allows putting the end-user in the driver seat of model development. It is impressive that [21] traced parallel coordinates to nineteenth century [22] to visualize the US Census data. See Fig. 29 reproduced from [21, 22].

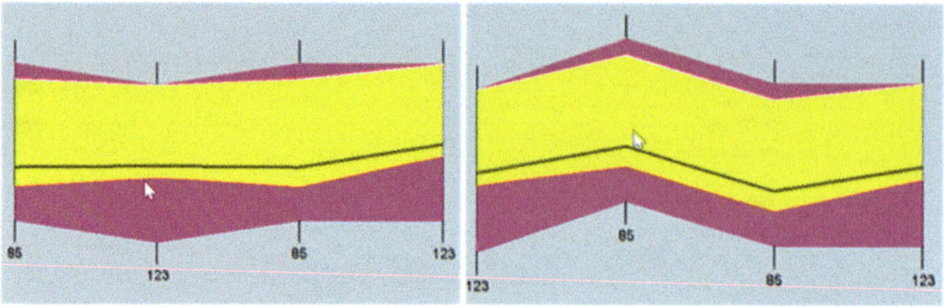

(a) Second coordinate shifted down.

(b) Second coordinate shifted up.

Fig. 28 Real-time Interaction with 250,000 4-D records with shifting coordinates up/down

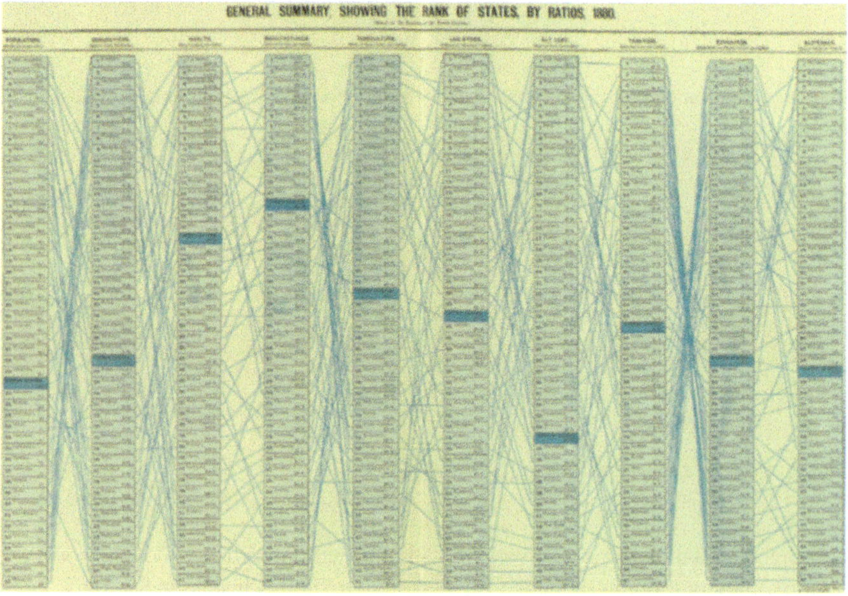

Fig. 29 Early parallel coordinates visualization [21, 22] showing the United States ranked by population, wealth, livestock, net debt, and others. The blue observation represents the U.S. average

Parallel coordinates were dormant and unknown for almost 100 years. Now parallel coordinates are actively used to explore n-D data. However, the use of parallel coordinates to improve supervised learning algorithms is still in the nascent stage.

The concepts of *hypercubes* and *hyperblocks* are used as the major concepts in this study to allow end-users to easily understand patterns on parallel coordinates. The base definitions of the hypercubes and hyperblocks are provided along with their very intuitive visualization in parallel coordinates. Then we described algorithms to discover HBs interactively and automatically in multiple HB settings: individual and multiple, overlapping, and non-overlapping, and in combination of hyperblocks with linguistic description of visual patterns.

This paper discussed challenges of classification algorithms based on hyperblocks and proposed the Hyper algorithm for classification with mixed and pure hyperblocks. It is shown that Hyper models generalize decision trees. The Hyper algorithm was tested on the benchmark Wisconsin Breast Cancer data. It allowed discovering pure and mixed HBs with all these data and then with tenfold cross validation. We established a link between hyperblocks, dimension reduction and visualization of lower dimensional HBs.

Hyperblocks in other general line coordinates. The HBs are simple and interpretable units in supervised machine learning. It creates an opportunity to use them in other General Line Coordinates beyond parallel coordinates. Below Figs. 30, 31 shows an example of a two hyperblocks (blue and red) in the Shifted Paired Coordinates (SPC) and Collocated Paired Coordinates (CPC) defined in [2]. Advantage of the visualization in SPC is that these blue and red rectangles do not overlap in contrast with them in CPC and parallel coordinates (see Fig. 31). SPC and CPC require only three simple rectangles for each HB but in parallel coordinates requires five more complex 4-gones.

Major benefits of hyperblock technology and Hyper algorithm are the abilities to discover and observe hyperblocks by end-users. It includes observing all HBs together side by side (see. Fig. 12) making patterns visible for all classes. Also, it allows detailed visual analysis of individual HBs of individual classes. HBs are well understood by the end-users in the original attributes with the ability to cut down

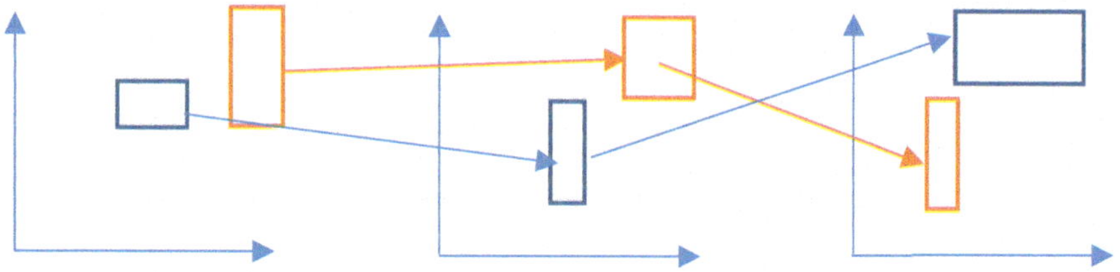

Fig. 30 Red and blue hyperblocks in shifted paired coordinates as connected rectangles

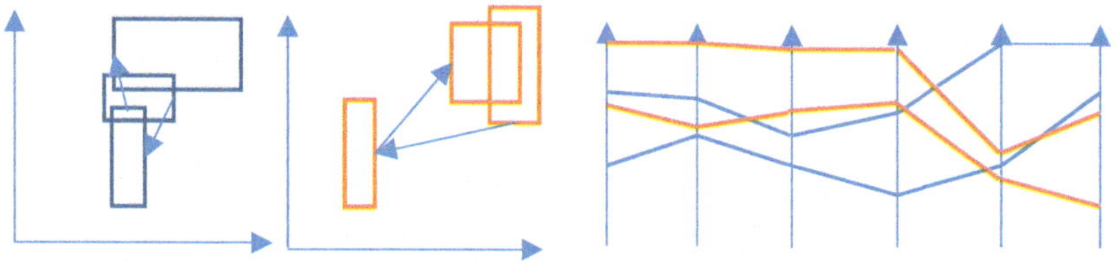

(a) Blue HB in CPC. (b) Red HB in CPC. (c) Red and blue HBs in parallel coordinates.

Fig. 31 Red and blue hyperblocks in collocated paired and parallel coordinates

unimportant attributes. Another advantage of HBs relative to DTs is abilities to avoid overgeneralization of data. DTs set up thresholds in each node that can be far away from actual data of classes. Figure 14 for HBs and Fig. 33 in Appendix for DT illustrate this situation.

Empty Cell Visualization. The paper also presented features of VisCanvas 2.0 software that implement Hyper algorithms and other HB functionality including Empty Cell Visualization method and dealing with large datasets. The conducted studies have shown that we can successfully deal with empty cells and large datasets. It also opens an opportunity to generalize the empty cell visualization method to other General Line Coordinates, like Shifted Paired Coordinates. The idea is that empty cell marker can be located at the beginning and/or end of each line coordinate, as it is shown in Fig. 32 for Shifted Paired Coordinates. Future studies include maturing the algorithms presented to the full scope of machine learning in 2-D parallel coordinates instead of only analytically in n-D. The VisCanvas software system will be expanded to support larger datasets using GPU and multithreading.

Another aspect of the future work is usability studies. Multiple prior studies with parallel coordinates already had shown that users can easily capture simple pattern in parallel coordinates. Hyperblocks belong to the category of simple patterns. This is a major reason why we concentrate this work on hyperblocks. Therefore, a traditional usability study where hyperblocks will be shown to the users with asking if a user considers them simple or not will not be much informative. We plan a more informative usability study where a domain expert without advanced knowledge of machine learning will discover and interpret pure enough hyperblocks using VisCanvas 2.0 tool.

Running Hyper algorithm of large datasets requires significant runtime. The future work will be on parallelization of the computational processes, efficient data

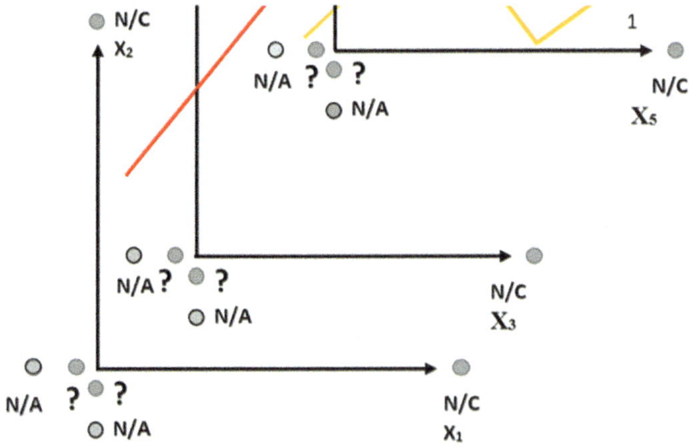

Fig. 32 Visualization of 6-D data with empty cells in shifted paired coordinates

buffering, using multi-threading, powerful GPU and supercomputer resources. It will allow for operations to take place in the background while the user is still able to interact with the UI. The challenges involved are identifying tasks, balance, data splitting, and data dependency. Both data and task parallelization are future challenges. To help shortening the run time big datasets can be split into smaller sets handled by multiple threads while the work done on them can also be parallelized. The use of thread pools, where threads are already instantiated and waiting to be requested, will be another way to explore and deal with the big data.

Appendix

See Table 8 and Fig. 33.

Table 8 Branches of ID3 decision tree for the WBC data (class 2–B, class 4–M)

<table>
<tr>
<td rowspan="2">|— **x2 < = 2.50**
||— **x6 < = 5.50**
|||— x 8 < = 9.00
||||— x1 < = 7.50
|||||— class: 2
||||— x 1 > 7.50
|||||— x 6 < = 2.00
||||||— **class: 2**
|||||— x 6 > 2.00
||||||— class: 4
|||— x 8 > 9.00
||||— class: 4
||— **x6 > 5.50**
|||— x 1 < = 2.50
||||— **class: 2**
|||— x 1 > 2.50
||||— class: 4</td>
<td>|— **x 2 > 2.50**
||— **x3 < = 2.50**
|||— x1 < = 5.50
||||— class: 2
|||— x 1 > 5.50
||||— class: 4
||— **x3 > 2.50**
|||— **x2 < = 4.50**
||||— x 6 < = 2.50
|||||— x 5 < = 3.50
||||||— class: 2
|||||— x 5 > 3.50
||||||— x 5 < = 7.00
|||||||— class: 4
||||||— x 5 > 7.00
|||||||— class: 2</td>
<td rowspan="2">|— **x2 > 2.50**
||— **x3 < = 2.50**
|||—**x2 > 4.50**
||||— x 4 < = 1.50
|||||— x 1 < = 7.00
||||||— x 9 < = 2.00
|||||||— class: 2
||||||— x 9 > 2.00
|||||||— class: 4
|||||— x 1 > 7.00
||||||— class: 4
||||— x 4 > 1.50
|||||— x 8 < = 2.50
||||||— x 8 < = 1.50
|||||||— class: 4
||||||— x 8 > 1.50
|||||||— x 6 < = 9.00
||||||||— class: 2
|||||||— x 6 > 9.00
||||||||— class: 4
|||||— x 8 > 2.50
||||||— class: 4</td>
<td rowspan="2">|— **x2 > 2.50**
||— **x3 > 2.50**
|||—**x2 < = 4.50**
||||— **x6 > 2.50**
|||||— x 1 < = 6.50
||||||— **x1 < = 5.50**
|||||||— x 6 < = 6.00
||||||||— x 4 < = 3.50
|||||||||— x3 < = 3.50
||||||||||— class: 4
|||||||||— x 3 > 3.50
||||||||||— class: 2
||||||||— x 4 > 3.50
|||||||||— class: 4
|||||||— x 6 > 6.00
||||||||— x7 < = 3.50
|||||||||— x 5 < = 6.50
||||||||||— class: 4
|||||||||— x 5 > 6.50
||||||||||— class: 2
||||||||— x 7 > 3.50
|||||||||— class: 4
||||||— x 1 > 5.50
|||||||— class: 2</td>
</tr>
<tr>
<td>|— **x2 > 2.50**
||— **x3 > 2.50**
|||— **x2 < = 4.50**
||||—**x6 > 2.50**
|||||— **x1 > 6.50**
||||||— x 8 < = 7.00
|||||||— class: 4
||||||— x 8 > 7.00
|||||||— x 8 < = 8.50
||||||||— x 1 < = 7.50
|||||||||— class: 4
||||||||— x 1 > 7.50
|||||||||— class: 2
|||||||— x 8 > 8.50
||||||||— class: 4</td>
</tr>
</table>

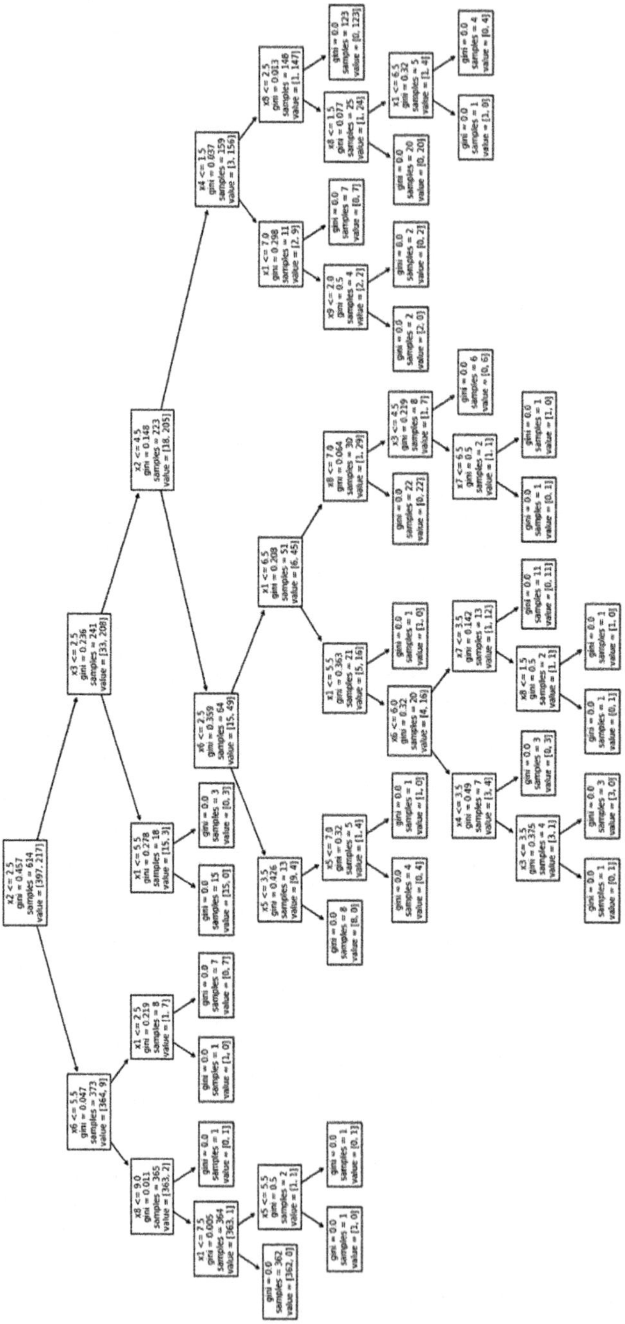

Fig. 33 ID3 Decision tree for WBC data with multiple "mini" terminal nodes

References

1. Kovalerchuk B, Ahmad MA, Teredesai A (2021) Survey of explainable machine learning with visual and granular methods beyond quasi-explanations. In: Pedrycz W, Chen SM (eds) Interpretable artificial intelligence: a perspective of granular computing. Springer, pp 217–267. https://arxiv.org/abs/2009.10221
2. Kovalerchuk B (2018) Visual knowledge discovery and machine learning, Springer
3. Kovalerchuk B, Nazemi K, Andonie R, Datia N, Bannissi E (Eds) (2022) Integrating artificial intelligence and visualization for visual knowledge discovery. Springer
4. Kovalerchuk B, Grishin V (2019) Adjustable general line coordinates for visual knowledge discovery in n-D data. Inf Vis 18(1):3–32
5. Kovalerchuk B, Hayes D (2021) Discovering interpretable machine learning models in parallel coordinates. In: 2021 25th International conference information visualisation (IV) 2021 Jul 5. IEEE, pp 181–188. arXiv:2106.07474
6. Martinez AR, Wegman EJ (2002) Encoding of text to preserve meaning. In: Proceedings of the army conference on applied statistics
7. Estivill-Castro V, Gilmore E, Hexel R (2020) Constructing interpretable decision trees using parallel coordinates. In: Intern. Conf. on Artificial Intelligence and Soft Computing 2020. Springer, pp 152–164
8. Inselberg A (1998) Visual data mining with parallel coordinates, Computational Statistics, vol 13, No.S 1
9. Inselberg A (2009) Parallel coordinates: visual multidimensional geometry and its applications. Springer Science & Business Media
10. Schneider B, Keim DA, El-Assady M (2020) Datashiftexplorer: visualizing and comparing change in multidimensional data for supervised learning. In: 15th Intern. joint conference on computer vision, imaging and computer graphics theory and applications, VISIGRAPP 2020, pp 141–148
11. Gilmore E (2021) Learning interpretable decision tree classifiers with human in the loop learning and parallel coordinates, Doctoral dissertation, Griffith University
12. Sansen J, Richer G, Jourde T, Lalanne F, Auber D, Bourqui R (2017) Visual exploration of large multidimensional data using parallel coordinates on big data infrastructure. In: Informatics, vol. 4, No. 3, p 21. Multidisciplinary Digital Publishing Institute
13. Tam GK, Kothari V, Chen M (2016) An analysis of machine-and human-analytics in classification. IEEE Trans Visual Comput Graph 23(1):71–80
14. Dua D, Graff C (2019) UCI machine learning repository. University of California, School of Information and Computer Science, Irvine, CA. http://archive.ics.uci.edu/ml
15. VisCanvas 2.0 (2023) https://github.com/CWU-VKD-LAB
16. Kovalerchuk B, Dunn A, Worland A, Wagle S (2023) Interactive decision tree creation and enhancement with complete visualization for explainable modeling. https://arxiv.org/pdf/2305.18432
17. Zhuravlev Y, Nikiforov V (1971) Recognition algorithms, based on the computation of estimates. Kibernetika 3:1–11
18. Qin Y, Cheng X, Li X, Leng Q (2020) An effective multi-label classification algorithm based on hypercube. In: International Conference on Intelligent Computing 2020. Springer, pp 476–483
19. Ohmaid H, Eddarouich S, Bourouhou A, Timouya M (2020) Comparison between SVM and KNN classifiers for iris recognition using a new unsupervised neural approach in segmentation. IAES Int J Artif Intell 9(3):429
20. VisCanvas 1.0 (2018) https://github.com/CWU-VKD-LAB

21. Weidele DK (2019) Conditional parallel coordinates. In: 2019 IEEE visualization conference (VIS) 2019 Oct 20. IEEE, pp 221–225
22. Gannett H, Hewes F (1883) General summary showing the rank of states by ratios 1880. In: Hewes FW (ed) Scribners statistical atlas of the United States showing by graphic methods their present condition and their political, social and industrial development. US Census Office, p 151
23. Xu Y, Hong W, Chen N, Li X, Liu W, Zhang T (2007) Parallel filter: a visual classifier based on parallel coordinates and multivariate data analysis. In: International Conference on Intelligent Computing. Springer, pp 1172–1183

Visual Knowledge Discovery with General Line Coordinates

Lincoln Huber, Boris Kovalerchuk, and Charles Recaido

1 Introduction

1.1 Motivation and Goal

While many powerful Machine Learning (ML) and visualization techniques exist for exploring n-D data, these techniques often lack explainability and/or are lossy [1–12]. This makes it difficult to recommend such ML models in high-risk classification scenarios where a misclassification may lead to disastrous results. Developing new ways to explain and interact with ML models will enhance the usability of these models.

All traditional ML methods are computational, where the predictions are produced by computations. Visualization is used to represent these results. The idea of *visual ML* and *Visual Knowledge Discovery* (VKD) is getting the actual prediction by visual methods or with a significant contribution of the visual method to the core of the prediction process.

One emerging n-D data visualization technique is General Line Coordinates (GLC). GLC are a category of visualization techniques specializing in reversible lossless n-D data visualization in 2-D and 3-D [1–4]. GLC-Linear (GLC-L) is a type of GLC specializing in solving supervised learning classification tasks in 2-D [1, 2]. While limited to discovering only *linear models*, GLC-L can be made *more*

L. Huber · B. Kovalerchuk (✉) · C. Recaido
Department of Computer Science, Central Washington University, Ellensburg, USA
e-mail: Boris.Kovalerchuk@cwu.edu

L. Huber
e-mail: Lincoln.Huber@cwu.edu

C. Recaido
e-mail: Charles.Recaido@cwu.edu

B. Kovalerchuk et al. (eds.), *Artificial Intelligence and Visualization: Advancing Visual Knowledge Discovery*, Studies in Computational Intelligence 1126,
https://doi.org/10.1007/978-3-031-46549-9_5

explainable than traditionally assumed for linear models as shown in [2]. Dynamic Scaffolding Coordinates (DSC) is another type of GLC introduced in [13]. When based on Parallel Coordinates (DSC1), it can be used to increase explainability like GLC-L. When based on Shifted Paired Coordinates (DSC2), it can be used to increase explainability and create worst-case validation datasets. Together, extensions of GLC-L, DSC1, and DSC2 increase the set of methods for VKD based on the lossless GLC.

The goal of this paper is to expand [13] and branch off from the original GLC-L algorithm to create new variations capable of further enhancing the explainability of ML models that can be built, visualized, and explored. We call the new branches of the GLC-L algorithm as GLC non-Linear (GLC-nL), GLC Interactive Rules Linear (GLC-IRL), GLC Hyperblock Rules Linear (GLC-HBRL), and GLC Worst Case Linear (GLC-WCL). The similarly purposed algorithms DSC1 and DSC2 in [13] will be used for comparison with the new GLC-L variations.

Major benefits from these new GLC-L branches are improved interpretability and accuracy in both linear and non-linear classification problems. This is done through the ability to accurately classify non-linear data using GLC-nL and the highly interpretable rules from hyperblocks from GLC-IRL and GLC-HBRL. GLC-WCL also gives the ability to analyze interpretability weaknesses in a model by constructing worst-case validation sets. The interactive nature of all these expansions reinforces these improvements with the added benefits of expert knowledge in human-guided VKD methods.

This paper is organized as follows. First, we analyze the interpretability of Linear Discriminant Functions (LDFs) in Sect. 1.2. Then we present the base GLC-L, GLC-nL, GLC-IRL, GLC-HBRL, GLC-WCL, DSC1, and DSC2 algorithms in Sect. 2. Next, the results of several case studies using these algorithms are shown in Sect. 3. A brief description of a new software system using each algorithm follows in Sect. 4. Last, the pros and cons of each algorithm are summarized, and an outline for future work is discussed in Sect. 5. The structure of the chapter can be seen visually in Fig. 1.

1.2 Analysis of Interpretability of Linear Discriminant Functions

Linear models play a very important role in current machine learning explainability studies. These models form the core of most popular LIME [5] and SHAP [6] explainability methods and, in general, are a preferred way to explain more complex models. Unconditional statements that linear models are intrinsically interpretable are common, e.g., [7–11] based on abilities to analyze fitted coefficients and the confidence bounds. Related models like piecewise linear functions to approximate complex nonlinear ML models are similarly discussed [9].

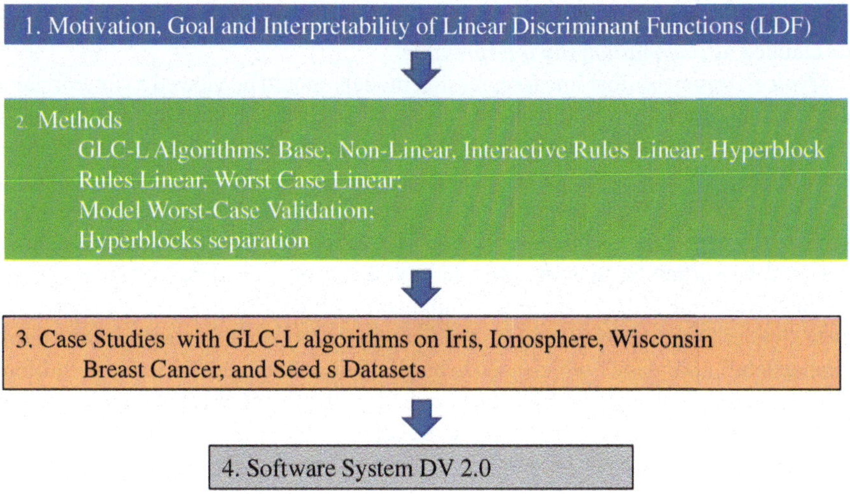

Fig. 1 Visual outline of paper

Unfortunately, these unconditional statements lack scientific ground. Below we provide arguments against such unconditional interpretability of linear models. The unconditional interpretability of linear models is an *incorrect statement* often based on a narrow definition of interpretability as *computational simplicity.* Considering linear models as always interpretable, and moreover as a "gold standard" of interpretability, is risky and has negative consequences.

An example of such *unconditional interpretability* is the interpretation of the *weights* of linear regression and discrimination models as *effects* of the features on prediction [7, 8]. Statements like the following quotes can often be found in the literature: "The linearity of the learned relationship makes the interpretation easy" [10], "...a simple linear function, i.e., a surrogate model which is *easy to interpret*" [12], and "...the elegant simplicity of linear models makes the results they generate easy to interpret" [9]. These statements reference business analysts, doctors, and industry researchers who use these models. Notwithstanding the decades of publications questioning linear models in econometric literature [14].

Statements about the unconditional interpretability of linear models have expanded to *alternative regression techniques* like *Generalized Additive Models* (GAM), naming them trusted linear models but used in new and different ways [9]. GAM generalize linear models to a weighted sum of functions for more complex tasks. Basically, *interpretability for GAM is equated with model simplicity* where simpler GAM are considered as more interpretable [9]. Thus, all critique about insufficiency of simplicity as a measure of interpretability is applicable to GAM too. *Piecewise linear functions* have the same interpretability issues as GAM. *Quantile regression* builds sets of linear models to different percentiles (subsets) of the training data. While these models are again claimed to be interpretable [9], each has the same interpretability issues as the linear models discussed above. They cannot be claimed as *unconditionally interpretable.*

Typical arguments that linear regression models should be viewed as interpretable are [15]: (1) linear models have *simple model structure*, (2) linear *structure* is easy to understand, (3) the output of the linear model is *easy* to understand, (4) the *weight* of each feature represents the mean change in the prediction given a one unit increase of the feature, and (5) the features with larger weights have more *effect* on the result.

Below we present *counterarguments.* The simplicity of (1) is not sufficient. If it would be sufficient, then any simple model can be considered as interpretable. A spurious linear model can be as simple as a model with important attributes or even simpler. Arguments (2) and (3) fundamentally rely on the meaning of the terms "understand" and "easy", which need to be clarified and defined. Different definitions of these terms can lead to acceptance or rejection of a linear model's interpretability.

For (4) and (5), one of the counterarguments is that different types of variables (e.g., categorical features vs. numerical features) have *different scales* [15]. The change of the scales changes the weights and effects of the features on the output. This results in contradictory interpretations of the model. Remedies like t-statistics and chi-square score [16] suggested in [15] do not resolve the issue fully. Typically, deep analysis of such remedies shows that they bring their own assumptions that are difficult to justify for given data. Moreover, discovering such assumptions and limitations often is difficult itself because they are not clearly stated.

A related counterargument is about *mutual dependence.* Correlation of the variables makes their weights confusing even when they are measured in the same scale. Remedies like using penalty [9] also often bring their own assumptions that are difficult to justify for given data. Below we follow [9] to summarize and analyze the current penalized regression alternatives to ordinary least squares regression. These methods combine L1/LASSO penalties for variable selection and Tikhonov/L2/ridge penalties. They make assumptions about data, but less than ordinary least squares regression. The minimization of the constrained objective functions penalizes for assigning large weights to correlated or meaningless variables. L1/LASSO penalties bring some weights to zero, selecting a small, *representative* subset of weights. Tikhonov/L2/ridge penalties help to get stable weights for *correlated variables*, while they *may not* create confidence intervals, t-statistics, or p-values for regression parameters.

The selection of variables to be suppressed in the penalty function is *nontrivial.* Doing this without *causal domain knowledge* can easily produce a meaningless result. It is also non-trivial for correlated variables. We can suppress the *cause* attribute instead of the dependent one.

Another counterargument is related to *human limitations* in understanding concepts with multiple variables [17]. In general, the actual situation with interpretability of linear models is quite complex and is not limited by the counterarguments listed above [18]. Linear models can be interpretable for *homogeneous features* like pixel intensities in an image or a time series, where all features measure the same property, like temperature, but at the different moments. However, even for such homogeneous data, the impact of the features expressed by the weights for heavily correlated features is quite confusing and can be misleading as we discussed above. Another example of explanation difficulties of linear models for homogeneous attributes is about the meaning of a weighted sum of systolic and diastolic blood pressure measurements. While both measure blood pressure inside the arteries, systolic one measures it when the heart is pumping, but the diastolic one measures it when the heart is resting between beats making the meaning of a weighted sum at least unclear.

For *heterogeneous features*, the situation is even more challenging to call linear models unconditionally interpretable. In heterogeneous data one attribute can measure temperature, another blood pressure, and another the size of a tumor. This makes the interpretation of the weighted sum of these features fundamentally challenging: it has no physical meaning. How many doctors are willing to make life-critical decisions based on such weighed sums? How many doctors are willing to explain the decision to a patient in these terms?

Consider another example of a linear function: 5(blood pressure) + 3(body temperature) + 7(BMI) to be a basis of the diagnostics and the treatment. Is it interpretable and explainable for a patient and a doctor? This example shows that the narrow definition of interpretability as abilities of the user to easily compute the output (*model computational simplicity*) cannot serve the domain experts who are actual end-users of the models.

Quasi-explainable weights. Below is an example that illustrates quasi-explainable weights. Consider a linear model: If [0.3*(tumor size X sq. mm) + 0.4(tumor shape measure Y) + 0.5(% of tumor growth from the last test Z)] > 10 then cancer. Even if it was 100% accurate on the given data, would anybody go to a cancer surgery based on such a model? Would a doctor accept the cancer conclusion based on this model? What is the meaning of the weighed summation of such heterogeneous values such as size, shape and % in oncology? Can we say that the size is less important than % because 0.3 is less than 0.5? If we measure size in sq. cm the coefficient for size will be 30 instead of 0.3. Will it mean that the importance of the size and % is reversed because 30 > 0.5? Thus, we are getting very different relative importance of these attributes. In both cases we get *quasi-explanation.* In contrast, if X_1 and X_2 were homogeneous attributes, then their weights could express the importance of the attributes meaningfully and contribute to the *actual not quasi-explanation.* This example adds doubts to the frequent claim that *weights* in linear models are a major and efficient tool to provide model interpretation to the user, e.g., [19] with multiple AutoML systems implementing it as a model interpretation tool, e.g. [20].

Thus, an explanation that uses *summation of heterogeneous attributes* can be a *quasi-explanation*, but hardly a *true explanation*. The *lightweight concept* of intrinsic interpretability equated to *simplicity* of the linear model computation misses the goal of all interpretability studies to convince the end-user that the model is good enough to be used by this user. In essence, it claims that the model is intrinsically interpretable if the user understands how to compute the output from the input. For a linear model everybody can do this simple computation, but it is not intrinsically interpretable.

When we try to sum up three apples and four oranges, we cannot say that we have seven apples or seven oranges. We need to construct a new item called fruit, then we can say that we have seven fruits or even more generally seven items [21]. While it was easy for apples and oranges making the same approach work for blood pressure and temperature is much more difficult when keeping a medical meaning intact for diagnostics.

Therefore, it is very difficult to justify that linear models are unconditionally intrinsically interpretable. While linear regression models are often interpretable for homogeneous data, these data are not typical in ML problems like healthcare where interpretability is paramount. Therefore, linear models cannot be claimed unconditionally interpretable.

This conclusion brings the immediate important consequence. It requires reexamining and *limiting the unconditional use* of popular *linear methods* and associated methods like LIME and SHAP and paying more attention to *decision trees* (DT) and *logic decision rules* in propositional or First-order Logic (FoL) as interpretable methods, which are available in analytical, computational, and visual forms, e.g., [22–29]. These methods allow both (a) converting linear models to interpretable DTs and logic rules or (b) construct them from the data directly similarly to general non-linear models. We will present methods specific to linear models in Sect. 2.

2 Methods

This section describes the main idea of general line coordinates (GLC) and specifically the main algorithm we use in this chapter. GLC represent n-D data in 2-D or 3-D *without loss* of information. It means that we want to be able to restore fully each n-D data point from its 2-D or 3-D representation. The traditional dimension reduction and visualization methods like Principal Component Analysis are *lossy* because they convert say a 10-D point to a 2-D point. The abilities to restore from a 2-D point 10-D points are very limited. Such dimension reduction also leads to corruption of the n-D distances as it is proven in the Johnson-Lindenstrauss lemma [1].

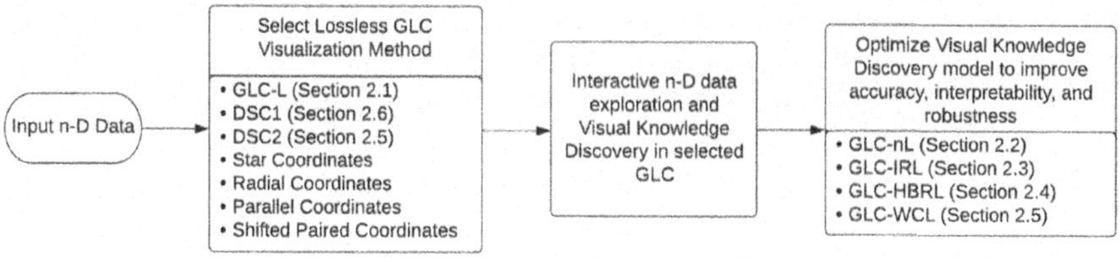

Fig. 2 Generic GLC based Visual Knowledge Discovery roadmap

The actual GLC approach is converting each n-D point **x** to a directed graph **x*** in 2-D/3-D. Typically these graphs are polylines. First, GLC were proposed in 2014 and summarized in [1]. There are an infinite number of possible GLC by locating n coordinates in 2-D in a variety of ways: curved, parallel, collocated, disconnected, etc. Before GLC were proposed only parallel coordinates and radial coordinates were known as lossless visualization methods. The advantage of expanding the class of lossless visualization methods is in the fact that different data sets may require different types of visualization to make patterns visible and discoverable visually. A roadmap for how GLC are generically used can be seen in Fig. 2.

This rest of this section presents the visualization algorithm GLC-L [1, 2] and its generalizations: GLC-nL, GLC-IRL, GLC-HBRL, and GLC-WCL. GLC-L deals with a linear function

$$F(\boldsymbol{x}) = c_1x_1 + c_2x_2 + c_3x_3 + \ldots c_nx_n$$

$F(\mathbf{x})$ serves as a Linear Discriminant Function (LDF), where threshold T is used to set up a classification rule.

$$\text{if } F(\mathbf{x}) < T \text{ then } \mathbf{x} \text{ belongs to class 1, else } \mathbf{x} \text{ belongs to class 2.}$$

The generalizations of GLC-L deal with more complex *non-linear* functions: GLC-nL visualizes general weighted sum of functions, GLC-IRL is capable of interactively creating hyperblocks based on an LDF, and GLC-HBRL automatically creates hyperblocks based on an LDF. In contrast, GLC-WCL is used to expose interpretability weaknesses in all types of GLC-L visualizations.

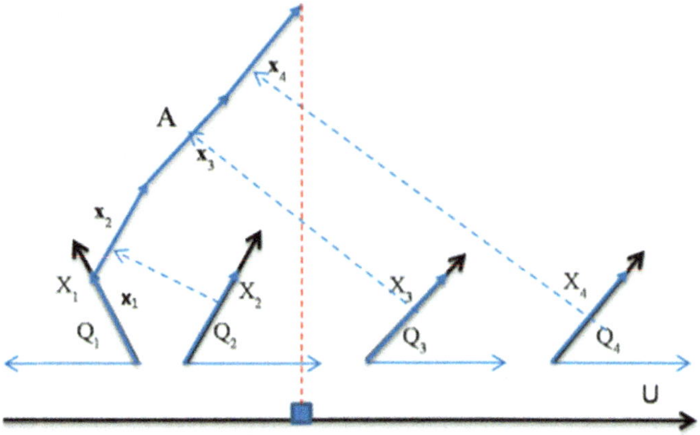

Fig. 3 4-D point $A = (1, 1, 1, 1)$ in GLC-L coordinates $X_1 - X_4$ with angles (Q_1, Q_2, Q_3, Q_4). Vectors $\mathbf{x}_i$ are shifted to connect one after another and the end of the last vector is project to the black line [1, 2]

2.1 Base GLC-L Algorithm

This section presents the base algorithm for GLC-L following [1, 2]. This algorithm is the base for all other GLC-L algorithms in this paper. The idea of GLC-L is illustrated by an example in Fig. 3, where there are four vectors $\mathbf{x}_i$ shifted to connect one after another.

To produce the GLC-L visualization we need to normalize coefficients c_i of $F(\mathbf{x})$ to produce a new linear function $G(\mathbf{x}) = k_1x_1 + k_2x_2 + \ldots + k_nx_n$ as described below. Let $K = (k_1, k_2, \ldots, k_n)$ and $k_i = c_i / |c_{max}|$, where $|c_{max}| = \max_{i=1:n}|(c_i)|$. Here all k_i are normalized to the interval $[-1, 1]$. The following property is true for F and G:

$$F(\boldsymbol{x}) < T \text{ if and only if } G(\boldsymbol{x}) < T/|c_{\max}|$$

Thus, F and G are equivalent linear classification functions. Below we present steps of the GLC-L algorithm for a given linear function $F(\mathbf{x})$ with coefficients $C = (c_1, c_2, \ldots, c_n)$.

Steps of the *GLC-L algorithm* are below:

(1) *Normalize* $C = (c_1, c_2, \ldots, c_n)$ by creating a set of normalized parameters $K = (k_1, k_2, \ldots, k_n)$ with normalized rule: if $y_n < T/|c_{max}|$ then $\mathbf{x}$ belongs to class 1, else $\mathbf{x}$ belongs to class 2, where y_n is a normalized value, $y_n = F(\mathbf{x})/|c_{max}|$.
(2) *Compute all angles* $Q_i = arcos(k_i)$ of the absolute values of k_i and locate coordinates $X_1 - X_n$ in accordance with these angles as shown in Fig. 3.
(3) *Draw vectors* $\boldsymbol{x}_1, \boldsymbol{x}_2, \ldots, \mathbf{x}_n$ *one after another*, as shown in Fig. 3. Then *project* the last point for $\boldsymbol{x}_i$ onto the horizontal axis U (see red dotted line in Fig. 3).

(4) *Choose classification task*

   a. For a two-class classification task, repeat step 3 for all n-D points with classes 1 and 2 drawn in different colors. Move points of class 2 by mirroring them to the bottom.
   b. For a multi-class classification task, combine all, but one class into a super class then repeat steps 3 for all n-D points with classes 1 and the super class drawn in different colors. Move points of the super class by mirroring them to the bottom.

This algorithm uses the property that $cos(arccos\ k) = k$ for $k \in [-1, 1]$. The projection of vectors $\boldsymbol{x}_i$ to axis U will be $k_i x_i$ and with consecutive location of vectors $\boldsymbol{x}_i$, the projection from the end of the last vector $\boldsymbol{x}_n$ gives a sum $k_1x_1 + k_2x_2 + \ldots + k_nx_n$ on axis U.

2.2 *Non-linear Algorithm GLC-nL*

As a linear classification method, GLC-L is limited to discovering only linear models, while often data are not linearly separable. To expand GLC-L to non-linear models, *GLC-nL (General Line Coordinates non-Linear)* was developed.

Non-linear models have multiple forms which include polynomial, weighted sums of functions, and kernel-based models. The expansion of GLC-L to GLC-nL allows to visualize them as follows.

Consider a *quadratic* function $G(\mathbf{x}) = k_{11}x_1 + k_{12}{x_1}^2 + k_{21}x_2 + k_{22}{x_2}^2 \ldots + k_{n1}x_n + k_{n2}{x_n}^2$. We add all ${x_i}^2$ to n-D point $\mathbf{x} = (x_1, x_2, \ldots, x_n)$ to produce a new n-D point and visualize this quadratic function in the GLC-L expansion. Similarly, any polynomial function can be visualized in GLC-nL.

A *general weighted sum of functions*, $G(\mathbf{x}) = k_1G_1(\mathbf{x}) + k_2G_2(\mathbf{x}) + \ldots + k_mG_m(\mathbf{x})$ is visualized in GLC-nL similarly, where the original n-D point $\mathbf{x} = (x_1, x_2, \ldots, x_n)$ is substituted by m-D point $P(\mathbf{x}) = (G_1(\mathbf{x}), G_2(\mathbf{x}), \ldots, G_m(\mathbf{x}))$ and visualized.

The non-linear classifier works as follows with a threshold *T.*

$$\text{if } G(\mathbf{x}) < T \text{ then } \mathbf{x} \text{ belongs to class 1, else } \mathbf{x} \text{ belongs to class 2.}$$

$G(\mathbf{x})$ serves as a *non-Linear Discriminant Function (nLDF).*

We consider *kernel-based* non-linear models as a form of this general weighted sum of functions [30, 31]. Here each $G_j(\mathbf{x})$ is a kernel. The GLC-nL algorithm uses SVM Support Vectors along with a polynomial or radial basis function (RBF) kernel to bolster the accuracy of non-linearly separable data.

Let, **x** be an n-D point and $\mathbf{y}_i = (y_1, y_2, \ldots, y_n)$ be a SVM support vector. For a polynomial kernel, the base equation of the kernel for producing $F(\mathbf{x})$ is as follows:

$$p_i(\mathbf{x}, \mathbf{y}_i) = (\gamma(\mathbf{x} \cdot \mathbf{y}_i) + 1)^3,$$

where $\gamma = 1/n$ and n is the number of dimensions in a dataset and $\mathbf{x} \cdot \mathbf{y}_i$ is a dot product of **x** and $\mathbf{y}_i$ [31].

For a RBF kernel, the base equation is as follows:

$$p_i(\mathbf{x}, \mathbf{y}_i) = e^{(-\gamma||\mathrm{x}-\mathrm{y}_i||^2)},$$

where γ and n are as above [31].

Below we present the steps of the *GLC-nL algorithm* for kernel-based models:

(1) *Run SVM and get the SVM support vectors* for a given dataset.
(2) Use either the polynomial or RBF kernel on vectors **x** and $\mathbf{y}_i$ to get new value p_i, where **x** is a vector from the original dataset and $\mathbf{y}_i$ is a SVM support vector.
(3) Repeat step 2 for all m support vectors $\{\mathbf{y}_i\}$ of SVM.
(4) Add m-D point $\mathbf{p} = (p_1, p_2, p_3, \ldots, p_m)$ to new dataset D.
(5) Repeat steps 2–4 for all n-D points $\{\mathbf{x}\}$ in a dataset.
(6) Perform the base GLC-L algorithm on new dataset D with newly generated coefficients in it or exported from the respective Kernel algorithms.

Note that with the newly generated coefficients in Step 6 we modify the original kernel algorithm and resulting models.

2.3 *Rules from Linear Discriminant Function*

In Sect. 1.2 we presented the deficiencies of the Linear Discriminant Functions (LDF) interpretability. In this section, to help to make LDF more interpretable, we have adapted the GLC-L algorithm to help produce *interpretable logical rules* $\{R\}$ for a given LDF. As we analyzed in Sect. 1.2, often linear models are considered *unconditionally interpretable* in machine learning. That analysis has shown that this is not the case. The same is true for more complex *non-linear models* which also will benefit from being interpreted by logical rules.

2.3.1 Rules for Linear Discriminant Function for a Given Case

First, we want to build an interpretable logical rule R for a given n-D point $\mathbf{x} = (x_1, x_2, \ldots, x_n)$ that belongs to class 1 (C_1) and Linear Discriminant Function (LDF) G,

$$G(\mathbf{x}) = k_1x_1 + k_2x_2 + \ldots + k_nx_n > \mathrm{T}$$

Assume that we have two n-D points **b** and **d** such that

$$\forall i = 1 : n\, b_i \leq x_i \leq d_i => \mathbf{x} \in \mathrm{C}_1$$

i.e., we have a rule

$$R: \text{ If } \mathrm{d}_1 \geq x_1 \geq b_1 \,\&\, d_2 \geq x_2 \geq b_2 \,\&\ldots\&\, d_n \geq x_n \geq b_n \text{ then } \boldsymbol{x} \text{ is in class } 1. \quad (1)$$

Assume also that any other n-D point **y** from a given dataset such that

$$d_1 \geq y_1 \geq b_1 \,\&\, d_2 \geq y_2 \geq b_2 \,\&\ldots\&\, d_n \geq y_n \geq b_n$$

also belongs to class C_1, i.e., we do not have any counterexample for given **b** and **d**. These two n-D points **b** and **d** form a *pure hyperblock*, where all n-D points belong to a single class C_1. The larger the difference between **b** and **d** means a stronger generalization of n-D point **x** by the rule *R*. In Fig. 4a GLC-L visualizes LDF *G*(**x**) by a blue polyline, *G*(**b**) by a green polyline, and *G*(**d**) by a black polyline. These polylines provide much more information than just the values of the function G on these n-D points. We call the n-D point **b** a *lower bound* and n-D point **d** an *upper bound.*

The main steps of the algorithm to find them is as follows:

(1) Order all attributes of all given n-D points according to the values of the coefficients k_i of *G* starting with the negative coefficients in increasing order. Denote reordered n-D points as $\mathbf{x}'$, $\mathbf{y}'$ and so on.
(2) Find a set of values $\{y'_{\min i}\}$, which are the smallest values of each y'_i and form $\mathbf{b}'$ from them, $b'_I = y'_{\min i}$.

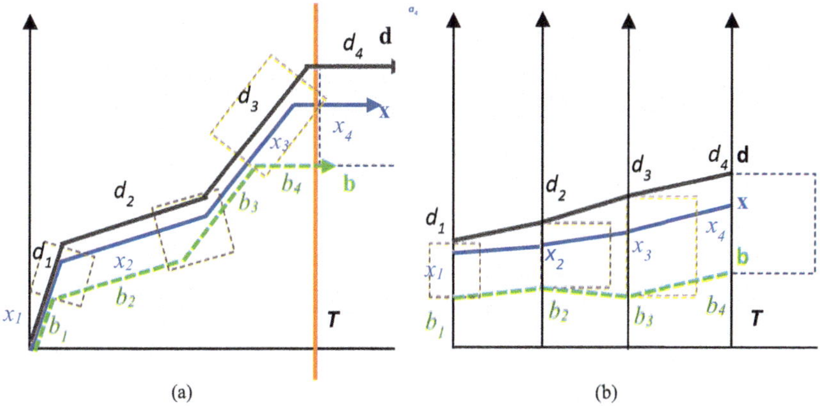

Fig. 4 LDF based interpretable rule in GLC-L **a** and in Parallel Coordinates **b** for 4-D point **x**

(3) Find a set of values $\{y'_{\text{miax}i}\}$, which are the largest values of all y'_I and form $\mathbf{d}'$ from them, $d'_I = y'_{\text{min}i}$.
(4) Draw $\mathbf{x}'$, $\mathbf{b}'$ and $\mathbf{d}'$ in GLC-L.

An alternative way to assign **b** and **d** is using domain knowledge from a domain expert/end-user. The dotted rectangles indicate the allowable areas of each attribute. A domain expert/end-user can interactively assign them in GLC-L using the domain knowledge, e.g., temperatures below 35 °C and above 38 °C are not allowed for the class of healthy people. The rule (1) can also be visualized in the Parallel Coordinates as a hyperblock as Fig. 4b shows.

2.3.2 Rules for Linear Discriminant Function for All Cases

The main idea of the algorithm to represent the whole LDF by a set of rules is as follows. We built a set of hyperblocks (HBs) [25] that represent a LDF by explicitly using the LDF. Some HBs can cover many cases and some HBs can be unique to individual cases. We designed several versions of the algorithm to start with the LDF. The main idea is exploiting the projection of the endpoint of each LDF polyline.

The first algorithm is *GLC Interactive Rules Linear (GLC-IRL)*. The idea of GLC-IRL is that a user can interactively select user's *areas* of interest and build rules for that area of interest in GLC-L.

The steps of the *GLC-IRL algorithm* are as follows:

(1) Interactively select some rectangular area *A* by selecting two 2-D points to outline the area. One 2-D point will establish the upper right or upper left corner of the rectangular area. The other 2-D point will establish the lower left or lower right corner of the rectangular area.
(2) Select a set of n-D points, with their end points located within the area *A* selected by step 1.
(3) For each attribute x_i find the smallest and largest values within the set of n-D points selected in step 2.
(4) Using the smallest values for each attribute x_i, create a n-D point **b** and using the largest values for each attribute x_i, create a n-D point **d**.
(5) Using **b** and **d**, create some rule *R* such that $d_1 \geq x_1 \geq b_1$ & $d_2 \geq x_2 \geq b_2$ & …& $d_n \geq x_n \geq b_n$ for any n-D point **x** within the set created in step 2.
(6) Interactively repeat step 1 for however many hyperblocks are needed.

2.4 *Rules from Hyperblocks for Linear Discriminant Function*

An opposite way to interpret Linear Discriminant Functions is building interpretable rules by constructing hyperblocks *independently* from a LDF and then matching those rules with the LDF and adapting them to the LDF. The advantage of this approach is in potentially building hyperblocks with higher accuracy than the LDF. If this happens then there is no reason to use the LDF with lower accuracy and interpretability than the rules from HBs.

If the result is *mixed*, i.e., for some cases the LDF has advantages in accuracy but some hyperblocks have advantages in accuracy, then there are options to integrate them to the models. When some *hyperblocks are more accurate* we obviously should use them because they are more interpretable. When the LDF is more accurate we can attempt to modify the hyperblocks to meet the accuracy of the LDF. It is always possible by building hyperblocks for each such case individually. However, it can end up with the large number of individual hyperblocks, which is obviously not desirable. In actual experiments with real data, it did not happen as we report in the case study section. Below we present three algorithms for independent hyperblock creation and one algorithm to match the independently created hyperblocks to a given LDF.

2.4.1 Hyperblock Algorithms

To create HB based rules for a given LDF, algorithms that create HBs are necessary. Below we present three algorithms used to independently create HBs: Interval Hyper (IHyper), Merger Hyper (MHyper), and Interval Merger Hyper (IMHyper).

Interval Hyper

The first algorithm is *Interval Hyper (IHyper)*. The idea of IHyper is that a hyperblock can be created by repeatedly finding the largest interval of values for some attribute x_i with a **purity** above a given threshold.

The steps for *IHyper algorithm* [32] are as follows:

(1) For each attribute x_i in a dataset, create an ascending sorted array containing all values for the attribute.
(2) *Seed* value a_1, the first value in the first sorted array and *compute* LDF $G(\mathbf{a})$ for the n-D point **a**, which corresponds to a_1. The first sorted array is an array of values of the first attribute. Note, instead of a_1 any value of any sorted array of x_i can be taken.
(3) *Initialize* $b_i = a_i = d_i$ for a_i
(4) *Create* HB for **a** such that

$$b_1 \leq a_1 \leq d_1 \;\&\; b_2 \leq a_2 \leq d_2 \;\&\ldots\&\; b_n \leq a_n \leq d_n$$

(5) Use the *next value* e_i in the same sorted array to expand the interval on the same attribute if the n-D point **e** that corresponds with e_i is either of the same class as **a** or that the interval on this attribute will remain above some purity threshold T despite adding e_i.

(6) Repeat step 4 until there are no more values left in the sorted array or adding e_i to the interval will drop it below some purity threshold T.
   a. If there are no more values left in the sorted array, save the current interval.
   b. If the interval will drop below some purity threshold, remove all values equal to e_i from the current interval and save what is left. If possible, repeat step 2 with the same attribute but use a seed value greater than e_i.
(7) For all saved intervals for attribute x_i, save the interval with the largest number of values.
(8) Repeat step 2 with the next sorted array.
(9) For all saved intervals for all attributes, save the interval with the largest number of values.
(10) Using the saved interval from step 7, create a hyperblock.
(11) Repeat step 1 with all n-D points not in a hyperblock until all n-D points are within a hyperblock or no new more intervals can be made with any attribute.

Merger Hyper

The second algorithm is *Merger Hyperblock (MHyper)* [25]. The idea for MHyper is that a hyperblock can be created by merging two overlapping hyperblocks.

The steps for the *MHyper algorithm* are as follows:

(1) Seed an initial set of *pure* HBs with a single n-D point in each of them (HBs with length equal to 0).
(2) Select a HB **x** from the set of all HBs.
(3) Start iterating over the remaining HBs. If HB_i has the same class as **x** then attempt to combine HB_i with **x** to get a pure HB.
   a. Create a joint HB from HB_i and **x** that is an envelope around HB_i and **x** using the minimum and maximum of each attribute for HB_i and **x**.
   b. Check if any other n-D point **y** belongs to the envelop of HB_i and **x.** If **y** belongs to this envelope add **y** to the joint HB.
   c. If all points **y** in the joint HB are of the same class, then remove **x** and HB_i from the set of HBs that need to be changed.
(4) Repeat step 3 for all remaining HBs that need to be changed. The result is a *full pure HB* that cannot be extended with other n-D points and continue to be pure.
(5) Repeat step 2 for n-D points that do not belong to already built full pure HBs.
(6) Define an *impurity threshold* that limits the percentage of n-D points from opposite classes allowed in a *dominant* HB.
(7) Select a HB **x** from the set of all HBs.
(8) Attempt to combine **x** with remaining HBs.
   a. Create a joint HB from HB_i and **x** that is an envelope around HB_i and **x**.
   b. Check if any other n-D point **y** belongs to the envelop of HB_i and **x.** If **y** belongs to this envelope add **y** to the joint HB.

c. Compute impurity of the HB_i (the percentage of n-D points from opposite classes introduced by the combination of **x** with HB_i.)
d. Find HB_i with lowest impurity. If this lowest impurity is below predefined impurity threshold create a joint HB.

(9) Repeat step 7 until all combinations are made.

Interval Merger Hyper

The third algorithm is *Interval Merger Hyper (IMHyper)*. The idea for IMHyper is to combine the IHyper and MHyper algorithms. The steps for the IMHyper algorithm are as follows:

(1) Run the IHyper algorithm.
(2) Create a set of any n-D points not within the HBs created in step 1.
(3) Run the MHyper algorithm on the set created in step 2 but add the HBs created in step 1 of this algorithm to the set of pure HBs created in step 1 of the MHyper algorithm.

2.4.2 Hyperblock Rules for Linear Discriminant Function

Below we present an algorithm denoted as *GLC Hyperblock Rules Linear (GLC-HBRL)*. This algorithm constructs hyperblocks for a given LDF, which has *better accuracy than exactly replicating the LDF*.

The steps for the *GLC-HBRL algorithm* are as follows:

(1) Run the IHyper algorithm, but, on step 8, before adding some HB **x** to an interval, confirm that the classification of the HB within the interval matches that of a given LDF. If the interval and LDF classifications do not match, run step 8a of the IHyper algorithm.
(2) Create a set of any n-D points not within the HBs created in step 1.
(3) Run the MHyper algorithm on the set created in step 2 but add the HBs created in step 1 of this algorithm to the set of pure HBs created in step 1 of the MHyper algorithm. For the MHyper algorithm, only join HBs if all points classified by the joined HB match all LDF classifications. Note, this step ensures that the resulting HBs will at least as accurate as LDF.

This algorithm does not allow hyperblocks to misclassify cases LDF classified correctly. It alters the algorithm from [25] and takes into account already created HBs. It creates HBs by combining [25] algorithm and the interval algorithm. As can be seen in Fig. 5a HB created using this algorithm misclassifies the same case as the LDF. This can be seen on the lower graph of Fig. 5 in which the misclassified cases of the HB are shown. This means we can explain the behavior of LDF using hyperblocks, including its misclassifications.

Although, it should be noted that because of the combination of the two algorithms, the number of cases within the blocks will often exceed that of the dataset. For

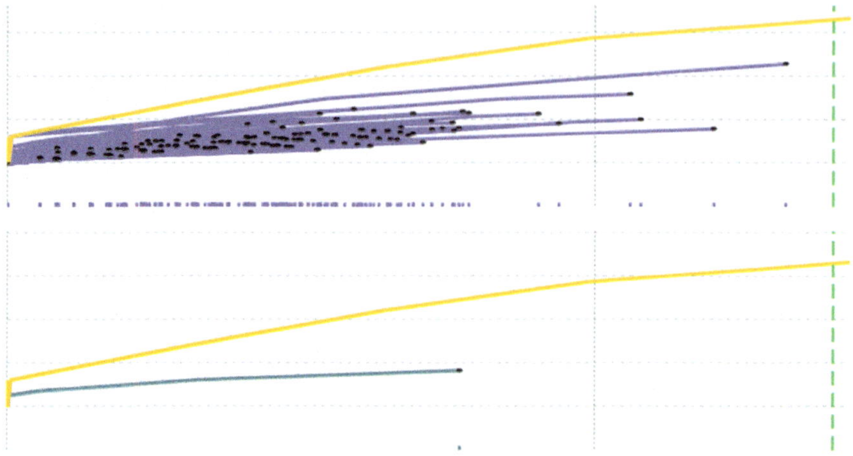

Fig. 5 Hyperblock containing 345 cases and an accuracy of 99.71%. The one misclassified case in this hyperblock is also misclassified by the LDF. Dotted green line—class discrimination line from LDF

instance, the Wisconsin Breast Cancer dataset has 683 valid cases, but the set of blocks which contain the hyperblock from Fig. 5 contain 732 cases.

2.5 Model Worst-Case Validation

While many ML models are powerful, understanding their interpretability weaknesses is crucial for high-risk classification tasks. One such interpretability approach that requires attention is getting a surrogate linear model like LIME.

If such surrogate model has a high accuracy, then it is considered as a good explanation for the original black box model. We critically analyzed this approach in Sect. 1.2 showing that it is not sufficient to claim that we get a good explanation in this way. Now we will consider a high accuracy requirement deeper.

A common way to evaluate the accuracy of the model is using k-fold cross validation [33] with $k = 10$ commonly used. In fact, tenfold cross validation builds 10 models with different data splits to training and validation and averages the accuracy of these 10 models. If that average accuracy is high enough and its standard deviation is small, then it is claimed that we have a high accuracy. Then one of those models is recommended to be used. It is either a model which has the highest accuracy or close to average. For tasks where the cost of each individual error is high, like medicine, this approach can produce exaggerated expectations of the model success. Therefore, for such tasks worst-case estimates have advantages over this one. It is based on the Shannon function, which search for an algorithm and its ML model that produce the highest accuracy on the worst data split to training and validation sets. For details of this approach see [13, 33].

To help estimate the worst-case accuracy of a ML model, we have made an algorithm to find the worst-case validation split for any model visualized in GLC-L. This is done by comparing the GLC-L projections on horizontal axis U to create an upper and lower bound of the worst cases by observing the areas of heavy overlap of projections of cases from opposite classes. Several figures in the case studies in Sects. 3.7 and 3.8 show these bounds as yellow dotted vertical lines.

This algorithm is called *GLC Worst-Case Linear (GLC-WCL)* and is as follows:

(1) Find the *lower bound* of the worst-case validation split by using the projections from GLC-L to locate the leftmost misclassified n-D point. If no point is misclassified set the lower bound to the threshold *T*.
(2) Find the *upper bound* of the worst-case validation split by using the projections from GLC-L to locate the rightmost misclassified n-D point. If no point is misclassified set the upper bound to the threshold *T*.
(3) If the range between the upper and lower bounds exceeds 90% of the total range of GLC-L projections, then find another worst-case range that will be not greater than 90% of the total range by excluding most extreme projection points to reach 90%. A user can change 90% to another value for the task at hand.
(4) Store all given n-D points with GLC-L projections between the upper and lower bounds as a worst-case validation split.

Note that this algorithm assumes the most complete worst-case validation set contains all data in the area where cases from opposite classes overlap.

Another algorithm called *Worst-Case estimates with Dynamic Scaffold Coordinates based on Shifted Paired Coordinates (WC-DSC2)* is presented below. It is a fully interactive process using Dynamic Scaffolding Coordinates based on Shifted Paired Coordinates (DSC2) illustrated in Fig. 6. This interactive process consists of finding areas in the DSC2 visualization where the cases of alternative classes are most heavily overlapped.

The *DSC2 graph construction algorithm* is as follows:

(1) Set up dataset sample coordinates in the same manner as a SPC plot.
(2) Create a scaffold from the origin to the attribute-pair point for each attribute-pair and for all samples.
(3) The first attribute-pair scaffold position is left untouched; however, the tail of the first scaffold is removed, making the tips of the first attribute-pair the "origin" of the polyline.
(4) Translate the remaining scaffolds, to the tip of the preceding scaffold.

Figure 7 shows multiple n-D cases visualized in DSC2 and a worst-case set created using WC-DSC2. This algorithm uses interactively created hyperblocks to identify n-D cases most likely to be misclassified. For Fig. 7, two attributes were created with Principal Component Analysis (PCA) and appended to the WBC dataset as the first two attributes. Then, graphically linear scalars of 150% were placed on the PCA attributes as they are they are most separating attributes (attributes of interest). The remaining attributes were then scaled to 5% each. 50 samples were then interactively selected with hyperblocks to create a worst-case validation set. A comparison between both methods can be found in Sect. 3.7 for the case study of GLC-WCL with the WBC dataset.

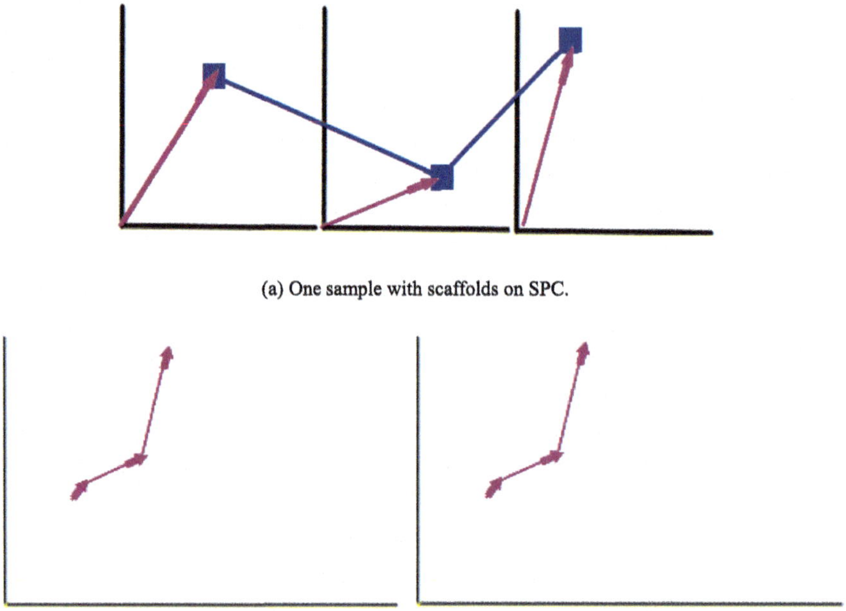

(a) One sample with scaffolds on SPC.

(b) Connecting the scaffolds.

(c) Removing the first scaffold.

Fig. 6 Visual steps for construction of the DSC2 plot

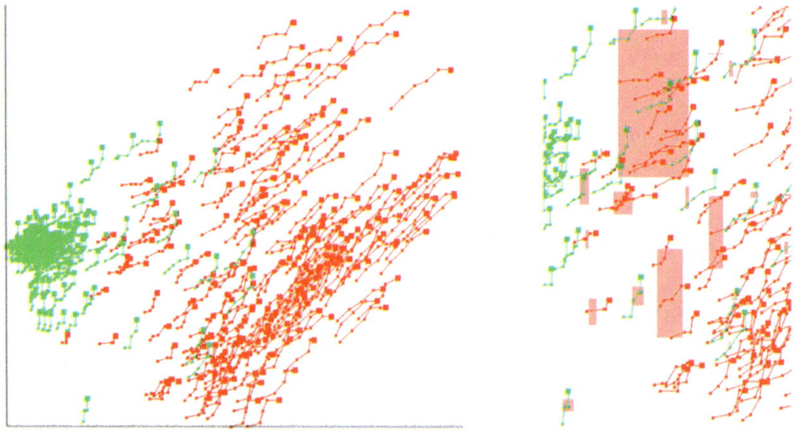

(a) WBC dataset with PCA attributes.

(b) Zoomed hyperblock selections for worst-case set for (a)

Fig. 7 Finding regions of heavy overlap in the WBC dataset

2.6 Automatic Separating Hyperblocks by Scaling Attribute

We have an interesting visualization challenge in General Line Coordinates. Two hyperblocks, which do not overlap in n-D space can overlap in 2-D GLC space including GLC-L [13]. The task is to develop a method to make them non-overlapping in GLC.

Below we described the algorithm for this. The main idea of this algorithm is first to identify an attribute X_i where those hyperblocks HB1 and HB2 to do not overlap. The existence of such attribute is the mathematical condition of non-overlap. In [13] it is done by selecting that attribute X_i as the first attribute to be visualized in the Dynamic Scaffold Coordinates based on Parallel Coordinates (DSC1).

The DSC1 graph construction algorithm (Fig. 8) is as follows:

(1) Set up dataset sample coordinates in the same manner as a PC plot.
(2) Apply a rotation transformation for each individual attribute axis with pre-defined angles.
(3) Create a scaffold from the origin to the attribute point for each attribute and for all samples.
(4) The first attribute scaffold position is left untouched; however, the tail of the first attribute scaffold is removed, making the tips of the first attribute the "origin" of the polyline.
(5) Translate the remaining scaffolds to the tip of the preceding scaffold.

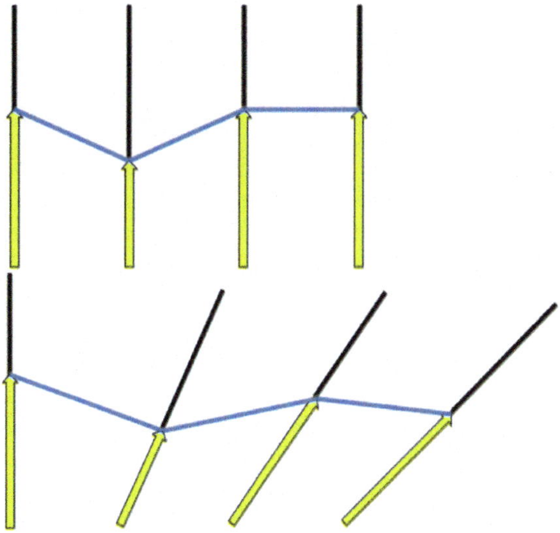

(a) Sample on parallel coordinates with and without rotation.

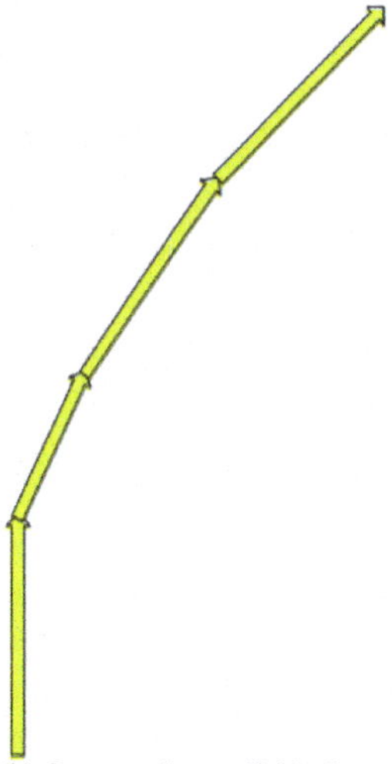

(b) Connect the scaffolds from tip-to-tail after rotation.

Fig. 8 Visual steps for construction of the DSC1 plot

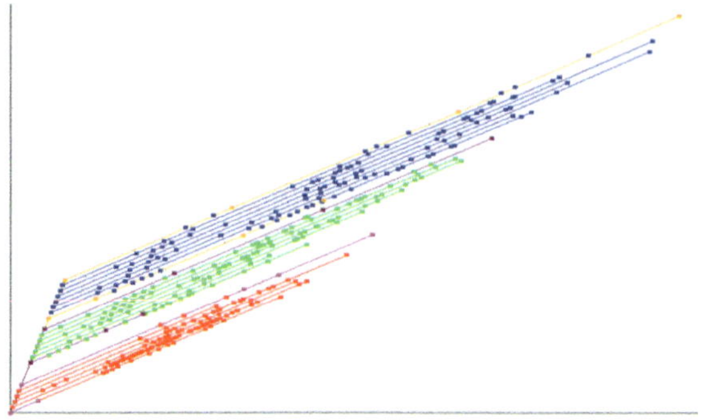

Fig. 9 Three non-overlapping pure HBs from the Iris dataset on DSC1

For DSC1, non-overlapping hyperblocks are guaranteed to be separated on at least one attribute, which is referred to as the *attribute of separation*. This attribute of separation is placed first in the order of attributes and given the steepest angle to emphasize its importance. In the case of multiple attributes of separation between any two hyperblocks only one is chosen randomly from them if no additional information is provided. The order for the remaining attributes does not matter in the DSC1. In Fig. 8 the scaffold tips are shown to retain all information of the sample.

Figure 9 demonstrates three hyperblocks in DSC1 with the Iris dataset where the red, green, and blue lines each represent cases from different classes in the dataset. The three hyperblocks are all separated on the fourth attribute. The fourth attribute was given the first spot in the attribute order followed by attributes 2, 3, and 1. Only one DSC1 plot is required to demonstrate these three hyperblocks as they share the same attribute of separation. DSC1 is an excellent tool for the Iris dataset as 140 samples can be separated with only the petal width attribute. However, separating the remaining 10 samples (not shown in Fig. 9) requires a DSC1 series with three attributes of separation.

The angles in the DSC1 graph construction algorithm are chosen to visually show separation of classes that separate on one attribute known as the attribute of separation. The attribute of separation is placed first in the order of attributes and given the steepest angle to emphasize its importance and the order for the remaining attributes sharing the same angle however the possibility exists to change the angle of each attribute as shown in Fig. 10.

Unfortunately, the technique shown in Fig. 10 does not work consistently when separating hyperblocks. Overlaps can happen because DSC1 and GLC-L visualizations rely on horizontal separation and do not consider vertical overlap. To get around this problem, another technique known as non-linear scaling can be applied to an artificial attribute projected vertically in DSC1 or GLC-L.

Figure 11a demonstrates how non-linear scaling can be applied on the first attribute-pair of the Iris dataset. The classes are pushed in the direction of the corresponding color arrows. The Virginica class (blue) is pushed up because it is above the black horizontal line and the Versicolor class (green) and Setosa class (red) are pushed down as they are below the black horizontal line. The red class is pushed to

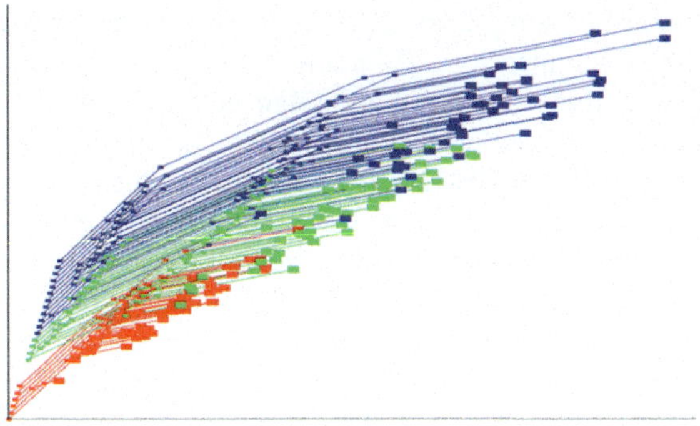

Fig. 10 DSC1 with a different rotation for each attribute

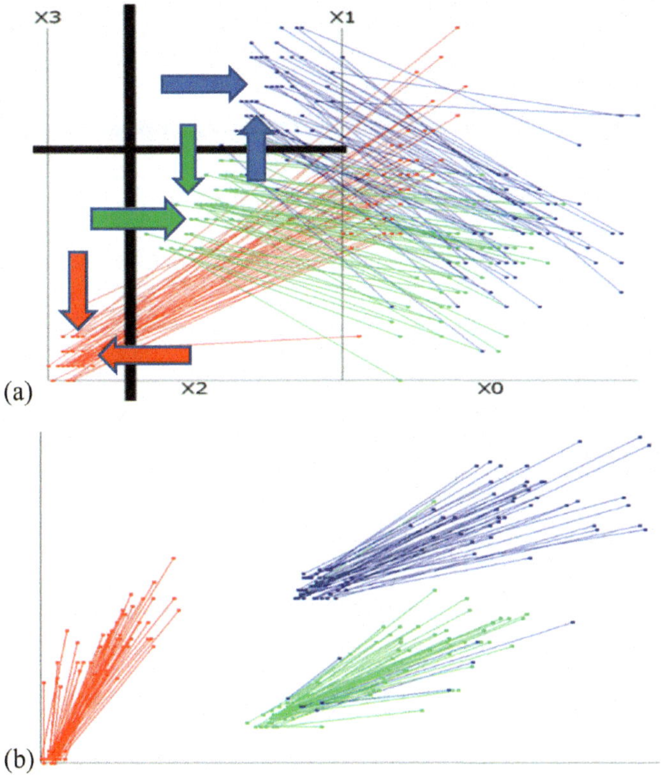

Fig. 11 Non-linear scaling technique on SPC for Iris Dataset. **a** Before scaling. **b** After scaling

the left because it is on the left side of the black vertical line whilst the green and blue classes are pushed right as they are right of the black vertical line. Figure 11b shows the same data with non-linear scaling applied.

This technique can be applied to GLC-L by picking up that attribute x_i and duplicate it to create (n + 1)-dimensional point **h** from the original n-D point **x**, $\mathbf{h} = (x_i, \mathbf{x})$ with x_i as the first attribute of h. Next, we assign an angle of 90° to x_i in GLC-L.

This is equivalent to its coefficient equaling zero and means it will not make any contribution to the value of LDF. However, we will be able to exploit this attribute to separate HBs. Since the intervals for HB1 and HB2 in x_i do not overlap (one is above another one) and the contribution of x_i to LDF is *zero*, we can exaggerate the distance between these intervals by disproportional scaling [34]. If needed, this will make the difference between HBs more visible.

An example of this is Fig. 12 which shows two HBs separated in n-D space, but heavily visually overlapped in all attributes but x_1.

Similarly, Fig. 13 shows the same two HBs even more heavily overlapped in 2-D GLC-L space. Applying the algorithm described above we get Fig. 14. As can be seen, disproportional scaling allows for the complete separation of both HBs without effecting the LDF classification. While the actual HBs in Fig. 14 occupy more space than in Fig. 13, we represent them zoomed out for comparison with Fig. 13.

The detailed steps of this algorithm are as follows.

(1) For two HBs which do not overlap in n-D space, find all attributes x_i where the HBs do not overlap. Such attributes will be denoted as *separating attributes*.
(2) For both HBs, calculate the height of upper and lower bounds of these HBs to be visualized in GLC-L.
(3) For both HBs, sum the values of the separating attributes for the upper and lower bounds.
(4) Calculate a scaling value.

    a. Find the HB with the greater sum of separating attributes for the upper bound. This HB will be denoted as HB1. The other HB will be denoted as HB2.
    b. Calculate the scaling value by subtracting the GLC-L height of the upper bound of HB2 by the GLC-L height of the lower bound for HB1.
    c. Divide the difference found in (b) by the sum of separating attributes for the lower bound of HB1.

(5) When running the GLC-L algorithm, duplicate the values of the separating attributes and multiply the duplicated values by the scaling value to create a new value h_i.
(6) Draw values h_i at a 90° angle for all n-D points within HB1 before any other values.

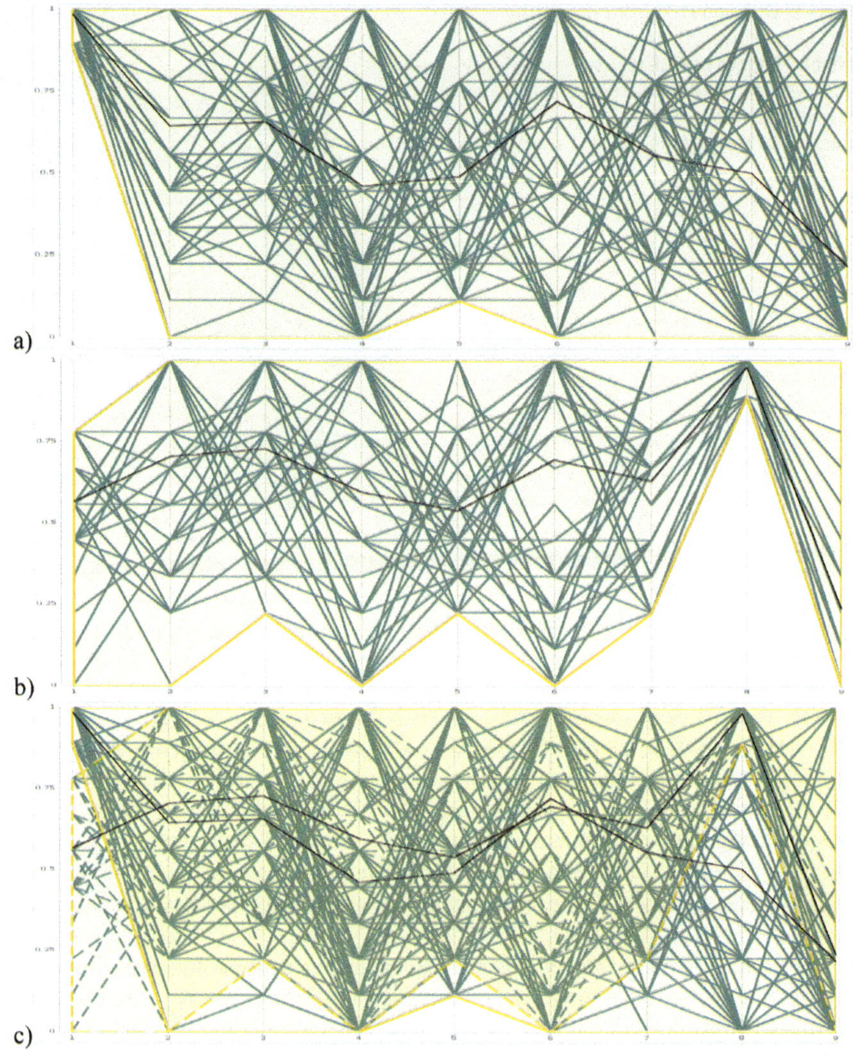

Fig. 12 Two HBs separated in n-D space but visually overlapped in parallel coordinates, where **a** shows HB1. **b** shows HB2, and **c** shows **a** and **b** combined with separation visible only in attribute x_1

3 Case Studies

Below are results from 8 case studies for the GLC-nL, GLC-IRL, GLC-HBRL and GLC-WCL algorithms. These case studies use the Iris, Ionosphere, Wisconsin Breast Cancer (WBC), and Seeds datasets from the UCI Machine Learning Repository. The Iris dataset has four attributes and three classes: setosa, versicolor, and virginica. Each

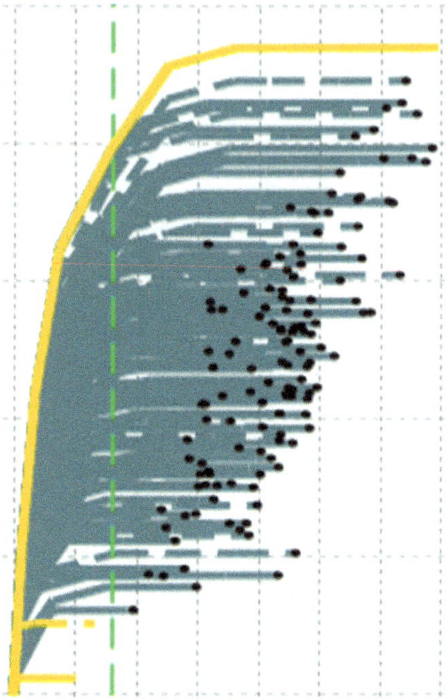

Fig. 13 HB1 and HB2 from Fig. 12 heavily overlapped when visualized in GLC-L space. Yellow solid lines are edges of HB1 and yeallow dotted lines are edges of HB2

class has 50 cases for a total of 150 cases. The Ionosphere dataset has 34 attributes and two classes: good and bad. The good class has 126 cases, and the bad class has 225 cases for a total of 381 cases. The WBC dataset has nine attributes and two classes: benign and malignant. The benign class has 444 cases, and the malignant class has 239 cases for a total of 683 cases. The Seeds dataset has seven attributes and three classes. Each class corresponds with a different type of wheat and has 70 cases for a total of 210 cases.

For each case study, accuracy is used as the primary means for assessing model quality. This is because accuracy has a direct meaning as a general quality of a model in contrast with other metrics such as F-score which is a combination of other metrics.

The first two case studies discuss GLC-nL and use the Iris and Ionosphere datasets. In both cases using GLC-L to create a LDF resulted in poor accuracy due to poor linear separation. But, by using GLC-nL, the performance was able to be improved. For the third and fourth case studies, GLC-IRL was discussed using the WBC and Seeds datasets. In both cases, highly interpretable and interactive hyperblocks were able to be created from the original GLC-L visualizations. In the fifth and sixth case studies, GLC-HBRL was discussed using the WBC and Ionosphere datasets. In both cases the LDF created by the initial GLC-L visualizations was able to be automatically improved on by a set of highly interpretable hyperblocks. Although, the increase of accuracy is mainly due to a significant set of hyperblocks containing only an individual case. Most of these types of hyperblocks contain cases that were originally misclassified by the LDF and were secondly singled out by the hyperblock

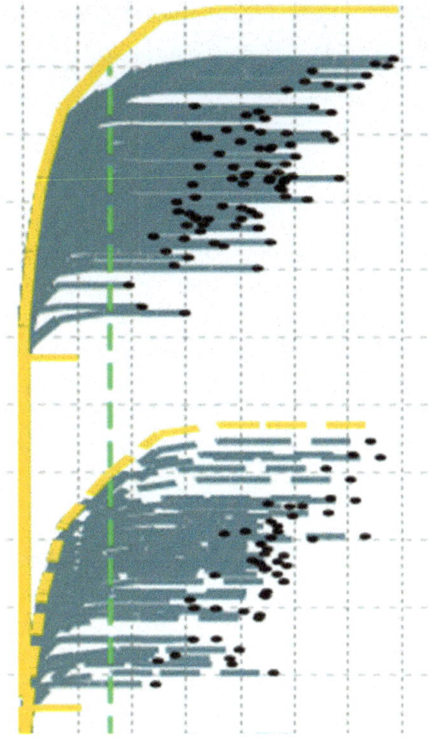

Fig. 14 HBs from Fig. 12 fully separated in GLC-L space

algorithm. Then, for the last two case studies, GLC-WCL was discussed using the WBC and Ionosphere datasets. In both cases GLC-WCL was able to create a worst-case set with significantly lower accuracy than the set with all data. These worst-case sets performed similarly across multiple ML models.

3.1 GLC-nL with Iris Dataset

For the first case study, we present results with the Iris dataset. For this study we combined the setosa and virginica classes into one super class named "combined class." We do this because of software limitations during classification problems with more than two classes on a single dimension linear separation threshold.

In this case, the setosa and virginica classes were combined because linear classification methods provide poor results when classifying the versicolor class and the combined class. For instance, the average accuracy for a tenfold cross validation is 70.67% for Linear Discriminant Analysis (LDA) and 64.67% for Logistic Regression (LR). See Table 1.

Table 1 10-fold cross validation (2 class Iris data)

Model	Fold 1 (%)	Fold 2(%)	Fold 3 (%)	Fold 4 (%)	Fold 5 (%)	Fold 6 (%)	Fold 7 (%)	Fold 8 (%)	Fold 9 (%)	Fold 10 (%)	Avg. (%)
DT	100	93.33	100	100	93.33	86.67	86.67	93.33	93.33	100	94.67
SGD	53.33	66.67	80	73.33	86.67	53.33	60	66.67	73.33	33.33	64.67
NB	86.67	93.33	93.33	100.00	93.33	86.67	93.33	86.67	73.33	93.33	90
SVM	100	100	100	100	93.33	86.67	86.67	100	86.67	100	95.33
KNN	100	100	100	100	93.33	86.67	86.67	100	86.67	100	95.33
LR	73.33	66.67	80	73.33	60	53.33	60	66.67	53.33	60	64.67
LDA	73.33	86.67	86.67	86.67	73.33	40	66.67	60	73.33	60	70.67
MLP	73.33	66.67	80	66.67	66.67	66.67	66.67	73.33	60	86.67	70.67
RF	100	100	100	100	93.33	86.67	86.67	93.33	80	100	94
Metrics											
Avg	84.44	85.93	91.11	88.89	83.70	71.85	77.04	82.22	75.56	81.48	82.22
St. Dev	17.00	15.07	9.43	14.14	13.38	18.79	13.38	15.63	12.91	24.44	15.42

Similarly, one of the LDF versions implemented in GLC-L produced 79.33% accuracy (see Fig. 15). Table 1 also shows that LDA accuracy varies widely between folds from 40% accuracy to 86.67%. These 40% serve as an *estimate of the worst-case accuracy* of LDA on Iris data. In contrast for SVM and Decision Tree (DT), this estimate of the worst-case accuracy is 86.67%, which is best-case estimate for LDA on these data.

In general, for non-linear classification methods the results are much better. For instance, the average accuracy for a tenfold cross validation is 95.33% for Support Vector Machines (SVM) and 95.33% for K-Nearest Neighbor (k-NN). See Table 1. The same applies for our GLC-nL algorithm with a radial basis function kernel which achieved an accuracy of 96% as Fig. 16 shows.

GLC-L outputs like Fig. 15 allow a user to analyze a poor and good performance of the model. It includes identifying and observing misclassified cases, marked with the red dots. It also allows finding and observing the most confidently classed cases of the classes (see black ovals for the violet class). The yellow vertical dotted lines outline the overlap areas selected by the user for the further analysis. Analysis of the angles and their values allow a user to see most contributing attributes. The segments of the polyline that are more horizontal correspond to an attribute with a larger contribution to LDF.

In GLC-nL, a user can modify already produced automatically high-quality rules to meet user's needs by dragging the threshold shown as a dotted green vertical line in Fig. 15. The analytic output shown in Table 1 allows a user comparing accuracy of different automatic ML algorithms and to see where the major differences between the very good and relatively low quality of the results are located.

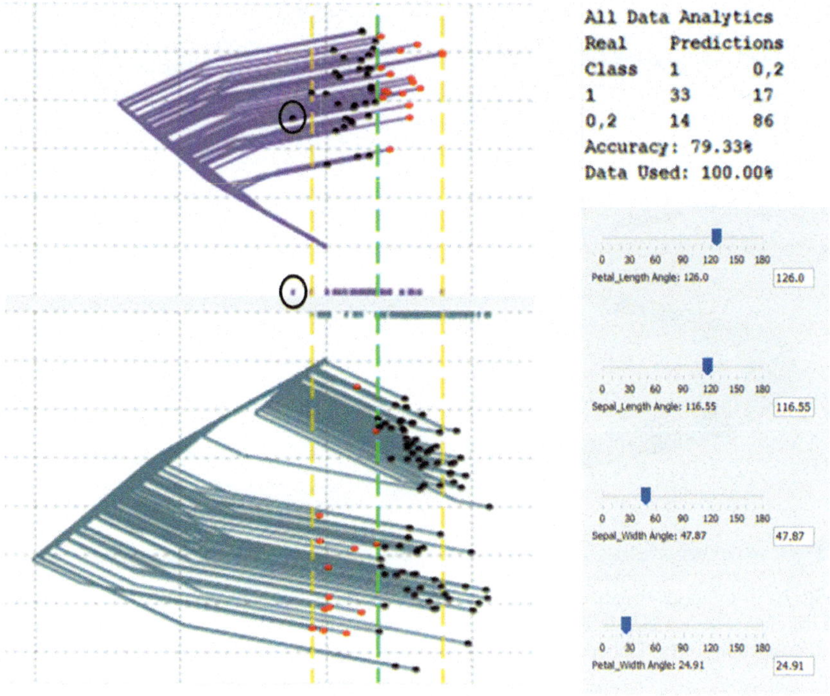

(a) Iris classes setosa and virginica (green), class versicolor (violet) in GLC-L. Dotted green line – class discrimination line. Dotted yellow lines – bounds for worst-case validation split.

(b) Analytics and angles.

Fig. 15 Iris dataset visualized with 79.33% accuracy with GLC-L

3.2 GLC-nL with Ionosphere Dataset

For the second case study, we present the results with the Ionosphere dataset. Normally, linear classification methods provide poor results when classifying the two classes in this dataset. For instance, the average accuracy for a tenfold cross validation is 85.77% for Linear Discriminant Analysis and 87.19% for Logistic Regression. See Table 2.

While slightly improved, an LDF for GLC-L performs similarly to other linear models with an accuracy of 91.74% (see Fig. 17).

For non-linear classification methods the results are much better. For instance, the average accuracy for tenfold cross validation is 93.43% for Random Forest (RF) and 93.44% Support Vector Machines (SVM). See Table 2. The same applies for the GLC-nL algorithm with a polynomial kernel, which gives accuracies up to 96.87% (see Fig. 18).

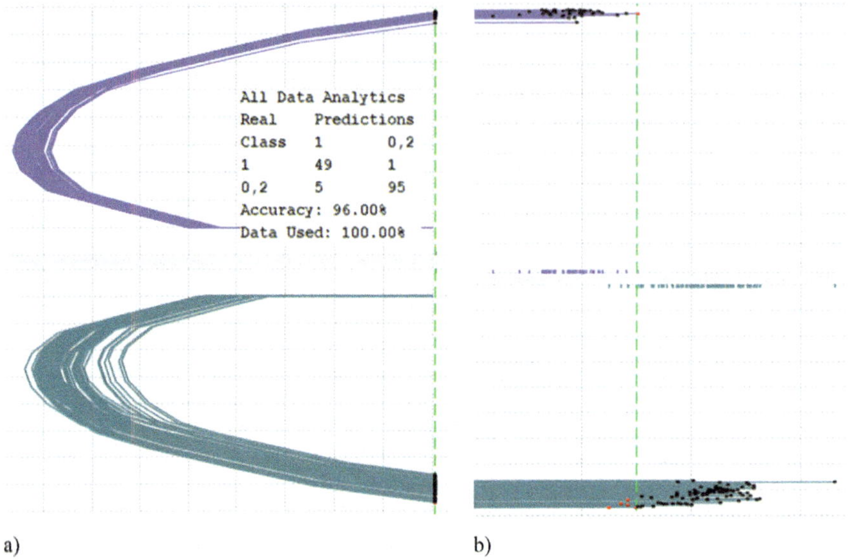

Fig. 16 Iris dataset visualized with 96% accuracy with GLC-nL with a radial basis function kernel. **a** Iris classes setosa and virginica (green), class versicolor (violet) in GLC-nL and analytics. **b** GLC-nL projections of (**a**) Dotted green line—class discrimination line

Table 2 10-fold cross validation (ionosphere data)

Model	Fold 1 (%)	Fold 2 (%)	Fold 3 (%)	Fold 4 (%)	Fold 5 (%)	Fold 6 (%)	Fold 7 (%)	Fold 8 (%)	Fold 9 (%)	Fold 10 (%)	Avg. (%)
DT	88.89	80	91.43	91.43	88.57	85.71	82.86	91.43	94.29	88.57	88.32
SGD	83.30	88.57	88.57	77.14	88.57	80	85.71	100	97.14	88.57	87.76
NB	86.11	88.57	94.29	82.86	82.86	85.71	80	97.14	94.29	85.71	87.75
SVM	94.44	94.29	97.14	85.71	91.43	85.71	97.14	100	100	88.57	93.44
KNN	83.33	85.71	88.57	71.43	77.14	77.14	91.43	88.57	91.43	82.86	83.76
LR	83.33	88.57	88.57	80	88.57	77.14	91.43	97.14	94.29	82.86	87.19
LDA	80.56	85.71	88.57	77.14	88.57	77.14	88.57	97.14	88.57	85.71	85.77
MLP	86.11	91.43	91.43	82.86	94.29	82.86	97.14	100	91.43	94.29	91.18
RF	100	88.57	94.29	82.86	88.57	88.57	100	100	100	91.43	93.43
Metrics											
Avg	87.35	87.94	91.43	81.27	87.62	82.22	90.48	96.83	94.60	87.62	88.73
St. Dev	6.23	3.98	3.19	5.73	4.95	4.47	6.85	4.15	3.90	3.78	4.72

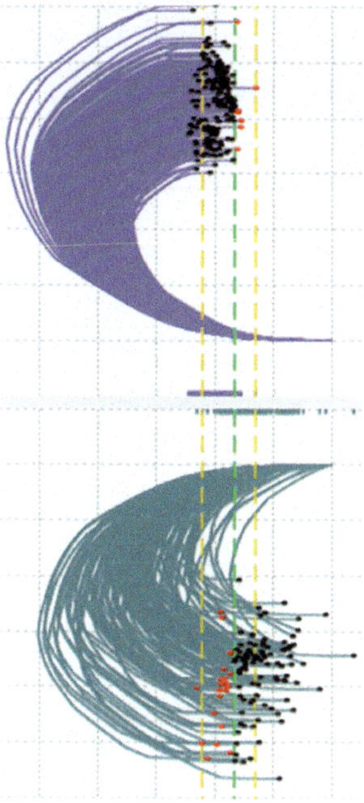

(a) Ionosphere classes: bad (green), good (violet) in GLC-L.
Dotted green line – class discrimination line.
Dotted yellow lines – bounds for worst-case validation split.

All Data Analytics

Real Class	Predictions 0	1
0	218	7
1	22	104

Accuracy: 91.74%
Data Used: 100.00%

(b) Analytics.

Fig. 17 Ionosphere dataset visualized with 91.74% accuracy with GLC-L

3.3 GLC-IRL with Wisconsin Breast Cancer Dataset

For the third case study, we present the results with the WBC dataset. The benign class is visualized in purple, and the malignant class is visualized in cyan. Figure 19 shows how three classification areas can be created with the GLC-IRL algorithm. These three areas classify all 683 cases with accuracies of 95.24, 89.83, and 100%.

While not necessary as accurate as more analytical methods, the GLC-IRL algorithm presents a domain expert a chance to create highly interpretable rules that may be very useful covering specific areas of interest.

3.4 GLC-IRL with Seeds Dataset

For the fourth case study, we present the results with the Seeds dataset. For this study, classes two and three have been combined into one superclass named "combined class." We do this because of software limitations during classification problems with more than two classes on a single dimension linear separation threshold. Class one is visualized in purple, and the combined class is visualized in cyan. Figure 20

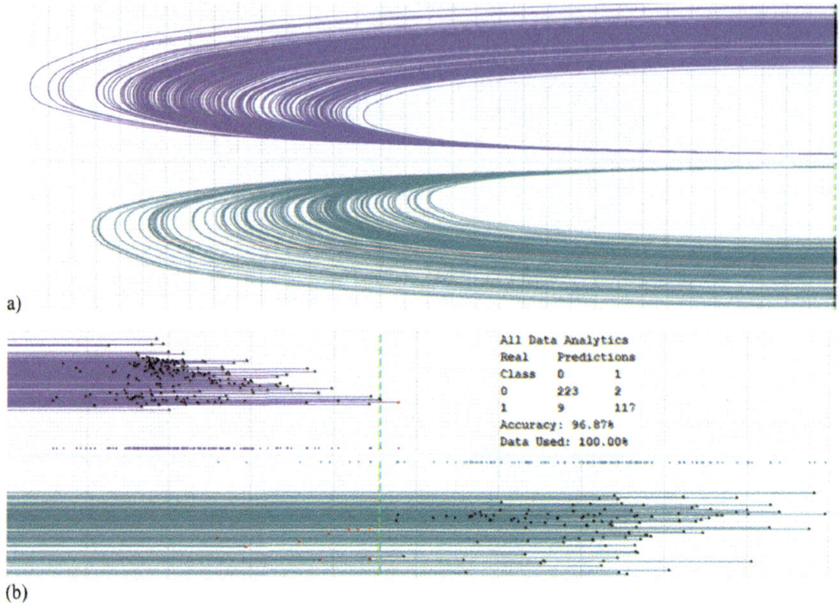

Fig. 18 Ionosphere dataset visualized with 96.87% accuracy with GLC-nL with a polynomial kernel. **a** Ionosphere class bad (green), class good (violet) in GLC-nL. **b** GLC-nL projections of (**a**) and analytics. Dotted green line—class discrimination line

shows how two classification areas can be created with the GLC-IRL algorithm. These two areas classify all 210 cases with accuracies of 95.71 and 99.27%.

3.5 GLC-HBRL with Wisconsin Breast Cancer Dataset

For the fifth case study, we present the results with the WBC dataset. In this instance, the hyperblocks created from the GLC-HBRL algorithm match up precisely with the cases classified correctly and incorrectly the GLC-L algorithm. All cases that were classified correctly by GLC-L are also classified correctly by the hyperblocks and all cases that were misclassified by GLC-L are either classified incorrectly by the hyperblocks or have been given their own hyperblock for independent analysis. The bounds of each hyperblock give a rule to which the linear discriminant function created by the original GLC-L visualization can be interpreted.

Figure 21 shows all WBC data visualized and classified with 98.1% accuracy by the visualized LDF.

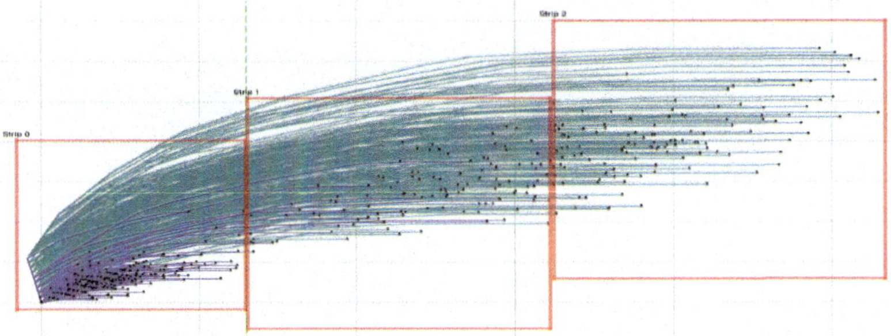

a)

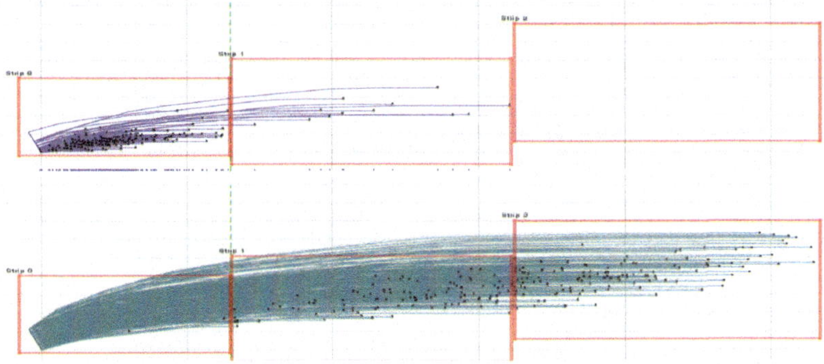

b)

Fig. 19 WBC dataset visualized with 3 generalized rules with GLC-L. Strip 0's rule has 95.24% accuracy, Strip 1's rule has 89.83% accuracy, and Strip 2's rule has 100% accuracy. **a** WBC class malignant (green), class benign (violet) in GLC-L with a combined graph. **b** WBC class malignant (green), class benign (violet) in GLC-L with separate graphs. Dotted green line—class discriminator. Red box—interactive selection box for generalized rule creation

Now our goal is to build a set of rules in the forrm of HBs that will mimic and interpret this LDF. Figure 22 shows these HBs built by the algorithm described above. In total we have 26 HBs with 13 HBs with individual n-D points. These HBs cover all 683 cases and exactly reproduce the LDF, i.e., for each HB they classify n-D points that are within that HB exactly as LDF for these n-D points both correctly and incorrecty. The first HB in Fig. 22 perfectly illustrates this situiation, where one case is misclassified by both HB and LDF.

Using both the HBs and LDF we can pick up a new case which is not a part of the existing cancer data and then we can identify a hyperblock where this case is located, and we can compute the value of the LDF for this case. This new case will then get a predictive explanation by both the HBs and LDF. Confirmation from both the HBs and GLC-L visualization will add user confidence in this predictive explanation.

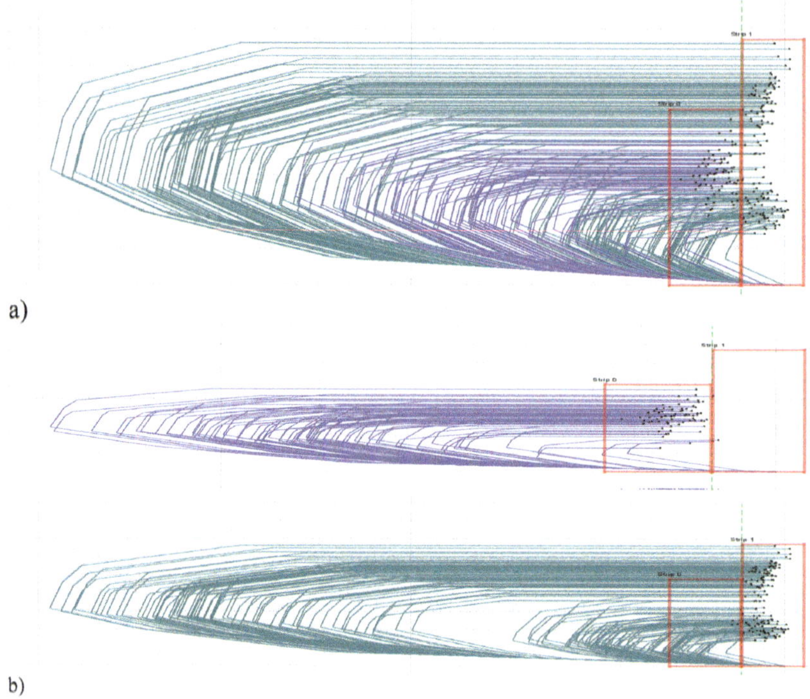

Fig. 20 Seeds dataset (2 class) visualized with 2 generalized rules with GLC-L. Strip 0's rule has 95.71% accuracy and Strip 1's rule has 99.27% accuracy. **a** Seeds class 2 and 3 (green), class 1 (violet) in GLC-L with a combined graph. **b** Seeds class 2 and 3 (green), class 1 (violet) in GLC-L with separate graphs. Dotted green line—class discriminator. Red box—interactive selection box for generalized rule creation

3.6 GLC-HBRL with Ionosphere Dataset

For the sixth case study, we present the results with the Ionosphere dataset. In this example, the hyperblocks created from the GLC-HBRL algorithm match up precisely with the cases classified correctly and incorrectly by the GLC-L algorithm. All cases that were classified correctly by GLC-L are also classified correctly by the hyperblocks and all cases that were misclassified by GLC-L are either classified incorrectly by the hyperblocks or have been given their own hyperblock for independent analysis. The bounds of each hyperblock give a rule to which the linear discriminant function created by the original GLC-L visualization can be interpreted.

Figure 23 shows all Ionosphere data visualized and classified with 90.88% accuracy by the visualized LDF. Figure 24 shows these HBs built by GLC-HBRL. In total 45 HBs were created with 27 HBs containing only an individual n-D point. These HBs cover all 224 cases and exactly reproduce the LDF. Hyperblock 1 in Fig. 24 gives an example of the HBs matching the classifications of LDF. In this hyperblock

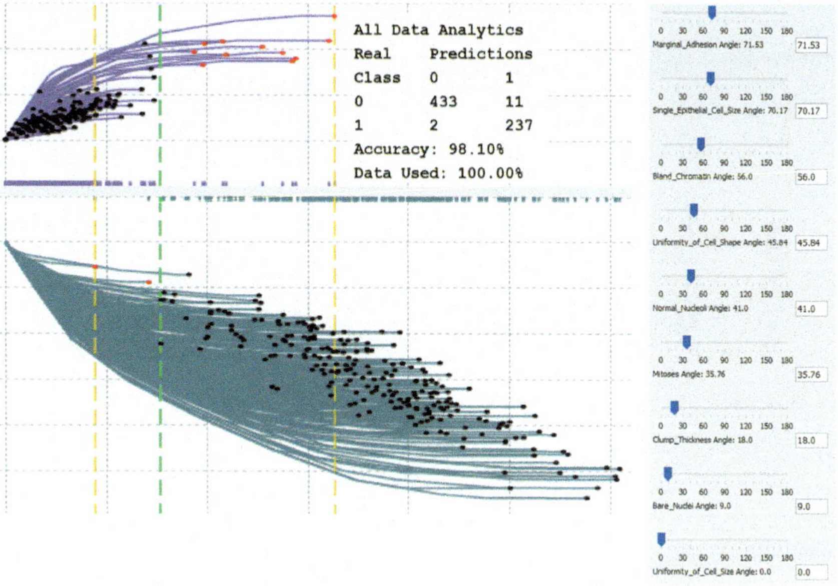

Fig. 21 Wisconsin breast cancer dataset visualized with 98.1% accuracy with GLC-L

there are 100 total n-D points and 5 misclassifications. It can be seen in (b) that all 5 misclassified cyan cases are also misclassified by LDF.

Another example of the HBs matching LDF is the 27 HBs containing only a single n-D point. In all 27 of these HBs, each n-D point is misclassified by the LDF. During the creation of the HBs these cases were particularly singled out due to their misclassifications.

3.7 *GLC-WCL with Wisconsin Breast Cancer Dataset*

For the seventh case study, we present the results with the WBC dataset. The benign class is visualized in purple, and the malignant class is visualized in cyan. Figure 25a shows this dataset visualized in GLC-L with 98.1% accuracy. Figure 25b shows the worst-case validation set for this LDF created with GLC-WCL and using 22.4% of the total data.

Using this worst-case validation set, analysis can be done on the overlapping data. This can be done by creating two additional LDF: a LDF without overlapping data and a LDF for only overlapping data. The first LDF describes how the dataset would perform when the worst-case validation set is removed. In this case, without overlapping data an LDF can be created with 100% accuracy. The second LDF describes how the dataset would perform with only the worst-case validation set. In this case, the created LDF only has 87.58% accuracy with only the overlapping data.

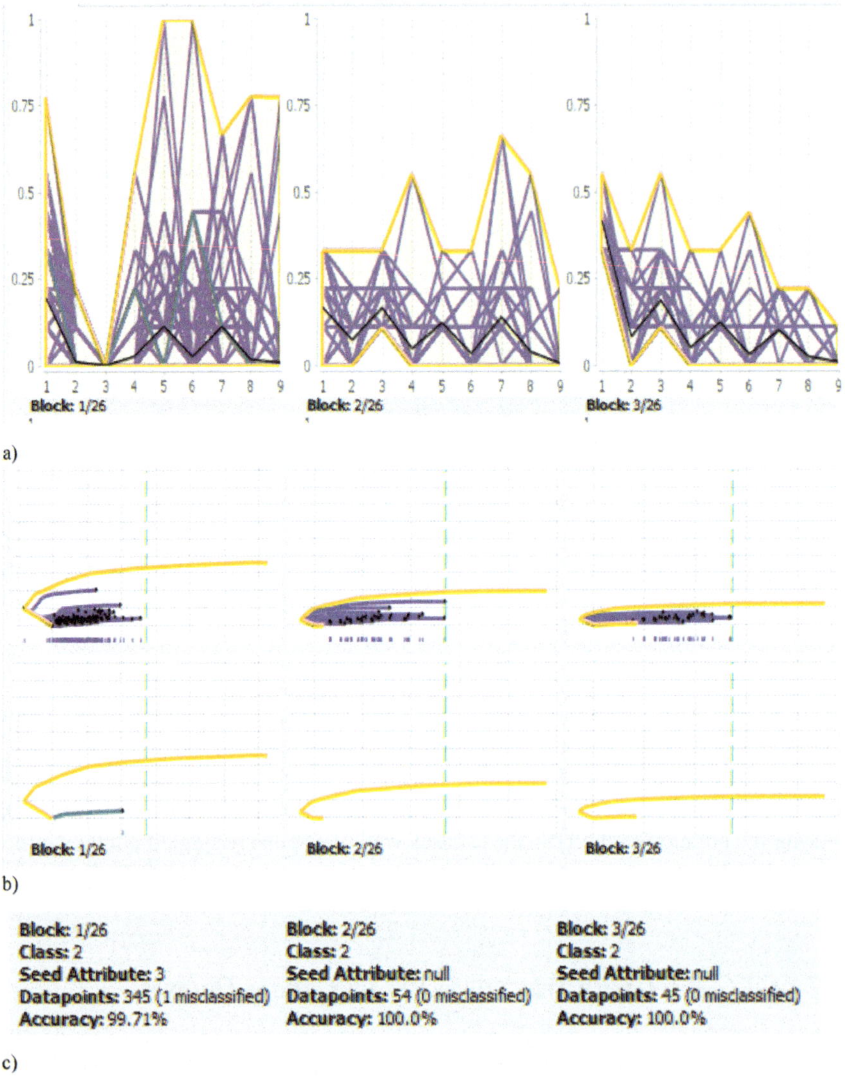

Fig. 22 First 3 HBs created with GLC-HBRL, **a** HBs visualized in parallel coordinates, **b** HBs visualized in GLC-L, **c** Analytics for each HB (see Fig. 28 in the Appendix for all HBs)

Then, a final analytic can be produced using the first LDF. By taking the LDF created without overlapping data and applying it to the worst-case validation set, we can get an estimate of the worst-case scenario for this dataset. In this case, the worst-case scenario accuracy was 79.74%.

In a high-risk classification task, knowing the worst-case scenario accuracy is of great benefit. For instance, in this case, despite having an overall accuracy of

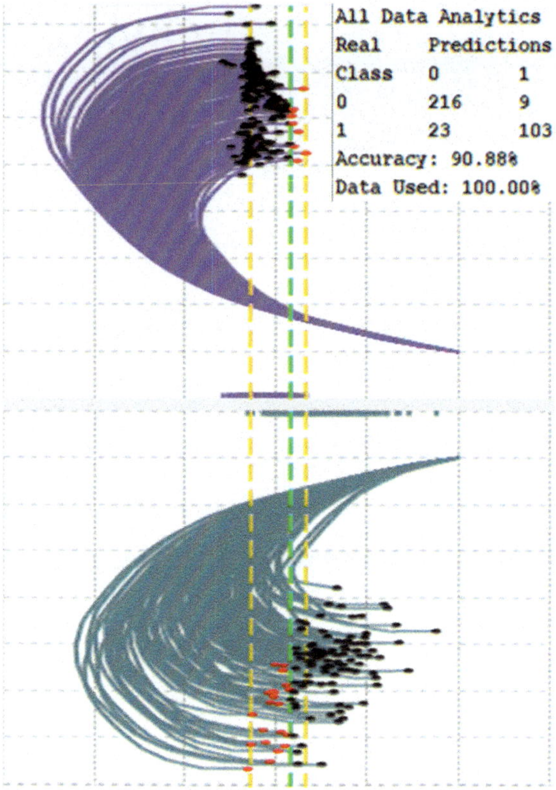

Fig. 23 Ionosphere dataset visualized with 90.88% accuracy with GLC-L

98.1%, we know that the worst-case scenario accuracy is 79.24% and the worst-case validation set contains 22.4% of the data. This means in a particularly bad scenario, 4.65% of this data will be misclassified. This reduces the accuracy from 98.1% to 95.35% and may make the model unusable if a particular accuracy is needed.

Beyond GLC-L classifications, the worst-case validation set also performs significantly worse for other ML models. Table 3 shows similar average worst-case classification accuracies for eight additional ML models: Decision Tree (DT), Support Vector Machine (SVM), Random Forest (RF), k-Nearest Neighbors (KNN), Logistic Regression (LR), Naïve Bayes (NB), Stochastic Gradient Descent (SGD), and Multilayer Perceptron (MLP). The worst performing being DT with an average of 77% accuracy and the best performing being LR and MLP with average accuracies of 85%.

Similarly, worst-case validation sets can be created with the interactive algorithm Worst-Case estimates with Dynamic Scaffold Coordinates based on Shifted Paired Coordinates (WC-DSC2). In Table 3, the worst-case validation set for WC-DSC2 was made by interactively selecting 50 cases likely to be misclassified from the WBC dataset. Unfortunately, due to poor visual separation this worst-case set did not reach its goal for validation with most ML models. The average ML model achieving 98% accuracy compared to the 82% average accuracy of the GLC-WCL worst-case set.

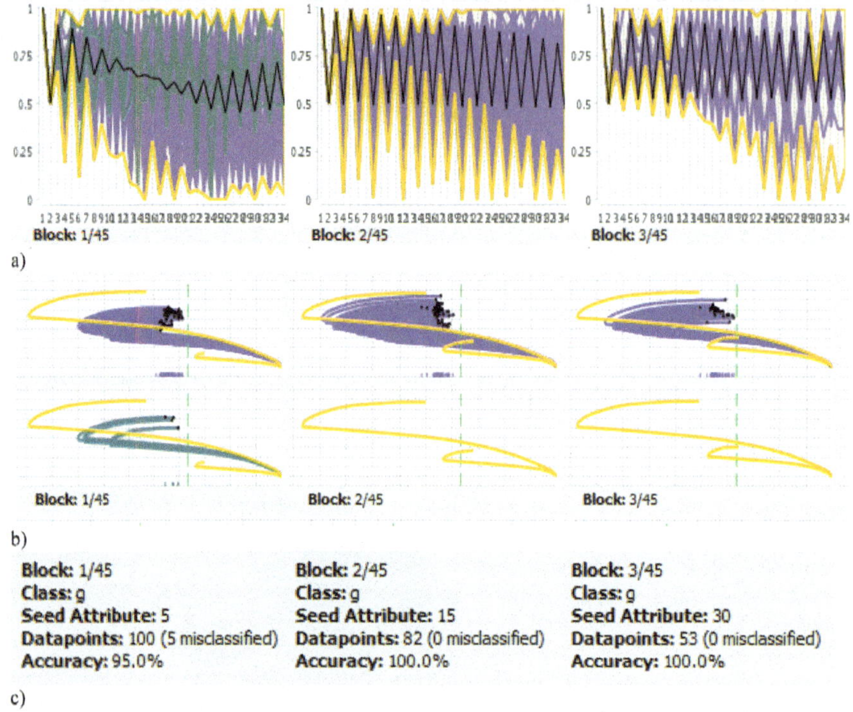

Fig. 24 First 3 HBs created with GLC-HBRL. **a** HBs visualized in parallel coordinates. **b** HBs visualized in GLC-L. **c** Analytics for each HB (see Fig. 29 in the Appendix for all HBs)

To combat this visualization issue with DSC2, techniques such as reordering attributes based on a DT, adding attributes created with Principal Component Analysis (PCA), and adding attributes created with t-Distributed Stochastic Neighbor Embedding (t-SNE) can be used to improve visualizations. In Table 3, PCA was used to create two new attributes for the WBC dataset before interactively selecting 50 cases likely to be misclassified using WC-DSC2. Using this enhanced visualization, WC-DSC2 was able to create a worst-case set with an average accuracy of 51%. The worst performing ML model being SGD with an average of 42% accuracy and the best performing ML model being NB with an average accuracy of 60%.

In this case, while WC-DSC2 can achieve a strong worst-case validation set by increasing manual effort, GLC-WCL can achieve lower worst-case estimates on the validation set completely automatically.

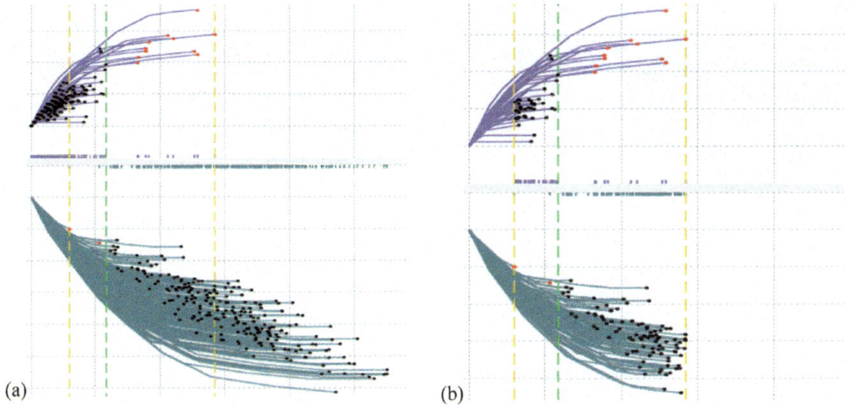


```
All Data Analytics          Data Without Overlap Analytics    Overlap Analytics          Worst Case Data Analytics
Real      Predictions       Real      Predictions             Real      Predictions      Real      Predictions
Class     0        1        Class     0        1              Class     0        1       Class     0        1
0         433      11       0         390      0              0         44       10      0         48       6
1         2        237      1         0        140            1         9        90      1         25       74
Accuracy: 98.10%            Accuracy: 100.0%                  Accuracy: 87.58%           Accuracy: 79.74%
c) Data Used: 100.00%       Data Used: 77.60%                 Data Used: 22.40%          Data Used: 22.40%
```

Fig. 25 Wisconsin Breast Cancer dataset visualized with 98.1% accuracy with GLC-L. **a** Wisconsin Breast Cancer class malignant (green), class benign (violet) in GLC-L. **b** Wisconsin Breast Cancer worst-case validation split class malignant (green), class benign (violet) in GLC-L. **c** Analytics for (**a**). Worst-case validation set using 22.4% of the dataset with 79.74% accuracy. Dotted green line—class discriminator. Dotted yellow lines—bounds for worst-case validation split

Table 3 Results for worst-case validation sets for WBC dataset on 8 ML Models

Model	Accuracy		
	GLC-WCL (%)	WC-DSC2 (%)	WC-DSC2 With PCA (%)
DT	77	100	46
SVM	84	96	52
RF	81	100	50
KNN	84	91	54
LR	85	100	56
NB	82	100	60
SGD	81	100	42
MLP	85	98	44
Avg	82	98	51

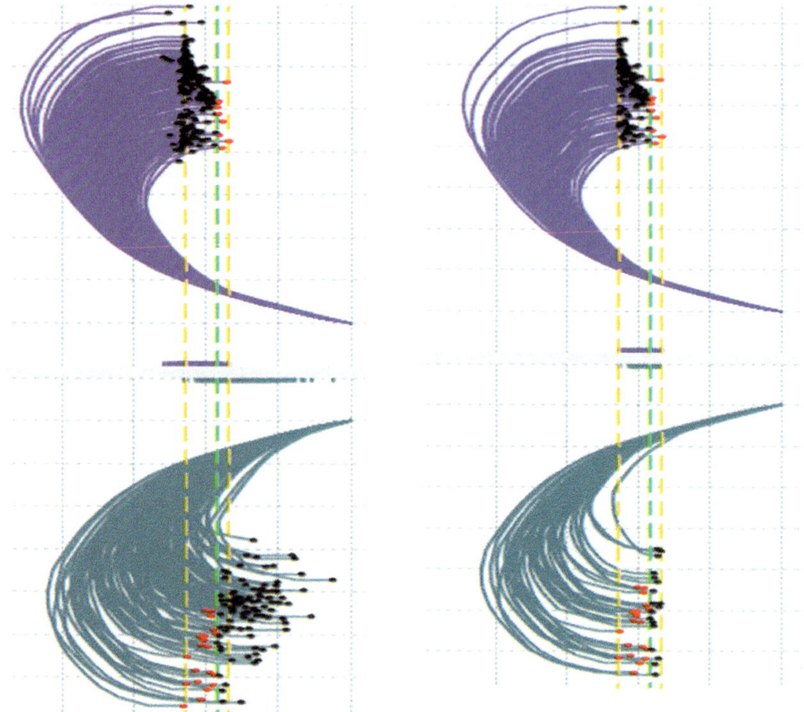

(a) Ionosphere class bad (green), class good (violet) in GLC-L.

(b) Ionosphere worst-case validation split class bad (green), class good (violet) in GLC-L

All Data Analytics

Real Class	Predictions 0	1
0	216	9
1	23	103

Accuracy: 90.88%
Data Used: 100.00%

Data Without Overlap Analytics

Real Class	Predictions 0	1
0	78	0
1	1	85

Accuracy: 99.39%
Data Used: 46.72%

Overlap Analytics

Real Class	Predictions 0	1
0	143	4
1	14	26

Accuracy: 90.37%
Data Used: 53.28%

Worst Case Data Analytics

Real Class	Predictions 0	1
0	132	15
1	25	15

Accuracy: 78.61%
Data Used: 53.28%

(c) Analytics for (a) Worst-case validation set using 53.28% of the dataset with 78.61% accuracy.

Fig. 26 Ionosphere dataset visualized with 90.88% accuracy with GLC-L. Dotted green line—class discrimination line. Dotted yellow lines—bounds for worst-case validation split

3.8 GLC-WCL with Ionosphere Dataset

For the eighth case study, we present the results with the Ionosphere dataset. The bad class is visualized in purple, and the good class is visualized in cyan in Fig. 26.

Figure 26a shows this dataset visualized in GLC-L with 90.88% accuracy. Figure 26b shows the worst-case validation set for this LDF created with GLC-WCL using 53.28% of the total data.

We can then create two more LDF with this worst-case validation set. The first LDF describes how the dataset would perform when the worst-case validation set is removed. In this case, without overlapping data an LDF can be created with 99.39% accuracy. The second LDF describes how the dataset would perform with only the worst-case validation set. In this case, the created LDF only has 90.37% accuracy with only the overlapping data.

Then, a worst-case analytic can be produced using the first LDF. By taking the LDF created without overlapping data and applying it to the worst-case validation set, we can get an estimate of the worst-case scenario for this dataset. In this case, the worst-case scenario accuracy was 78.61%.

4 Software System DV 2.0

Figure 27 illustrates the DV 2.0 software developed and used for all experimentation in this paper. There are several important labels to note: label (1) shows the location of the field where a user can change angles interactively; label (2) is where the analytics of the dataset is displayed; label (3) shows the slider locations for range, overlap, and threshold control; and label (4) indicates where the data will be visualized.

All algorithms, visualizations, and GUI were implemented in Java and JFreeChart on the Windows 10 Operating System using the DV 2.0 software. Additionally, many Machine Learning (ML) methods were used. Linear Discriminant Analysis (LDA) was used to get angles and threshold values for General Linear Coordinates Linear (GLC-L) visualizations. Support Vector Machines (SVM) was used to get the Support

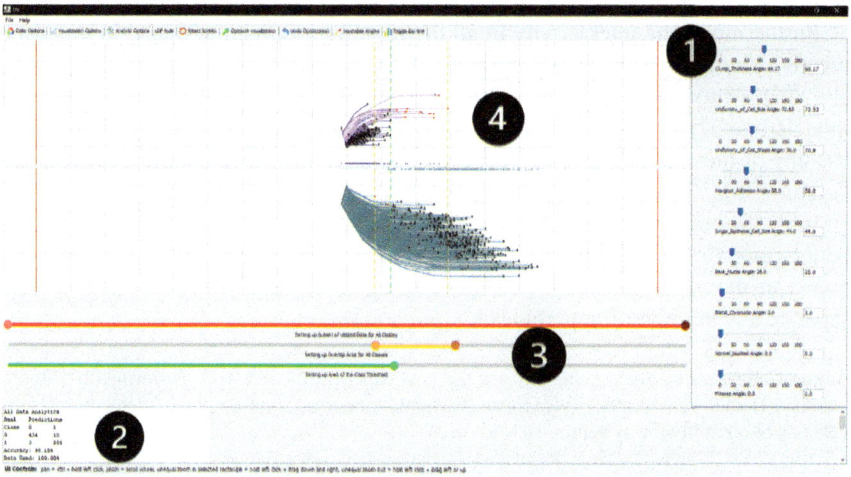

Fig. 27 DV 2.0. Label (1)—location of the field angles for user interaction. Label (2)—location of the analytics of the dataset. Label (3)—slider locations for range, overlap, and threshold control. Label (4)—data visualization location

Vectors used in the GLC non-Linear (GLC-nL) algorithm. Then, the ML methods LDA, SVM, Decision Tree, Stochastic Gradient Descent, Naïve Bayes, k-Nearest Neighbors, Linear Regression, Multilayer Perceptron, and Random Forest were all used to create k-Fold cross validation comparison tables. All these ML methods were implemented in Python using scikit-learn.

5 Conclusion

In Machine Learning understanding block-box methods on multidimensional data is a key challenge. Powerful ML methods often are unexplainable and weaker more explainable methods often perform poorly on complex data. In this chapter, a Visual Knowledge Discovery approach to General Line Coordinates (GLC) was introduced as a potential solution. Specifically, the previously introduced GLC-Linear (GLC-L) and Dynamic Scaffolding Coordinates (DSC) were expanded to produce, explain, and visualize non-linear classifiers with explainable rules. This was done through the algorithms GLC non-linear (GLC-nL), GLC Interactive Rules Linear (GLC-IRL), GLC hyperblock rules linear (GLC-HBRL), and DSC based on Parallel Coordinates (DSC1).

Additionally, GLC-L was expanded to interactively find worst-case validation splits with visual knowledge discovery algorithms. This was done through GLC worst-case linear (GLC-WCL), and DSC based on Shifted Paired Coordinates (DSC2), to ensure the accuracy and interpretability of these non-linear models and rules. In our case studies, experiments with the Iris, Ionosphere, Wisconsin Breast Cancer, and Seeds datasets showed that these visual knowledge discovery methods could compete with other ML algorithms.

Furthermore, the interactivity in all the new algorithms introduced in this paper greatly reinforces both the interpretability and accuracy of these ML models by allowing for expert input to help drive the model creation process. These new ML models majorly benefit human-guided Visual Knowledge Discovery methods. To use any of these new methods, the developed experimental software is available at GitHub [35].

In the future, further expansions to the GLC-HBRL algorithm will be needed to better generalize linear and non-linear discriminant functions. Moreover, advancements to the separation of hyperblocks in GLC-L space will be needed to fully separate any amount of hyperblocks.

Acknowledgements We are thankful for the past experiments and work performed by Morgan Leblanc, Daniel Van Houten, Tyler Swan, Fawziah Alkharnda, and Stephan Adams on previous versions of our software system.

Appendix

See Figs. 28 and 29.

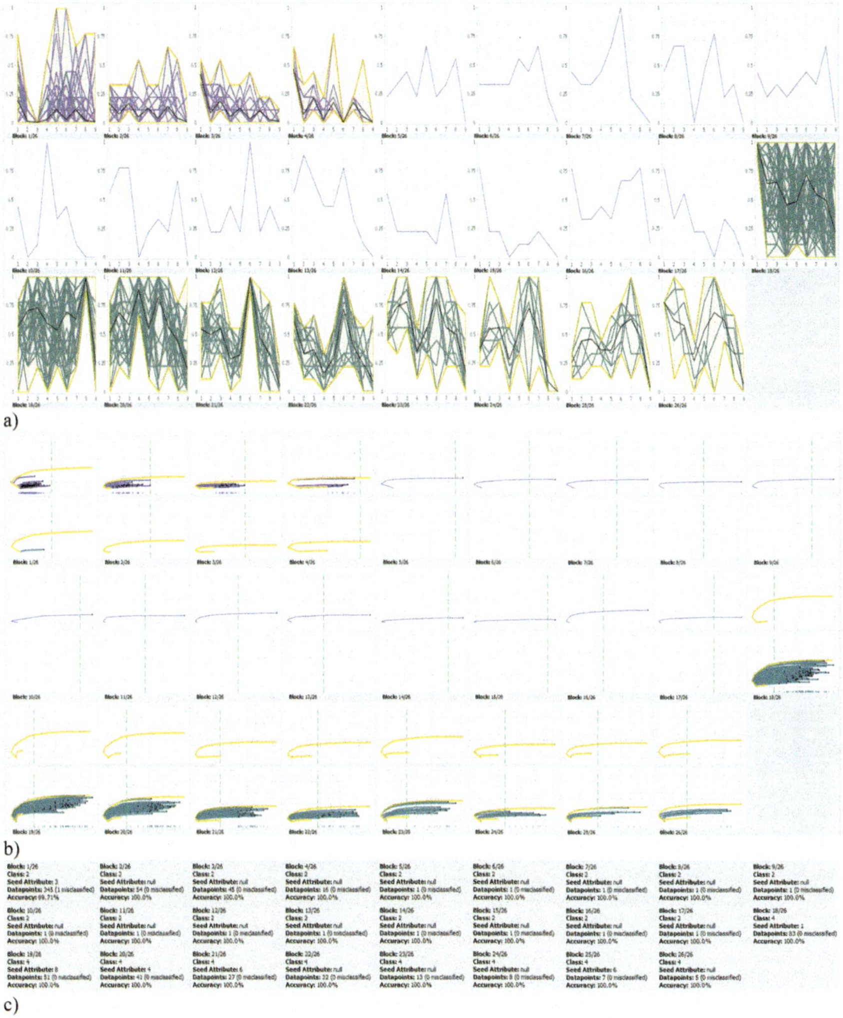

Fig. 28 HBs created with GLC-HBRL with Wisconsin Breast Cancer Dataset. **a** HBs visualized in parallel coordinates, **b** HBs visualized in GLC-L, **c** Analytics for each HB

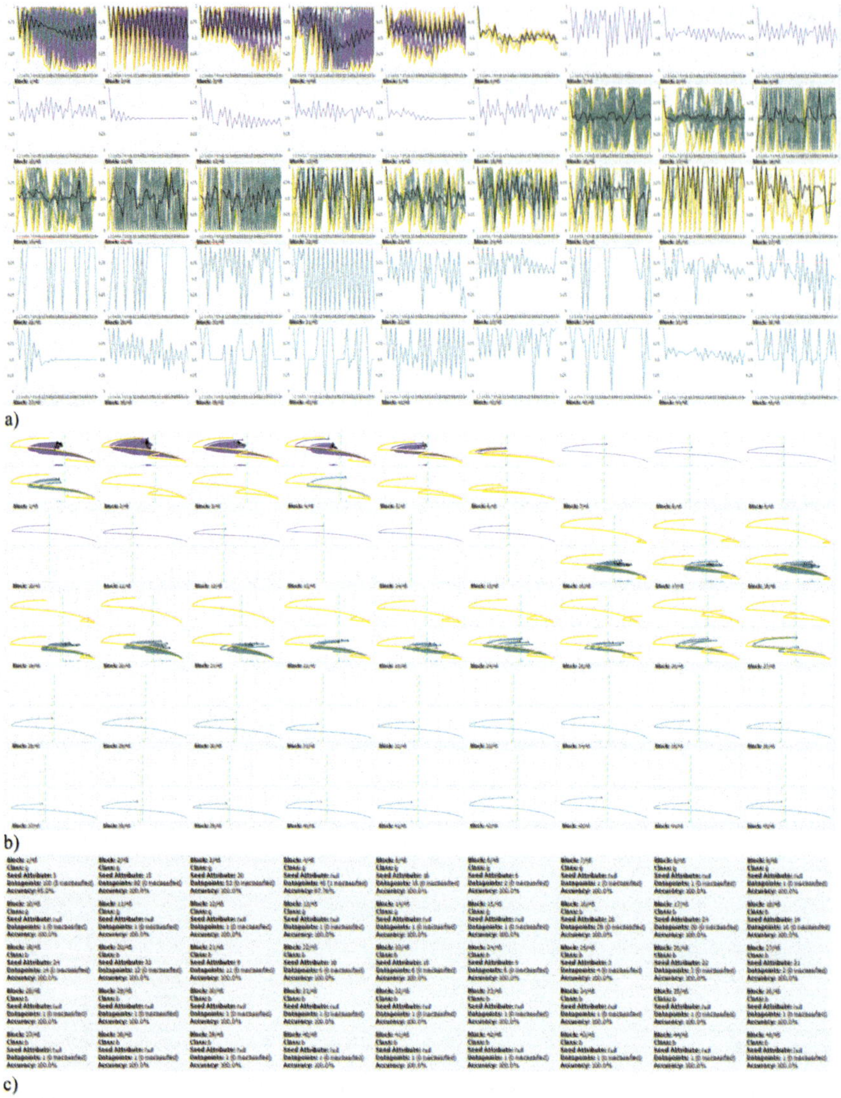

Fig. 29 HBs created with GLC-HBRL with Ionosphere Dataset. **a** HBs visualized in parallel coordinates. **b** HBs visualized in GLC-L. **c** Analytics for each HB

References

1. Kovalerchuk B (2018) Visual knowledge discovery and machine learning. Springer
2. Kovalerchuk B, Dovhalets D (2017) Constructing Interactive Visual classification, clustering, and dimension reduction models for N-D Data. Informatics 4(3):1–27. https://doi.org/10.3390/informatics4030023

3. Kovalerchuk B, Ahmad MA, Teredesai A (2021) Survey of explainable machine learning with visual and granular methods beyond quasi-explanations. Interpretable artificial intelligence: a perspective of granular computing. Springer, pp 217–67. https://arxiv.org/pdf/2009.10221
4. Kovalerchuk B, Nazemi K, Andonie R, Datia N, Bannissi E (eds) (2022) Integrating artificial intelligence and visualization for visual knowledge discovery. Springer
5. Ribeiro MT, Singh S, Guestrin C (2016) Why should I trust you?. Explaining the predictions of any classifier. In: Proceedings of the 22nd ACM SIGKDD. pp 1135–1144
6. Lundberg SM, Lee SI (2017) A unified approach to interpreting model predictions. Advances in neural information processing systems, pp 1–10. https://proceedings.neurips.cc/paper/2017/file/8a20a8621978632d76c43dfd28b67767-Paper.pdf
7. Molnar C, Casalicchio G, Bischl B (2020) Interpretable machine learning–a brief history, state-of-the-art and challenges. In: Joint European conference on machine learning and knowledge discovery in databases. Springer, Cham, pp 417–431
8. Freitas AA (2014) Comprehensible classification models: a position paper. ACM SIGKDD explorations newsletter 15(1):1, 10
9. Hall P, Phan W, Ambati S (2017) Ideas on interpreting machine learning. https://www.oreilly.com/radar/ideas-on-interpreting-machine-learning/
10. Molnar C (2020) Interpretable machine learning. https://christophm.github.io/interpretable-ml-book/
11. Lundberg SM, Erion G, Chen H, DeGrave A, Prutkin JM, Nair B, Katz R, Himmelfarb J, Bansal N, Lee SI (2020) From local explanations to global understanding with explainable AI for trees. Nat Mach Intell 2(1):56–67. https://doi.org/10.1038/s42256-019-0138-9
12. Holzinger A, Saranti A, Molnar C, Biecek P, Samek W (2022) Explainable AI methods-a brief overview. In: International workshop on extending explainable AI beyond deep models and classifiers. Springer, pp 13–38
13. C Recaido B (2022) Kovalerchuk, interpretable machine learning for self-service high-risk decision-making. In: 26th international conference information visualization. IEEE, pp 322–329. arXiv:2205.04032
14. Criticisms of econometrics (2022) https://en.wikipedia.org/wiki/Criticisms_of_econometrics
15. Moraffah R, Karami M, Guo R, Raglin A, Liu H (2020) Causal interpretability for machine learning-problems, methods and evaluation. ACM SIGKDD Explorations Newsl 22(1):18–33
16. Li J, Cheng K, Wang S, Morstatter F, Trevino RP, Tang J, Liu H (2018) Feature selection: a data perspective. ACM Comput Surv (CSUR) 50(6):94
17. Miller GA (1956) The magical number seven, plus or minus two: some limits on our capacity for processing information. Psychol Rev 63(2):81–97. https://doi.org/10.1037/h0043158
18. Lasso G, Khan S, Allen SA, Mariano M, Florez C, Orner EP, Quiroz JA, Quevedo G, Massimi A, Hegde A, Wirchnianski AS, Bortzrd Malonis RH, Georgiev GI, Tong K, Herrera NG, Morano NC (2022) Longitudinally monitored immune biomarkers predict the timing of COVID-19 outcomes. Plos Comput Biol 19(3)
19. Doshi-Velez F, Kim B (2017) Towards a rigorous science of interpretable machine learning. https://arxiv.org/abs/1702.08608
20. Xanthopoulos I, Tsamardinos I, Christophides V, Simon E, Salinger A (2020) Putting the human back in the AutoML loop. In: CEUR workshop proceedings. http://ceur-ws.org/Vol-2578/ETMLP5.pdf
21. Kovalerchuk B, Schwing J (eds) (2005) Visual and spatial analysis: advances in visual data mining, reasoning and problem solving. Springer
22. Ribeiro MT, Singh S, Guestrin C (2018) Anchors: high-precision model-agnostic explanations. Proc AAAI Conf Artif Intell 32(1)
23. Guidotti R, Monreale A, Ruggieri S, Pedreschi D, Turini F, Giannotti F (2018) Local rule-based explanations of black box decision systems. arXiv preprint arXiv:1805.10820.
24. Marques-Silva J, Ignatiev A (2022) Delivering trustworthy AI through formal XAI. Proc AAAI 3806–3814. https://www.aaai.org/AAAI22Papers/SMT-00448-Marques-SilvaJ.pdf
25. Kovalerchuk B, Hayes D (2021) Discovering interpretable machine learning models in parallel coordinates. In: 2021 25th International conference information visualisation (IV). IEEE, pp 181–188. arXiv:2106.07474

26. Mitchell T (1997) Machine learning, McGraw-hill
27. Muggleton S (ed) (1992) Inductive logic programming. Morgan Kaufmann
28. Džeroski S (2009) Relational data mining. In: Data mining and knowledge discovery handbook. Springer, Boston, MA, pp 887–911
29. Kovalerchuk B, Vityaev E (2000) Data mining in finance: advances in relational and hybrid methods. Kluwer
30. Kernel method. Wikipedia. https://en.wikipedia.org/wiki/Kernel_method
31. Support Vector Machines. scikit. (n.d.). https://scikit-learn.org/stable/modules/svm. Accessed 20 Feb 2023
32. Kovalerchuk B, Neuhaus N (2018) Toward efficient automation of interpretable machine learning. In: 2018 IEEE international conference on big data. Seattle, IEEE, pp 4933–4940. 978-1-5386-5035-6/18
33. Kovalerchuk B (2020) Enhancement of cross validation using hybrid visual and analytical means with Shannon function. In: Beyond traditional probabilistic data processing techniques: interval, fuzzy etc. methods and their applications. Springer, Cham, pp 517–543
34. Wagle SN, Kovalerchuk B (2022) Self-service data classification using interactive visualization and interpretable machine learning. In: Integrating artificial intelligence and visualization for visual knowledge discovery. Springer, Cham, pp 101–139
35. GitHub: https://github.com/CWU-VKD-LAB, DV2.0, DSCVis

Unveiling Insights: Empowering AI/ML Through Visual Knowledge Discovery

Search Space Visualizations in the Context of Optimization Problems: A Literature Review

Ennio W. L. Silva, Hugo A. D. do Nascimento, Juliana P. Felix, Humberto J. Longo, Bernd Scheuermann, and Les Foulds

1 Introduction

Conceptually, the *search space* of an optimization problem consists of the set of all possible combinations of values of its variables, which may include both feasible and infeasible solutions. The space may be finite and bounded, small or extremely large, or infinite and unbounded, among other classifications [12]. Other similar terms to the concept of a search space are *solution space* and *design space*. The solution space usually relates to the set of all feasible solutions [16], and, therefore, represents a subset of the search space. The design space has been more frequently applied when referring to design processes that involve the exploration of parameters in either engineering or art.

Visualizing the search spaces of optimization problems, particularly in complex real-world scenarios, such as engineering design problems, provides an important aid

E. W. L. Silva (✉)
Instituto Federal de Educação, Ciência e Tecnologia do Tocantins, Tocantins, Brazil
e-mail: ennio.silva@ifto.edu.br

H. A. D. do Nascimento · J. P. Felix · H. J. Longo · L. Foulds
Instituto de Informática, Universidade Federal de Goiás, Goiás, Brazil
e-mail: hadn@ufg.br

J. P. Felix
e-mail: julianafelix@ufg.br

H. J. Longo
e-mail: longo@ufg.br

L. Foulds
e-mail: fouldsles92@gmail.com

B. Scheuermann
Hochschule Karlsruhe, University of Applied Sciences, Karlsruhe, Germany
e-mail: bernd.scheuermann@h-ka.de

B. Kovalerchuk et al. (eds.), *Artificial Intelligence and Visualization: Advancing Visual Knowledge Discovery*, Studies in Computational Intelligence 1126,
https://doi.org/10.1007/978-3-031-46549-9_6

for monitoring, analyzing, and guiding the execution of search algorithms. In general, these problems have many variables and constraints and may contain conflicting objectives [3]. The intrinsic dependencies among these elements are often hard to understand and treat algorithmically.

Many solution visualizations focus on showing the structure of a single solution and highlighting its strengths and weaknesses with respect to the objective functions and constraints of a given problem. In contrast, the present chapter surveys work to date reported in the scientific literature on search space visualizations, abbreviated here by *SSV*, which aim at a larger picture. They provide visual elements that can help an observer grasp how well algorithms used explore opportunities to move towards the optimum of the problem. Executing a search algorithm typically involves continuously computing new solutions, at which each solution represents a position in the search space. Such solutions computed shall henceforth be denoted as *visited*. Search space visualization can draw an overall picture of the set of visited solutions and provide insights into the neighborhood structures, emphasize feasible and infeasible solution regions, and, in the case of dynamic optimization problems, show changes in the objective function values over the solution space.

Standard visualization techniques can be used for search space visualization, such as scatter plots, parallel coordinates, tree maps, and heat maps. These techniques, alone or in combination with other approaches, provide different levels of abstraction and detail depending on the characteristics of the given problem and the user's needs.

Furthermore, search space visualizations can be an important asset for supporting visual analytics and knowledge discovery. They open opportunities for new high-level interactive resources that can make visual analytics more effective for solving complex real-world problems. In fact, this kind of integration is in line with the development of other works that use standard visualizations for enhancing human-guided optimization processes [5, 8, 9, 18].

Search space visualization is mentioned in many scientific articles. In our previous work [38], we conducted a literature review in this field with over 65 articles filtered from a set of 264 related works included that were included from an initial set of 4005 articles. This provided quantitative and comparative results that address some previously-posed research questions that are discussed in the next section of the current chapter.

In the present chapter, we extend our previous work [38] by providing a more insightful discussion on the findings of that literature review in the context of search space visualization for optimization problems. Furthermore, we provide an advanced analysis considering the growth of the area over the years, the types of techniques applied to handle multidimensional data, and the difference between strategies used for visualizing both single- and multi-objective problems.

The remainder of this work is organized as follows: Sect. 2 describes the methodology adopted for the literature review and their quantitative results regarding the accepted and the filtered articles; Sect. 3 presents the analysis of the filtered articles after data extraction; Sect. 4 draws our conclusions and proposes ideas for future work.

2 Literature Review Methodology

We followed a methodology for the literature review based on the guidelines provided by Kitchenham and Charters [17] for Systematic Literature Reviews (SLRs). SLRs are widely used to identify, evaluate and interpret relevant pieces of research for a specific area of interest. We now briefly describe this methodology. For a detailed description of the protocol developed, the reader is kindly referred to [38].

The goal of the review was to find and analyze works in the scientific literature that present SSVs of either continuous or discrete optimization problems. Therefore, in order to guide this process we defined a list of research questions to be covered. The questions were targeted first at obtaining a panorama of previous contributions in that research field and its changes over the years, the variety of information visualization techniques that were applied to SSVs, the kind of optimization problems studied, how high-dimensional problems were supported by SSV techniques, and how these visualizations differed for both single- and multi-objective problems. Additionally, as a way to better understand how this field has evolved, other information such as the main academic publication vehicles and their public availability were considered.

A set of keywords were collaboratively defined after the analysis of several "control" articles that were known to address the defined questions. The final search string was defined as: ("search-space" OR "search space" OR "space of solution" OR "space of search" OR "solution space" OR "design space") AND ("visualization" OR "visualisation" OR "graphic representation" OR "visual representation" OR "dimensional reduction" OR "dimension reduction").

The search string was applied to the fields "title", "abstract", and "keywords" (usually referred to as TAK) of eight distinct digital libraries: ACM Digital Library, Elsevier, IEEE Digital Library, Scopus, Springer Link, Google Scholar, ISI Web of Science, and arXiv. Additionally, a backward snowballing approach [2, 45] was used, where the references of included articles were also studied to identify other articles of interest, resulting in a few more to be analyzed. This process resulted in a total of 4005 articles and was carried out in August 2021.

The abstracts of all the articles previously selected were analyzed in order to identify those that satisfied the inclusion criteria: "works in which information visualization techniques are applied for the understanding or exploration of the search-space of a continuous or combinatorial optimization problem". On the other hand, works that were not related to optimization problems, those that did not present an approach to understanding the search space, or works for which a full article was not available or was not written in English, Portuguese, or German were excluded from the future evaluation (exclusion criteria).

Table 1 summarizes the number of articles retrieved from each database. Many articles (1235) appeared duplicated in more than one search database, and, when removed, reduced the number of found articles to 2770. After reading all abstracts and analyzing them according to the inclusion and exclusion criteria mentioned previously, 264 articles remained.

Table 1 Number of articles found on each search database

Database	Total
ACM digital library	51
Elsevier	937
IEEE digital library	284
Springer link	10
Scopus	1711
Google scholar	153
ISI web of science	352
arXiv	138
Snowballing	369
Sum	4005

Finally, all 264 articles passed through a quality assessment stage to rank papers in order of relevance. An integer score ranging from 0 to 10 was assigned to each article based on 5 quality questions defined. Each question could be answered with one of the following options and its respective score—No (0), Partially (1), or Yes (2). The questions used are: "Do 'visualization' and 'solution space' or their synonyms appear in the title or abstract?", "Was the article published in an Information Visualization conference or journal?", "Was the article published in a well-known conference or journal?", "Is the optimization problem explicit in the title or abstract?", "Is the optimization problem well-known?". In this study, we restricted our analysis to articles that achieved a minimum score of 8, representing such a condition a *Performance Filter* (PF). By applying the PF, the set of articles was reduced to 65, which were then finally studied and had information extracted from them in order to answer the research questions. A synthesis of the data extracted from these remaining articles is publicly available as a supplementary material.[1] To support this literature review, the collaborative online tool Parsif.al (https://parsif.al/) was employed.

3 Research Analysis

In this section, we discuss our findings related to SSVs in the context of optimization problems. The information presented here is based on the articles that passed the PF described in the previous section. Among other issues, we discuss the growth of the field over the years, the existing information visualizations (IV) techniques for search-space understanding, how these visualizations differ between single and

[1] Link to supplementary material: https://drive.google.com/drive/folders/1iEbEDJEEBXJw76tNZiytyWjhOndEaEVs.

multi-objective problems, and also the classes of optimization problems that are mainly used to exemplify the visualization techniques.

3.1 Overview of SSV Articles

Information Visualization (IV) is a rapidly evolving research area that has grown in popularity in recent years. The growth of IV is evidenced by the increasing number of publications in the area. In order to understand how this growth has affected subfields of IV such as search space visualization (SSV), we created statistics of the articles included in the SLR.

The results can be seen in Fig. 1, where the number of included articles per year is represented by the line graph, and the number of filtered articles that passed the PF is illustrated by the bar graph. The considerable growth in the number of articles over the years confirms that the topic has attracted increasing interest within the scientific community. Please note that the numbers for the year 2021 are preliminary as the database queries took place in August 2021.

The subsequent statistics provide an overview of the distribution of articles across countries. A summary of this information is presented in Table 2, which shows the dominance of authors based in the USA (16) and the UK (14). Some countries share the same number of publications, such as Brazil and Canada, each with 4 articles. Several other countries are listed, including some from Europe, and also from Asia and Oceania, however in significantly lower numbers. We note that a

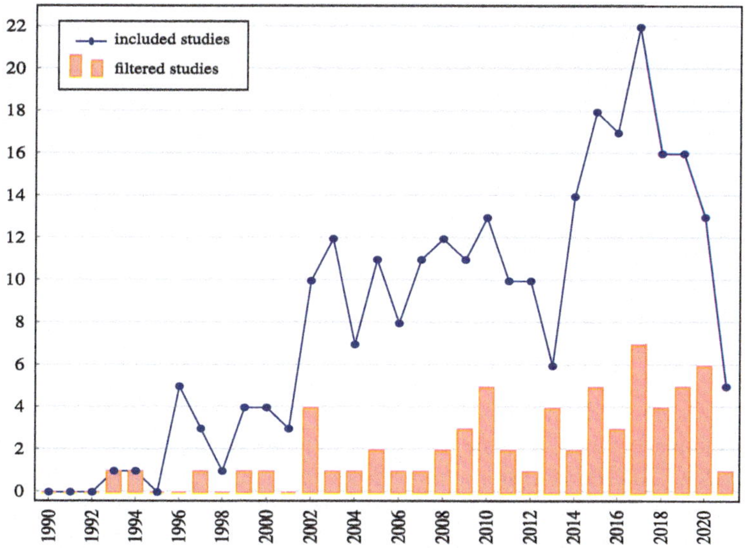

Fig. 1 Number of articles published per year

Table 2 Leading countries by the number of publications

Country	Quantity of publications
USA	16
UK	14
Germany	6
Switzerland	5
Brazil, Canada	4
Austria, China, France, Poland	3
Australia, Netherlands, Republic of Korea	2
Finland, Greece, India, Ireland, Italy, Malaysia, Portugal, Singapore, Slovak Republic, Spain	1

single article could have been carried out by authors from different institutes or countries, contributing to more than one country in the table. Therefore, the sum of publications by countries does not reflect the number of articles analyzed in this literature review. Despite the collaboration between authors from different countries, we observed that the cases where an author would appear to collaborate in more than one research group are rare. This fact leads us to believe that it makes sense to expand international collaboration to promote knowledge sharing and leverage synergies in the SSV field.

Finally, regarding the use of SSV in science, we were further interested in identifying in which other research areas it is being used. Although SSV techniques are naturally applied to computer science and optimization problems, the literature review made it possible to observe its application to other major fields. Using the Science of Science (Sci2) Tool [37] to analyze the journal articles among the filtered articles, we were able to identify the other main disciplines in which SSV has been applied, which are: mathematics, physics, electrical, chemical, and mechanical engineering, and social sciences.

3.2 Dimensionality Reduction

Representing high-dimensional data is a frequent challenge. When there are a large number of variables involved in a problem, it can be difficult to represent the nature of the problem in a clear and objective way. A common strategy for dealing with this issue is through dimensionality reduction, which consists of transforming the representation of high-dimensional data into a lower-dimensional representation, such as 2D or 3D. This transformation allows for a clearer and more understandable representation of the data, making it easier to identify important patterns and trends.

There are several techniques to reduce the dimension of a given problem, whether through attribute selection or feature extraction. In the literature review, only a few articles performed dimensionality reduction for feature extraction. Instead, most works that performed some kind of dimensionality reduction aimed at the attribute extraction phase. Among the techniques used, those that stand out are Multidimensional Scaling (MDS) [10, 32], Self-Organizing Map (SOM) [13, 23, 24, 36], ISOMAP (Isometric Mapping) [21, 43] and t-SNE (t-Distributed Stochastic Neighbor Embedding) [40, 41].

3.3 SSV Techniques for Search-Space Understanding

In the construction of a visualization tool, one of the main tasks is the development of the technique to be used in a given application or context. This choice is usually based on the type of information to be treated, the type of values, the nature and size of the domain, and the operations the user intends to carry out. Thus, it is up to specialists and designers to identify the best technique(s) to be employed in their application.

In this sense, there are several common visualization techniques that are applied to search space visualization, such as scatterplots, parallel coordinates, graphs, heatmaps, and other types of charts. In our review, we identified the information visualization techniques present in the articles and classified them by dimension, i.e. Bidimensional (2D) or Tridimensional (3D).

Most visualization techniques found are 2D, with 86% of the articles analyzed using them. The application of scatterplots stands out, appearing in 26 articles, followed by parallel coordinates and graph drawing (each in 11 articles), heatmap (in 10 articles), line chart (in 9 articles), and bar chart (in 7 articles). The visualization techniques employed in each article are described in the supplementary material mentioned in Sect. 2.

Some examples of these main techniques are illustrated in Figs. 2 and 3. Namely, in Fig. 2, the authors in [41] were interested in verifying the representation of the search space under a genetic improvement bug-fixing scenario using different dimensionality reduction techniques. Their visualizations are scatterplots, with each solution being represented as a point and colored according to its fitness. The image on the left was generated by a projection based on Multiple Correspondence Analysis (MCA), while the one on the right was obtained by the t-Distributed Stochastic Neighbour Embedding (t-SNE) method. Although MCA can demonstrate some structure in the data, particularly by displaying solutions with the same fitness level in close proximity, the t-SNE can reveal and exhibit further patterns. For example, it can unveil worm-like artifacts, which are actually parts of the search trajectories, to show that very similar solutions exist.

The second example of a 2D visualization technique is a type of parallel coordinates presented in [15], which the authors called *search space coverage plot*. The visualization, illustrated in Fig. 3, aggregates all solutions from all iterations of a particle swarm optimization method into a single chart. The vertical parallel axes represent the dimensions (variables) of the solutions, while the solutions are drawn

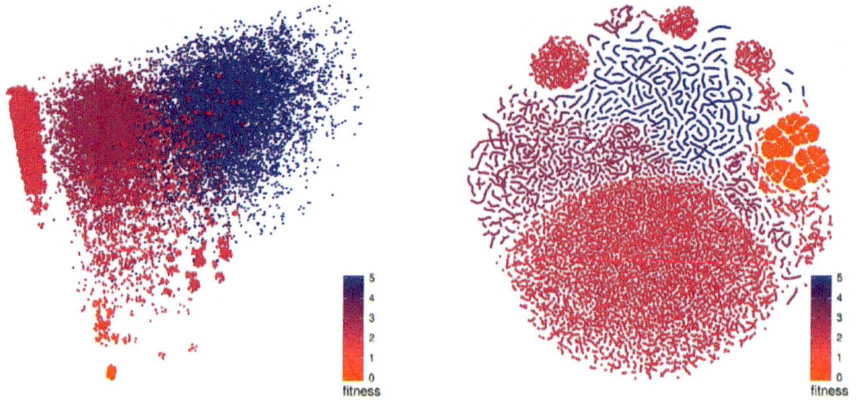

Fig. 2 Search space represented by the scatterplot technique. Figure by [41], licensed under CC BY (https://creativecommons.org/licenses/)

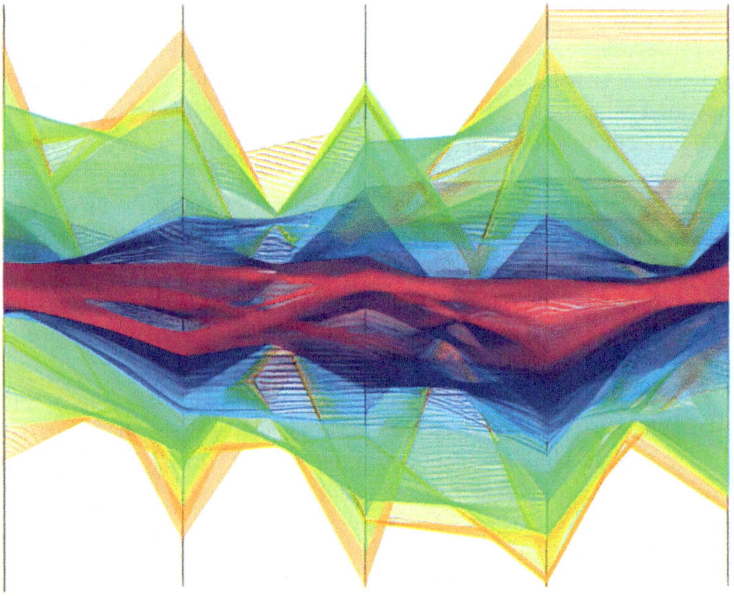

Fig. 3 Search space represented by parallel coordinates technique. Reprinted from [15] © 2009, with permission from ACM (Association for Computing Machinery)

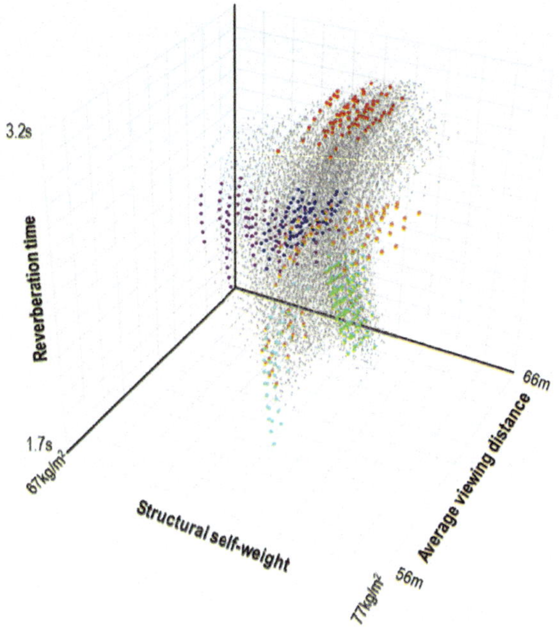

Fig. 4 Search space represented by 3D scatterplot technique. Reprinted from [28], © 2020, with permission from Elsevier

as lines connecting all axes. Colors and an overlaid order (from back to front) are employed to highlight the sequence in which the solutions were produced, with lighter colors (in the back) indicating initial iterations of the search, and darker colors (in the front) representing the last iterations.

The most commonly used 3D techniques are 3D Scatterplot, appearing in 3 articles [25, 26, 28], and 3D Surface, appearing in 2 articles ([7, 35]). Despite the fact that there are fewer 3D occurrences than 2D representations, they have promising potential for usage in immersive interactive environments.

Some visual examples of 3D techniques are represented in Figs. 5 [28] and 4 [35]. In Fig. 5 we can see the representation of the search space for the vehicle routing problem and how it facilitates the analysis of the entire topological structure of this space. The solutions are not visible, but it is possible to perceive, through the colors, the regions that contain solutions of greater or lesser quality. In Fig. 4, the space presents solutions to a design arena problem [35]. The solutions are presented by points mapped on three axes representing performance indicators, that is, the objectives to be optimized.

3.4 Single-Objective Versus Multi-objective Visualizations

For single-objective and multi-objective problems, the goal is to find an optimal solution that minimizes or maximizes a single objective function.

Fig. 5 Search space represented by 3D surface technique. Image by [35], used under the Creative Commons license, © 2013 reserved to LUT Scientific and Expertise Publications

In contrast, multi-objective problems have two or more objectives that can be conflicting. The goal is to find a set of non-dominated solutions, known as the Pareto-optimal front, that represents a trade-off between the different objectives.

In the study reported here, 33 of the articles analyzed dealt with single-objective optimization problems, 31 with multi-objective problems, and only one explored both types of optimization problems. The reader is referred to the supplementary material (see footnote 1) where this information can be verified. A curious fact is that, for both kinds of problems, scatterplot was the most used technique for visualizing the search space. However, how its visual mapping was defined usually changed from one case to the other. The second most frequent visualization techniques were, equally, heatmaps and parallel coordinates for single-objective problems, and only parallel coordinates for multi-objective problems. This may suggest that heatmaps can be more challenging to use for showing the search space of multi-objective problems.

A general strategy for presenting the search space of multi-objective problems has been to assign one or more objective functions to the spatial axes of the visualization. For instance, a 2D scatterplot of known solutions with axes representing two objective measures was employed in [6, 33]. A set of non-dominated solutions were shown in some of the visualizations, which represent a good compromising solution to the problem.

When there are more than three objective functions under analysis, visualizing the search space is more complex. In this case, a parallel coordinate technique can be used. Coordinates may be assigned to objective functions as well as to the values of the variables that describe the problem, if necessary. An interesting variation of the parallel coordinates technique is presented in [6, 20] and is illustrated in Fig. 6. A chord diagram, as this technique is called, maps every objective of the problem to a segment of an arc, while solutions are drawn as lines intersecting all arcs.

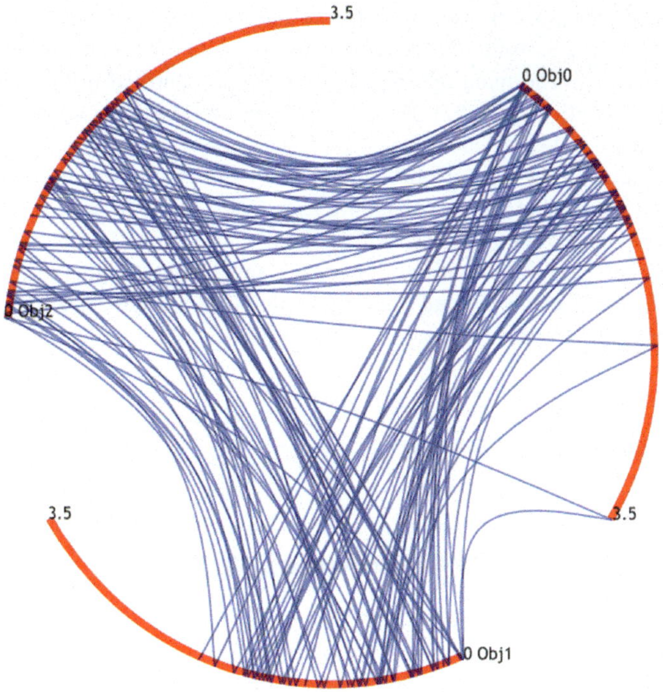

Fig. 6 A search space visualization for the DTLZ problem with chord diagram. Reprinted from [20], ©2017, with permission from Elsevier

3.5 Optimization Problems Investigated Using SSV

Search Space Visualization (SSV) techniques can be particularly useful in analyzing optimization problems as they allow decision-makers to quickly and easily visualize and understand potential solutions to a problem. SSVs are especially useful in high-dimensional optimization problems, where it can be difficult to grasp the relationship between variables and solutions.

In order to better understand how SSV techniques have been explored to analyze or solve optimization problems, we have identified and divided the optimization problems appearing in the articles into three categories: Benchmark Functions (BF), Classical Problems (CP), and Specific Problems (SP). As it is well-known, BF are functions that can be tested to measure the performance of optimization problems; CP are optimization problems that are widely studied in the literature, such as the Traveling Salesman Problem (TSP), the Vehicle Routing Problem (VRP), and the 2-D Bin Packing Problem. Finally, SP are special problems with peculiar characteristics that do not fit into either of the other two categories.

Each problem in the studied articles can be assigned to at least one of the 3 categories (BF, CP, SP), sometimes to more than one category, even if the problem has already been investigated elsewhere. In our analysis, we identified that optimization problems that fit in the BF category correspond to 37% of the total number of problems studied. The main optimization problems identified in this set were the Rosenbrock Function [30, 32], Rastrigin's Function [14, 23, 24], and the DTLZ class [10, 11, 43, 44].

For problems in the CP category, a total of 34 were identified, corresponding to 30% of the total. In this set, we can highlight the VRP [4, 35], the TSP [19, 42], and the Graph Layout Problem (GLP) [18, 22]. Finally, we found 37 problems associated with the SP category, representing 32% of the total. Arena Design [28], Conceptual Design [33], Car Design [1], and Material Design [31] are some examples of works that were assigned to this category.

3.6 Supported Interactions

Identifying the interactions supported by the visualizations is an important activity in order to understand how the visualizations have been explored by the different optimization problems.

In our analysis, we observed that 37/65 articles used interactive forms of visualizations or implementation systems. The implemented interactive actions vary from adjusting the level of detail (such as zooming in/out), to more advanced optimization methods that guide the problem resolution. For instance, in the work of [34], the users can select the solution, zoom in/out, and make inclusion and exclusion visualization models. Furthermore, in [35] the authors implemented a tool that allows interactive examination with free-form rotation around the visualizations.

Regarding more advanced interactions involving exploration or integrating visualization with optimization, two works stand out, namely [27] and [7].

In [27], the authors present an interactive visualization system for solving an engineering design problem. The system uses a genetic algorithm to generate many solutions and provide an interface allowing the user to interact with the generated data. Figure 7 illustrates an overview of the developed system, which enables the verification of the data in different perspectives, be it 2D or 3D.

Several interactive approaches were proposed in this work. One concerns the selection of an area to: (a) identify clusters; (b) focus the algorithm on the selected region or outside it; or (c) examine a region in more detail through zoom. For example, in Fig. 7, the identification of *clusters* in different visualization techniques can be seen, giving the user a wider range of clusters in relation to the search space.

Other interactive forms adopted by the authors include the selection of a region for interaction with the genetic algorithm and zoom for closer observation of a specific region. The genetic algorithm can produce solutions both inside and outside the selected area, the choice of which, is up to the user. These strategies allow the users to explore new solutions under different aspects and choose solutions to compare them.

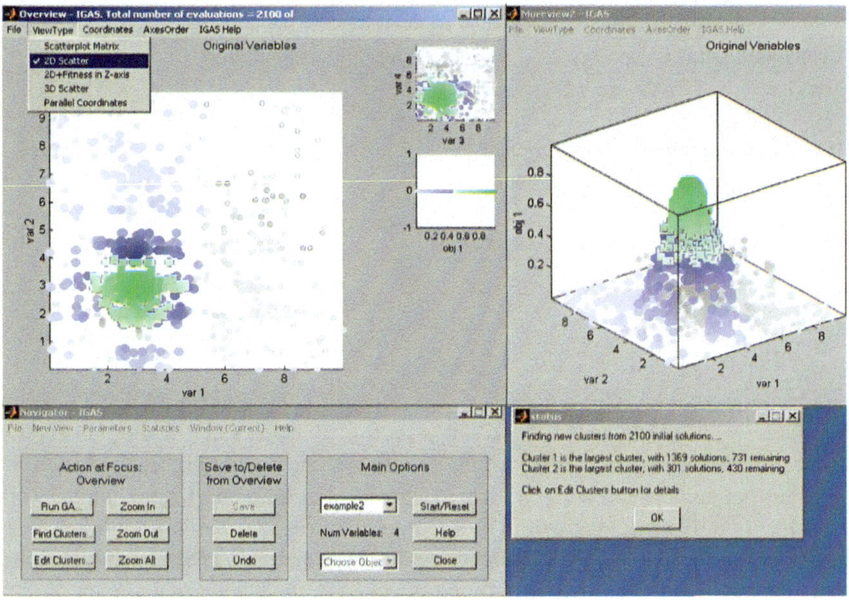

Fig. 7 System interface overview for solving an engineering design problem using a genetic algorithm. Reprinted from [27], © 2005, with permission from Elsevier

In [7], the authors also present an interactive system for engineering design optimization problems, which, in this case, involves both a discrete and a continuous examination of the search space. In the discrete aspect, a workspace is created in which one has a broad view of the search space through various visualization techniques. For example, a graph of parallel coordinates is constructed, which allows the user to filter the solutions according to the modification of the thresholds of decision variables. Furthermore, scatter plots facilitate a relational analysis between two variables, and if it is 3D, the space can be rotated for viewing at different angles; and a histogram is used to display the distribution of solutions over space according to the decision variable and its respective value.

3.7 *Main Publication Vehicles*

Research on search space visualization can be found in various journals and conference proceedings, depending on the specific application area or method used, such as computer graphics, human-computer interaction, and operational research. To better identify what those publishing means are, we collected quantitative information on all filtered articles. Table 3 shows the top 5 publication vehicles based on the number of articles analyzed.

Table 3 Leading publication vehicles based on the number of publications

Conference/Journal	Quantity of publications
The Genetic and Evolutionary Computation Conference	8
IEEE Congress of Evolutionary Computation	5
IEEE Transactions on Visualization and Computer Graphics	4
IEEE Congress on Evolutionary Computation	4
arXiv	3

From the total of analyzed articles, we observed that 32 were published in conferences, 24 in journals, and there was a total of 9 articles that were published in different vehicles, such as arXiv or books, for instance. Among these vehicles, we went further and observed that 31% are related to Evolutionary Computation, while 21% are directly associated with IV or Computer Graphics and 45% in other areas. This fact implies that many works that involve information visualization techniques to understand the search space are found in publication vehicles that are not directly related to Information Visualization. An interesting fact arising from the list of the main vehicles is the presence of a conference in the first position of the ranking. This may be due to the speed with which conference articles are evaluated and published compared to journals, which normally have a slower response time.

3.8 *Availability of the Proposed IVs*

An important part of disseminating knowledge and contributing to any field is by making works and the proposed solutions publicly available. In order to identify how this might apply to the field of Search Space Visualization (SSV), we investigated articles found in this literature review and tried to identify which of the systems or SSVs developed were made available so that it could be reused in future studies.

Our results showed that, of all articles analyzed in this review, only three authors made their solutions available to the general public [13, 29, 39]. The authors shared their solution by publicly providing the source code of their systems. In [13], for example, the authors created an open-source visualization called SwarmViz. In that system, users are able to set up Particle Swarm Optimization experiments through canonical benchmark functions or from external input data. In [29], the authors described a web-based visual analytic system termed HYperTendril, that allows users to refine the search space and to obtain insights into the complex behavior of hyperparameter search algorithms, as well as to diagnose their configurations. Lastly, in [39], the authors present a modified Self-Organizing Map (SOM) algorithm that enables users to select regions of interest in the design space using as examples a benchmark nonlinear analytical, as well as engineering design problems.

Considering that a great number of classic optimization problems have been studied, as well as some widely known benchmark functions, the availability of visualization interfaces or developed systems might possibly be of great use in science, especially for comparative purposes.

4 Conclusions

In this chapter, we provided an overview of Search Space Visualizations (SSVs) in the context of optimization problems. A literature review was carried out and several works were identified that apply Information Visualization techniques to promote the understanding and exploration of search spaces. Among the topics discussed in this chapter, we covered the growth of the SSV field over the years in terms of the number of publications, countries, and areas where this approach has been studied, as well as the variety of information visualization techniques used for dealing with high-dimensional problems, and to present the structure of search spaces. It was noted that there is a predominance of 2D visualizations, particularly those based on scatterplots and parallel coordinates. Few 3D visualizations were identified, and we believe that they may become more common with the popularization of Virtual Reality (VR) and Augmented Reality (AR) devices.

For a better understanding of SSVs applied to both single-objective and multi-objective problems, a deeper analysis was carried out, identifying those visualization techniques that most appeared in the articles investigated. Over the several optimization problems analyzed in each of the works, we could observe the high usage of benchmark functions, such as Rastrigin's Function and the DTLZ problem, and also classical optimization problems, such as the Vehicle Routing Problem (VRP) and the Traveling Salesman Problem (TSP). An analysis according to the publication vehicles of the selected articles was also carried out, which implied that many of the works were published in conferences or journals related to Evolutionary Computation. Finally, the supported interactions of the works were verified, and a discussion on the public availability of the proposed solutions was carried out.

There is a continuous expansion of the field, confirmed by the growth in the number of articles over the years. Researchers from all over the world have been exploring visualization approaches for solution-spaces, led by those affiliated with the USA, followed by the UK and Germany. There is still a lack of specific programming libraries and tools to support investigations on solution-space visualization, given that only a few articles released source code for such an aim.

Finally, we observe that search space visualization is also a valuable tool in visual analytics and knowledge discovery. It enables users to explore and analyze complex datasets, identify patterns and relationships, and discover new insights that may not have been apparent through traditional data analysis techniques. With the use of visualization tools, users can gain a deeper understanding of the search space, which can help in decision-making and problem-solving based on optimization processes.

Acknowledgements This study was financed in part by the Coordenação de Aperfeiçoamento de Pessoal de Nível Superior—Brasil (CAPES)—Finance Code 001, and by the Instituto Federal de Educação, Ciência e Tecnologia do Tocantins.

References

1. Abi Akle A, Minel S, Yannou B (2017) Information visualization for selection in design by shopping. Res Eng Design 28:99–117. https://doi.org/10.1007/s00163-016-0235-2
2. Badampudi D, Wohlin C, Petersen K (2015) Experiences from using snowballing and database searches in systematic literature studies. In: Proceedings of the 19th international conference on evaluation and assessment in software engineering, pp 1–10. https://doi.org/10.1145/2745802.2745818
3. Cibulski L, Mitterhofer H, May T, Kohlhammer J (2020) Paved: Pareto front visualization for engineering design. Comput Graph Forum 39(3):405–416. https://doi.org/10.1111/cgf.13990
4. Daneshpajouh H, Zakaria N (2017) A clustering-based visual analysis tool for genetic algorithm. In: VISIGRAPP. https://doi.org/10.5220/0006135902330240
5. do Nascimento HA, Eades P (2005) User hints: a framework for interactive optimization. Future Gener Comput Syst 21(7):1177–1191. https://doi.org/10.1016/j.future.2004.04.005
6. Fieldsend J, Everson R (2013) Visualising high-dimensional pareto relationships in two-dimensional scatterplots. In: International conference on evolutionary multi-criterion optimization. Springer, pp 558–572. https://doi.org/10.1007/978-3-642-37140-0_42
7. German BJ, Feigh KM, Daskilewicz MJ (2013) An experimental study of continuous and discrete visualization paradigms for interactive trade space exploration. J Comput Inf Sci Eng 13(2):021004. https://doi.org/10.1115/1.4023857
8. Hägele D, Abdelaal M, Oguz OS, Toussaint M, Weiskopf D (2022) Visual analytics for nonlinear programming in robot motion planning. J Vis 25(1):127–141. https://doi.org/10.1007/s12650-021-00786-8
9. Hakanen J, Miettinen K, Matković K (2021) Task-based visual analytics for interactive multiobjective optimization. J Oper Res Soc 72(9):2073–2090. https://doi.org/10.1080/01605682.2020.1768809
10. He Z, Yen GG (2016) An improved visualization approach in many-objective optimization. In: 2016 IEEE congress on evolutionary computation (CEC), pp 1618–1625. https://doi.org/10.1109/CEC.2016.7743982
11. He Z, Yen GG (2017) Comparison of visualization approaches in many-objective optimization. In: 2017 IEEE congress on evolutionary computation (CEC), pp 357–363. https://doi.org/10.1109/CEC.2017.7969334
12. Hernandez A, Pollman A (2021) Bounding the solution space of complex systems in terms of non-numeric and\or uncontrollable scenario variables. Systems 9(4). https://doi.org/10.3390/systems9040089
13. Jornod G, Di Mario E, Navarro I, Martinoli A (2015) Swarmviz: an open-source visualization tool for particle swarm optimization. In: 2015 IEEE congress on evolutionary computation (CEC), pp 179–186. https://doi.org/10.1109/CEC.2015.7256890
14. Keshk M (2017) Semantic evolutionary visualization. In: Tan Y, Takagi H, Shi Y, Niu B (eds) Advances in swarm intelligence. Springer International Publishing, Cham, pp 624–635
15. Khemka N, Jacob C (2009) VISPLORE: a toolkit to explore particle swarms by visual inspection. In: Proceedings of the 11th annual conference on genetic and evolutionary computation, GECCO '09. Association for Computing Machinery, New York, NY, USA, pp 41–48. https://doi.org/10.1145/1569901.1569908
16. Khosrow-Pour M (2015) Encyclopedia of information science and technology. Information Science Reference

17. Kitchenham BA, Charters S (2007) Guidelines for performing systematic literature reviews in software engineering. Tech. Rep. EBSE 2007-001, Keele University and Durham University Joint Report. https://www.elsevier.com/__data/promis_misc/525444systematicreviewsguide.pdf
18. Klau GW, Lesh N, Marks J, Mitzenmacher M (2010) Human-guided search. J Heuristics 16(3):289–310. https://doi.org/10.1007/s10732-009-9107-5
19. Kong X, Schunn CD (2007) Global versus local information processing in visual/spatial problem solving: the case of traveling salesman problem. Cogn Syst Res 8(3):192–207. https://doi.org/10.1016/j.cogsys.2007.06.002
20. Koochaksaraei RH, Meneghini IR, Coelho VN, Aes FGG (2017) A new visualization method in many-objective optimization with chord diagram and angular mapping. Knowl-Based Syst 138:134–154. https://doi.org/10.1016/j.knosys.2017.09.035
21. Kramer O (2016) Solution space visualization. Studies in Big Data. Springer International Publishing. https://books.google.com.br/books?id=nHJBDAAAQBAJ
22. Kwon O, Ma K (2020) A deep generative model for graph layout. IEEE Trans Vis Comput Graph 26(01):665–675. https://doi.org/10.1109/TVCG.2019.2934396
23. Lee HC, Kim YH (2019) Visualizing swarm behavior with a particle density map. In: Proceedings of the genetic and evolutionary computation conference companion, GECCO '19. Association for Computing Machinery, New York, NY, USA, pp 415–416. https://doi.org/10.1145/3319619.3321903
24. Lotif M (2014) Visualizing the population of meta-heuristics during the optimization process using self-organizing maps. In: IEEE congress on evolutionary computation (CEC), pp 313–319. https://doi.org/10.1109/CEC.2014.6900265
25. Lotif M, Coelho ALV (2011) Visually inspecting the search behavior of harmony search and its variants with viz3d. In: IEEE congress of evolutionary computation (CEC), pp 616–623. https://doi.org/10.1109/CEC.2011.5949676
26. Nassersharif B, Ence D, Au M (1994) Visualization of evolution of genetic algorithms. In: Proceedings of world congress on neural networks WCNN'94, vol 1. Lawrence Erlbaum Associates, Hillside, NJ, USA, pp 560–565
27. Packham ISJ, Rafiq MY, Borthwick MF, Denham SL (2005) Interactive visualisation for decision support and evaluation of robustness—in theory and in practice. Adv Eng Inf 19:263–280. https://doi.org/10.1016/j.aei.2005.07.006
28. Pan W, Sun Y, Turrin M, Louter C, Sariyildiz S (2020) Design exploration of quantitative performance and geometry typology for indoor arena based on self-organizing map and multi-layered perceptron neural network. Autom Constr 114:103163. https://doi.org/10.1016/j.autcon.2020.103163
29. Park H, Nam Y, Kim JH, Choo J (2021) Hypertendril: visual analytics for user-driven hyperparameter optimization of deep neural networks. IEEE Trans Visual Comput Graph 27(2):1407–1416. https://doi.org/10.1109/TVCG.2020.3030380
30. Parsopoulos K, Georgopoulos V, Vrahatis M (2008) A technique for the visualization of population-based algorithms. In: IEEE congress on evolutionary computation (IEEE world congress on computational intelligence), pp 1694–1701. https://doi.org/10.1109/CEC.2008.4631018
31. Peerless JS, Sevgen E, Edkins SD, Koeller J, Kim E, Kim Y, Garg A, Antono E, Ling J (2020) Design space visualization for guiding investments in biodegradable and sustainably sourced materials. MRS Commun 10(1):18–24. https://doi.org/10.1557/mrc.2020.5
32. Pohlheim H (2006) Multidimensional scaling for evolutionary algorithms—visualization of the path through search space and solution space using sammon mapping. Artif Life 12(2):203–209. https://doi.org/10.1162/artl.2006.12.2.203
33. Pu P, Lalanne D (2000) Interactive problem solving via algorithm visualization. In: IEEE symposium on information visualization (INFOVIS 2000), pp 145–153. https://doi.org/10.1109/INFVIS.2000.885103
34. Pu P, Melissargos G (1997) Visualizing resource allocation tasks. IEEE Comput Graph Appl 17(4):6–9. https://doi.org/10.1109/38.595258

35. Rasku J, Kärkkäinen T, Hotokka P (2013) Solution space visualization as a tool for vehicle routing algorithm development. In: Proceedings of the Finnish operations research society 40th anniversary workshop (FORS40), vol 13, pp 9–12
36. Richardson T, Winer E (2011) Visually exploring a design space through the use of multiple contextual self-organizing maps. In: 37th Design automation conference, parts A and B, international design engineering technical conferences and computers and information in engineering conference, vol 5, pp 857–866. https://doi.org/10.1115/DETC2011-47944
37. Sci2 Team (2009) Science of Science (Sci2) Tool. https://sci2.cns.iu.edu. Indian a University and SciTech Strategies. Accessed 27 Jan 2023
38. Silva EW, do Nascimento HA, Felix JP, Longo HJ, Scheuermann B (2022) A systematic literature review of solution-space visualization approaches in the context of optimization problems. In: 2022 26th International conference information visualisation (IV). IEEE, pp 48–53. https://doi.org/10.1109/IV56949.2022.00017
39. Thole SP, Ramu P (2020) Design space exploration and optimization using self-organizing maps. Struct Multidiscip Optim 62(3):1071–1088. https://doi.org/10.1007/s00158-020-02665-6
40. Tyagi A, Xie C, Mueller K (2020) Visual steering for one-shot deep neural network. Synthesis. https://doi.org/10.48550/ARXIV.2009.13008
41. Veerapen N, Ochoa G (2018) Visualising the global structure of search landscapes: genetic improvement as a case study. Genet Program Evolvable Mach 19(3):317–349. https://doi.org/10.1007/s10710-018-9328-1
42. Volke S, Zeckzer D, Scheuermann G, Middendorf M (2015) A visual method for analysis and comparison of search landscapes. In: Proceedings of the 2015 annual conference on genetic and evolutionary computation, GECCO '15. Association for Computing Machinery, New York, NY, USA, pp 497–504. https://doi.org/10.1145/2739480.2754733
43. Walker D, Fieldsend J, Everson R (2012) Visualising many-objective populations. In: Proceedings of the 14th annual conference companion on genetic and evolutionary computation, GECCO '12. Association for Computing Machinery, New York, NY, USA, pp 451–458. https://doi.org/10.1145/2330784.2330853
44. Walker DJ (2015) Visualising multi-objective populations with treemaps. In: Proceedings of the companion publication of the 2015 annual conference on genetic and evolutionary computation, GECCO Companion '15. Association for Computing Machinery, New York, NY, USA, pp 963–970. https://doi.org/10.1145/2739482.2768445
45. Wohlin C (2014) Guidelines for snowballing in systematic literature studies and a replication in software engineering. In: ACM international conference proceeding series, pp 1–10. https://doi.org/10.1145/2601248.2601268

Relative Confusion Matrix: An Efficient Visualization for the Comparison of Classification Models

Luc Etienne Pommé, Romain Bourqui, Romain Giot, and David Auber

1 Introduction

Machine Learning (ML) and Deep Learning (DL) algorithms are widely used to solve detection [1], segmentation [3] or classification [2] tasks. In the literature, an instance of such algorithms is called a *model*. To be able to efficiently solve a task, a model needs its parameters to be set up correctly, through an iterative process called the *training phase*. When a model with good performance is found, other models are built and tested to attempt to outperform the first one. Then, all the built models can be benchmarked to measure their accuracy, their behavior, their commonalities and differences. The focus of this paper is to present a comparison and a user evaluation of different visualization techniques that enable to compare results obtained by two different trained models on the same classification task.

A classification task consists in assigning a label called *class* (for instance "cat") from a bounded dictionary (i.e. alphabet) to an input data (for instance, an image). The resolution of the task is entrusted to a model, which produces the output class label, called a *prediction*. To evaluate the performance of the model in charge of the task, one computes the predictions for each input sample of a dataset for which classes are already known (*ground truth*) and then, predictions are compared to ground truths.

L. E. Pommé (✉) · R. Bourqui · R. Giot · D. Auber
University of Bordeaux, CNRS, Bordeaux INP, LaBRI, UMR 5800, 33400 Talence, France
e-mail: luc.pomme-cassierou@labri.fr

R. Bourqui
e-mail: romain.bourqui@labri.fr

R. Giot
e-mail: romain.giot@labri.fr

D. Auber
e-mail: david.auber@labri.fr

B. Kovalerchuk et al. (eds.), *Artificial Intelligence and Visualization: Advancing Visual Knowledge Discovery*, Studies in Computational Intelligence 1126,
https://doi.org/10.1007/978-3-031-46549-9_7

The performance of a model can be evaluated at different levels. (1) At a global level, measures such as the accuracy (correct predictions / total number of samples) provide a score for the entire test set. (2) At a class level, measures are used to assess the performance of the model through the lens of specific classes. For instance, the recall of a class indicates whether the samples of that class were correctly classified.

Several works [4–7] contain a detailed analysis of their models. These analyses are often based on class-level measures and rely on a visualization called *Confusion Matrix* (CM) [8]. A CM is a square matrix in which the rows represent all possible ground truths of the input samples. Its columns represent all possible predictions of a model, in the same order. The cell (row, col) counts the percentage of samples of ground truth row predicted as class col. In standard CM representations, a gradient of colors maps the percentage in each cell to a color. Some variants display both the percentages and the mapped color in each cell. A dark color denotes a high percentage while a light color denotes a low percentage.

Confusion matrices are designed to efficiently illustrate the class performance of a single model. However, when designing a new model for a new classification task, a common approach consists in training several models before selecting the best one. While the accuracy can provide some cues about the overall behaviors of the compared models, it does not support the comparison of models at a class level. Especially when two models have a similar global performance, analyzing their performance at a thinner level (class level for instance) becomes crucial. In such case, one confusion matrix per model may be required. However, this visualization may not be sufficient to efficiently compare the model's performance at a class level. In particular, when comparing two models, it is important to assess (1) whether one of the models achieves better performance for all classes or (2) one model is better for a subset of classes but worst for another subset. Identifying these classes can also be important to the application domain. Some classes may be more critical than others, and any mistake in these classes may be prejudicial, as in the medical field, for example.

In this work, different strategies are studied to compare two models solving the same classification task. The contribution of this paper is twofold. First, we present the Relative Confusion Matrix (RCM) which enables a performance comparison of two models at a class level. Then, we conducted a user evaluation to compare RCM with two existing visualization methods to compare two models at a class-level. This paper is an extension to our previous work [9]. It includes more details on the evaluation design and on the user's feedback, and incorporates a real application of RCM in a context of deep neural networks pruning.

The next section (Sect. 2) provides a brief overview of existing model comparison techniques. Then, we present the new RCM visualization (Sect. 3) and how to interpret it. Then, we describe the user evaluation protocol (Sect. 4), the visualizations that are compared to RCM, and the results (Sect. 5) of the evaluation. Finally, the results are discussed (Sect. 6), and a brief use case of RCM is presented (Sect. 7) before drawing the conclusions (Sect. 8).

2 Related Work

In this paper, we are interested in the visual comparison of the performances of two models and this section discusses about existing techniques for that purpose.

The most common way to evaluate the performance of individual classification models is to compute measures such as the accuracy, the recall or the precision. It is well established that position-based [10] visualizations like bar charts are efficient approaches to represent these kinds of quantitative information, for a comparison purpose. These visualizations are particularly suited to compare one or two characteristics over several models [11]. For example, to compare the mean accuracy percentage, the mean error percentage or mean response times over several views during a user evaluation, simple position-based visualizations like bar charts or more sophisticated ones like box plots are commonly used.

For binary classification problems, the output probabilities of the models are sometimes considered instead of the output class. In these cases, different thresholds can be applied on the probabilities to determine one prediction class per input. For each threshold, it is possible to estimate a measure of how much a model benefits, making correct predictions for a class, compared to how much it costs, incorrectly classifying input samples from another class into that class. Then, a Receiver Operating Characteristic (ROC) [12] curve in a form of a line chart can be drawn where each point of the curve is one of the measures associated with a threshold. However, it is less suitable for problems with more than two classes, or when several models are compared together because of overlapping curves.

The visualizations are more complex when the compared characteristics of the models are distributions of values. In Squares [7], Ren et al. vertically juxtapose representations of individual classification models. Each row contains one horizontal histogram per class showing the distribution of the highest output probability of the samples. Each histogram shows the true positives, the false positives and the false negatives of its class. In each bin, the samples of the same ground truth are displayed with the same color, so the number of required colors matches the number of classes and the number of rows of the visualization is the number of models being compared.

Alsallakh et al. [6] spread the classes along a confusion wheel to display the distributions of the samples. For each class, they use colors to differentiate performance measures such as True Positives (TP), False Positives (FP), False Negatives (FN) and True Negatives (TN) in histograms of classification probabilities. In this representation, the diameter of the wheel is strongly dependent on the number of classes and the resolution of curved histograms. This may be problematic when comparing the characteristics of two spatially distant classes, especially when the number of classes is large.

In general, a confusion matrix is used [5, 13] to compare the performance of different classes in a model. The literature establishes a few variants of confusion matrices that share the advantage of being model-agnostic visualizations. For example, NEO [14] fills each matrix cell with a colored rectangle whose size is proportional to the percentage in that cell, instead of using a gradient of colors to encode percentages.

As Hinterreiter et al.[4] stated, very few works address the problem of efficiently comparing the performance of two models at a class level. Most of the time, the model's quality is not questioned at a class level. Though, when models comparison is available at a class level, the CM of two distinct models are not even displayed at a time, which does require memory attention from the user to compare both models. For example, Li et al. [15] display one CM at a time per neural network obtained through several iterations of pruning.

For visual comparison problems in general, Gleicher [16] suggested three approaches to design components. The first and most common way to compare different objects of the same type is the *juxtaposition* of instances of the same visualization. Talbot et al. [5] juxtapose multiple confusion matrices for class-level comparisons. The second way consists in the *superimposition* of several instances of the same visualization. Basak et al. [17] establish a parallel between superimposed directed weighted graphs and superimposed adjacency matrices. In each cell, they display two colored concentric rectangles to encode the edges weights of two graphs. A parallel can be established between this adjacency matrix design and a confusion matrix design where the two colored concentric rectangles in each cell represent the percentage produced by two classification models. The third way uses a *explicit encoding* to combine the information of the two objects. In this last category, ConfusionFlow [4] suggest comparing multiple confusion matrices in a single visualization using one polyline in each cell. Each point on a polyline represents the percentage of samples classified in that cell by one model. This representation allows a global comparison of several models at a class-level. This method seems well suited when the polylines in the cells are almost monotonic, for example in the case of checkpoint models taken from the same learning process, because the performance tends to improve in each cell. In case of a comparison of multiple independent models, this visualization might be difficult to apprehend.

3 Relative Confusion Matrix

The goal of this work is to establish an efficient visualization to compare at a class scale two distinct models A and B that solve the same classification task. To make this comparison fair, we consider from now on that both models are fed with the same input data. They also necessarily have the same amount of output classes. This section presents our first contribution: the Relative Confusion Matrix (RCM), a visualization designed to efficiently compare two models at a class level.

The design of the RCM is inspired by the design of standard confusion matrices: it is a square table where rows represent all possible ground truths. Columns represent all possible output predictions, in the same order.

The RCM falls in the *special encoding* category as all of its cells encode the cell-by-cell absolute percentage difference between models B and A. In standard confusion matrices, the gradient of colors represents the percentages in each cell. In RCM, the gradient of colors represents the variation of the percentage in each cell between B and A. A dark color indicates a large variation between B and A while a light color indicates a small variation between B and A. This variation can be either positive if B performs better than A, negative if A performs better than B or zero if both perform the same. In practice, the gradient scales on the opacity color channel, not on the RGB channels. Moreover, the opacity range is fixed to [0.1, 1] either for positive or negative variations. When there is no variation, the opacity is set to 0. Even if this approach reduces the number of different colors that can be used in the gradient, it has the advantage to make a clear separation between cells with any variation (light blue) and cells with no variation (white color).

To indicate which model performs the best on each cell, when a difference exists between the two models, a symbol is displayed in the center of the cell on top of the background color. Three cases are then distinguished, so only two distinct symbols are required. The absence of variation between the two models in a cell is indicated with an absence of symbols and a white color (i.e. a blue color with an opacity of 0, on a white background). This absence of symbols is designed to increase the contrast between cells with any variation and cells with no variation, to ease the readability of the matrix. As the symbols are drawn in the center of the cells, they hide some pixels of the background color of the cells in which they are drawn. The more pixels that are hidden, the fewer pixels that display the background color of the cell, which can potentially alter their perception, particularly for models with many classes or when the matrix contains an agglomeration of symbols in neighbor cells. To limit the number of hidden colored pixels in the cells, the chosen symbols must have a simple shape, which means no loop. For efficiency, the two symbols must be dissimilar to be easily distinguishable. Ideally, they also are semantically opposed to each other. The $+$ and $-$ symbols satisfy all of these constraints. They are well suited to denote positive or negative differences between A and B. They have opposite meanings. They also require a different number of pixels to draw, since a $+$ requires almost twice as many pixels. Other symbols like letters (A or B for example) could have been reasonable choices but these letters contain both loops and sharp borders which together add complexity to the symbol. Also, for the same font size and thickness, $+$ and $-$ are drawn with fewer pixels. Thus, a $+$ (resp. $-$) is used in the cells where B (resp. A) shows better performance than A (resp. B). Besides, $+$ and $-$ symbols are particularly appropriate in a context of model B being a model derived from A. In that case, a $+$ (resp. $-$) can be seen as B improving (resp. deteriorating) the performance compared to A. Moreover, one could have used two gradients of colors to distinguish diagonal cells from non-diagonal cells or to differentiate positive variations from negative variations. To avoid the problem of lightness comparison of several gradients of colors in case positive variations need to be compared to negative variations, we only used one gradient of colors for the whole RCM. Figure 1 shows an example of RCM on had-crafted data and illustrates how to read the RCM on symbols on both diagonal and non-diagonal cells.

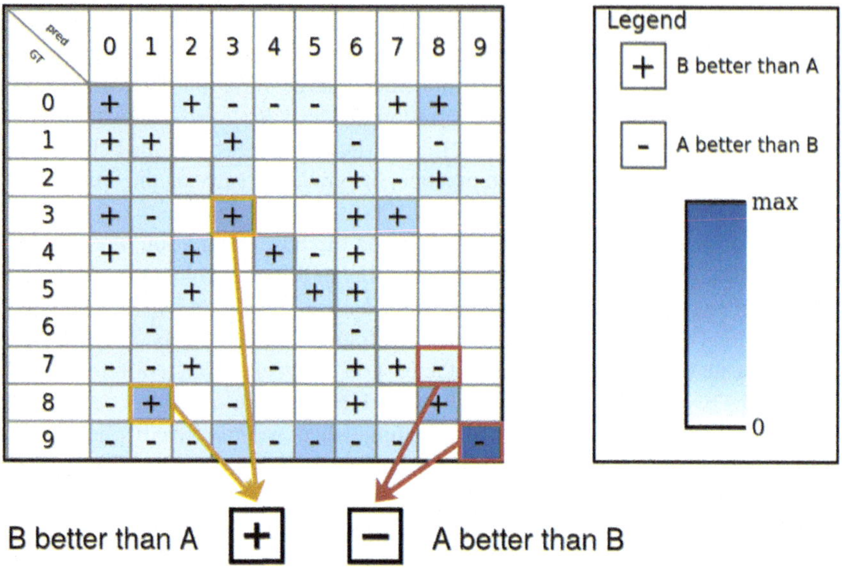

Fig. 1 Example of Relative Confusion Matrix (RCM) representation. Symbols (if any) designate the best model while the gradient indicates how large the difference between A and B is

4 User Evaluation

To measure the efficiency of this new visualization, the evaluation described in this section follows the recommendations given by the Purchase protocol [11]. We compare RCM, a visualization falling in the special encoding category, to two visualizations inspired from the literature. We define four tasks to solve on each visualization. We also define how data have been generated for each task, and how the sessions with users were organized.

4.1 Visualizations

This work focuses on how a model performs relatively to another and we are not interested in the performance of models alone. We estimate the efficiency of RCM by comparing it to existing visualizations. We decided to evenly represent the three categories mentioned in Sect. 2 (juxtaposition, superimposition, special encoding) in the comparison, RCM belonging to the last one.

The Twin Confusion Matrix (TCM) represents the juxtaposition category and is a special case of the EnsembleMatrix [5]. This visualization is composed of two standard confusion matrices displayed side by side. The two standard confusion matrices use the same design with only a gradient of colors to represent the percentages in the cells, as shown in Fig. 2.

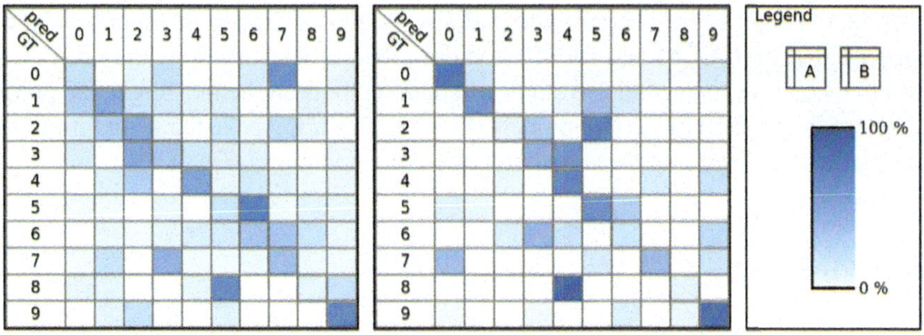

Fig. 2 Example of Twin Confusion Matrix (TCM) representation. Each matrix represents one of the two models *A* and *B*. A blue gradient encodes the percentages in the cells

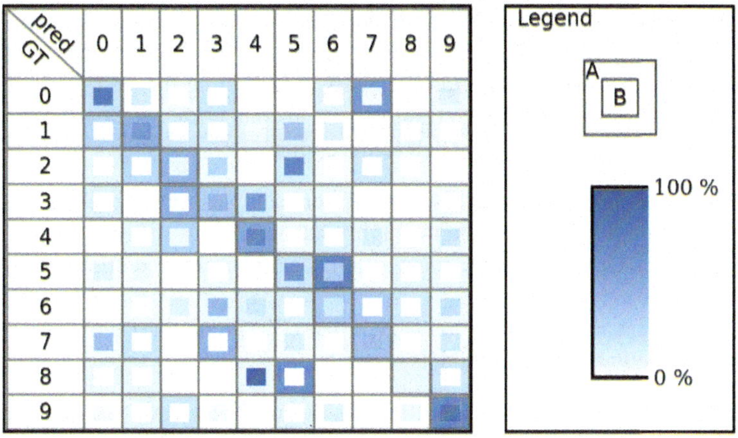

Fig. 3 Example of Combined Confusion Matrix (CCM) representation. The matrix displays a superimposed version of the two confusion matrices: each cell contains two squares. The outer one represents *A* while the inner one represents *B*

The Combined Confusion Matrix (CCM), which represents the superimposition category, is based on Basak et al. [17] adjacency matrix. This visualization is a confusion matrix whose cells encode the corresponding cells of both *A* and *B*. Each cell contains two nested colored squares: an outer square to represent the cell in *A* and an inner square to represent the same cell in *B*. The same gradient of colors is used for all the squares in all the cells, as shown in Fig. 3.

4.2 Hypotheses

In this evaluation, we consider TCM as a baseline visualization and we make the assumption that comparing two models with a superimposed visualization or a special encoding visualization is better than a side-by-side visualization. In other words, we suppose that both CCM and RCM are more efficient visualizations than the TCM baseline ($H_{baseline}$). Second, we study whether symbols are better suited than colors to determine the best of the two models (H_{symb}). Third, we make the assumption that finding the darkest color is easier than comparing contrasts (H_{dark}).

4.3 Tasks

To fairly measure the efficiency of the three visualizations, we designed four tasks to be solved by users. These tasks are based on the literature and focus on the performance difference between the two models. In Squares [7], two complementary options are suggested to compare models. The first one consists in detecting the classifier with the largest number of errors. The second one consists in comparing distributions of classes. We have adapted these two ideas to the three confusion matrix designs, so the first two tasks consist in indicating the best model under a given condition while the last two tasks consist in comparing two distributions. This led to the following tasks, formulated as questions:

- T1: Which model does correctly classify more samples for the highlighted class?
- T2: Which model is the best on the highlighted cell?
- T3: On which class is there the most difference of correct predictions between the two models?
- T4: Among incorrect predictions, which pair (GT, prediction) shows the greatest difference between the two models?

For T1 (resp. T2), the given condition is a diagonal cell (resp. non-diagonal cell) and the user is asked to select the best of the two models on that cell. In both cases, the suggested cell is chosen randomly with a uniform probability distribution. Both the line and the column of that cell are framed with a distinct color, directly on the matrix (RCM, CCM) or on the matrices (TCM).

To solve T3 (resp. T4), the user must select the diagonal (resp. non-diagonal) cell with the greatest difference between models A and B. The difference between T1 and T2 (resp. T3 and T4) is that the reasoning to solve the task is inverted. Indeed, diagonal cells indicate correct predictions while non-diagonal cells indicate incorrect predictions. The best model on a diagonal cell is the one that makes the most correct predictions (T1), while the best model outside of the diagonal is the model that makes the least classification mistakes (T2).

4.4 Dataset

The users were asked to solve the four tasks on a dataset of matrices represented with one of the three visualizations introduced in Sect. 4.1. To ensure the generalizability of the results [17], we randomly generated the matrices for three sizes (2×2, 10×10

or 30×30). The sizes would correspond to classification problems with respectively 2, 10 and 30 classes. A pair of matrices of equal sizes was generated for each trial, one matrix for A and one for B. In total, two pairs of matrices per task (T1–T4), per visualization (TCM, CCM, RCM) and per size (2×2, 10×10 or 30×30) were generated. The generator implements two constraints to produce models with relatively average or high performance, to try to mimic real classification models. First, we empirically fixed the density of all matrices to 0.3. This means exactly 30% of the nondiagonal cells of each matrix in a pair count a minimum of one sample. For each pair of matrices, the 30% filled cells in A are likely not the same cells in B. Thus, the percentage of nondiagonal cells that show a difference between A and B are in [30%, 60%]. The second constraint implemented in the generator forces all classes of each model to have a minimal recall of 0.3. This means each model is guaranteed to correctly classify in the diagonal a minimum of 30% of the samples of each class.

For every pair of matrices generated for any task, the right answer is determined by construction. The uniqueness of the answers is also verified during the generation process. Several additional pairs of matrices with various sizes were generated for the tutorial, and for training [11].

4.5 Evaluation Protocol

During the entire evaluation, the same person supervises all the participants. Every participant begins the evaluation with a hand-crafted tutorial designed specifically for this evaluation. The tutorial has the objective to limit the expertise gap that may exist between expert participants and participants that were unfamiliar with confusion matrices before the evaluation. This tutorial explains the interest of confusion matrices, how TCM, CCM and RCM are read and interpreted, and how to solve the four tasks on specific examples. An illustration of the tasks description of this tutorial is given in Fig. 4. For each task and for each visualization in the tutorial, the participant is given one answered example and one example to train. The correction is given after each training examples and the participant is free to retry or come back at any point in the tutorial to ask as many questions as necessary to the supervisor until the end of the tutorial. Once the participants have finished the tutorial, they are all given the same instructions and indications for the evaluation part. Both the instructions given in the tutorial and the instructions given by the supervisor insisted on not answering randomly or too fast, but correctly as fast as possible.

For the whole evaluation, the resolution of all matrices in pixels are the same. 2×2 matrices are displayed in 300×350, 10×10 matrices are displayed in 400×450, and 30×30 matrices are displayed in 500×550. These dimensions are a trade-off between several constraints illustrated in Fig. 5.

During the evaluation, the trials are grouped by task. The idea is to study how well participants solve tasks individually, not how quickly they can adapt to a task change. For generalizability, two latin squares are used to ensure the participants do not face either the same tasks or the same trials in the same order. All participants

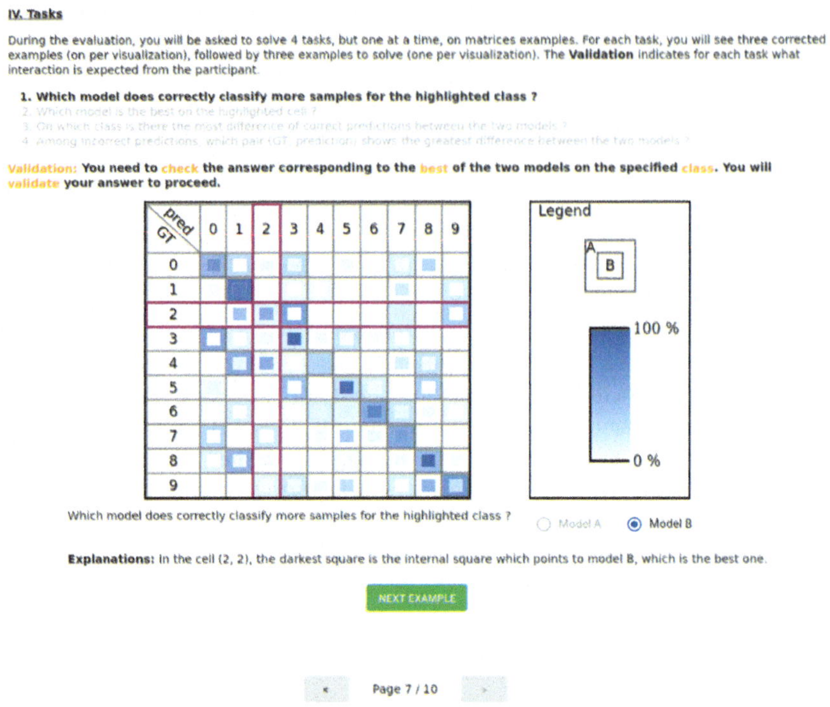

Fig. 4 Screenshot of the tutorial segment where T1 is detailed on a pair of matrices of size 10×10 with the CCM visualization

have one minute at their disposal to solve each trial, and three seconds break between each trial. Another break is scheduled after each task. In the end, the participants are asked to give their feedback on the visualizations and tasks through a short survey.

5 Results

This section presents the results of the user evaluation. These results are investigated under three conditions which are the three visualizations (TCM, CCM, RCM).

Participants. 26 volunteers in total performed the evaluation with a mean age of 31.5 years. The measures of three participants were excluded due to a lack of focus during the evaluation and because some trials were answered randomly, according to both the participants and the supervisor. We collected the answers to each trial for each participant to compute Error Percentages (EP), with a Response Time (RT) for each trial. One participant was considered an outlier based on its mean response time over all the trials, despite having the best mean error rate over all trials. Among the remaining 22 participants, some of them did not provide an answer to specific trials, within the time limit. The only conclusion that can be established in these cases is

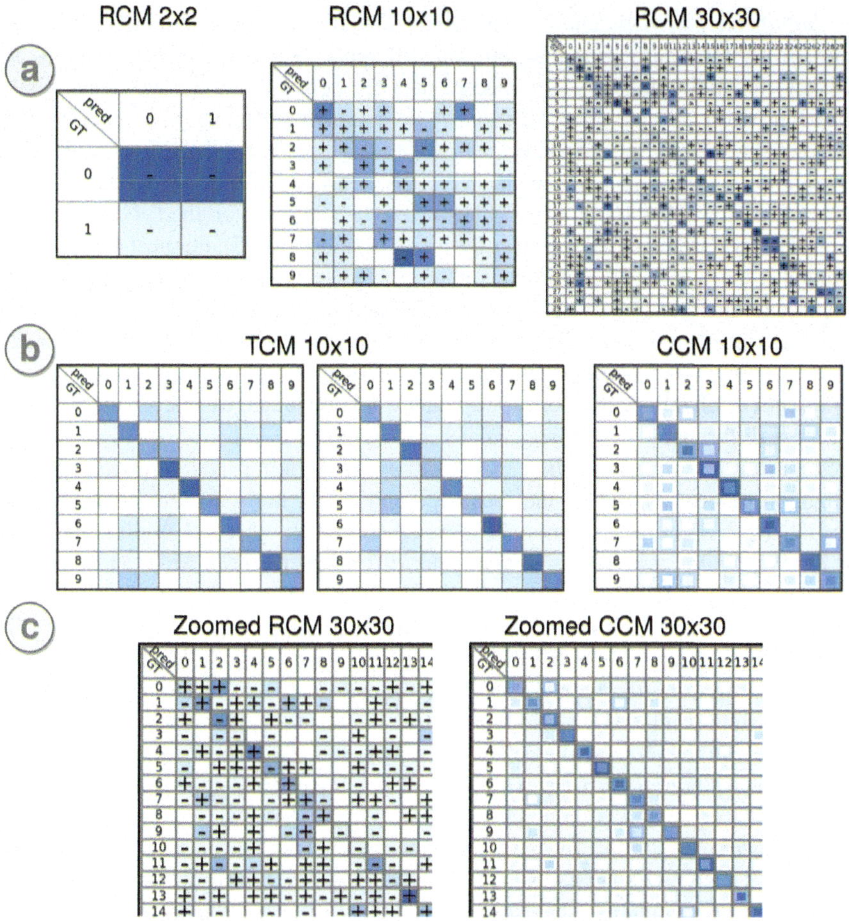

Fig. 5 Illustration of the constraints to fix the dimensions of the matrices in pixels. (a) First, to enable comparisons in the results analysis, the matrices of different sizes (2 × 2, 10 × 10 and 30 × 30) should approximately have the same dimension in pixels. (b) Second, for comparison purpose, the dimensions of each matrix should take into account that TCM requires the display of two matrices. Those two matrices should fit on screen without zooming or panning. It was decided during the pre-pilot tests [11] not to include such interactions on the visualizations to limit the number of studied parameters. (c) Third, the internal cells should not be too small because two visualizations require the content of the cells to be visible (two concentric squares in CCM and a symbol in RCM). (a)(b)(c) Last, for each visualization and each size, the matrix headers should have larger width/height than the internal cells to anticipate situations in which the classes are alphanumeric labels instead of indices

that one minute was not enough for these participants to answer these trials. Over all trials for all participants, only seven trials are concerned. Because no answer was given to these trials, the response times to these trials were considered meaningless and they were not included in the following statistical analysis.

Significance of visualization. This detailed analysis starts with a high-level comparison to detect any global significant difference between the visualizations for EP and RT, individually. The collected data for both EP and RT are not normally distributed for at least one condition. A nonparametric statistical approach, using the Kruskal-Wallis independent measures test was then used, with a significance threshold of 0.05. This test detects an overall significant difference on the mean RT ($p\text{-}value_{RT} = 4.9 * 10^{-8} < 0.05$). However, the same test on the EP data does not reveal any significant difference ($p\text{-}value_{EP} = 0.17 > 0.05$). For post-hoc pairwise comparison on RT, a Conover test with corrected $p\text{-}values$ was run and revealed a significant difference between all pairs of conditions. In Fig. 6, an orange bar indicates a significant pairwise difference to all other conditions. According to the mean response times of all conditions, one can conclude that globally, *RCM* is the most efficient visualization. Further experiments were conducted to analyze the reason(s) of these differences.

Significance of visualization, per task. By aggregating data per task and visualization before computing the mean RT and Error, the aim is to enhance some visualization properties that may be more efficient than others. Overall, at this level, a Kruskal-Wallis test detects a significant difference for both MEP and the mean RT per task. The post-hoc pairwise comparisons between each pair of visualization is depicted in Fig. 7. However, the only noticeable significant differences for T1 and T3 are between TCM and the other visualizations, and TCM shows the worst response times and the worst error percentages. According to the participants, the spatial distance induced by the juxtaposition of the two matrices of A and B forces to memorize the colors of the cells, which increases the response time of each trial and the risk of confusion between neighbor cells. Interestingly, the post-hoc pairwise tests do not detect any significant difference either for EP or RT when comparing all nondiagonal cells for T4, when the participants were asked to find the greatest amount of difference between incorrect predictions. This result is analyzed later with further investigation. The second significant difference concerns T2. There is a significant mean RT difference between CCM and RCM. In that case, RCM is more efficient than CCM. The participants are also more confused with CCM than for other visualizations. In this visualization, participants answer a little slower than

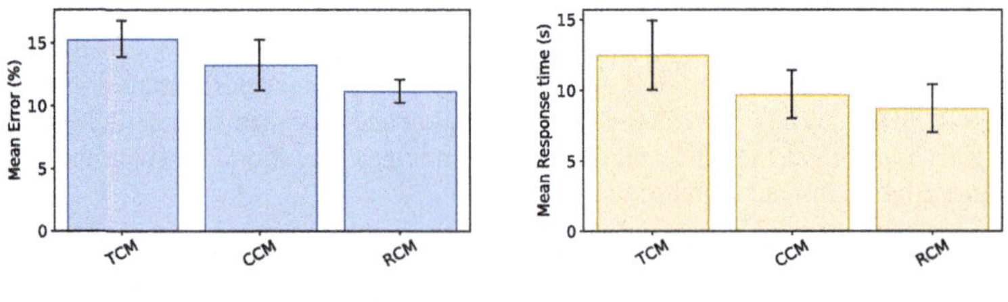

Fig. 6 Mean Error Percentage (MEP) and Mean Response Time (MRT) with standard error bars per visualization. Statistical tests showed that the visualization had a significant effect on the MRT, but not on the MEP. An orange bar indicates that the post-hoc pairwise comparisons of this visualization against all others is significant ($p\text{-}value < 0.05$) according to Conover test [18]. A blue bar alone indicates no significant difference using the same test

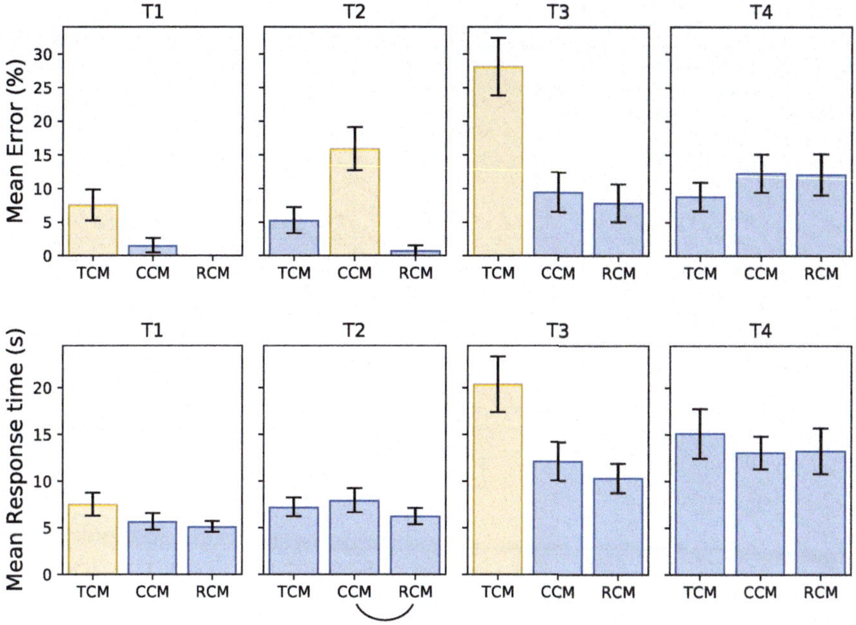

Fig. 7 Mean Error Percentage (MEP) and Mean Response Time (MRT) with standard error bars per task, per visualization. Post-hoc comparisons are run with the Conover test [18] when the Kruskal-Wallis tests show any significant difference. An orange bar indicates that the post-hoc pairwise comparisons of this visualization against all others is significant ($p\text{-}value < 0.05$). A blue bar alone indicates no significant difference using the same test. An additional arc connecting two blue bars indicates a significant posthoc difference between the two visualizations

RCM on average but tend to be misled by the color inversion in CCM between diagonal and nondiagonal cells. However, we observed that, at some point, some of the participants realized their mistake and changed their behavior for the next trials.

Significance of visualization, per task, per matrix size. Based on this analysis, we investigate further to detect whether the size of matrices impact the participants' performance on the different visualizations. The statistical results are illustrated in Fig. 8. In this figure, the phenomenon observed before between CCM and RCM appears with T2. As the matrix size grows, the EP difference is more significant. With an increasing matrix size, the CCM visualization has fewer pixels to display each matrix cell. Then, fewer pixels are available to display inner and outer squares which increases the difficulty to estimate small differences between close colors. With comparable matrices size, participants make fewer errors in less time with an RCM than with a TCM and a CCM. As shown in Fig. 8, statistical tests also reveal significantly faster response times for RCM without committing significantly more or less mistakes. The only case CCM outperforms RCM is on large matrices (30×30) for the comparison that must be done for T4, on all nondiagonal cells. The fact that RCM is less efficient than CCM could be explained by the density of black symbols in the matrix that may affect the perception of light colors. Nonetheless,

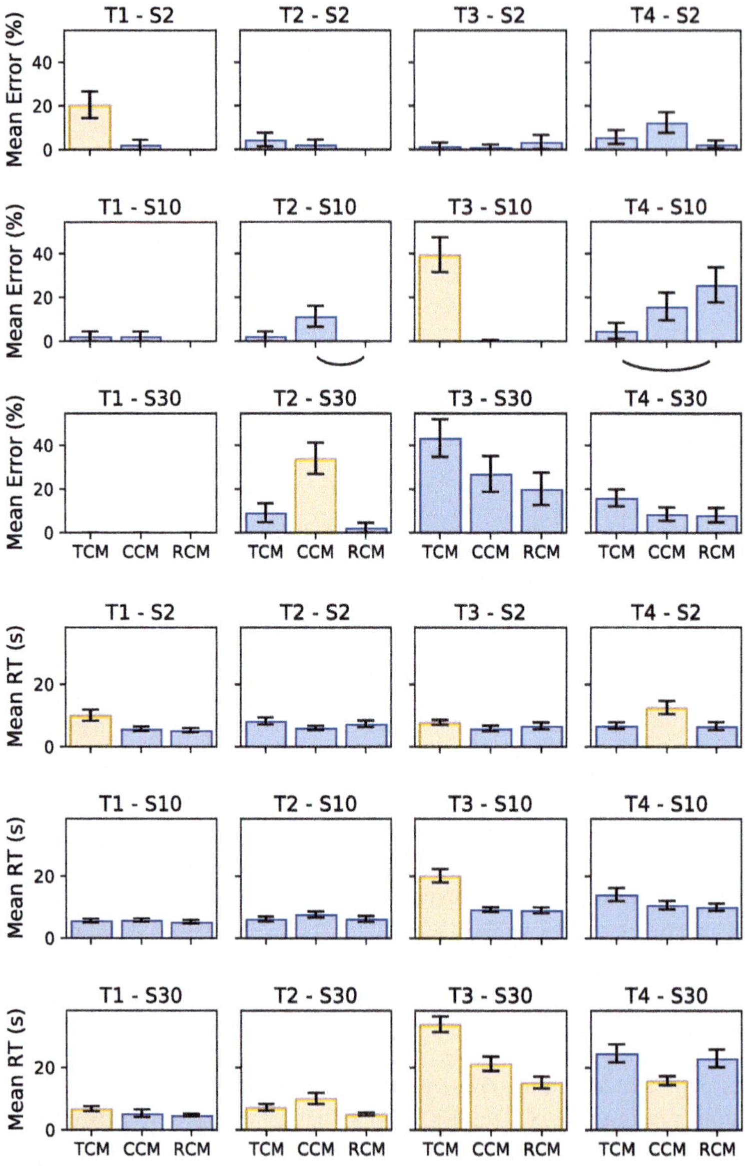

Fig. 8 Mean Error Percentage (MEP) and Mean Response Time (MRT) with standard error bars per task, per matrix size, and per visualization. Post-hoc comparisons are run with the Conover test [18] when the Kruskal-Wallis tests show any significant difference. An orange bar indicates that the post-hoc pairwise comparisons of this visualization against all others is significant (p-$value < 0.05$). A blue bar alone indicates no significant difference using the same test. An additional arc connecting two blue bars indicates a significant posthoc difference between the two visualizations

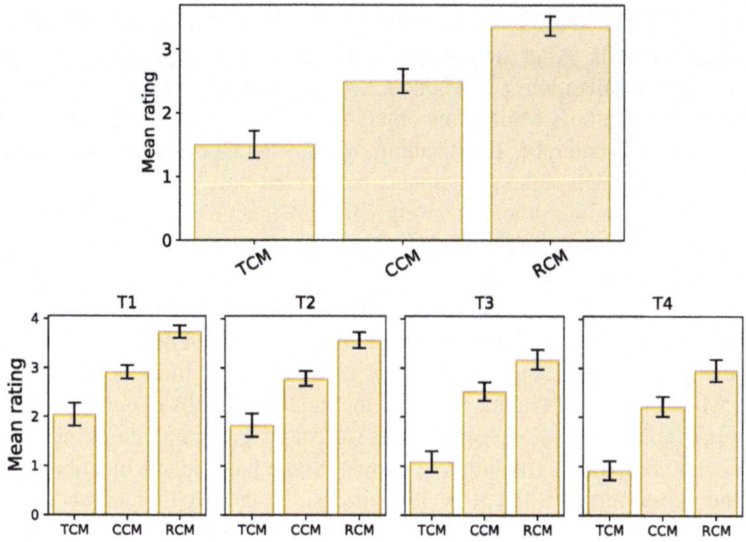

Fig. 9 Mean ratings per visualization (top) and Mean rating per task, per visualization (bottom) with standard error bars. All bars are orange, indicating that all the rating differences between pairs of visualizations are significant. The means and standard errors indicate RCM is the most preferred visualization

TCM almost never outperforms neither RCM nor CCM, which illustrates the problem of juxtaposing visualizations.

Participants' preference. All participants were asked to rate the three visualizations between 0 and 4, 0 being the worst mark and 4, the best mark. We collected their rates globally, and for each individual task. We analyzed these preferences with the Kruskal-Wallis test to have a qualitative evaluation of the three visualizations. The results indicate a significant difference in the participants' preferences for at least one of the visualizations. Post-hoc comparisons using the Conover were then conducted. They revealed a significant difference between all pairs after *p-values* correction. The ranks of the three visualizations showed that, most of the time, users preferred RCM to CCM to TCM. This tendency is exactly the same for each individual task (see Fig. 9).

6 Discussion

The user evaluation reveals that most users are not comfortable with manipulating side-by-side visualizations to compare models. Such comparisons are relatively slow and require from the user to memorize several colors at a time to make comparisons, especially when comparing distributions of colors (T3–T4). As expected, the participants had difficulties to solve the tasks when the elements to compare are spatially distant from each other. This validates our hypothesis $H_{baseline}$.

Surprisingly, the tasks T1 and T2 are not solved any faster with or without the symbols in the cells. After surveying our participants, it appeared that, unlike for the two other visualizations (TCM and CCM), some of the participants learned the meaning of the symbols while some others needed to systematically check the legend on the left to remember the meaning of the + and − symbols. This additional validation step, which some participants performed only for the RCM, artificially increased their response time. We strongly believe that an expert of the three visualizations that would be able to learn the meaning of the symbols without constantly using the legend would answer significantly more efficiently on RCM (H_{symb}).

The H_{dark} hypothesis concerns the tasks T3 and T4. The results on T3 tend to validate this hypothesis since participants found the darkest diagonal cells of RCM faster than the greatest contrasts of TCM or CCM. This difference is particularly visible when the size of the matrices and the number of cells to compare increases. However, to solve task T4, the participants were not significantly faster at finding the darker color compared to the higher contrast. Some participants mentioned that for large and dense matrices like 30 × 30 matrices, the density of symbols in the cells with small differences tend to disturb the perception of the colors in neighbor cells. To solve this problem, we suggest an interactive version of RCM with a threshold that determines the minimum amount of difference in a cell, from which a symbol needs to be displayed (see Fig. 10). Otherwise, this threshold could control the percentage

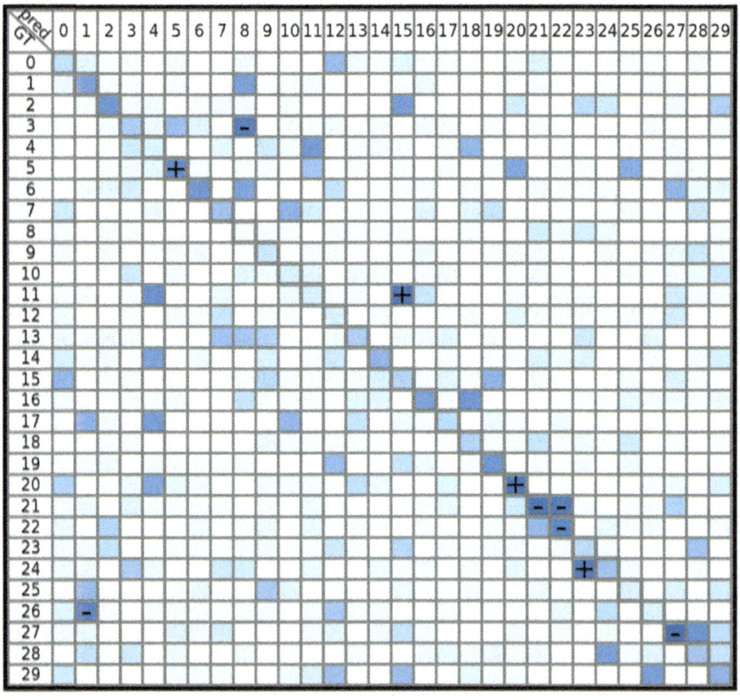

Fig. 10 Example of 30 × 30 Relative Confusion Matrix with a threshold at 90%: the symbols in the colored cells are only displayed if the difference is superior to 90% of the maximum value in the matrix in absolute value

of cells that show the greatest difference between the two models in which symbols have to be displayed.

Participants also mentioned the neighborhood of a cell can falsify the perception of its color(s), especially for TCM and CCM, when the aforesaid cell is close to the diagonal dark cells. To solve this problem, a participant suggested encoding the diagonal and non-diagonal cells with two distinct gradients of colors. Perhaps a more in-depth evaluation could help determine the trade-off between a clear distinction of the diagonal, especially with large matrices, and the comparison of different gradients of colors.

7 Real Application of RCM

The results of the user evaluation demonstrated that most of the time, RCM either outperforms or is equivalent to TCM and CCM. This section illustrates the benefits of RCM in a real use case like deep neural networks pruning.

Deep neural networks are modern model architectures that reach state-of-the-art performance on many tasks, particularly classification tasks [3]. These models easily reach millions of parameters and require billions of operations to process a single input data which makes them expensive to use in production and hard to embed in small IoT devices [15]. The performance of such models is usually measured on a global level with metrics such as the accuracy while the class-level performance is studied through standard confusion matrices. The idea behind network pruning is to reduce the resources requirements (computational cost and memory usage) to run a network A, pre-trained on a dataset of interest, while maintaining as much as possible its performance.

In a previous work, we presented NetPrune [19], a visualization tool that enables to analyze the importance of all parameters in the network. This tool also provides means to remove some parameters based on their importance with regard to the proposed metrics, to obtain a small-sized neural network. However, removing too many parameters at a time may irreversibly damage the performance of A. For that reason, the pruning process used in NetPrune is exploratory and iterative: each step, some parameters are removed from the network architecture of A which is then fine-tuned to try to recover the potential performance losses. A new network B is produced as a result, in the end of a step. This network has fewer parameters than its parent network A and needs to process fewer operations for each input data. Each new network solves the same classification problem its parent solves.

Removing parameters during an iteration tends to damage the global network performance, but not necessarily all the classwise performance are damaged. Moreover, the fine-tuning steps sometimes produce networks with better performance than their parent. In that case, after each iteration, it is important to determine whether the child network has better or worse performance globally, but also locally (i.e. for each class). This is where the Relative Confusion Matrix comes into play. It enables a user to efficiently evaluate the classwise performance of a network B, relatively to its parent network A.

For this use case, we were interested in the classification of small-sized grayscale images representing ten different fashion articles, taken from the Fashion-MNIST dataset [20]. An instance A of VGG-16 [21] was trained on this dataset for 50 epochs and reached 88.21% of accuracy. This instance of VGG-16 contains 14.98M of parameters and 627.33M FLOPs (operations to process one input image). We made the assumption that A was oversized compared to the difficulty of the classification task and used NetPrune to iteratively remove some of its parameters. We consider the screenshots of two networks B and C, derived from A, where C is also derived from B, which means B (resp. C) is the result of several pruning iterations from A (resp. B). Network B reaches 91.46% of global accuracy, with 166.26k parameters and 52.29M FLOPs, while network C reaches 91.47% (+0.01%) with 70.47k parameters and 21.22M FLOPs. As both networks reach similar performance on the global level, the network with the lowest number (*i.e.* C) of parameters should be preferred. However, in practice, one may need to analyze the differences between B and C with more precision to make a decision related to the field of application. This information can be determined with the help of the confusion matrices of B and C displayed in Fig. 11. In this figure, the diagonal cells confirm that both networks demonstrate high performance for all classes even though the diagonal cell of the "Shirt" class has a lighter color for both B and C. However it is difficult to determine with precision which network performs the best on each class, mainly because the two cells to be compared, each time, in the two matrices, are spacially distant from each other. Figure 12 shows the same information with a Relative Confusion Matrix that displays the performance of C relatively to B. Since network B is an ancestor of network C, a $+$ can be interpreted as an improvement while a $-$ can be interpreted as a deterioration of B. Thus, it is clear that network C has improved the performance of class "Coat" (+1.40%) at the expense of class "Shirt" ($-$2.10%). At the same time, a $+$ in the dark blue diagonal cell of the "Ankle boots" class indicates that network C makes more correct predictions on that class. The $+$ in the cell ("Ankle boot", "Sneaker") indicates that most of the improvements of

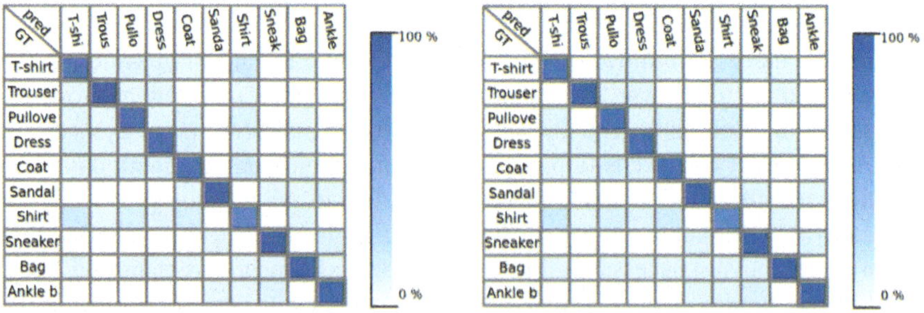

Fig. 11 Confusion matrices of two networks B (left) and C (right), both derived from A, an instance of VGG-16 [21], trained on the Fashion-MNIST [20] dataset. Here, C is also derived from B. The screenshots of the two standard confusion matrices are taken with NetPrune [19]

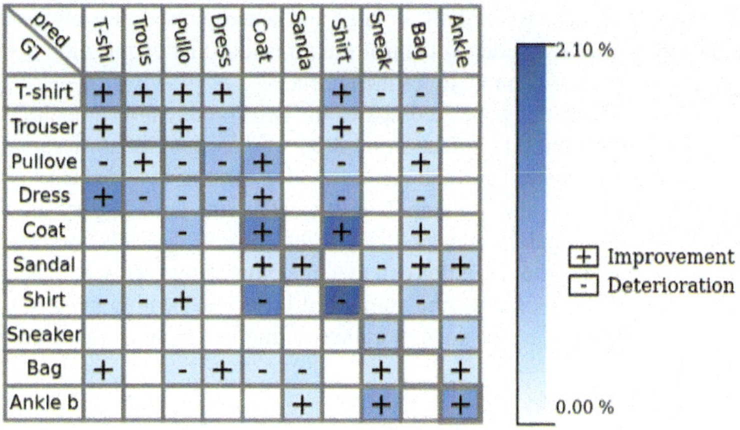

Fig. 12 RCM of the pruned network C against its ancestor B. Here, the symbols + (resp. −) indicate that C (resp. B) is better than B (resp. C). Since network B is an ancestor of network C, a + can be interpreted as an improvement while a − can be interpreted as a deterioration of B. The screenshot of RCM is taken with NetPrune [19]

C over B in the "Ankle boots" class is due to samples that are less misclassified as "Sneakers". The same remark can be made for "T-shirts" that are less misclassified as "Shirts" by C, or samples of class "Dress" that are less misclassified as "T-shirts" but more misclassified as "Shirts" or "Trousers" for example. In that case, depending on the field of application, network C is not necessarily preferable to network B.

Depending on the context or the requirements, looking into the classwise performance of a model may become crucial, and we show with this practical example that the Relative Confusion Matrix is well suited for the comparison of two classification models.

8 Conclusion

In this paper, we address the problem of effectively comparing the classwise performance of two classification models. We propose a new visualization called Relative Confusion Matrix (RCM) to encode the differences between two models with a gradient of colors and symbols. We validate the effectiveness of the RCM to compare two classification models through a user evaluation. This paper includes a real use case of RCM in a context of pruning deep neural networks.

This work shows the comparison results and an analysis of RCM, up to 30 classes. To address more classes and reduce the number of rows and columns of the confusion matrices, we suggest to establish a hierarchy that groups similar classes if it does not exist, and apply hierarchical interactions, as described by Görtler et. al. [14]. To enhance the most important symbols for large and dense matrices, we suggest adding an interaction tool like a slider to control the threshold that enables the display of symbols and reduce the complexity of the matrix.

Eventually, one may not want to compare only two classification models. To adapt our visualization to more classification models, we can imagine to designate each model with a unique symbol, which can be written with as few pixels as possible. Instead of displaying $+$ and $-$ symbols in the cells, we could display the symbol representing the best model in this cell. Then, the color of the cells could represent how much this model is better in this cell than the other. However, when the number of models to compare increases, the number of cells filled with a symbol tends to the total number of cells, hence the importance of finding symbols taking little space such as $+$ and $-$.

References

1. Ren S, He K, Girshick R, Sun J (2016) Faster R-CNN: towards real-time object detection with region proposal networks. IEEE Trans Pattern Anal Mach Intell 39(6):1137–1149
2. Girshick R, Donahue J, Darrell T, Malik J (2014) Rich feature hierarchies for accurate object detection and semantic segmentation. In: Proceedings of the IEEE conference on computer vision and pattern recognition, pp 580–587
3. Krizhevsky A, Sutskever I, Hinton GE (2017) Imagenet classification with deep convolutional neural networks. Commun ACM 60(6):84–90
4. Hinterreiter A, Ruch P, Stitz H, Ennemoser M, Bernard J, Strobelt H, Streit M (2020) Confusion flow: a model-agnostic visualization for temporal analysis of classifier confusion. IEEE Trans Vis Comput Graph
5. Talbot J, Lee B, Kapoor A, Tan DS (2009) Ensemble matrix: interactive visualization to support machine learning with multiple classifiers. In: Proceedings of the SIGCHI conference on human factors in computing systems, pp 1283–1292
6. Alsallakh B, Hanbury A, Hauser H, Miksch S, Rauber A (2014) Visual methods for analyzing probabilistic classification data. IEEE Trans Visual Comput Graph 20(12):1703–1712
7. Ren D, Amershi S, Lee B, Suh J, Williams JD (2016) Squares: Supporting interactive performance analysis for multiclass classifiers. IEEE Trans Visual Comput Graphics 23(1):61–70
8. Powers DM (2020) Evaluation: from precision, recall and F-measure to ROC, informedness, markedness and correlation. Preprint at arXiv:2010.16061
9. Pommé LE, Bourqui R, Giot R, Auber D (2022) Relative confusion matrix: efficient comparison of decision models. In: IV2022-IVE-information visualization evaluation and user studies
10. Mackinlay J (1986) Automating the design of graphical presentations of relational information. ACM Trans Graph 5(2):110–141
11. Purchase H (2012) Experimental human-computer interaction: a practical guide with visual examples. Cambridge University Press
12. Fawcett T (2006) An introduction to ROC analysis. Pattern Recogn Lett 27(8):861–874
13. Amershi S, Chickering M, Drucker SM, Lee B, Simard P, Suh J (2015) Model tracker: redesigning performance analysis tools for machine learning. In: Proceedings of the 33rd annual ACM conference on human factors in computing systems, pp 337–346
14. Görtler J, Hohman F, Moritz D, Wongsuphasawat K, Ren D, Nair R, Kirchner M, Patel K (2021) Neo: generalizing confusion matrix visualization to hierarchical and multi-output labels. In: CHI conference on human factors in computing systems, pp 1–13
15. Li G, Wang J, Shen HW, Chen K, Shan G, Lu Z (2020) CNN pruner: pruning convolutional neural networks with visual analytics. IEEE Trans Vis Comput Graph 1

16. Gleicher M (2017) Considerations for visualizing comparison. IEEE Trans Visual Comput Graph 24(1):413–423
17. Basak A, Bach B, Henry Riche N, Isenberg T, Fekete JD (2013) Weighted graph comparison techniques for brain connectivity analysis. In: Proceedings of the SIGCHI conference on human factors in computing systems, pp 483–492
18. Conover WJ, Iman RL (1979) On multiple-comparisons procedures, Technical Report LA-7677-MS, Los Alamos Scientific Laboratory
19. Pommé LE, Bourqui R, Giot R, Vallet J, Auber D (2023) NetPrune: a sparklines visualization for network pruning. Vis Inf
20. Xiao H, Rasul K, Vollgraf R (2017) Fashion-mnist: a novel image dataset for benchmarking machine learning algorithms. Preprint at arXiv:1708.07747
21. Simonyan K, Zisserman A (2014) Very deep convolutional networks for large-scale image recognition. Preprint at arXiv:1409.1556

Computation of Pixel-Oriented Grid Layout for 2D Datasets Using VRGrid

Adrien Halnaut, Romain Giot, Romain Bourqui, and David Auber

1 Introduction

Data visualization is an efficient way to analyze and communicate numerical data [1] that are intractable in tabular or textual formats. Such numerical data vary in size (i.e., number of elements in the dataset) and dimensionality (i.e., their number of features). Many visualization techniques [2] have been designed to get insights from high-dimensional datasets such as, for example, histograms and scatter plots [3]. However, all these techniques are constrained by the resolution of the end-user screen (i.e., number of available pixels). This constraint makes visualization of high-dimensional data challenging, especially when the number of dimensions is higher than the number of available visual attributes [4]. Tackling these challenges is crucial as usage of high-dimensional data is now common in many research and industrial fields, including bioinformatics [5] and artificial intelligence analyzis [6].

High dimensional data visualization is usually done by reducing the number of dimensions until it fits into the number of available visual attributes. Various dimension reduction techniques exist for different purposes [7]; they focus on optimizing different criteria, and their visualization method is usually a scatter plot, where x and y coordinates are mapped to the 2D projection and both size and color of the points depend on two other criteria. In this chapter, we assume data-points are dis-

A. Halnaut (✉) · R. Giot · R. Bourqui · D. Auber
University of Bordeaux, CNRS, Bordeaux INP, LaBRI, UMR 5800, 33400 Talence, France
e-mail: adrien.halnaut@labri.fr

R. Giot
e-mail: romain.giot@labri.fr

R. Bourqui
e-mail: romain.bourqui@labri.fr

D. Auber
e-mail: david.auber@labri.fr

B. Kovalerchuk et al. (eds.), *Artificial Intelligence and Visualization: Advancing Visual Knowledge Discovery*, Studies in Computational Intelligence 1126,
https://doi.org/10.1007/978-3-031-46549-9_8

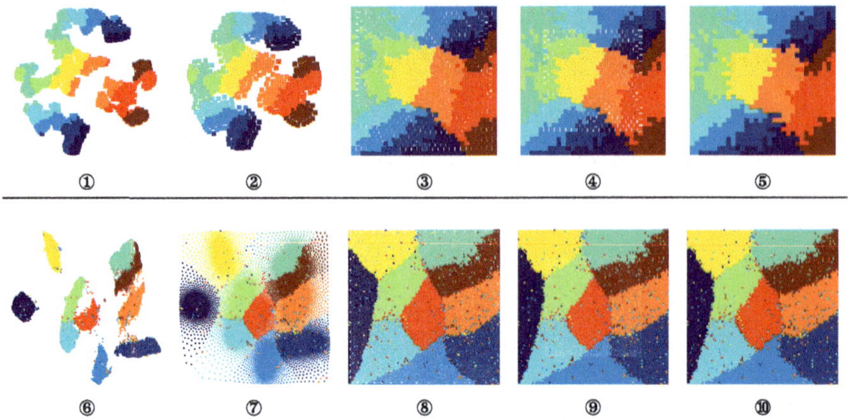

Fig. 1 Swissroll dataset (1 024 samples with 3 dimensions) projected with t-SNE in $\mathbb{R}^2$ ①, and MNIST dataset (10 000 samples with 768 dimensions) projected with UMAP in $\mathbb{R}^2$ ⑥. Both projections represent data in a lower dimensional space having many overlapping elements which hinder their interpretation. VRGrid is applied to both projections (⑤ and ⑩) to compute a shape-preserving, grid layout arrangement solving the overlapping issue. This method can be applied on any 2D dataset, or any $\mathbb{R}^N$ dataset projected into a $\mathbb{R}^2$. VRGrid consists in relaxing elements (② and ⑦) until a stable state is reached, and iteratively place elements into pre-computed grid emplacements, starting from the outer edges (③ and ⑧) toward the center (④ and ⑨)

played with the same size and their color attribute has a specific semantic. High-dimensional data visualization techniques are efficient in many fields [8], but the larger the dataset is, the larger the drawing space needs to be [9]. Popular modern high-dimension reduction techniques such as t-SNE [10] or UMAP [11] produce non-uniform data projections that group together similar elements of the dataset; as the size of the high-dimensional dataset increases, those groups get denser, and the visual occlusion of individual elements is globally increased. If not handled properly, this visual occlusion results in wrong assumptions about the dataset's nature [12]. For example, in Fig. 1, the MNIST dataset [13] is visualized using UMAP dimension reduction technique (⑥), and each element is colorized according to its class. The yellow cluster on the top-left part of the projection seems to be composed of only about 5 to 10 outliers, while in reality there are more than 20 outliers hidden behind other elements which are properly displayed with VRGrid (⑩).

Several methods exist to increase the perceptual scalability of dimension reduction methods. Here, we make the choice to design a pixel-oriented method that relies on a grid layout. Pixel-oriented methods aim to provide informative data visualization while prioritizing perceptual scalability over accurate spatial fidelity. This perceptual scalability is achieved by removing visual occlusion and maximizing drawing space usage, resulting in a compact visualization technique of data. Grid-layout arrangements are a subset of pixel-oriented visualization methods which are scaled to the pixel size, which is the smallest object the display monitor can draw, free of any overlapping data. The state-of-the-art method Self-Sorting Map [14] provides compact results with good information retention compared to regular non-uniform data projections. This method consists in populating a grid with high-dimensional data and performing permutations to optimize a correlation metric based on the dissimilarity between elements in both the high-dimensional space and in the grid's 2D space. Despite the good results, focusing on a single comparison criterion seems limited compared to the numerous regular data projection methods existing in the dimension reduction field. DGrid [15, 16] is a two-step process that takes as input a dataset alreadt projected in R2 and performs arrangements on projected elements to form a grid arrangement. Those arrangements are processed by first partitioning the drawing space and then performing arrangements locally. It prioritizes neighborhood conservation and is designed to scale to larger datasets with a complexity smaller than $O(N^2)$. Both methods are discussed in Sect. 2.

We propose a novel method that produces a grid arrangement of any 2D data, Voronoi Relaxation Grid (VRGrid). Compared to other arrangement methods, VRGrid takes 2D data as input (which may be projected high-dimensional data) and makes use of a relaxation algorithm to uniformly scatter the 2D points into a finite space. Points are progressively assigned into grid cells while respecting the configuration computed by the relaxation method. The relaxation is applied multiple times to progressively refine the placements of points that are still unassigned to grid cells. This process is done until all points are assigned to a grid cell. VRGrid can be used as a complementary visualization method of already projected high-dimensional data.

The relaxation method used by VRGrid here is Lloyd's algorithm [17], which is used to compute Centroid Voronoi Tessellations [18, 19] (CVT) configurations that is helpful in uniformly scattering a set of coordinates in a 2D space. We conduct a quality evaluation of VRGrid arrangements using various datasets, against state-of-the-art methods. The contributions of this chapter are a novel iterative method able to arrange any set of 2D points into a grid with minimal deformation, a public implementation of the algorithm and an exhaustive comparison between VRGrid and the state-of-the-art methods SSM and DGrid. This chapter extends a previous communiucation [20] by giving deeper information on VRGRid algorithm, alternative configurations and more details on the evaluation.

The outline of this chapter is as follows. Section 2 presents state-of-the-art methods to produce grid-based arrangements. Section 3 presents the VRGrid algorithm, its limitations and an analysis of its computation cost. Section 4 presents how the evaluation protocol has been conducted and Sect. 5 its results. Section 7, presents several VRGrid variations to further improve the quality/computation time ratio of the method. Finally, Sect. 8 summarizes the paper and highlights how VRGrid can be improved.

2 Related Works

Regular data projections are efficient in representing data on a screen, but their construction requires empty drawing space to represent differences between two elements and often results in overlapping data when the drawing space is too small. Various techniques exist to project methods that aim to optimize readability over accurate distance encoding in data representation, and are presented in this section.

General Space-Filling Visualization [21] is a way to improve this drawing space usage by encoding input data into a structure that fills a squared drawing space. *Fractal Curves* are an efficient way to represent data organization. It consists in ordering high-dimensional data along a fractal curve using a dissimilarity metric. Fractal curves such as Hilbert's [22] are designed to preserve the proximity of their elements once they are drawn in a 2D space. Keim's work on pixel-oriented visualization [23] shows that this technique is used in several application domains. Those fractal curves are not limited to orthogonal structures; GosperMap [24] makes use of Gosper curves to present hierarchical data in an undetermined visual structure, resulting in a geographic map-looking visualization. However, while being efficient in their application domain, the necessary reduction of data to the 1D-space (i.e., the order on the fractal curve) implies a large information loss with complex data.

Grid Arrangement Methods can be categorized into two subsets, the first one projects data directly into a grid layout starting from high-dimensional space, while the other makes use of an intermediate dimensional reduction method before organizing the lower-dimensional result into a grid.
IsoMatch [25] makes use of the IsoMap [26] dimension reduction technique to project high dimensional data into a 2D space, followed by the Hungarian algorithm [27] to fit projected data into a specific structure (usually a grid). The Hungarian algorithm performs combination optimizations over an assignment problem. In compact data projection, the assignment problem is presented as positioning elements into a grid with each possible assignment associated with a cost function, usually representative of the data deformation it would create. While the Hungarian algorithm is designed to find the best possible configuration, its high computational cost of $O(n^4)$, with n being the dataset's size, makes it hardly usable for large dataset, and by extension the IsoMatch method itself. VRGrid uses a similar approach as it consumes projected data as an input and performs arrangement from it to build a grid. However, VRGrid's computational complexity can be adapted to scale to large datasets at the cost of arrangement quality, as explained in Sect. 3.2.
DGrid (Distance-preserving Grid) [15, 16] is a grid arrangement method designed to be a post-processing step of reduction dimension method. It consumes projected data positions as input and partitions the drawing space according to their sorting order on both x and y axes. Partitions are then progressively split in half until they can fit in a predetermined grid cell. This method is shown to be very efficient in both neighborhood preservation and processing time, and its complexity of $O(N.logN)$ makes it scalable to large problems. However, due to the partitioning strategy, deformation of data topology can occur on unevenly scattered data (e.g., t-SNE). In comparison,

VRGrid strategy is not based on partitioning data and thus its grid arrangement does not produce such deformations, but at the cost of a higher computational cost.

Self-Sorting Map (SSM) [14] takes high-dimensional data and arranges them onto a grid such that the distance between arranged elements represents dissimilarity between them. Elements are first randomly disposed into a grid, then smartly swapped to maximize the correlation between distances and a given dissimilarity metric. SSM progressively split the grid and manages to produce structured layouts in a relatively short time frame with a given complexity of $O(L(logN)^2)$ and can be lowered using parallel computing. This method is proven to be efficient, but only dissimilarity information is used to arrange the input data. Topology data such as the positions of the input dataset are unused. VRGrid strongly differs from this approach as the input data coordinates are taken into account during the grid construction, and relies on already projected data to work. Here, we focus primarily on methods retaining as much information from the original projection as possible, so methods relying on fractal curves (projecting into $\mathbb{N}$) or sampling the data (*e.g.*, Self-Organizing Maps [28] or OoDAnalyzer [29]) are not included in the scope of this chapter.

3 VRGrid: Voronoi Relaxation Grid

The VRGrid method (summarized in Fig. 2) aims at computing a square grid-layout arrangement for a given dataset P (but can be easily generalized for rectangles). There is no restriction on the nature of the dataset if each element of the dataset is a 2D coordinate. VRGrid relies on the application of a relaxation algorithm able to evenly scatter 2D points into a convex and bounded 2D space. It first initializes a structure for the 2D points of the input dataset to be placed into, then, for every iteration of the method, applies the relaxation algorithm and iteratively assigns scattered 2D points to the grid cells. Only a subset of points is assigned at each VRGrid iteration, starting from the edges of the grid toward its center, to prevent distortion of the configurations computed by the relaxation method. The relaxation method is then reapplied on unassigned points in a smaller space and the process goes on until all points are assigned to a grid cell. This section describes each step of VRGrid's process and explains the design choices.

3.1 Algorithm

In this section, we detail each step of the algorithm which is depicted in Algorithm 1. The algorithm has an initialization step, and then cycles between the application of the relaxation methods and the point assignation to the grid cells.

Initialization. For a given dataset P composed of n 2D points, we note G the grid structure where P will be arranged.

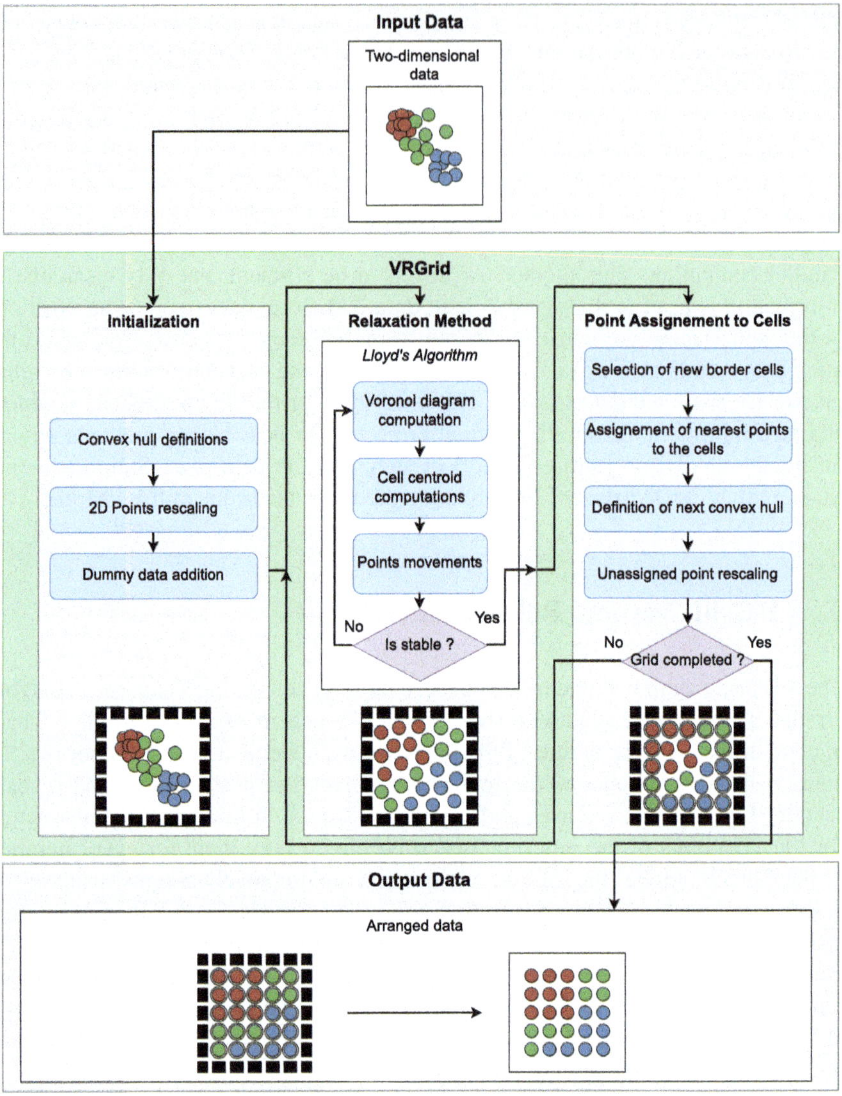

Fig. 2 VRGrid consumes a set of 2D points and outputs a grid-based arrangement of it. It first initializes the grid structure and prepares the data to be processed. Then, VRGrid iteratively scatters data using a relaxation method and assign points in the current border cells until all points are assigned

G is a square grid composed of $(\lceil\sqrt{n}\rceil + 2) \times (\lceil\sqrt{n}\rceil + 2)$ cells: the $\lceil\sqrt{n}\rceil \times \lceil\sqrt{n}\rceil$ cells where P points will be assigned during the VRGrid application, and $4 \times (\lceil\sqrt{n}\rceil + 1)$ additional cells around it to facilitate the algorithm execution. However, these cells will not contain points. We note $\mathcal{B}_0, \mathcal{B}_1, ...\mathcal{B}_k$ the convex hulls centered

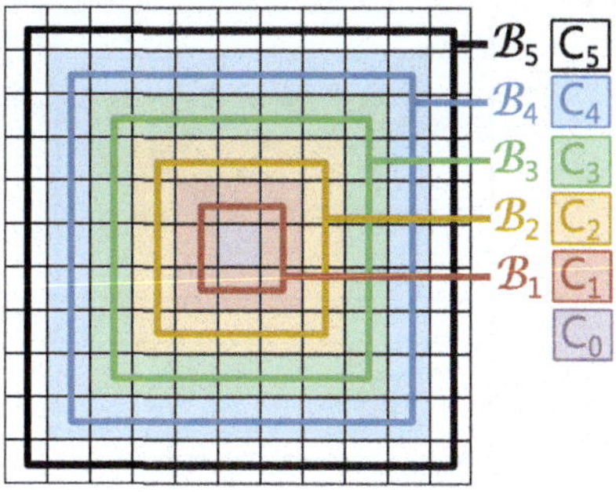

Fig. 3 Definition of convex hulls ($\mathcal{B}_1$ to $\mathcal{B}_5$) and associated delimiting cells (C_0 to C_5, colorized according to their appropriated convex hull) for a 11×11 sized grid used for a 9×9 sized dataset

around the middle of G, with $k = \lceil \sqrt{n}/2 \rceil + 1$. Each convex hull $\mathcal{B}_i$ corresponds to a square with a width of w_i cells such as $w_i = (2i - 1)$ if $\lceil \sqrt{n} + 1 \rceil$ is odd and $w_i = 2(i + 1)$ otherwise.

Figure 3 illustrates the construction of the borders of the convex hulls and their appropriate representing cells for an 11×11 sized grid. We note C_i the set of cells of G that represents $\mathcal{B}_i$ for each convex hull. If $\lceil \sqrt{n} + 1 \rceil$ is odd, C_0 is composed of the middle cell of G. These sets are chosen in a way that at the i-th iteration of VRGrid, 2D points are relaxed inside the space delimited by $\mathcal{B}_{k-i}$ and are assigned to the cells in the C_{k-i-1} set. Next, 2D coordinates of the points of P are rescaled to fit into $\mathcal{B}_k$; this ensures that there is enough grid cells inside $\mathcal{B}_k$ for the 2D points to be assigned to: $|C_{k-1} \cup C_{k-2} \cup ... \cup C_0| \geq |P|$. Furthermore, as the next parts of the algorithm requires that there is no 2D points having exactly the same coordinates, each coordinate is being slightly displaced by randomly chosen 2D vectors. This displacement is small enough to differenciate two coordinates but not enough to alter the original meaning of the dataset. The last initialization step of VRGrid aims to handle where $|P| < \lceil \sqrt{n} \rceil^2$, which is when there are more grid cells than points to assign. This configuration would result in empty cells and because of VRGrid's iterative nature, these empty cells would be located toward the middle of the grid and hinder its visualization. To prevent this configuration from happening, we add "dummy" data into P. These "dummy" points are processed in the same way as the rest of P but are removed from the result to not be confused with the actual data. We chose to uniformly place them in B_{k-1}, the relaxation algorithm can still change their coordinates in next step. To illustrate an iteration of VRGrid step by step, Fig. 4 shows complete initialization of VRGrid on a randomly generated dataset of 78 points, distributed as three clusters.

Relaxation Method. VRGrid makes use of relaxation algorithm to scatter 2D points in a finite and convex space. At the i-th iteration of VRGrid starting with $i = 0$, this space is defined by the $\mathcal{B}_{k-i}$ convex hull, and unassigned points P_i are processed by the relaxation algorithm such as:

$$P_i = P_{i-1} \backslash C_{k-i} \quad \text{with } P_0 = P \tag{1}$$

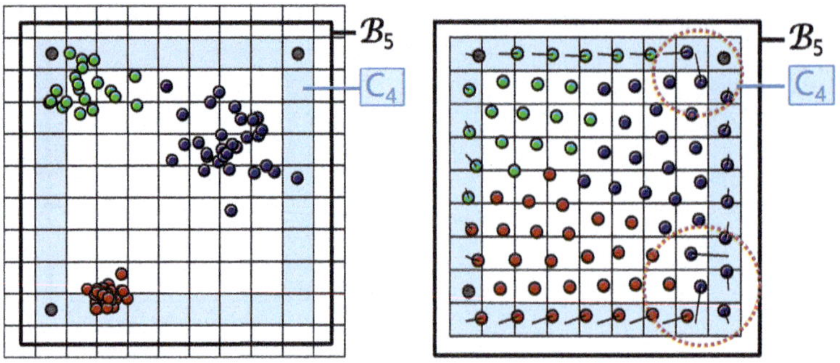

(a) VRGrid initialization

(b) Result of Lloyd's algorithm after 223 iterations

Fig. 4 Overview of VRGrid's first iteration on a sample dataset. In (a), current convex hull is $\mathcal{B}_5$ and C_4 (light blue cells) is the first set of cells to be assigned P points to, as defined in Fig. 3. Points have been rescaled to fit into the convex hull and $9 \times 9 - 78 = 3$ dummy points (grey circles) have been placed in C_4 to use all grid's cells at the end of the algorithm. In (b), segments represent the nearest point of P_0 for every cell of C_4. Red dashed circle highlights assignments that may seem unintuitive due to the assignment order. This is due to how the order of assignments is computed which results in farther candidates as the assignments go

The relaxation algorithm needs to evenly scatter P_i points toward the cells in C_{i-1}.

As mentioned earlier, we chose to use Lloyd's algorithm [17] to compute the relaxation of 2D points. It takes as input a set of coordinates D and iteratively processes them to compute a Centroid Voronoi Tesselation [19] (CVT) configuration. This configuration implies that for a computed Voronoi diagram based on a set of 2D points, each generating point of a cell is also its centroid. This disposition results in evenly scattered points into the finite space. While its convergence has been proven in 1D spaces [18], the convergence of the algorithm in higher dimensional spaces, including 2D spaces, has yet to be proven. However, it has been proven [30] that in a bounded convex space, the algorithm always converges toward a CVT configuration, which is the case here because of the usage of convex hulls defined earlier.

Lloyd's algorithm consists in repeating the following steps until stabilization: (1) The Voronoi diagram V is computed based on D. Each point $d \in D$ is associated with its corresponding Voronoi cell $v_d \in V$. (2) The centroid $C(v_d)$ of the cell v_d is computed. (3) Each point d is moved to their corresponding $C(v_d)$.
Stabilization is reached when the distance between each point d and their respective centroid $C(v_d)$ is smaller than a given ϵ. Smaller ϵ values give a more precise coordinate disposition and need more algorithm iterations. For exemple, Fig. 4b shows stabilization reached after 223 iterations of Lloyd's algorithm.

Since VRGrid only focuses on the position assignment of the points closest to the convex hull edges, we can impose a fixed number of iterations to Lloyd's algorithm to separate these candidates from the other points without waiting for complete stabilization. Doing so reduces the number of iterations and thus the overall computing time, but potentially lowers the final arrangement quality.

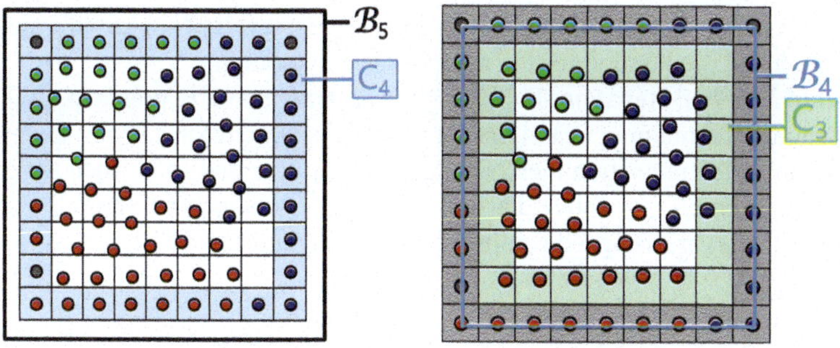

(a) Assignation to C_4 (b) Next iteration preparation

Fig. 5 Point assignation step (a) and preparation for next VRGrid iteration (b). Convex hull is changed to $\mathcal{B}_4$ and C_3 is the next set of cells to be assigned points to. Other P points are rescaled to a smaller scale to stay inside of $\mathcal{B}_4$

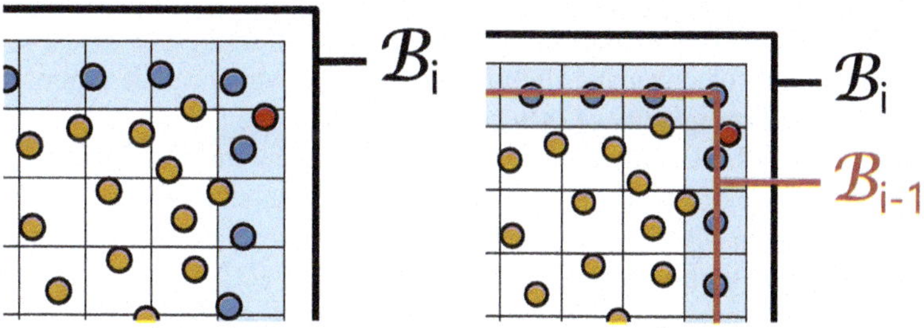

(a) Assignation to C_{i-1} (b) Next iteration preparation

Fig. 6 Problematic configuration where a point (red circle) is not chosen to be assigned in C_{k-i-1} cells (blue circles) despite being close to $\mathcal{B}_{k-i}$. Because of that, when defining the new convex hull $\mathcal{B}_{k-i-1}$, the point is outside of it, and cannot be processed by the relaxation method as intended on the next VRGrid iteration. Rescaling unassigned data solves this issue

Point Assignation to Grid Cells. After the relaxation step (i.e., once points are scattered enough into the current convex hull $\mathcal{B}_{k-i}$), we compute for each cell $c \in C_{k-i-1}$ its closest point $p \in P_i$. p is then assigned to c's position. In Fig. 4b, the assignment of closest points is shown by a segment toward their respective cell (Fig. 5).

We then prepare the next iteration $i + 1$ of VRGrid; convex hull $\mathcal{B}_{k-i-1}$ defines the area for the next application of the relaxation method and C_{k-i-2} the next grid cells to assign 2D points to. However, there is no guaranty that every point of P_{i+1} is inside $\mathcal{B}_{k-i-1}$ after C_{k-i-1} assignment. Indeed, the relaxation step is applied on the whole space delimited by $\mathcal{B}_{k-i-1}$, so any point of P_i could be close to the hull border and still not be assigned to a C_{k-i-1} cell. Since only $|C_{k-i-1}|$ points are assigned to grid cells, one or more points can be left out between the border of $\mathcal{B}_{k-i}$ and $\mathcal{B}_{k-i-1}$. Figure 6 presents an example where such a situation occurs.

To ensure this does not happen before the next relaxation algorithm application, we apply a rescale of P_{i+1} points to fit into the newest convex hull at the end of each

VRGrid iteration. Figure 5 shows the assignment result following Lloyd's algorithm application and preparation for the next VRGrid's iteration.

End of VRGrid Iteration. Following the assignment of P_i points in C_{k-i-1} and the rescaling of remaining unassigned points, the next iteration $i + 1$ is applied (i.e., application of the relaxation method in the $\mathcal{B}_{k-i-1}$ bounded space then assignment of P points into C_{k-i-2} cells). This iterative process is applied until all grid cells (except C_k cells) are assigned a point of the input dataset.

3.2 *Complexity Analysis*

This section details the computational complexity of VRGrid by inspecting each step of the algorithm. For better readability, we name n the size of the input dataset including eventual dummy points.

Initialization. VRGrid initialization step consists in building a set of convex hulls, rescaling input data of size n and slightly displacing it, which can be summarized as the computational complexity of $O(n)$.

Relaxation Method. This part differs depending on the chosen relaxation method for VRGrid. In case of Lloyd's algorithm, which consists of three steps: computing the Voronoi diagram ($O(n.log(n))$), computing the centroid of each cell ($O(n)$) and moving coordinate to the corresponding cell's centroid ($O(n)$). The computational complexity of each iteration of the algorithm is thus $O(n.log(n))$. The number of iterations i needed for Lloyd's algorithm depends on the density of the processed data (denser data needs more iterations than already evenly scattered data), and the stabilization factor ϵ (lower values and needs more iterations than higher values). As presented in the previous section, i can also be fixed to not wait for stabilization and thus lower the overall complexity of the relaxation method. For this relaxation method, the resulting computational complexity is $O(i.n.log(n))$.

Point assignment to cells. VRGrid's point assignment steps consist in looking in P for the closest points of each cell in C_i, assigning them and scaling the remaining unassigned P points to the next convex space. The nearest element of any coordinate in a 2D space can be efficiently computed using a quad-tree structure of $O(n.log(n))$ complexity. Assigning points and the scaling process are both $O(n)$. The point assignment step results in the computational complexity of $O(n.log(n))$.

VRGrid iterations. VRGrid re-applies the relaxation method until all $G \backslash C_k$ cells are used. The number of iterations is defined by the number of convex hulls k built during the initialization step. VRGrid thus needs to iterate $\lceil \sqrt{n}/2 \rceil + 1$ times.
As a result, VRGrid's complexity is bound to:

$$O\left(\sqrt{n}.i.n.log\left(n\right)\right) = O\left(i.n^{\frac{3}{2}}.log(n)\right) \tag{2}$$

Algorithm 1: The full VRGrid algorithm.

Data: Set of N points $\mathcal{D} \subset \mathbb{R}^2$;
Stability indicator ϵ
Result: Set of N points $\mathcal{D} \subset \mathbb{N}^2$
▶ Initialization
$k \leftarrow \left\lceil \sqrt{N}/2 \right\rceil + 1$;
$stable \leftarrow \bot$;
$doRescale \leftarrow \bot$;
if $\left\lceil \sqrt{N} \right\rceil + 1 \mod 2 = 0$ **then**
  $C_0 \leftarrow \{(-1, 1), (1, 1), (1, -1), (-1, -1)\}$;
else
  $C_0 \leftarrow \{(0, 0)\}$;
for $i \in \{1, \cdots, k\}$ **do**
  $C_i \leftarrow \{(-i, i), (i, i), (i, -i), (-i, -i)\}$;
  for $x \leftarrow \{-i + 1, \ldots, i - 1\}$ **do**
    $C_i \leftarrow C_i \cup \{(-i, x), (i, x), (x, i), (x, -i)\}$;
$\mathcal{F} \leftarrow C_k; \mathcal{D} \leftarrow rescale(\mathcal{D}, k)$;
$\mathcal{D} \leftarrow displace(\mathcal{D})$;
▶ Main loop
$iter \leftarrow 1$;
while $|\mathcal{D} \backslash \mathcal{F}| > 0$ **do**
  while $\neg stable$ **do**
    $d \leftarrow 0$;
    $V \leftarrow voronoiDiagram(\mathcal{D})$;
    for $u \in \mathcal{D}$ **do**
      $c_u \leftarrow computeCentroid(V_u)$;
      ▶ d allows to check if current configuration is stable, and thus may be computed differently if needed.
      $d \leftarrow d + ||u - c_u||$;
      $u \leftarrow c_u$;
      $doRescale \leftarrow doRescale \vee isOutside(C_{iter}, u)$;
    if $doRescale$ **then**
      $\mathcal{D} \leftarrow \mathcal{F} \cup rescale(\mathcal{D} \backslash \mathcal{F}, k)$;
    $stable \leftarrow d \leq \epsilon$;
  ▶ Snap point to grid positions
  $qt \leftarrow quadtree(\mathcal{D} \backslash \mathcal{F})$;
  for $c \in C_{k-iter}$ **do**
    $tgt \leftarrow closestInQuadtree(qt, c)$;
    $\mathcal{D}[tgt] \leftarrow c$;
    $\mathcal{F} \leftarrow \mathcal{F} \cup \{\mathcal{D}[tgt]\}$
  $iter \leftarrow iter + 1$;

4 Experimental Protocol

We evaluate how VRGrid competes against state-of-the-art methods by looking at how well data topology is represented by the method and how different are the proposed arrangements from their input data. Five aspects are evaluated: pairwise distance preservation, neighborhood preservation, pairwise relative positioning, global positioning, and computation time. In the evaluation, we focus on VRGrid, and the baselines Self Sorting Maps [14] and DGrid [15, 16]. We perform the evaluation in a context of large dataset processing ($\geq$ 1 000 elements). Self-Sorting Maps (SSM) is used and evaluated as a post-processing method, meaning its input data is already projected, 2D data.

VRGrid varies in its accuracy by changing the ϵ value and its number of fixed iterations i. Smaller ϵ values would result in more precise data representation but slower processing time while higher ϵ values would result in faster processing time but imprecise data positioning. Similarly, higher i values would result in better data representation at the expense of longer processing time, while lower i values would result in faster processing time for worse data positioning.

The number of iterations before obtaining a stable configuration varies from one dataset to another, but most of the node movements are computed during the first iterations of the algorithm, as shown for example in Fig. 7. In this evaluation, we test two configurations of VRGrid: VRGrid$_P$ is computing Lloyd's algorithm until stable configuration with an ϵ value fixed to 10^{-3}. VRGrid$_F$ is using a fixed number of iterations i which results in faster processing times at the cost of quality degradation.

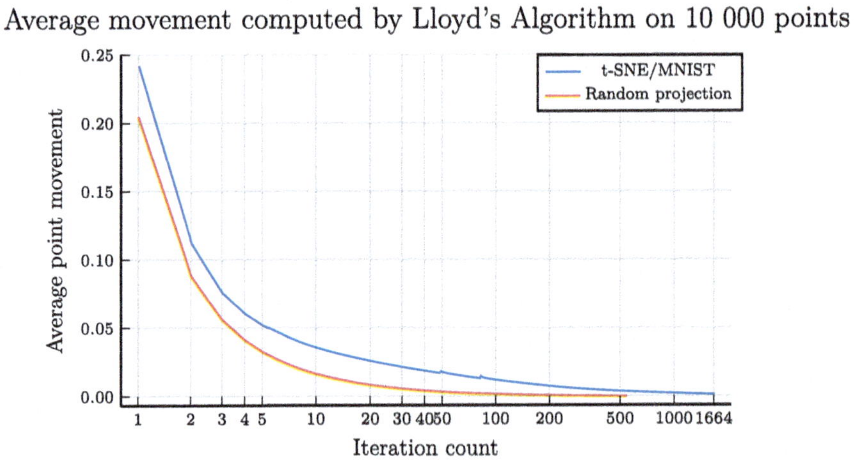

Fig. 7 Average distance traveled by each point of datasets of varying density when applying the Lloyd's algorithm. The t-SNE projection of MNIST has several dense areas, while the randomly scattered dataset has uniform density. The Lloyd's algorithm needs 1644 iterations to reach stable configuration in the first dataset but only needs 544 iterations on the second one. It should be noted that most of the traveled distances happen in the first iterations of the algorithm

We fix i to 10 for Lloyd's algorithm to limit the relaxation mainly on the closest points of the convex hull border instead of the most denser point area.

4.1 Evaluation Metrics

The evaluation focuses on manifold understanding (reflected by pairwise distance preservation and neighborhood preservation) and drawing deformation minimization (depicted by relative and global positioning movements). We measure these aspects for each scenario using the following metrics:

Distance Preservation (DP) is evaluated using the Cross-Correlation metric to compare the distance between the grid arrangement and the 2D input data spaces. It is used by SSM as a loss function [14]. This comparison is done by computing pairwise distances in the input data and how they are correlated in the arrangement. Distances that are not correctly represented in the grid arrangement bring an empirical penalty to the evaluation:

$$DP = \frac{1}{|\mathcal{D}|^2} \sum_{(s,t)\in\mathcal{D}^2} \frac{\left(\mathcal{G}(s,t) - \overline{\mathcal{G}}\right)\left(\delta(s,t) - \overline{\delta}\right)}{\sigma_{\mathcal{G}}\sigma_{\delta}} \tag{3}$$

where $\mathcal{D}$ is the dataset, $\overline{\mathcal{G}}$ and $\overline{\delta}$ are respectively the mean distances between elements in the grid and $\mathbb{R}^2$. $\sigma_{\mathcal{G}}$ and σ_{δ} are standard deviations of the distance respectively in the grid and in $\mathbb{R}^2$. Higher DP values mean better distance preservation.

Neighborhood Preservation (NP). As scatter plots are usually interpreted by how elements are positioned compared to the others, this metric rates how similar is each element's neighborhood in the grid arrangement to the input data.

Usually, neighborhood preservation is evaluated by computing the ratio of the k-nearest neighbors for each element between the input dataset and the projection. As this evaluation is focused on grid arrangements, we consider the neighborhood $knn_{\mathcal{G}}(k, i)$ of the element i to be composed of the elements around i within a square area of k cells size, including diagonals. This also means that elements close to the border of the grid have a smaller neighborhood than others. This neighborhood is then compared with the input data using the same number of neighbors:

$$NP = \frac{1}{2 - 2^{1-r}} \sum_{1\leq i\leq r} \sum_{\forall s\in\mathcal{D}} \frac{|knn_{\delta}(s, f_i(s)) \cap knn_{\mathcal{G}}(s, f_i(s))|}{|\mathcal{D}| \times f_i(s) \times 2^{i-1}} \tag{4}$$

with $\mathcal{D}$ the dataset, r being the maximal Chebyshev distance the neighborhood should be computed to, $knn_{\delta}(s, k)$ computing the k nearest neighbors of s in $\mathbb{R}^2$ and $knn_{\mathcal{G}}(s, k)$ the k nearest neighbors of s in the grid using Chebyshev distance, and $f_i(s)$ the function giving the number of neighbors to compute for $s \in \mathcal{D}$ depending of its position in the grid:

$$f_0(s) = 0, \quad f_i(s) = \begin{cases} f_{i-1}(s) + 1 + 2i & \text{if } s \text{ is in a corner} \\ f_{i-1}(s) + 1 + 4i & \text{si } s \text{ is on a border} \\ f_{i-1}(s) + 8i & \text{otherwise} \end{cases} \tag{5}$$

With this formula, higher values means the better the neighborhood is preserved.

Relative Positioning Preservation (RPP) is computed using the Orthogonal Ordering metric primarily designed to evaluate shape preservation of overlap solvers in graph drawing [31]. We use this metric to evaluate the preservation of each element relatively to others before and after computing the grid arrangement:

$$RPP = \frac{ooni_x + ooni_y}{N(N-1)} \tag{6}$$

with N being the dataset size, $ooni_x$ and $ooni_y$ the axis inversion count on both axes:

$$ooni_x = \sum_{\substack{(u,v)\in\mathcal{D}^2 \\ x_\delta(u) > x_\delta(v)}} \begin{cases} 1 & \text{if } x_\mathcal{G}(u) < x_\mathcal{G}(v) \\ 0 & \text{otherwise} \end{cases} \tag{7}$$

$$ooni_y = \sum_{\substack{(u,v)\in\mathcal{D}^2 \\ y_\delta(u) > y_\delta(v)}} \begin{cases} 1 & \text{if } y_\mathcal{G}(u) < y_\mathcal{G}(v) \\ 0 & \text{otherwise} \end{cases} \tag{8}$$

with $x_\mathcal{G}(s)$ and $y_\mathcal{G}(s)$ returning the x and y coordinate of s in the grid, and $x_\delta(s)$ and $y_\delta(s)$ doing the same but in $\mathbb{R}^2$.

Lower RPP values mean better preservation of elements relative to others and thus better results, while higher values mean higher deformation and thus worst results.

Global Positioning Preservation (GPP) is being computed using the Node Movements metric. It evaluates how coordinates moved from the original projection to the resulting arrangement [31]. While coordinate movements are unavoidable if the shape of the projection is altered, we assume the faithful grid arrangement of 2D data involves minimal movements:

$$GPP = \frac{1}{|\mathcal{D}|} \sum_{s\in\mathcal{D}} ||\mathcal{G}(s) - shift(s)||^2 \tag{9}$$

$$shift(s) = (x_\delta(s) + xbb_\mathcal{G} - xbb_\delta, y_\delta(s) + ybb_\mathcal{G} - ybb_\delta) \tag{10}$$

where $xbb_\mathcal{G}$ and $ybb_\mathcal{G}$ are the x and y coordinate of the convex hull's center in the grid, and xbb_δ and ybb_δ doing the same but for the $\mathbb{R}^2$ data. Lower GPP values mean smaller alterations from the original positions and thus better results, while higher values mean worst results.

4.2 Datasets and Visualization Methods

Datasets. In the evaluation, we use datasets varying in sizes, dimensions, and natures. Datasets "Swissroll", "Broken swissroll", "Helix" and "Twinpeaks", generated with the *Matlab Toolbox for Dimensionality Reduction* [32], have very specific shapes in a 3D space and are often tested in the dimension reduction research field, including SSM [14]. Their shape reveals the strategies of the projection methods and their limitations. To reflect actual real-life scenarios, we also proceed with our evaluation on real-life datasets MNIST [13] and Fashion-MNIST [33] (or F-MNIST) which are both commonly used in the machine learning field. Using t-SNE [34, 35], UMAP [11], MDS [7, 35] and IsoMap [26, 35], we then project the datasets in a 2D space to generate our evaluation datasets. Every combination result of these datasets are called "real datasets" in our evaluation.

We also proceed with our evaluation on 800 randomly generated 2D datasets varying in sizes (64, 100, 1 000 and 10 000 points) and scattering distribution methods (evenly scattered using Permuted Congruential Generator [36] and dense using normal distribution) to evaluate the stability of the grid arrangement methods.

Data Visualization. In this evaluation, we focus on how well evaluated methods can represent their 2D input data into a grid. Dimension reduction methods used to generate testing datasets are renowned for their efficiency, and it is common to colorize projected data using class information. This colorization usually reveals data clusters and outliers. However, well-clustered points can hardly be visually compared against each other since they have the same color. To visually evaluate how grid arrangement result differs from its input data, we colorize arranged points according to their initial 2D positions. Doing so shows initially close points with a similar hue while initially distant points will have a strongly different color. We can thus compare how points moved compared to their neighbors after the arrangement even if they were inside the same cluster.

Self Sorting Map and DGrid Implementations. At the time of the redaction of this chapter, the SSM algorithm implementation of its authors is no longer accessible [37]. The SSM algorithm used in this study is implemented in Rust following the directives of the authors' paper [14]. According to our evaluation of the Cross-Correlation Score, which is also used in the authors' paper, the quality of our produced results is very close to the authors' algorithm performances.

Both Orthogonal Ordering and Node Movement Minimization metrics are comparing the positioning of coordinates between the original projection and grid arrangement, which preservation is not guaranteed by the SSM method by design. To evaluate SSM with RPP and GPP, we automatically correct SSM results by applying all possible orthogonal transformations to the arrangement: flip of grid cells on X-axis and/or Y-axis and retaining the configuration which has the best GPP value. The Fig. 8 depicts an example of such correction.

DGrid has been implemented following the algorithm presentation given by the authors [16] in Python. Obtained results in both computation times and neighborhood preservation correlate with the performances presented in the authors' paper [15].

(a) Input data (b) SSM output (c) SSM output flipped on X-axis

Fig. 8 Example of SSM result correction on UMAP/Swissroll projection (**a**). The SSM arrangement (**b**) is not well aligned with (**a**), as red elements are on the right grid side although they are placed on the left in the input image. This result has a GPP value of 0.38, but if this arrangement is flipped on its X-axis as seen in (**c**), the GPP value is lowered to 0.08 which is better only requiring a trivial post-process

5 Results

We compare methods across the five metrics presented in Sect. 4.1 and present several results computed on real data, colorized accordingly to the original dataset labeling in Fig. 9 and their 2D position in Fig. 10. To evaluate whenever a method is significantly better than the other on a given metric, we make use of the Wilcoxon signed rank statistical test [38] computed on the random datasets. If the computed p-value is lower than 5×10^{-2}, then the results are significantly different. Wilcoxon signed rank test's p-values between VRGrid configurations and both SSM and DGrid are presented in Table 1 and are all lower than 5×10^{-2}. All differences between VRGrid_P, VRGrid_F, SSM and DGrid evaluations are thus statistically significant.

Visual Inspection. Input data are identifiable in VRGrid_P and VRGrid_F results, the initial positions of clusters and their composition are preserved in the grid arrangements. However, VRGrid_F has distortions along both diagonals especially noticeable using position-based colorization. VRGrid_P has no such visible distortions and has a better-looking separation between the different element classes. In SSM arrangements, the position of the different item groups is slightly different compared to the input data, while still being recognizable (e.g., UMAP projection of MNIST and t-SNE projection of Fashion dataset are slightly rotated clockwise by SSM compared to the input data). Aside from that, SSM arrangements look quite like VRGrid_P 's with sharper borders between each data class. However, when using position-based colorization, we can notice a few elements far from their original neighbors, notably in the UMAP/Swissroll scenario. In DGrid arrangements, the partitioning strategy is very noticeable on MNIST and Fashion results in the form of small square-shaped areas. In the Swissroll/UMAP arrangement, the placement of the red classes in the lower-left part of the grid does not seem intuitive. Position-based colorization reveals partitioning artifacts even more which can disturb data visualization interpretation.

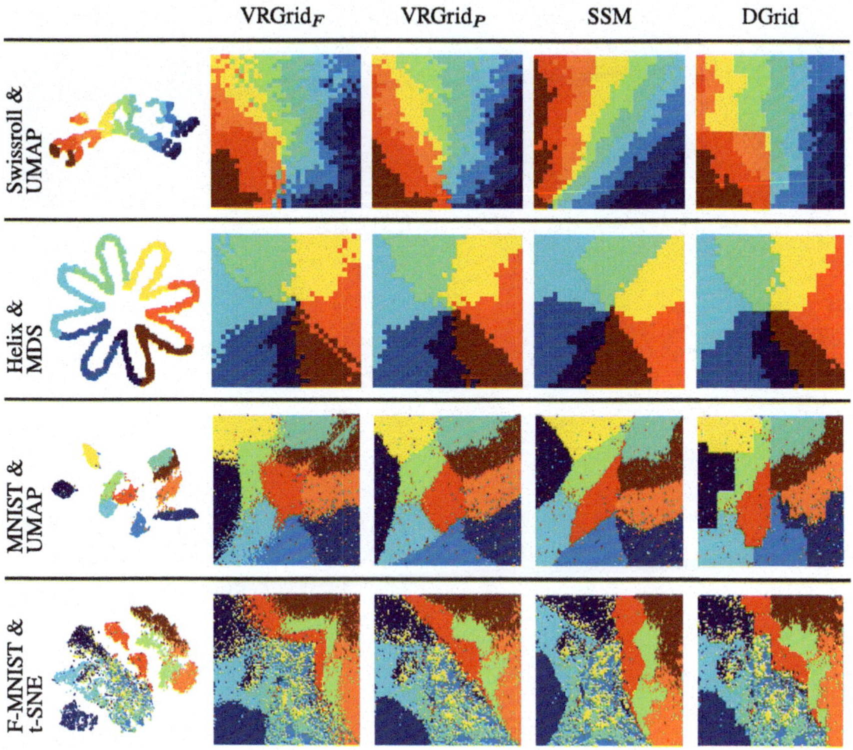

Fig. 9 Obtained results on real datasets, projected in $\mathbb{R}^2$ (left column) using UMAP, MDS and t-SNE. In this table, elements are colorized using their label. Input data is globally recognizable among all arrangements, but VRGrid$_F$ and DGrid arrangements show distortions and fragmented configurations

Distance Preservation (DP). Our Distance Preservation measure is computed using Cross-Correlation and results are depicted in Fig. 11, with higher values meaning better pairwise distance preservation. On real datasets, distances are well preserved across all methods. VRGrid$_P$ produces the best results overall with values about 2.57% better than SSM, 2.61% better than DGrid, and 5.41% better than VRGrid$_F$. On randomly generated datasets, VRGrid$_P$ is also better with results about 1.77% better than SSM, 6.73% better than DGrid, and 2.63% better than VRGrid$_F$. It can be concluded that all four methods manage to preserve distances when arranging data into a grid on real datasets, but VRGrid$_P$ performs better.

Neighborhood Preservation (NP). Neighborhood Preservation results are depicted in Fig. 12, with higher values meaning better neighborhood preservation across elements. On real datasets, DGrid produces the best results which are overall 8.94% better than SSM, 19.44% better than VRGrid$_P$, and 37.52% better than VRGrid$_F$. On random datasets, SSM produces the best arrangement with values

Fig. 10 Obtained results on real datasets, projected in $\mathbb{R}^2$ (left column) using UMAP, MDS and t-SNE. In this table, elements are colorized using their initial 2D position. This colorization rule allows to highlight disformations from the input datasets, such as VRGrid$_F$ diagonal distortions (1st column) or SSM's occasional misplacements (1st row, 3rd column)

Table 1 Computed p-values of the Wilcoxon signed rank test performed on VRGrid$_P$ and VRGrid$_F$ against SSM and DGrid. All values are lower than 5×10^{-2} making differences between all methods statistically significant in our evaluation

Method	Metric	SSM	DGrid	Method	Metric	SSM	DGrid
VRGrid$_P$	DP	3.52×10^{-56}	6.51×10^{-67}	VRGrid$_F$	DP	2.00×10^{-4}	1.66×10^{-28}
	NP	1.03×10^{-83}	7.20×10^{-34}		NP	7.71×10^{-114}	1.93×10^{-6}
	RPP	6.14×10^{-125}	1.37×10^{-27}		RPP	1.70×10^{-89}	6.17×10^{-19}
	GPP	2.55×10^{-125}	4.20×10^{-43}		GPP	2.01×10^{-104}	3.92×10^{-7}

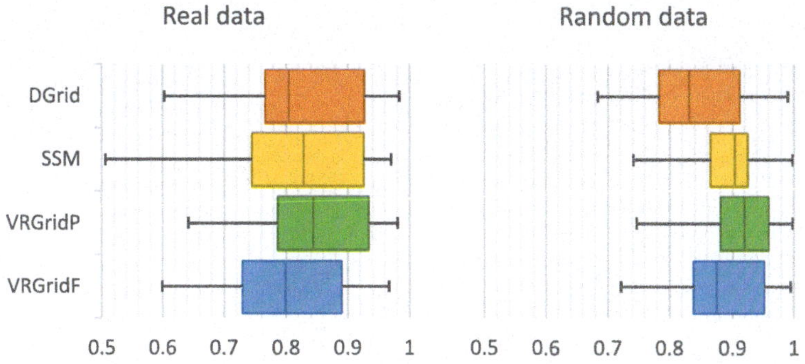

Fig. 11 Distance Preservation measure. Higher values are better. VRGrid$_P$ produces the best results overall, followed by SSM, then by DGrid or VRGrid$_F$ depending of the dataset's nature

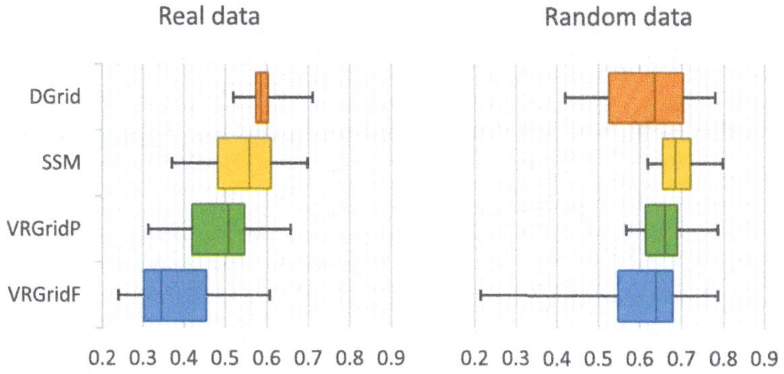

Fig. 12 Neighborhood Preservation measure. Higher values are better. Depending of the dataset's nature, DGrid and SSM are getting the best results, followed by VRGrid$_P$ and finally VRGrid$_F$

about 4.76% better than VRGrid$_P$, 12.16% better than DGrid, and 16.19% better than VRGrid$_F$.

SSM is better than VRGrid$_P$ at preserving neighborhoods on both real and randomly generated data. DGrid is better than VRGrid$_P$ on real data but not on randomly generated data. VRGrid$_F$ has the worst results on both real and random datasets.

Differences in quality between methods can be visually observed in proposed arrangements in Fig. 9, where SSM separates different data classes with sharp separations between groups while VRGrid$_P$ and VRGrid$_F$ slightly mix those separations. As for DGrid, despite its locally focused arrangements, the method preserves the neighborhood very well on real data but not as well on random data, mainly on random datasets of size 10 000 regardless of the distribution method. This behavior requires additional investigation as it seems that the partitioning technique of DGrid handles grids of sizes different than a power of 2 badly, yet the method still performs well with MNIST and Fashion-MNIST which are the same size.

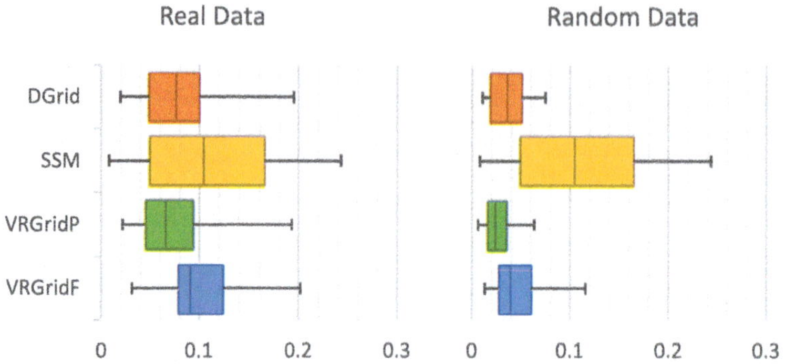

Fig. 13 Relative Position Preservation measures. Lower values are better. VRGrid$_P$ has the best results overall, followed by DGrid, then VRGrid$_F$ and finally SSM

Relative Position Preservation (RPP). Relative Position Preservation values are presented in Fig. 13, with lower values meaning lower local permutations in the grid arrangement resulting in better representation of input data. On real data, relative positioning in VRGrid$_P$ arrangements are conserved about 44.22% better than SSM, 6.53% better than DGrid, and 29.04% better than VRGrid$_F$. On randomly generated data, VRGrid$_P$ is about 62.04% better than SSM, 15.01% better than DGrid, and 34.04% better than VRGrid$_F$. SSM's poor performances can be visually observed on produced arrangements in Fig. 9 where arrangements are slightly rotated and distorted compared to the input data, when VRGrid$_P$, VRGrid$_F$ and DGrid manage to preserve the overall positioning of the elements in their arrangements.

By design, SSM does not preserve the input data positions, which can explain its poor performance in that aspect of the evaluation. DGrid has retention information but is outperformed by VRGrid$_P$.

Global Position Preservation (GPP). Global Position Preservation values are presented in Fig. 14, with lower values meaning less overall deformation of the original data projection. On real datasets, VRGrid$_P$ produces arrangement with 46.59% less movements than SSM, 14.83% less than DGrid, and 21.60% less than VRGrid$_F$. On random datasets, VRGrid$_P$ arrangements are also 46.47% better than SSM, 23.55% better than DGrid, and 17.66% better than VRGrid$_F$. Thus, VRGrid$_P$ produces an arrangement with fewer distortions than other methods.

Observed GPP values also show that the fragmented aspect of DGrid arrangements or the distortions in VRGrid$_F$ arrangements is as highly penalized as SSM.

Computation Times are presented in Table 2, with lower values meaning faster processing. When used as a post-processing method, SSM yields faster computation times than VRGrid. DGrid surpasses all other methods as it takes less than 0.02 second to process the largest datasets in our evaluation. VRGrid$_P$, despite its overall good quality results, except in neighborhood preservation, is about 270 times slower than SSM and cannot compete to DGrid, making it currently hardly practical for large datasets such as MNIST by taking around 90 min to process against 15 sec for SSM

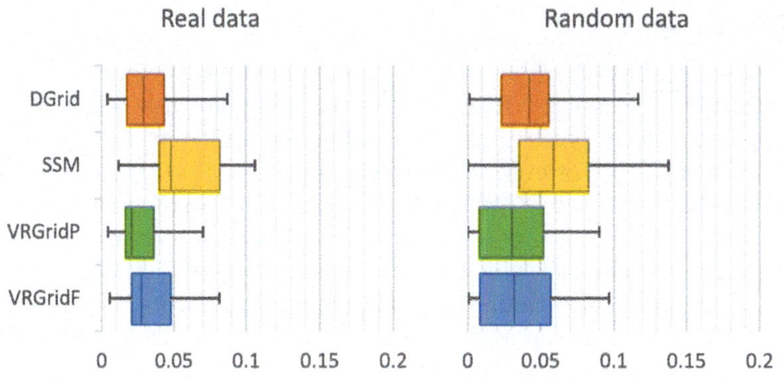

Fig. 14 Global position preservation measure. Lower values are better. VRGrid_P has the best scores overall, followed by VRGrid_F and DGrid depending on the nature of the data, and finally SSM has the worst results

Table 2 Average computation time in seconds on tested datasets sorted by their nature and size. DGrid is the fastest method, followed by SSM, then VRGrid_F and finally VRGrid_P

Dataset type	Dataset size	VRGrid_F	VRGrid_P	SSM	DGrid
Real	1 024	1.65	66.94	0.93	<0.01
	10 000	46.17	5 594.37	16.13	0.01
Random (dense)	64	0.04	0.46	0.02	<0.01
	100	0.07	1.01	0.03	<0.01
	1 000	1.65	68.53	0.52	<0.01
	10 000	46.94	5 628.95	15.92	0.01
Random (scattered)	64	0.05	0.40	0.03	<0.01
	100	0.07	0.98	0.04	<0.01
	1 000	1.64	63.10	0.55	<0.01
	10 000	46.57	4 999.92	9.46	0.01

and 0.015 second for DGrid. By not limiting the number of iterations, VRGrid_P can reach up to ten times the number of processed elements, which is hardly comparable to VRGrid_Fat larger scales. Furthermore, density of data seems to have a lesser impact on VRGrid_P total computation time with a 11% difference between dense and scattered data. VRGrid_F computation times are between up to 100 times faster than VRGrid_P, but are still slower than SSM.

6 Discussion

The evaluation shows that VRGrid$_P$ brings better results than SSM, DGrid and VRGrid$_F$ on distance preservation (DP) and minimizes deformation of the initial data (RPP and GPP). However, neighborhood preservation (NP) is a weak point of VRGrid, as both SSM and DGrid have better results in less processing time.

SSM is significantly weaker at staying close to the original data projection shape. Indeed, even using minimal deformation configuration of SSM's results explained in Sect. 4.2, both positioning metrics show that SSM arrangements are more distorted than VRGrid$_P$, sometimes producing sources of visual misinterpretation as shown in Fig. 10 in the UMAP/Swissroll scenario (1st row, 3rd column). While DGrid evaluation values are closer yet lower than VRGrid$_P$ in retaining input data shape, the visualization method proposed in Sect. 4.2 shows that DGrid's arrangements are locally very fragmented compared to other methods. However, this distortion is not very apparent in our evaluation results, notably by the RPP and GPP metrics, whereas SSM's outliers seem to be detected and penalized accordingly, as seen in the UMAP/Swissroll scenario. A dedicated novel metric that detects this issue should be designed in future works.

With those observations, we can affirm that all methods are useful in different usage scenarios. Scenarios only requiring compact projection of data (i.e., ignoring input data's shape) can already make good use of SSM as a post-processing method, since neighborhoods and distances are well preserved with reasonable processing time. However, VRGrid and DGrid are useful for producing compact projections as an additional tool to the regular data projection, such as data overlap solving tasks and/or numerous dataset projection navigation in a dynamic environment. In that aspect, we showed that, given a larger processing time, VRGrid brings significantly better results than DGrid, and does process its arrangement globally on its input data compared to DGrid's locally focused strategy. Dynamic or multi-view visualization tools would better benefit from VRGrid arrangements than other tested methods' as VRGrid produces overall fewer distortions.

Finally, our evaluation shows that VRGrid can compete against SSM and DGrid for the best arrangement of 2D dataset quality-wise, but falls short for the fastest processing method spot with its current implementation. The number of iterations needed for VRGrid$_P$ is of the same order as the dataset size n if not higher, which implies a computational complexity of $O\left(n^{\frac{5}{2}}.log(n)\right)$. The cut in computation time of VRGrid$_F$ by imposing a fixed number of iterations helps in reducing the computation time by up to 100 times. However, this modification is hindering the overall quality of the arrangement visually. Still, this tradeoff in quality is still insufficient to surpass SSM computation times on larger datasets.

As it stands now, VRGrid$_P$ would benefit from improvements on Lloyd's algorithm implementation, as cutting down iteration count results in a large decrease in overall arrangement quality. Lloyd's algorithm had several ways to be improved with algorithmic techniques such as over-relaxation [39] and hardware optimization [40, 41]. Those improvements led the computation of CVT to become at least 10 times faster than the classic Lloyd's algorithm running on CPU. This improvement potential could be a way to mitigate the high computational complexity of VRGrid_P and make it usable for bigger dataset processing.

7 Other Variants of VRGrid

Additionally to VRGrid_P and VRGrid_F, we also conducted experiments on two other VRGrid configurations. One of them is an attempt at finding a "sweet spot" between VRGrid_P and VRGrid_F called VRGrid_S (for *Stable*) for faster processing of data while keeping results of good quality. The other experiment is an attempt at improving the quality of the output while requiring even more computation time. VRGrid_S has a movement threshold as a stability factor like VRGrid_P , but with an ϵ value larger than VRGrid_P 's (10^{-2} instead of 10^{-3}). The obtained results are visually similar to VRGrid_P but with several misplaced elements, and the processing times are lower than VRGrid_P but still about 3 to 5 times greater than VRGrid_F . The obtained results are better than VRGrid_F , but there is a significant drop in RPP and GPP scores compared to VRGrid_P , as shown in Table 3.

The second configuration consists in improving the quality of VRGrid outputs by limiting the movements of the points during the first iterations of the relaxation algorithm. By fixing a maximum movement distance for each point instead of moving them directly to the centroid of their Voronoi cell, we can limit the risks of points of two dense areas next to each other to mix themselves and thus lowering the quality of the arrangement. This modification further increase processing time only marginaly, as it has effect only on the first iterations of the algorithm. We noticed a small improvement of the quality of VRGrid outputs, but this modification of the algorithm may not be compatible with the improvements mentionned in Sect. 6, and it is discouraged to pursue testing on this aspect of VRGrid for small quality improvement, instead work on the reduction of its computation time.

Table 3 Relative score obtained by VRGrid_S on evaluation metrics compared to VRGrid_P and VRGrid_F . Negative value means VRGrid_S is worse and positive values means VRGrid_S is better

Method	DP (%)	NP (%)	RPP (%)	GPP (%)
VRGrid_P	−1.33	−5.57	−22.53	−13.73
VRGrid_F	+1.35	+8.76	+16.61	+6.01

8 Conclusion

When visualizing projected data, overlap may occur and hinder the data analysis. Clustering methods such as t-SNE are popular high-dimension reduction techniques emphasizing even more visual occlusion in their results. Pixel-oriented visualization techniques, such as grid arrangement methods, are an efficient way to eliminate this visual occlusion and improve the perceptual scalability of these methods.

We presented VRGrid, a post-processing technique using relaxation methods to provide a grid arrangement for 2D data. VRGrid consumes the position of input data and incrementally transforms them into a grid disposition while limiting the deformation of the initial data shape. We evaluated this method against the state-of-the-art methods Self-Sorting Maps and DGrid on several aspects focused on data understanding and minimizing distortions. In this evaluation, we showed that VRGrid using Lloyd's algorithm surpasses SSM and DGrid with better distance preservation and minimal deformation but falls short in neighborhood preservation and more specifically in computing time. Limiting the number of Lloyd's algorithm iterations significantly reduces processing time but at the cost of an arrangement of lower quality across all four metrics.

These two observations showed that the usage of a relaxation method by VRGrid to converge toward a CVT configuration can be used to produce a grid arrangement for 2D data. To prevent deformations seen with VRGrid_F, the relaxation method must process enough iterations before VRGrid's iterative border-building step. This requirement may result in larger computation time, as seen with VRGrid_P, which is not desirable when working with large datasets. Lloyd's algorithm implementation can already be improved in several ways thanks to previous works in the domain. Other relaxation techniques can also be used with VRGrid, Broyden-Fletcher-Goldfarb-Shanno method (or BFGS) and its variants [42] are optimization methods that can be used to compute CVT configurations with fewer iterations than Lloyd's algorithm. Finally, RPP and GPP metrics were efficient to detect misarranged elements by SSM in the resulting grid but have not penalized VRGrid_F and DGrid as much for their visually different results from VRGrid_P. Further work on the evaluation of grid arrangement quality and practicability would be useful as the three methods may be interpreted differently by the end-user if given as is.

Acknowledgements We would like to thank the anonymous reviewers for their constructive comments to improve this chapter. This work has been carried out with financial support from the French State, managed by the French National Research Agency (ANR) in the frame of the Involvd project (ANR-20-CE23-0023-04). Experiments presented in this chapter were carried out using the Labo's in the Sky with Data (LSD), the LaBRI data platform partially funded by Region Nouvelle Aquitaine, and using PlaFRIM experimental testbed, supported by Inria, CNRS (LABRI and IMB), Université de Bordeaux, Bordeaux INP and Conseil Régional d'Aquitaine (see https://www.plafrim.fr/).

References

1. Friendly M (2008) A brief history of data visualization. Handbook of data visualization. Springer, pp 15–56
2. Liu S, Cui W, Wu Y, Liu M (2014) A survey on information visualization: recent advances and challenges. Vis Comput 30(12):1373–1393
3. Keim DA (2002) Information visualization and visual data mining. IEEE Trans Visual Comput Graph 8(1):1–8
4. Bertin J (1973) Sémiologie graphique: Les diagrammes-les réseaux-les cartes. Gauthier-VillarsMouton & Cie, Technical Report
5. Valdes-Mora F, Salomon R, Gloss BS, Law AMK, Venhuizen J, Castillo L, Murphy KJ, Magenau A, Papanicolaou M, de la Fuente LR et al (2021) Single-cell transcriptomics reveals involution mimicry during the specification of the basal breast cancer subtype. Cell Rep 35(2):108945
6. Rauber PE, Fadel SG, Falcao AX, Telea AC (2016) Visualizing the hidden activity of artificial neural networks. IEEE Trans Visual Comput Graph 23(1):101–110
7. Engel D, Hüttenberger L, Hamann B (2012) A survey of dimension reduction methods for high-dimensional data analysis and visualization. In: Visualization of large and unstructured data sets: applications in geospatial planning, modeling and engineering-proceedings of IRTG 1131 workshop 2011. Schloss Dagstuhl-Leibniz-Zentrum fuer Informatik
8. Liu S, Maljovec D, Wang B, Bremer P-T, Pascucci V (2016) Visualizing high-dimensional data: advances in the past decade. IEEE Trans Visual Comput Graph 23(3):1249–1268
9. Donoho DL et al (2000) High-dimensional data analysis: the curses and blessings of dimensionality. AMS Math Challenges Lecture 1(2000):32
10. Maaten LVD, Hinton G (2008) Visualizing data using t-sne. J Mach Learn Res 9:2579–2605
11. McInnes L, Healy J, Saul N, Grossberger L (2018) Umap: uniform manifold approximation and projection. J Open Source Softw 3(29):861
12. Wattenberg M, Viégas F, Johnson I (2016) How to use t-sne effectively. Distill 1(10):e2
13. LeCun Y, Bottou L, Bengio Y, Haffner P (1998) Gradient-based learning applied to document recognition. Proc IEEE 86(11):2278–2324
14. Strong G, Gong M (2014) Self-sorting map: an efficient algorithm for presenting multimedia data in structured layouts. IEEE Trans Multimedia 16(4):1045–1058
15. Hilasaca GM, Paulovich FV (2019) A visual approach for user-guided feature fusion. In: Anais Estendidos da XXXII conference on graphics, patterns and images. SBC, pp 133–139
16. Mamani GMH (2019) A visual approach for user-guided feature fusion. Ph.D. dissertation, Universidade de São Paulo
17. Lloyd S (1982) Least squares quantization in PCM. IEEE Trans Inf Theory 28(2):129–137
18. Du Q, Emelianenko M, Ju L (2006) Convergence of the Lloyd algorithm for computing centroidal Voronoi tessellations. SIAM J Numer Anal 44(1):102–119
19. Du Q, Faber V, Gunzburger M (1999) Centroidal Voronoi tessellations: applications and algorithms. SIAM Rev 41(4):637–676
20. Halnaut A, Giot R, Bourqui R, Auber D (2022) Vrgrid: efficient transformation of 2d data into pixel grid layout. In: 26th International conference information visualisation
21. Shneiderman B (1992) Tree visualization with tree-maps: 2-d space-filling approach. ACM Trans Graph (TOG) 11(1):92–99
22. Hilbert D (1935) Über die stetige abbildung einer linie auf ein flächenstück. In: Dritter Band: Analysis—Grundlagen der Mathematik—Physik Verschiedenes. Springer, pp 1–2
23. Keim DA (2000) Designing pixel-oriented visualization techniques: theory and applications. IEEE Trans Visual Comput Graph 6(1):59–78
24. Auber D, Huet C, Lambert A, Renoust B, Sallaberry A, Saulnier A (2013) Gospermap: using a gosper curve for laying out hierarchical data. IEEE Trans Visual Comput Graph 19(11):1820–1832
25. Fried O, DiVerdi S, Halber M, Sizikova E, Finkelstein A (2015) Isomatch: creating informative grid layouts. Comput Graph Forum 34(2):155–166. Wiley Online Library

26. Tenenbaum JB, De Silva V, Langford JC (2000) A global geometric framework for nonlinear dimensionality reduction. Science 290(5500):2319–2323
27. Kuhn HW (1955) The Hungarian method for the assignment problem. Naval Res Logist Q 2(1–2):83–97
28. Kohonen T (1990) The self-organizing map. Proc IEEE 78(9):1464–1480
29. Chen C, Yuan J, Lu Y, Liu Y, Su H, Yuan S, Liu S (2020) Oodanalyzer: interactive analysis of out-of-distribution samples. IEEE Trans Vis Comput Graph
30. Emelianenko M, Ju L, Rand A (2008) Nondegeneracy and weak global convergence of the Lloyd algorithm in r^d. SIAM J Numer Anal 46(3):1423–1441
31. Chen F, Piccinini L, Poncelet P, Sallaberry A (2020) Node overlap removal algorithms: an extended comparative study. J Graph Algorithms Appl
32. Van Der Maaten L, Postma E, Van den Herik J (2009) Dimensionality reduction: a comparative. J Mach Learn Res 10(66–71):13
33. Xiao H, Rasul K, Vollgraf R (2017) Fashion-mnist: a novel image dataset for benchmarking machine learning algorithms. Preprint at arXiv:1708.07747
34. Van Der Maaten L (2014) Accelerating t-SNE using tree-based algorithms. J Mach Learn Res 15(1):3221–3245
35. Pedregosa F, Varoquaux G, Gramfort A, Michel V, Thirion B, Grisel O, Blondel M, Prettenhofer P, Weiss R, Dubourg V, Vanderplas J, Passos A, Cournapeau D, Brucher M, Perrot M, Duchesnay E (2011) Scikit-learn: machine learning in Python. J Mach Learn Res 12:2825–2830
36. O'Neill ME (2014) PCG: a family of simple fast space-efficient statistically good algorithms for random number generation. ACM Trans Math Softw
37. Gong M (2019) Organize data into structured layout. http://www.cs.mun.ca/~gong/research/DataOrganization.html
38. Wilcoxon F (1992) Individual comparisons by ranking methods. In: Breakthroughs in statistics. Springer, pp 196–202
39. Xiao X (2010) Over-relaxation Lloyd method for computing centroidal Voronoi tessellations. University of South Carolina
40. Rong G, Liu Y, Wang W, Yin X, Gu D, Guo X (2010) GPU-assisted computation of centroidal Voronoi tessellation. IEEE Trans Visual Comput Graph 17(3):345–356
41. Zheng J, Tan TS (2020) Computing centroidal Voronoi tessellation using the GPU. In: Symposium on interactive 3D graphics and games, pp 1–9
42. Liu Y, Wang W, Lévy B, Sun F, Yan D-M, Lu L, Yang C (2009) On centroidal Voronoi tessellation-energy smoothness and fast computation. ACM Trans Graph (ToG) 28(4):1–17

A Method for Comparative Visualization of Labeled Multidimensional Data and Its Application to Machine Learning Data

Karen Kosaka and Takayuki Itoh

1 Introduction

Comparing training data for machine learning is an important issue. For example, it is important for transfer learning methods to compare source and target data to understand their differences. In other cases, training data may be compared and selected from multiple datasets. Therefore, it is important to compare training data not only quantitatively but also qualitatively. Visualization is an effective tool for this purpose.

Generally, training data for machine learning consists of a large number of samples. Here, samples are assumed to be image files, audio files, or document files, or other common types of files. The features of each sample are represented as a multidimensional vector, and one or more labels are assigned to each such sample. The above specifications suggest that the samples constituting the training data constitute labeled multidimensional data. Therefore, in this study, we visualize them on the same screen for the purpose of comparing multiple labeled multidimensional data. The requirements for this visualization are as follows.

Requirement 1: To represent differences in distribution among data by visualizing multiple data on the same screen.

Requirement 2: For each label assigned to a set of samples in a dataset, to realize an easy-to-understand representation of how samples with high similarity are concentrated, and how outlier samples are distributed.

Requirement 3: To realize a representation that makes it easy to compare how the same labeled sample groups have different distributions in each dataset.

To meet the above requirements, we propose the following visualization method.

K. Kosaka · T. Itoh (✉)
Ochanomizu University, Tokyo, Japan
e-mail: itot@is.ocha.ac.jp

B. Kovalerchuk et al. (eds.), *Artificial Intelligence and Visualization: Advancing Visual Knowledge Discovery*, Studies in Computational Intelligence 1126,
https://doi.org/10.1007/978-3-031-46549-9_9

- Apply the same dimensionality reduction method to all samples for all the datasets, and map all samples to the same screen space. This satisfies Requirement 1.
- For a set of samples that the same label is assigned in each dataset, densely distributed samples on the screen are surrounded by polygons. Other samples are highlighted as outliers. This satisfies Requirement 2.
- For multiple datasets, give the same color to groups of samples to which the same labels are assigned. This satisfies Requirement 3.

We can effectively represent the differences between multiple datasets with features and labels by using the visualization method described above. This chapter proposes an extension of a scatterplot drawing method by Nakabayashi et al. [10] to realize such a visualization. In addition, we present an application with typical machine learning sample datasets, MNIST[1] and USPS[2], and discuss the effectiveness of the proposed method.

This study started specifically with the comparison of training data for machine learning, and was presented as a conference paper [7]. This chapter reorganizes our conference paper for more general purposes on the comparison of multiple labeled multidimensional datasets. More specifically, we defined requirements for comparative visualization of general multiple labeled multidimensional datasets described in this section and reorganized the introduction of related work as described in Sect. 2. We also reorganized the description of implementation details as described in Sect. 3 and replaced all the examples with more effective ones as introduced in Sect. 4.

2 Related Work

2.1 Multidimensional Data Visualization

Visualization of multidimensional data has been actively discussed in many articles on information visualization. Scatterplots and Parallel Coordinate Plots (PCP) are especially famous and well-discussed methods for multidimensional data visualization. Scatterplot-based visualization methods can be divided into dimension-selection-based methods and dimension-reduction-based methods. The method presented in this chapter applies a dimension reduction method to represent multidimensional datasets.

Numeric evaluation of scatterplots has been an active research topic. Wilkinson et al. [21] proposed nine features of scagnostics, the metrics for the usefulness of scatterplot displays, based on the appearance of the scatterplots. Wang et al. [20] proposed the new scagnostics considering human perception. Several other studies focus on specific metrics of scatterplots, including correlation [3, 15] and class separation [14, 16]. Scagnostics can be applied to the method presented in this

[1] http://yann.lecun.com/exdb/mnist/.

[2] https://paperswithcode.com/dataset/usps.

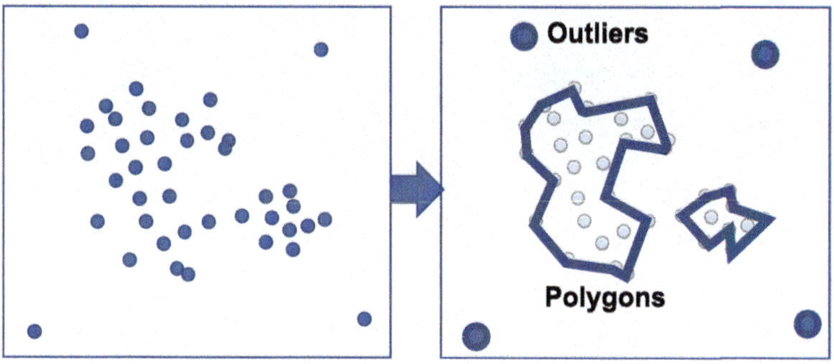

Fig. 1 Illustragion of the drawing process by Nakabayashi et al. [10]

chapter while drawing the dataset as polygons and outlier points as described in the next section.

Meanwhile, there have been several studies on drawing scatterplots that improve their visual comprehensibility. Bagplot [11] clearly divides outlier and other points by a simple visual representation; however, it often does not work well when the dense points form concave regions. Continuous Scatterplots [1] represents the two-dimensional spaces based on the density of points. The method may be time-consuming for some types of datasets. Schreck et al. [12, 13] presented methods for drawing the distribution of points by simpler objects such as convex hulls. The methods do not always display objects corresponding to all the points. Sunspot plots [19] can be used to combine the continuous rendering in dense regions and usual point drawing in sparse regions. This method applies colors to represent the density of points, and therefore, it is difficult to represent the distribution of categorical values in addition to the distributions of points in the screen space.

The scatterplot drawing presented in this chapter is based on an extension of a method proposed by Nakabayashi et al. [10]. This method displays a multidimensional dataset by representing the dataset as a selective set of scatterplots. The algorithm for the selection of scatterplots is inherited by another multidimensional data visualization method that displays a set of low-dimensional PCPs [5].

The processing procedure of this method consists of the following two steps.

Selecting the scatterplots: Select some important scatter plots with two axes for any two variables in the multidimensional data by a simple and interactive slider operation.

Drawing the scatterplots: Draw the point clouds displayed on the scatterplot as "outliers" and "polygons enclosing non-outlier clouds." Fig. 1 illustrates the concept of this process, and Fig. 2 shows the examples of the drawing of the scatterplots.

The study presented in this chapter extends the method by Nakabayashi et al. in such a way that the drawing method is applied independently to each label of the input dataset. Here, the color is established uniquely for each label and for each

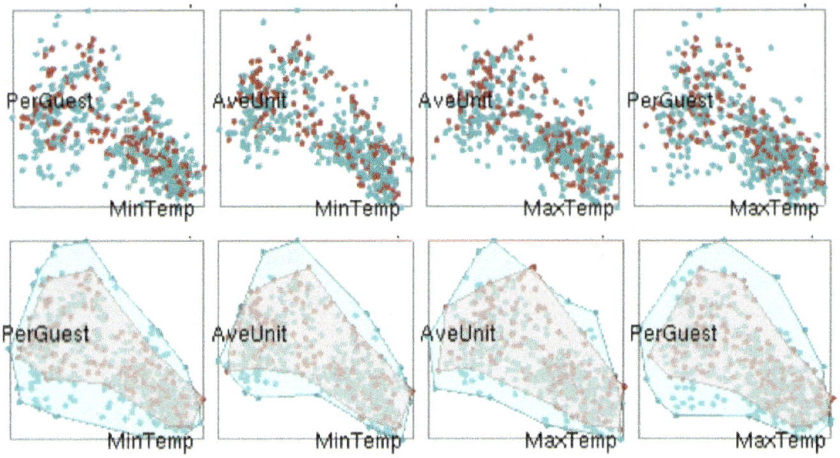

Fig. 2 (Upper) Before enclosing datapoints by polygons [10]. (Lower) After enclosing datapoints by polygons. While these examples represent a single dataset, the method proposed in this chapter extends this representation to display multiple datasets

dataset, and the threshold values of the "outlier group" and the "non-outlier group" are automatically calculated independently for each label and for each dataset.

2.2 Visualization of Labeled Data for Machine Learning

There have been various studies on visualization to analyze labeled data for machine learning. Ma et al. [8] presented a visualization to explain how knowledge already assigned to training data is transferred to destination data. This study is specialized for transfer learning and does not deeply discuss how to represent labeled multidimensional data. Swabha et al. [17] discussed the use of the training data to improve learning accuracy through their visualization method. This method does not implicitly represent label information. The study by Smilkov et al. [18] provided a visualization for local search, cluster discovery, and latent space semantic discovery. This study applies PCA (Principal Component Analysis) for scatterplots, but it does not represent multiple datasets in the same display space.

Multidimensional data visualization methods have been also applied to labeling processes of training data for machine learning. The visualization method by Moehrman et al. [9] showed that a user interface for sample selection contributes to label assignment. This method applied SOM (Self-Organizing Map), not scatterplots. Xiang et al. [23] visualized a hierarchical sample set by dimensionality reduction, and simultaneously visualizes a large number of representative samples belonging to a higher hierarchy by label correction. The system presented by Iijima et al. [4] (see Fig. 3) also visualized the results of automatic label assignment using small data

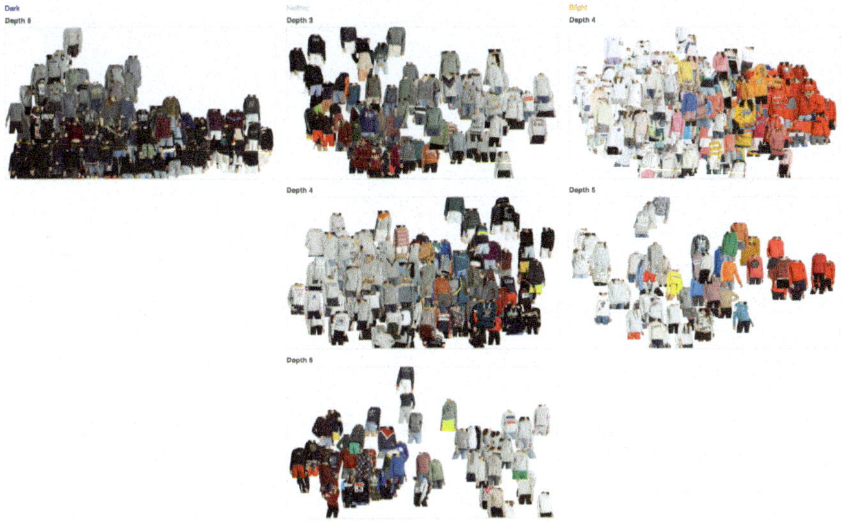

Fig. 3 Visualization of automatic label assignment for training data [4]. The method is specialized to image datasets

and fuzzy decision trees, and provides a user interface for correcting inappropriate label assignment results. The data representation of our study is different from these studies since they are specialized to image datasets and actually mapped images as the datapoints of scatterplots. Bernard et al. [2] demonstrated the possibility of supporting labeling by showing the separability of class distributions, and also shows that dimensionality reduction is effective for this purpose. This study represented dense regions of datapoints of a multidimensional dataset as polygons such as convex hull similar to our method, but the representations applied in this study do not distinguish outliers and non-outliers, as illustrated in Fig. 4. In addition, it does not aim to compare multiple training data.

Since visualization by dimensionality reduction has been effectively applied to the above methods, we adopt visualization by dimensionality reduction in this study as well. However, the above studies differ from the study presented in this chapter since they do not aim to compare multiple training data.

3 Comparative Visualization of Labeled Multidimensional Data

3.1 Overview and Processing Flow

The proposed method visualizes multiple multidimensional datasets on the same screen in order to ascertain differences in distribution among multidimensional

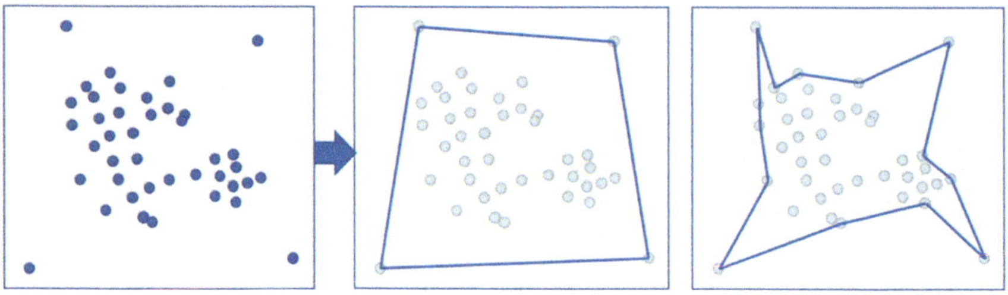

Fig. 4 Illustrations of representations of a set of datapoints as polygons [2]. This study does not distinguish outliers and non-outliers

datasets. The method applies the same dimensionality reduction to the features of all samples and projects them onto a two-dimensional screen space so that we can visualize multiple datasets on the same screen.

Our implementation uses t-SNE (t-Distributed Stochastic Neighbor Embedding) as the dimensionality reduction method. We would like to keep the aggregability of similar datapoints and the separateness of outlier datapoints of the visualization results in this study. Non-linear dimensionality reduction methods such as t-SNE or UMAP (Uniform Manifold Approximation and Projection) have better properties for this purpose. We applied both methods in the trial and subjectively determined that t-SNE had better visualization results. Also, several experiments [6, 22] demonstrated t-SNE had better results. As a result, we applied t-SNE as the dimensionality reduction method in this study.

The method displays the samples with the same label for each datapoint as polygons surrounding the density distributed samples, to make it easier to understand how similar samples are distributed and how outliers are distributed for each label. Each polygon is drawn in its own color. Polygons are created for each label, and a unique threshold and hue are assigned to each label. A unique saturation and brightness are assigned to each dataset. Figure 5 shows an illustration of this visual representation. Here, threshold values are automatically set for each label and each dataset to distinguish between outlier samples and the other samples.

3.2 *Process to Enclose Areas of High Point Cloud Density with Polygons*

The method applies the Delaunay triangulation method to enclose densely distributed samples, similar to the method used to draw scatterplots in Nakabayashi et al. [10]. The Delauney triangulation method generates a triangular mesh by connecting a given set of points so that the minimum angle of the triangles constituting the mesh is maximized. Once a triangulation mesh that connects all the points has been generated

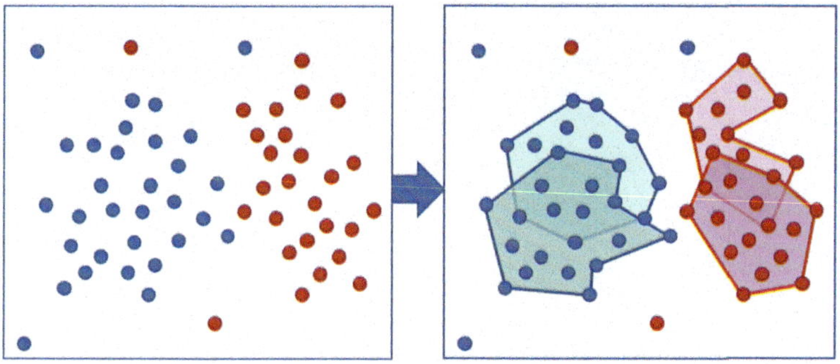

Fig. 5 A polygonal rendering with a hue assigned to each label and unique saturation and lightness assigned to each dataset. The same label is represented in dark and light blue and red

using this procedure, a triangulation mesh consisting only of points that are close together is generated by deleting triangles that have edges longer than a threshold value. Then, a polygon surrounding the outer frame is generated as the "inclusive area of non-outlier clouds," and the points outside the polygon are defined as outliers.

Following the above process, the method draws the following three types of shapes. Examples of Shape 1, Shape 2a, and Shape 2b are shown in Fig. 1.

Shape 1: Draw a small circle for each outlier.
Shape 2a: Draw the set of edges that form the boundary of the inclusive region as thick line segments.
Shape 2b: Paint a set of triangles that make up the inclusive region semi-transparently by alpha blending.

This method extends the above implementation to classify points in the same scatterplot by label and dataset, and generates a separate inclusion region for each label and dataset. The user must interactively specify the threshold for triangle deletion while using the above implementation, whereas the method presented in this chapter automatically sets the threshold for each label and dataset using the method described in the next section.

3.3 Automatic Threshold Setting

The proposed method removes triangles that contain edges whose lengths exceed the threshold value, and draws the outer boundary of the remaining triangles, as described in the previous section. The appropriate threshold value may vary from label to label and from dataset to dataset. It is cumbersome to manually set threshold values for each label and each dataset individually. Therefore, this method implements an automatic method for setting appropriate threshold values for each label by extending the scatterplot evaluation metrics [21].

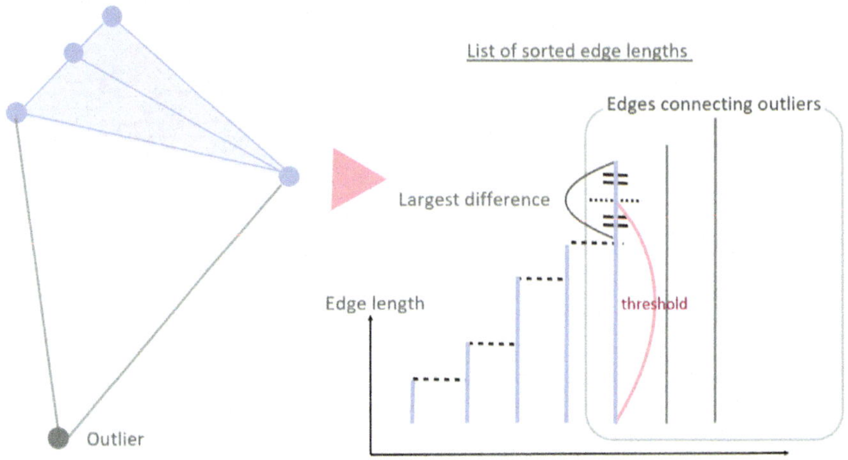

Fig. 6 Sort the edges of the left-hand figure. The average of the two edges with the largest difference between them is used as the threshold value

The procedure of the automatic setting method is illustrated in Fig. 6. First, the edges of each polygon are sorted in order of length. Then, the difference ($len_i - len_{i+1}$) between the lengths of the i-th and the $(i+1)$-th sides of the sorted polygon is calculated. The value of i with the largest difference is extracted and the value ($len_i - 0.5(len_i - len_{i+1})$) is set as the threshold value.

3.4 Color Specification

Our method specifies the color of each dataset and label by the following formula based on the HSB color system.

$$H = 2\pi \frac{i}{N}$$

$$S = B = \alpha \frac{j+1}{M} + (1.0 - \alpha)$$

Here, H represents hue, S represents saturation, and B represents lightness. N and M ($0 \leq i < N, 0 \leq j < M$) are the total number of labels and datasets, respectively, i and j are the serial numbers of labels and datasets, respectively, and α is a real number satisfying ($0 \leq \alpha \leq 1$). This formula assigns a unique saturation and lightness to each dataset and a unique hue to each label. Figure 7 shows an example of visualization in which each label is assigned a color to be displayed.

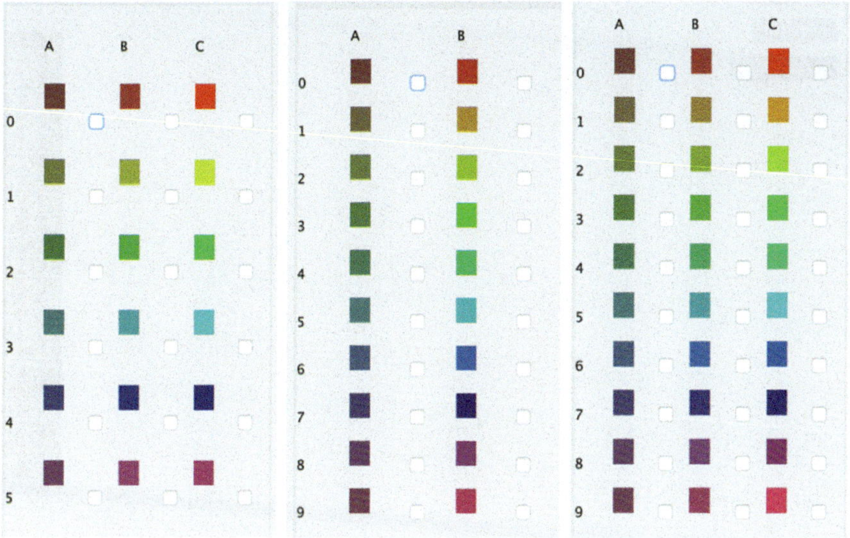

Fig. 7 An example of a visualization in which each label is assigned a color to be displayed. The rows are the datasets and the columns are the labels

3.5 Display of Images in the Input Dataset

The proposed visualization system simultaneously visualizes a set of images with features and labels. Specifically, the images corresponding to the scatterplots are displayed on the right side of the screen. The size of the images displayed on the screen is constant, but users can see all images by scrolling the screen. In the scatterplot of the visualization screen shown in Fig. 8, the samples labeled 0 in the MNIST data described below are displayed, and all the points are connected by lines. Therefore, all images of MNIST samples labeled 0 are displayed. The images also correspond to the samples selected on the label selection screen, and to the threshold slider. In other words, only images up to the label selected by the user and the threshold set by the user are displayed. Figure 9 displays samples labeled with MNIST 0 as in Fig. 8, but due to the low the threshold value, the number of samples surrounded by polygons is smaller than in Fig. 8. Therefore, the number of displayed images is also smaller than in Fig. 8.

3.6 User Interface

Figures 10 and 11 show the user interface of our implementation. By operating the control panel on the left side of the screen, the user can interactively change the contents of the panel on the right side of the screen. The operation panel on the left

Fig. 8 The image surrounded by polygons is displayed on the right side of the screen

side of the screen has a main tab and a color tab. The operation panel in Fig. 10 displays the main tab, and the operation panel in Fig. 11 displays the color tab. The file selected by pressing the UPLOAD button must describe the two-dimensional coordinate values of samples, their labels, and the absolute paths of image files associated with the samples. The COLOR tab displays a table of labels, their assigned hues, and checkboxes. The right panel displays only the images with the label selected by the user when marking a check box and pressing the SUBMIT button. This function is useful when there are many labels or when only images with a specific label are to be displayed.

The interface operation procedure is to select the UPLOAD FILE button and select the file containing the data information that is selected for display (Fig. 12).

Then, the presented method creates and displays a triangular mesh using the Delaunay triangulation method when a user presses the MAKE DELAUNAY button. The image of the dataset is also displayed on the right side of the screen, and the color tab shows which colors are assigned to each color. It is also possible to select the labels as shown in Figs. 13 and 14).

4 Examples

In this chapter, we discuss the effectiveness of the proposed method with examples using the MNIST and USPS datasets that provide handwritten text images and are commonly used as sample data for machine learning.

Fig. 9 Example of display with different threshold values for 8. Note that only the image in the area surrounded by the polygon is displayed

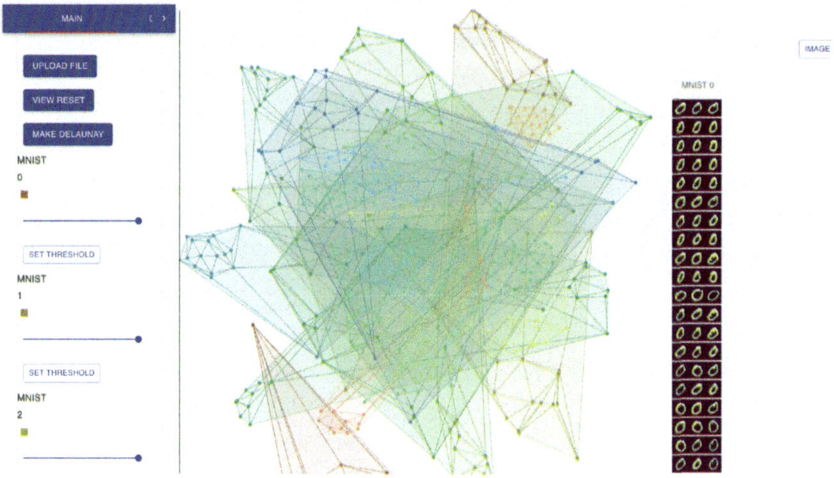

Fig. 10 Screen with the main tab displayed on the operation panel on the left side of the screen

4.1 Visualization of Single Dataset

Firstly, we visualized either of the two datasets independently and found how similar and exceptional images aredistributed in the datasets.

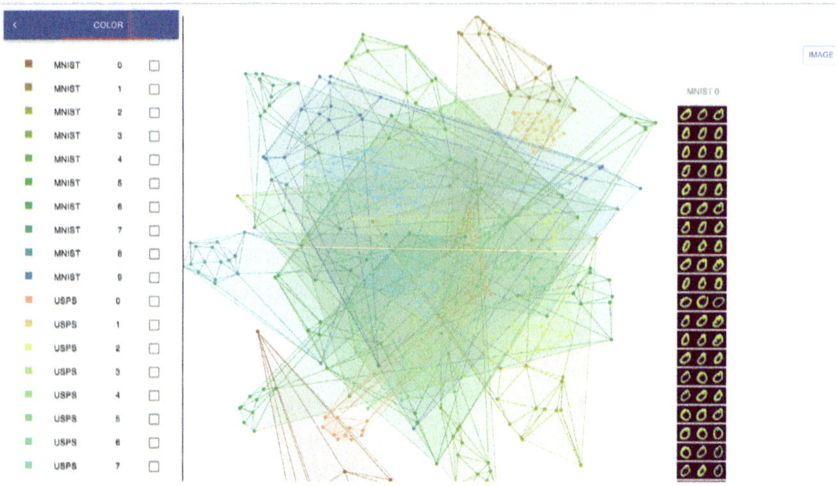

Fig. 11 Screen with the color tab displayed in the operation panel on the left side of the screen

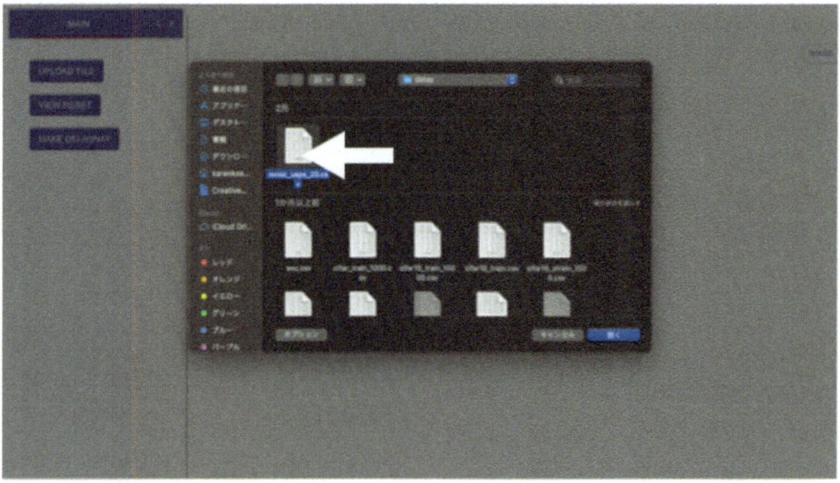

Fig. 12 The screen after the UPLOAD FILE button is pressed. The user selects the file to be displayed

Figure 15 shows the distributions of images labeled 0 or 9 in MNIST, where brown points correspond to images labeled 0 while blue points correspond to images labeled 9. Here, each blue point is enclosed within a triangle of brown points. The image corresponding to the enclosed blue point is shown at the right side of this figure. Actually, this image is confusing since it may look like 0 though the text is written as 9.

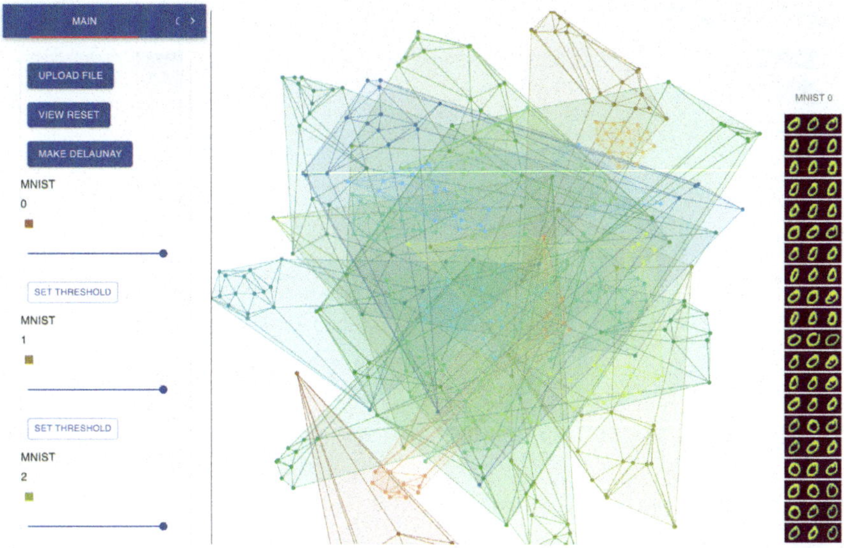

Fig. 13 Screen with all labels displayed

Fig. 14 The screen showing two labels selected and displayed from Fig. 13

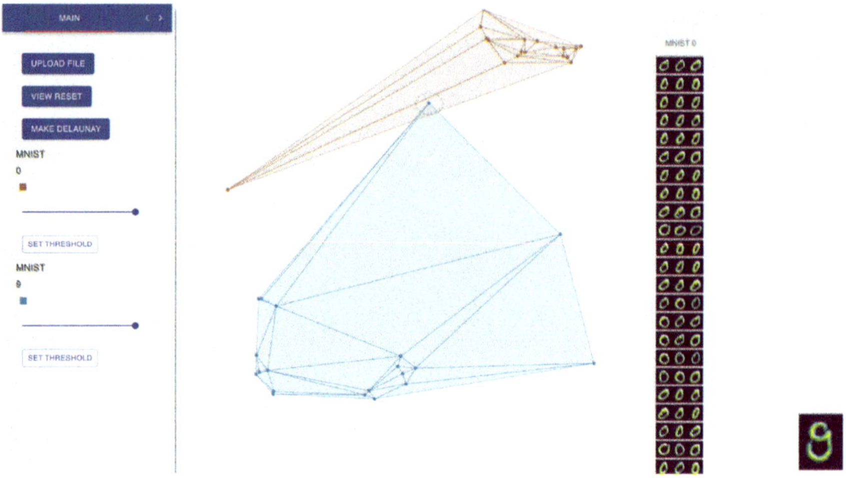

Fig. 15 Images labeled 0 or 9 in MNIST

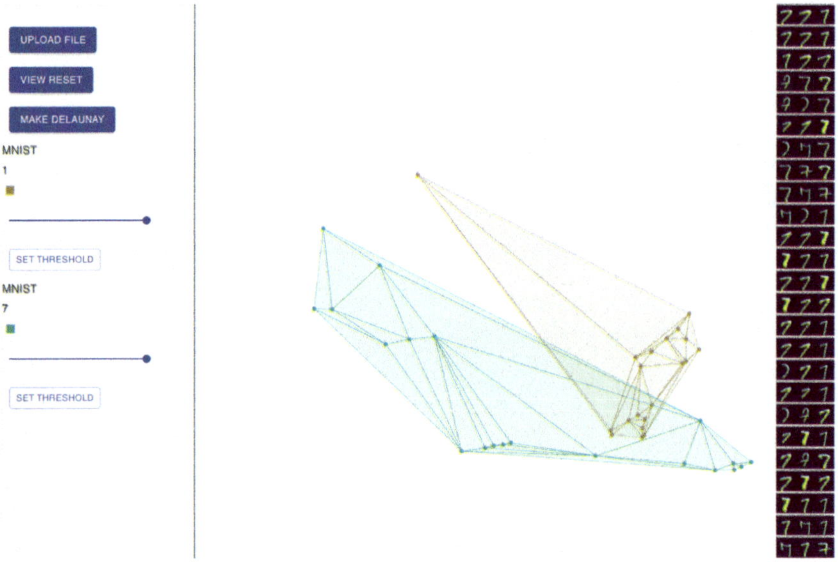

Fig. 16 Images labeled 1 or 7 in MNIST

Figure 16 shows the distributions of images labeled 1 or 7 in MNIST, where ochre points correspond to images labeled 1 while green points correspond to images labeled 7. This figure shows that a large part of the triangles of the two colors overlap and therefore features of the images are somewhat similar. Actually, some of the text images labeled 7 shown in Fig. 16 may look written as 1.

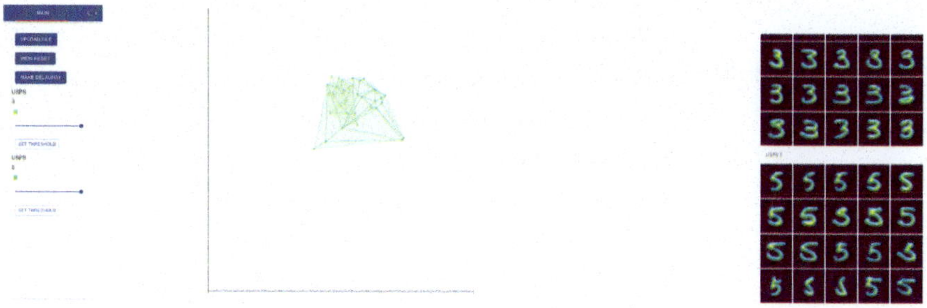

Fig. 17 Images labeled 3 or 5 in USPS

Figure 17 shows the distributions of images labeled 3 or 5 in USPS, where light-green points correspond to images labeled 3 while moss-green points correspond to images labeled 5. This figure shows that a large part of the triangles of the two colors overlap and therefore features of the images are somewhat similar. The text images labeled 3 are shown in the upper-right of the figure while the images labeled 5 are shown in the lower-right end. The figure shows some of the text images labeled 5 shown in Fig. 17 may look written as 3.

The above visualization examples demonstrate that the overlapped triangles depict the existence of similarly-looking but differently-labeled text images.

4.2 Visualization of Multiple Datasets

We applied t-SNE to MNIST and USPS simultaneously to project the datasets onto a two-dimensional space. Images labeled 6 and 9 were selected from each dataset, with 6 represented by brown and 9 by green for MNIST and 6 and 9 by red and light blue for USPS, as shown in Fig. 18. We observed how similar text images are distributed using the presented visualization.

Figure 19 shows that the point clouds that have the same label and belong to different datasets are closer; the point clouds labeled 6 and 9 are located farther apart and thus have different features.

While observing Fig. 20, we determined the cluster with a larger number of points as the cluster of regular handwriting of 9, and the points located in other places correspond to outlier handwriting images. We checked the outlier images and found that they were not mislabeled, but they looked as if they were written as 9.

Figure 21 shows that similar results were obtained for the samples labeled 6. This visualization displays 20 images for each label, for each dataset.

Next, we selectively displayed images labeled 1 in MNIST and USPS. Figure 22 shows the distributions of images labeled 1 in MNIST and USPS, where dark ochre points correspond to images in MNIST while light ochre points correspond to images in USPS. Two regions of polygons of MNIST and USPS are closely placed but they

Fig. 18 The 6 of MNIST is represented by brown, 9 by green, the 6 of USPS by red, and its 9 by light blue

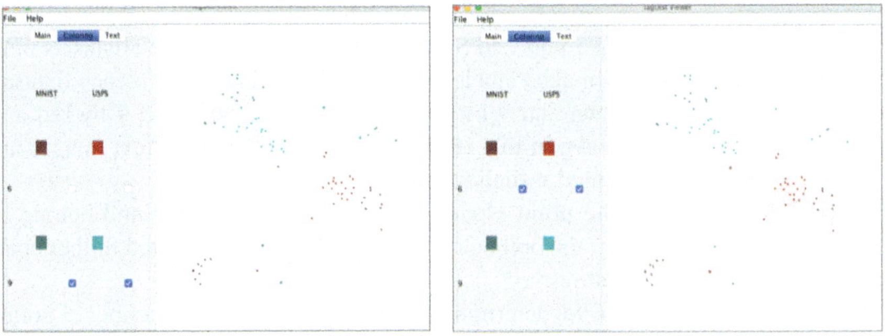

Fig. 19 The left side of the visualization result shows only images with 9 in each dataset, while the right side shows only images with 6

do not overlap. The visualization suggests that text images labeled 1 in both datasets may look somewhat similar, but in fact, are slightly different. Actually, images of MNIST displayed at the upper-right end and images of USPS displayed at the lower-right end are written as 1, but the shapes of the text in the two image groups look different since the texts in MNIST are more diagonal. Another interesting observation in this visualization is that an outlier point surrounded by a red circle corresponds

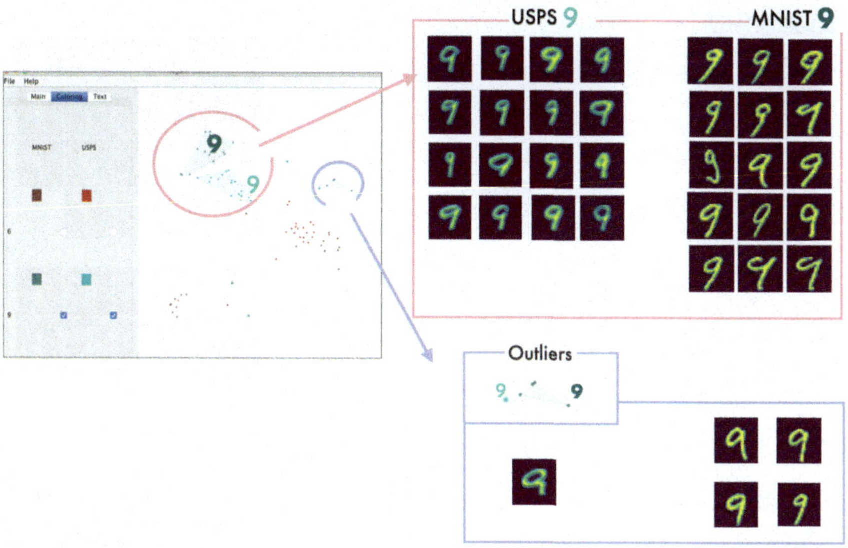

Fig. 20 Only images labeled as 9 in each dataset were displayed

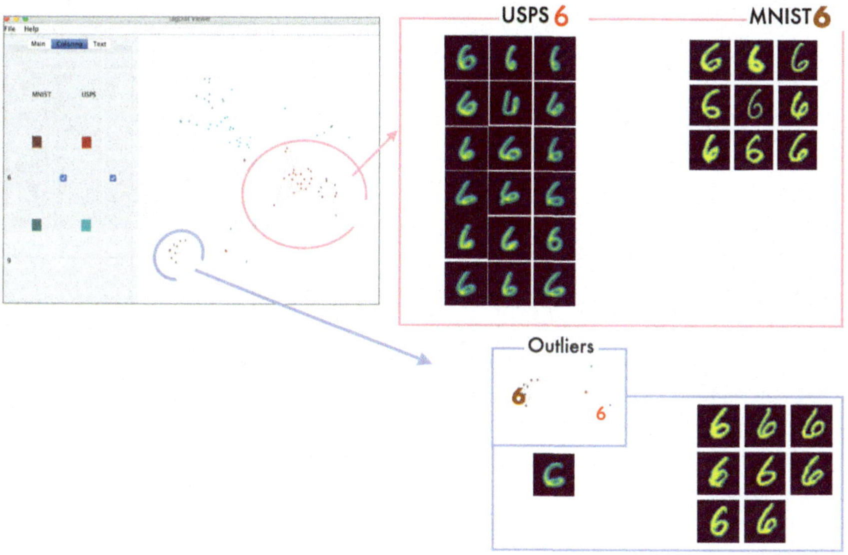

Fig. 21 Only images labeled as 6 in each dataset were displayed

to a text image shown under the red circle that looks much different from other text images.

Figure 23 shows the distributions of images labeled 2 in MNIST as well as images labeled 1 in MNIST and USPS, where dark ochre points correspond to images labeled 1 in MNIST, light ochre points correspond to images labeled 1 in USPS, and yellowish

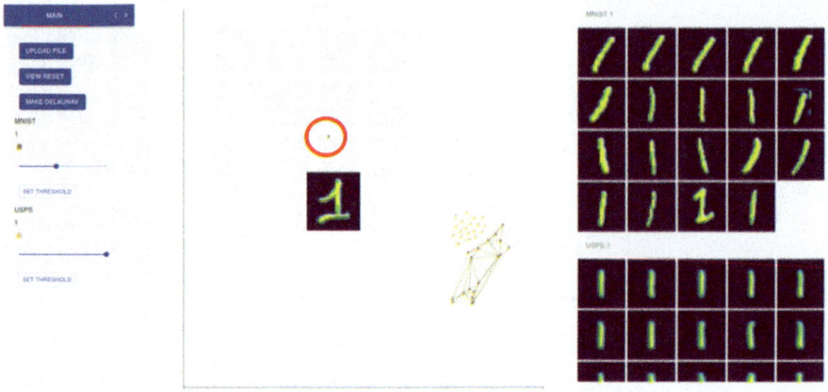

Fig. 22 Images labeled 1 in MNIST and USPS

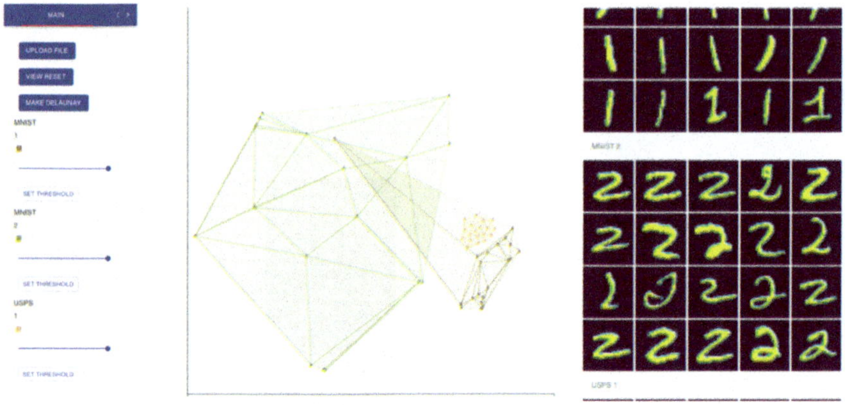

Fig. 23 Images labeled 1 in MNIST and USPS, and images labeled 2 in MNIST

green points correspond to images labeled 2 in MNIST. Here, a dark ochre point (the point surrounded by a red circle in Fig. 22) is enclosed by a triangle of yellowish-green points. This result suggests that the text image corresponding to the above-mentioned dark ochre point is labeled 1 but the feature of this image is closer to the images labeled 2.

The above visualization examples demonstrate that the presented visualization can suggest the similarity or difference among the images that have the same label but belong to different datasets. Also, the examples demonstrate that the outliers represented by this visualization actually depict the existence of different-looking text images.

5 Conclusion

In this chapter, we proposed a visualization method for comparing multiple multidimensional data, and reported an example of its application to the comparison of training data for machine learning. The proposed method applies the same dimensionality reduction to multiple multidimensional data, each of which is labeled, and displays them on a single screen. The method generates a triangular mesh connecting the points on the scatterplot, and by deleting triangles with long sides, the points located at close distances on the screen are represented as polygons. In this chapter, we propose a method to automatically set the threshold value of edge length for triangle deletion independently for each label. This method is expected to facilitate the understanding of differences in the distributions of multiple datasets.

The proposed method has several limitations. First, the numbers of labels and datasets are limited. In this chapter, ten different labels were represented by 10 different hues, but as long as labels are represented by hues, the limit is probably about 10. Also for data sets, this chapter only compared two data sets, and we have not applied to the comparison of three or more datasets. Furthermore, it is currently difficult to apply the method to datasets where each instance is assigned multiple labels.

The presented visualization can be applied to non-image datasets those datapoints have labels and multidimensional values. However, we need to modify the user interface to apply to non-image datasets. The current user interface displays a set of images corresponding to user-selected datapoints so that users can visually compare the images. We need to develop additional components to show the content of non-image data.

As future work, we would first like to expand the user interface and verify the operation on large-scale data with thousands or tens of thousands of instances. In addition, we would like to improve and conduct experiments for use in comparing training data for machine learning. We would also like to conduct user evaluation experiments on the effectiveness of this method.

References

1. Bachthaler S, Weiskopf D (2008) Continuous scatterplots. IEEE Trans Visual Comput Graph 14(6):1428–1435
2. Bernard J, Hutter M, Zeppelzauer M, Fellner D, Sedlmair M (2018) Comparing visual-interactive labeling with active learning: an experimental study. IEEE Trans Visual Comput Graph 24(1):298–308
3. Harrison L, Yang F, Franconeri S, Chang R (2014) Ranking visualizations of correlation using weber's law. IEEE Trans Visual Comput Graph 20(12):1943–1952
4. Iijima A, Itoh T (2021) Visualization for image annotations based on semantic differential. IEEE VIS Posters
5. Itoh T, Kumar A, Klein K, Kim J (2017) High dimensional data visualization by interactive construction of low dimensional parallel coordinate plots. J Vis Lang Comput 43:1–13

6. Kobak D, Linderman GC (2019) UMAP does not preserve global structure any better than t-SNE when using the same initialization. https://doi.org/10.1101/2019.12.19.877522
7. Kosaka K, Itoh T (2021) A visualization method for training data comparison. In: 25th International conference on information visualisation (IV2021), pp 205–211
8. Ma Y, Fan A, He J, Nelakurthi AR, Maciejewski R (2021) A visual analytics framework for explaining and diagnosing transfer learning processes. IEEE Trans Visual Comput Graph 27:1385–1395
9. Moehrmann J, Bernstein S, Schlegel T, Werner G, Heidemann G (2011) Improving the usability of hierarchical representations for interactively labeling large image data sets. In: 14th International conference on human computer interaction: design and development approaches volume part I (HCII'11), pp 618–627
10. Nakabayashi A, Itoh T (2019) A technique for selection and drawing of scatterplots for multi-dimensional data visualization. In: Proceedings of 23rd international conference on information visualisation (IV2019), pp 62–67
11. Rousseeuw PJ, Ruts I, Tukey JW (1999) The bagplot: a bivariate boxplot. Am Stat 53(4):382–387
12. Schreck T, Panse C (2007) A new metaphor for projection-based visual analysis and data exploration. In: IS&T/SPIE conference on visualization and data analysis
13. Schreck T, Schüßler M, Worm K, Zeilfelder F (2008) Butterfly plots for visual analysis of large point cloud data. In: International conference in central europe on computer graphics, visualization and computer vision (WSCG'08), pp 33–40
14. Sedlmair M, Tatu A, Munzner T, Tory M (2012) A taxonomy of visual cluster separation factors. Comput Graph Forum 31(3):1335–1344
15. Shao L, Mahajan A, Schreck T, Lehmann DJ (2017) Interactive regression lens for exploring scatter plots. Comput Graph Forum 36(3):157–166
16. Sips M, Neubert B, Lewis JP, Hanrahan PM (2009) Selecting good views of high-dimensional data using class consistency. Comput Graph Forum 28(3):831–838
17. Swayamdipta S, Schwartz R, Lourie N, Wang Y, Hajishirzi H, Smith NA, Choi Y (2020) Dataset cartography: mapping and diagnosing datasets with training dynamics. In: Proceedings of conference on empirical methods in natural language processing
18. Smilkov D, Thorat N, Nicholson C, Reif E, Viégas FB, Wattenberg M (2016) Embedding projector: interactive visualization and interpretation of embeddings. In: NIPS 2016 workshop on interpretable machine learning in complex systems
19. Trautner T, Bolte F, Stoppel S, Bruckner S (2020) Sunspot plots: model-based structure enhancement for dense scatter plots. Comput Graph Forum 39(3):551–563
20. Wang Y, Wang Z, Liu T, Correll M, Cheng Z, Deussen O, Sedlmair M (2020) Improving the robustness of scagnostics. IEEE Trans Visual Comput Graph 26(1):759–769
21. Wilkinson L, Anand A, Grossman R (2005) Graph-theoretic scagnostics. In: IEEE symposium on information visualization, pp 157–164
22. Xiang R, Wang W, Yang L, Wang S, Chaohan X, Chen X (2021) A comparison for dimensionality reduction methods of single-cell RNA-seq data. Front Genet 12:646936
23. Xiang S, Ye X, Xia J, Wu J, Chen Y, Liu S (2019) Interactive correction of mislabeled training data. In: IEEE conference on visual analytics science and technology (VAST), pp 57–68

Illuminating Insights: Visual Analytics Applications and Case Studies

Road Traffic Flow Prediction with Visual Analytics

Nuno Datia, Matilde P. M. Pato, João Vaz, and João Moura Pires

1 Introduction

Predicting road traffic flow is important for a number of reasons. First and foremost, it helps to alleviate congestion on roads, which can lead to improved travel times and reduced fuel consumption. This, in turn, helps to reduce air pollution and greenhouse gas emissions. In addition, predicting traffic flow can allow city planners to make more informed decisions about how to allocate resources. Furthermore, by providing drivers with accurate traffic predictions, it can assist them to plan their routes more efficiently, leading to fewer accidents and a safer overall driving experience. Overall, the ability to predict road traffic flow using machine learning can provide significant benefits for individuals, communities, and the environment [32].

Due to an increasing number of people living in cities, there is a greater likelihood of traffic jams on the roads [20]. The effects of traffic jam on people's lives are negative at various levels. Some of these include [21, 24]: (i) the car is wearing out and consuming more fuel; (ii) more time spent commuting; and (iii) higher levels

N. Datia (✉)
NOVA LINCS and Future Internet Technologies, ISEL-Lisbon School of Engineering, Politécnico de Lisboa, Lisboa, Portugal
e-mail: datia@isel.ipl.pt

M. P. M. Pato
NOVA LINCS & LAGIGE & Future Internet Technologies, ISEL-Lisbon School of Engineering, Politécnico de Lisboa, Lisboa, Portugal
e-mail: matilde.pato@isel.pt

J. Vaz
ISEL-Lisbon School of Engineering, Politécnico de Lisboa, Lisboa, Portugal
e-mail: A41920@alunos.isel.pt

J. M. Pires
NOVA LINCS, NOVA School of Science and Technology, Costa da Caparica, Portugal
e-mail: jmp@fct.unl.pt

B. Kovalerchuk et al. (eds.), *Artificial Intelligence and Visualization: Advancing Visual Knowledge Discovery*, Studies in Computational Intelligence 1126,
https://doi.org/10.1007/978-3-031-46549-9_10

of stress and anxiety. Monitoring and estimating traffic flow allows city officials to develop mobility policies that reduce traffic jams.

Traffic infrastructures are critical components of the planning process for a smart city [9]. Forecasting is considered the beginning of every planning process [10], as it allows planners to anticipate future issues and create solutions while reducing costs. It is critical to support decision makers in gaining more awareness of the interconnected factors that influence traffic, both quantitative and qualitative. For instance, learn more about the models and the parameters, their impact on predictions, and present results at multiple levels of granularity. By using visual means and interaction, Visual Analytics (VA) and Visual Knowledge Discovery (VKD) seek to discover new knowledge beyond visualising given data, machine learning algorithms and already known patterns [17].

In 2017, the Lisbon City Council (CML) implemented a management platform in the city [26]—Lisbon integrated management platform (PGIL). This platform integrates, compiles and presents various real-time data sources and supports the decision-making process of different actors through various thematic *dashboards*. Interactive dashboards are used to present data insights in a way that is easily understandable and actionable for humans. They provide a visual representation of data that can be easily understood by people from different backgrounds and with different levels of technical expertise. There are several reasons why interactive dashboards are effective for presenting data insights to humans:

help to clarify complex information Interactive dashboards can help to simplify and clarify complex data by using visualisations such as charts, graphs, and maps to communicate information in an easy-to-understand format;

allow users to explore data Interactive dashboards enable users to explore data by providing various filters, drill-down options, and other interactive features that allow them to focus on specific data sets or trends;

facilitate decision-making Interactive dashboards provide users with the ability to quickly identify trends and patterns in data, which can help them make informed decisions;

allow for real-time monitoring : Interactive dashboards can be configured to update automatically, providing users with real-time data and insights. This can be especially useful for monitoring the performance of a business or identifying potential issues.

Combining machine learning and interactive dashboards can be a powerful combination because they complement each other in a number of ways. For example, machine learning can help to identify patterns and trends in data that may not be immediately obvious to humans. Machine learning models could be trained to identify trends in real-time data and update the dashboard with the latest insights. Interactive dashboards can help visualise those trends in an easy-to-understand format. Visual analytics tools use sophisticated processes to analyse data sets using visual representations of the data. However, Lisbon city council still lacks the predictive models necessary to provide accurate estimates of many city operations and life-cycle, including of air pollution [29] and road traffic [31], to name a few.

This work extends and updates our previous work [31] and presents an Interactive Visual Analytical Dashboard (iVAD) to support the analysis of traffic jams using a proposed road traffic indicator. It displays the level of traffic flow within a city at different Levels-of-Detail (LoD), as patterns may appear in some LoD and not in others [28], including: (i) Street level, (ii) Area Level, and (iii) City level. iVAD displays road traffic indicators based on real and predicted data, and supports spatial and temporal data exploration. The uncertainty of predictions is visually communicated to analysts when using predicted data. The main contributions of this work are as follows:

1. a curated data set containing feature-engineered result on traffic jams and weather information;
2. a multi-level road traffic flow indicator definition, that incorporates a ML model;
3. an interactive dashboard that supports the road traffic flow indicator using real and predictive data.

The rest of the chapter is organised as follows. Section 2 describes the related work, followed by a discussion of the preparation and creation of the predictive model (Sect. 3). In Sect. 4, the interactive visual analytical dashboard is presented. Next, the user based evaluation is discussed (Sect. 5) and, finally, we draw some conclusions and point out directions for future work (Sect. 6).

2 Related Work

During the last years, researchers have used several techniques, including ML, to predict traffic inside cities. In particular, researchers try to create models for short-term traffic flow predictions. We organised a list of relevant references, related to our work, concerning three main topics: (i) the data used; (ii) regarding the ML algorithm's employed; and (iii) concerning the visual support given to the end-users.

2.1 Sources of Data

Most of the works rely on sensor data to model and predict the traffic behaviour [4, 6, 7, 22, 27, 30, 34, 35]. Bratsas et al.[6] use on-board probe car data, generated by a set of sensors, containing location (GNSS), orientation and speed. The data were acquired from a 1200 taxi fleet, every fifteen minutes for a two week period. Bachechi et al. [4] use real-time sensor data (the number of vehicles, their average speed and the type of vehicle) to produce a traffic flow model in road lanes of the urban area. Each sensor has a GPS coordinate and the identification of the lane where the data was collected. The traffic sensor data were collected using two different procedures: the regional sensor data are queried through a web service, while the urban sensor data are stored in a local PostgreSQL database and retrieved using SQL queries.

The 54 regional traffic sensors produce an average of 48900 observations per day, aggregated by 15 min. The 345 urban traffic sensors produce around 379204 observations every day. Tempelmeier et al. [30] use sensor data (average traffic speed) collected every 15 min in Hannover region. The traffic data was complemented with event data, selected between October 2017 and January 2018. Shahid et al. [27] used traffic intensity data combined with air pollution to predict traffic, collected by the city of Aarhus, Denmark, and make available as open data. Traffic data was collected by sensors spread throughout the city, which recorded how many vehicles passed every five minutes. It also includes longitude, latitude and timestamp. Zhang and Kabuka [34] create a traffic flow forecast model based on traffic data, collected in real time, from over 39,000 individual detectors in the state of California. The data was sampled hourly and combined with weather forecast data (precipitation, average wind speed, minimal and maximal temperature). A feature called *vehicle miles travelled* (VMT) indicates the number of miles driven along a specific freeway on an hourly basis. It is calculated by multiplying the number of hourly traffic on a freeway segment by its length. In their research, Chawla et al. [7] used sensor data from traffic counters and vehicle detector stations in California District 4 and the Bay Area California highways. The traffic data was collected on an hourly basis and included the flow of vehicles, vehicle speeds, vehicle occupancy, among other information. It comprises 5 years of data spanning 2015–2019, totalling about 75 million records. The traffic data was augmented with contextual information, namely incident and weather data. Ni et al. [22] uses the same traffic data as above. However, they combine it with twitter data collected by the same spatial and temporal window accordingly, through the Twitter Streaming API with geo-location filter. After data cleanse, a set of 5 444 527 valid tweets remains. The two data sets were cross-referenced taking into account the main access routes to the location where the event will take place. The authors shown the contextual data helps improving the prediction. Essien et al. [11] also uses Twitter data to improve traffic speed predictions. By mixing traffic and weather. related data sets (hourly observations of rainfall and temperature values) with traffic tendencies learned from Twitter users, prediction performance is increased. The traffic data set consists of historical 5-min observations of traffic flow characteristics (speed, flow, and density). Ten traffic sensors are located 0.3 mi apart in the study area. The data collection period is one year (from April 2016 to April 2017). Zaho et al. [35] use traffic data were collected from about 500 observation stations, mostly deployed within the fifth ring road of Beijing, with a frequency of 5 min. Each station provides vehicle volume, lane occupancy and average velocity. The data collection ranges from January 2015 to June 2015, and contains about 25.11 million validated records and 0.81 million missing or invalid data. The first five months of data are used as training data.

Some works rely on crowdsourced data [2, 12, 13]. Fernandes et al. [13] built a data set by collecting data from TOMTOM public API. The data set includes 80 d of data, starting from October 2018. It includes, among other information, the average speed at a selected point, the free flow speed expected under ideal conditions, the current travel time and the travel time under ideal free flow conditions (both in seconds). Farajiparvar et al. [12] used a deep learning approach to build a model that is able to forecast traffic speed. This model is based on Waze crowdsourced data, collected from 31 segments of an interstate road in Tennessee. Data range from June to November 2019. To complement the speed information, authors used Waze's spatial information to complement the speed data. Albuquerque et al. [2] uses data from traffic jams (Waze data) in the Lisbon metropolitan area, collected between January 2019 to July 2020, comprising 12 619 459 instances. The data was combined with road accident data in Lisbon involving vehicles and motorcycles. Do notice that when the traffic level was 5, representing a jam, the Delay column had the value -1 that was treated like missing data and replaced with a value prediction using a Pearson correlation and linear regression.

Other works rely on traffic images. Nubert et al. [23] present and discuss a potential application that estimates the traffic density at traffic lights/junctions using public cameras, then adjusts the traffic lights accordingly to get the most efficient result. They used 4582 labelled traffic images from 3 cameras at different junctions and angles. Each image is labelled with a traffic density level, a 5 class value, from *empty* to *traffic jam*.

2.2 ML algorithms

Researchers apply a wide range of algorithms when it comes to machine learning, depending on the tasks and data. In [6], the authors compared the Random Forests (RF), Support Vector Regression (SVR), Multilayer Perceptron (MLP) and Linear Regression (LR) to predict the speed of cars inside a city. They have used Mean Absolute Error (MAE) and Root Mean Squared Error (RMSE) to assess the quality of the models, and concluded the RF and LR were efficient when the pattern of mean speed was almost linear. However, when an abrupt change in traffic occurred, these models had large RMSE values. In those cases, Support Vector Regression (SVR), Multilayer Perceptron (MLP) achieved lower RMSE values, indicating that these models were more capable of following abrupt changes in speed.

To estimate the number of cars in a given place, Shahid et al. [27] compared a set of regression models, including ensemble regression. The list consists of regression algorithms which includes Decision Tree (DT), Gradient Boosting (GB), Ridge Regression (RR), RF, SVR, LR and MLP, alone. Authors also consider four ensemble models, which comprises two algorithms connected with boosting: (i) SVR–DT, (ii) SVR–GB, (iii) SVR–RF, and (iv) SVR–MLP. The experimental findings show the effectiveness of the proposed framework, which decreases the error rate by about 2.47% when compared with other non-ensemble techniques. Using as evaluation indicators RAE, MAE, RMSE, and Mean Squared Error (R^2), it was shown that SVR–MLP achieves the best results in all aspects.

Zhang and Kabuka [34] used a modified version of a Recurrent Neural Network (RNN), that combines ideas taken from LSTM. According to the authors, the best results were obtained using a GRU-based deep learning approach, with two hidden layers with 500 neurons each. They also used other ML algorithms for comparison, namely SVR, RF, ARIMA, to estimate the traffic low volume. Using MSE, MAE and RMSE, the results shown the proposed method yields the least error rate. Direct comparison of simplified RNN and LSTM RNN models, with the same hyperparameters, showed an improvement in prediction accuracy up to 5.8%, achieving over 94% accuracy.

In experiments, the authors concluded the SVR model performs best under stable conditions with minor variations, while the MLP model adapts better to situations with significant variations. Fernandes et al. [13] evaluates the performance of AutoRegressive Integrated Moving Average (ARIMA) and Long Short-Term Memory (LSTM) in traffic flow forecasting. They predict the mean velocity, concluding that LSTM outperforms ARIMA, considering MAE, RMSE, and Mean Squared Error (MSE). Nevertheless, both approaches were able to identified trends and the cyclic nature of traffic.

Ni et al. [22] build model for short-term traffic flow using four common methods, namely: (i) SVR, (ii) Autoregressive model (AR), (iii) Artificial Neural Networks (ANN), and (iv) K-Nearest Neighbour (KNN). The best results were reported for Support Vector Regression (SVR), showing a decrease of about 7% and 24% in Mean Absolute Percentage Error (MAPE) and RMSE, respectively. This is in relation to a data set without features that come from social network

Nubert et al. [23] use a Convolutional Neural Network (CNN) to classify the density of traffic in junctions. They used a InceptionV3 pre-trained model, removed the penultimate layer, and trained a new softmax layer on top of it. Since the number of images were small, a cross-validation was performed using a simple grid-search. The reported results indicated a Accuracy (ACC) 74.3% and an F1 Score (F1) 81.3%.

Bratsas et al. [6] assessed the effectiveness of their models via multiple comparisons in three scenarios: (I) Speed prediction at random dates and times on randomly selected roads, (II) Speed prediction on randomly selected roads, at eight consecutive dates and times, and (III) Speed prediction on randomly selected roads for the duration of a specific 24h time window. The authors useSVR, RF, MLP and LR to predict traffic speed in a region of interest, assessed by RMSE and MAE. In the first scenario the median RMSE was between 6–7 km/h for all models, with SVR performing better than the other models. In the second scenario, the minimum RMSE values occurred again for the SVR model, accompanied by the MLP. In the last scenario, the MLP model performed better during periods of larger changes in traffic flow. In contrast, the SVR model performed better during periods of smaller changes in traffic flow.

Farajiparvar et al. [12] used two deep learning methods, Seq2Seq model (S2S) and WaveNet model (WN), to forecast traffic speed in road segments. In different time steps, ranging from five minutes to two hours, WaveNet model (WN) with side information achieves the highest performance while exhibiting the lowest MAPE. S2S with side information outperforms ARIMA for long-term prediction and improves MAPE by 0.03%. With WN, MAPE is reduced by 0.9% and learning time is reduced by elimination of recurrent connections which is more efficient than Seq2Seq with LSTM. WN models show superior prediction accuracy when it comes to long-term forecasting.

Essien et al. [11] used a deep bi-directional LSTM autoencoder traffic flow prediction model implementation that uses tweet, weather, and traffic data sets for model training. In their study, the input multivariate time series data set is reconstructed using an overlapping sliding window approach. For this reason, a multi-step 1-h-ahead prediction model was developed to learn the historical time series data set features. The model was evaluated using a technique called walk-forward validation, using MAE, RMSE, Symmetrical Mean Absolute Percentage Error (sMAPE). Based on the empirical analysis, the MAE was reduced from 8 to 5.5 vehicles per hour when tweet data was included along with traffic, rainfall, and temperature data sets.

Zhao et al. [35] created a short-term traffic forecast, to estimate the traffic flow in 15, 30, 45 and 60 min. The main model is a LSTM network tested with 2, 3, 5 and 6 layers. The model was compared to other ML algorithms, namely Recurrent Neural Network (RNN), ARIMA, SVR, Radial Basis Function network (RBF) and Stacked Autoencoder model (SAE). Thr performances of the different algorithms for each time frame were evalluated using MAE, MSE, and MRE. The performance of the proposed LSTM network is better, especially when the forecast time is larger, i.e., longer than 15 m. SVR performed poorly, compared to more recent approaches. ARIMA has more forecast errors.

Chawla et al. [7] developed two models, one from a LSTM network and another using a Gated Recurrent Unit model (GRU) model, to estimate traffic congestion in any given area. The LSTM model is defined with 32 neurons in the first hidden layer and one neuron in the output layer. The model is fitted for 50 training epochs with a batch size of 32. The GRU model is defined with 32 neurons in the first hidden layer and one neuron in the output layer. The model fitted for 50 training epochs with a batch size of 100. Both models were used to predict January 2020 traffic speed on selected highways, based on the five-year historical training data. The authors compare both model with RF, XGBoost model (XGB), and Ensemble learning model. Using MAE, RMSE, MAPE and R^2, as the performance measure, results shows most effective model was LSTM, with very low bias and variance. This model achieved a RMSE of 6.736 and a MAPE of 2.09. In addition, XGboost also achieved positive results, with low bias and small variance too.

2.3 Visual Integration of Models

In the majority of research studies covered in this section, that model traffic flow, or related features to traffic, dashboards were not created to communicate their results to users. Albuquerque et al. [2] provides a set of static map visualisations, in the form of heatmaps, of congestion and road accidents. Traffic flow is depicted in three colours. Red corresponds to locations where more traffic congestion occurred and orange and yellow are locations where traffic congestion was less common. There are four colours that represent the distribution of accidents, with orange and yellow representing streets that have more accidents. Green and blue represent less common places for accidents to happen. However, the analysis is based on a static statistical analysis, where no form of interaction exists on the dashboards.

Some researches do create interactive dashboards. Tempelmeier et al. [30] developed an interactive dashboard to display spatio-temporal urban traffic patterns. The interaction is mainly focused on selecting the day displayed on the map. A tooltip for viewing historical data is available when clicking on an event on the map. Bachechi et al. [4] developed the Trafair Traffic Dashboard (TTD), a visual analytics tool that allows the exploration of real-time and historical data. The dashboard uses a traffic model created with the Simulation of Urban MObility (SUMO) simulator [19]. It makes available data about vehicle counts and their average speed in every road lane in the urban area. Notice there was no ML technique used to support the displayed data. Every 15 min a new simulation based on the traffic sensor data of the last 3 h is performed and combines traffic data with random and predefined vehicles' routes. The dashboard displays real-time traffic information, sensor measurements and derived statistics, traffic model historical data, and emerging trends. It is possible to choose a specific day to access more detailed insights, namely, the top 30 sensors in terms of vehicle count. Nevertheless, the contextual map only shows the current situation, colouring each road according to traffic congestion. It allows users to select a time frame to display a graphic of traffic conditions. However, such conditions apply to the whole region of interest (city of Modena).

Finally, Chawla et al. [7] integrates the ML model into an interactive dashboard that supports visual exploration of traffic flow in context. The user can select a particular freeway on a specific day and see the predicted speed results obtained from various models. The dashboard also provides users with both real-time and historical information, with the latter displayed in the form of an average value for each hour of a day.

2.4 Summary

From the literature review, summarised in Table 1, it is clear that sensor data is the most common source of data to analyse traffic behaviour. However, these sensors are only located in certain parts of the cities, making it difficult to make a ML model applicable throughout a large area. As pointed out by [2], at least for the city of Lisbon, the main road traffic problems occur on the principal roads inside the city. The works that use crowdsourced data overcome this problem [3, 14], relying on a collaborative effort to report traffic problems, with successful results. In addition, contextual data mixed with traffic data can benefit traffic predictions. Based on the many ML algorithms used, we can see that different algorithms are best suited to specific tasks and the given restrictions. Last, most of the works do not include ways of integrating models into user interfaces that will turn them into value for decision makers in cities, in particular. Thus, there is a need for ML models to be integrated into interactive dashboards.

Table 1 Summary of the related work discussed in this section organised into 5 categories: (i) *Year*, the year of publication; (ii) *Data used*, the sources of data used to predict traffic flow; (iii) *ML*, the Machine learning algorithms used, even if only for comparison purposes; (iv) *Evaluation criteria*, the performance metrics used to evaluate the quality of the model; (v) *Dashboard* indicates if the authors produce a Dashboard that integrates the ML model, towards a visual analytics solution; and (vi) *Dependant variable*, indicates which variable were prediction by the model

Reference	Year	Data used	ML	Evaluation criteria	Dashboard	Dependant variable
Ni et al. [22]	2014	Sensor data + Twitter data	AR, ANN, SVR, KNN	RMSE, MAPE	–	Vehicle miles travelled
Zhao et al. [35]	2017	Sensor data	LSTM, RNN, ARIMA, SVR, RBF, SAE	MAE, MSE, MRE	–	No. vehicles
Nubert et al. [23]	2018	Camera data	CNN	ACC, F1	–	Traffic density level
Zhang and Kabuka [34]	2018	Sensor data + Weather data	RNN, LSTM, SVR, RF, ARIMA	MAE, RMSE, MSE	–	Vehicle miles travelled
Bratsas et al. [6]	2019	Sensor data (on-board probe)	SVR, RF, MLP, LR	RMSE, MAE	–	Speed
Fernandes et al. [13]	2019	Crowdsourced data (TOMTOM)	ARIMA, LSTM		–	Speed
Farajiparvar et al. [12]	2020	Crowdsourced data (WAZE)	S2S, WN	MAPE		Speed
Tempelmeier et al. [30]	2020	Sensor data + event data	–	–	Yes, interactive	Speed
Albuquerque et al. [2]	2021	Crowdsourced data (WAZE)	–	–	Yes, static	Congestion level, no. Road Accidents
Bachechi et al. [4]	2021	Sensor data	–	–	Yes, interactive	No. vehicles, speed
Essien et al. [11]	2021	Sensor data + Weather data + Twitter data	LSTM	MAE, RMSE, sMAPE	–	No. vehicles
Shahid et al. [27]	2021	Sensor data + polution data	DT, GB, RR, RF, SVR, LR, MLP, SVR–DT, SVR–GB, SVR–RF, SVR–MLP	R^2, MAE, RMSE, RAE	–	No. vehicles
Chawla et al. [7]	2022	Sensor data + event data + weather data	LSTM, GRU, RF, XGB, Ensemble	MSE,RMSE, MAPE, R^2	Yes, interactive	Speed

In this work, weather information will be used in conjunction with traffic flow data, to create a traffic flow ML model integrated with an interactive dashboard.

3 Data Preparation and Modelling

For creating the model to forecast traffic flow data, Waze and meteorological data were used. Additional data sources were identified as potential solutions (e.g. traffic sensors derived from electromagnetic coils and events), but they were not able to be collected in real-time, which was one of the requirements of this study. The resulting data set was derived from data collected between November and December 2020, during the COVID-19 pandemic. The collection takes place, using two open data Web APIs: (i) EMEL Open Data[1] (waze powered), and (ii) IPMA Forecast,[2] in Portuguese, both reporting *json* data. Different frequencies were used to fetch and store data. Weather data is fetched every hour, while traffic data is fetched every five minutes. Frequency is based on the rate of change of the data sources.

Weather information is collected by three stations inside the city. Although there seems to be no significant fluctuation among stations for each hour, it was decided to use the maximum, minimum and average values of temperature, wind speed and accumulated precipitation. Waze data is a subset of the standard data under the "*Waze for Cities data program*".[3] The retained features are illustrated in Table 2.

3.1 *Dependent Variable*

Many of the works in the literature use the speed of vehicles passing on a road to assess traffic fluidity. Even though this is a good indicator of road congestion, the sensitivity of the human brain to different speeds [16] does not necessarily translate into a realistic estimate of a precious commodity these days—time. "*How long will I take to get home?*" is a more common question than "*How fast can I get there?*". People place significant importance on time, on its singularity, and on its impact on the rhythms of social life [33]. Therefore it was decided to use the `delay` as the dependent variable to consider on the definition of the Road Traffic flow Indicator (RTI). Of course the faster we travel, the less time is needed to get from A to B. Waze data seems to confirm that, as we can see from the correlation between some of the variables (Table 3).

However, there is singularity in the data. If the road congestion `level` is 5 (meaning a jam), the speed goes to zero and the `delay` becomes -1. Since Waze reports abnormal traffic situations and not normal conditions, around 50% of the

[1] Available at https://opendata.emel.pt/index.html.

[2] Available at https://api.ipma.pt.

[3] Available at https://www.waze.com/wazeforcities/.

Table 2 Waze data kept for data processing and feature engineering. Not all features are used for model creation, and contextual attributes are missing from this table. For a complete list of attributes, please use the following DOI: 10.13140/RG.2.2.22001.35689 to get access to the data set

Name	Description
datetime	Date and time of the jam
city	The city where the jam took place. Maybe empty
street	The name of the jammed street
level	The level of congestion, [0 .. 5] (0 = lane completely free, 5 = lane completely congested)
length	The length of the jam, in meters
end_node	The nearest exit after the jam
speed	Average traffic speed in meters per second
road_type	Type of the road (1 = Street, 2 = Main road, 3 = Motorway, 4 = Ramp, 5 = Path, 6 = Main street, 7 = Secondary street, 8 = Driveway, 9 = Walkway, 10 = Pedestrian way, 11 = Exit, 14 = Driveway, 15 = Boat crossing, 16 = Stairs, 17 = Private way, 18 = Railway, 19 = Exclusive lane (taxi/bus), 20 = Driveway or inside parking lot, 21 = Service road)
delay	The expected delay to travel through the jam
position	Set of coordinates representing the congestion location (Object 'GeoJSON')

Table 3 Correlation coefficients between characteristics `speed`, `length` and `level` in relation to the dependent `delay`

Correlation coefficient	Speed	Length	Level
Pearson	−0.748	−0.472	0.881
Spearsman	−0.917	−0.704	0.949

records have a `delay` equalling -1 for the period examined. From a mathematical perspective, there is a discontinuity in the `delay` value when the `level` passes from 4 to 5. Therefore, care must be taken when dealing with those records. The value of `level` will be changed from -1 to a positive value based on the temporal context.

3.2 *Feature Engineering*

The data were assumed to have a normal distribution (based on those different from -1 in the `delay` feature). A function based on the maximum delay in each day was used to change the -1, described by (1).

$$\max(\texttt{delay}) + n \times \texttt{stdev}. \tag{1}$$

To determine which is the "best" value for n, a set of tests are carried using $n \in [1, 3]$ and comparing the results with the original data set using MAE, `rmse` and R^2. A non-optimised regression model is used to predict the delay. The results show that

R^2 is very similar (differences < 0.01), while the MAE and `rmse` increase with n. In order to maintain the distribution tidier, $n = 1$ is chosen, although the distribution is generally skewed right.

As a result, the MAE found is about 2 min, which is appropriate for the human scale and the problem at hand. To the best of our knowledge, this is the first study to transform the Wazed-based delay into a prediction variable.

The next step was to include the recent past in the data set. Taking into account the results shown in Table 3, there seems to be a high degree of correlation between a set of features, which are not available at the present time when the delay need to be predicted. Thus, any model that uses them to estimate the dependent variable will perform worse. One possibility to overcome this situation is to encode the recent past into a set of new features. This assumes the actual delay is the result of the accumulation of traffic, in the case of a jam, or the reduction of cars, otherwise. The idea used here is similar to attenuation-accumulation maps [1], where the observation an event now depends on the cumulative effect of a set events in the past. Short-term historical data is incorporated by transforming `delay`, `length`, and `speed` into nine new features, representing the maximum, median, and average value over the previous three 30-minute periods, spanning 1h30m. The data set has also been updated with contextual data, namely features related to time, such as rush hours and time of day, along with the classic textual description of time and weather features.

The final data set includes 65 features, and is publicly available (DOI: 10.13140/RG.2.2.22001.35689).

3.3 Predictive Model

The goal of the model is to predict the value of the delay for the next 30 minutes. The regression model was built using XGBoost [8], in the H2O tool.[4] The XGBoost algorithm is a distributed gradient boosting supervised learning algorithm designed to be highly efficient. It implements a parallel tree boosting algorithm based on Gradient Boosting. This is a numerical optimisation problem where the objective is to minimise the loss function of the model by adding weak learners using gradient descent. It was also tried with lower success with other algorithms (RF, SVR). Although the algorithms used produce difficult interpretable models, the goal is not to explain why the model produces a specific response and communicate that to the end users. Nevertheless, based on feature importance, we can interpret the model in the first place, giving us an insight on the usability of the model in real-life. Figure 1 shows the most important features. As such, the features `level` and `speed`, and `length` were removed because they share a high correlation with the dependent variable, some are the strongest predictors, and they are unavailable in "real-time" to be useful in practice.

The train set was taken from the 49th week of 2020. It covers the whole city of Lisbon, and has a total of 86532 records. The test set contains the data for the 50th

[4] https://www.h2o.ai.

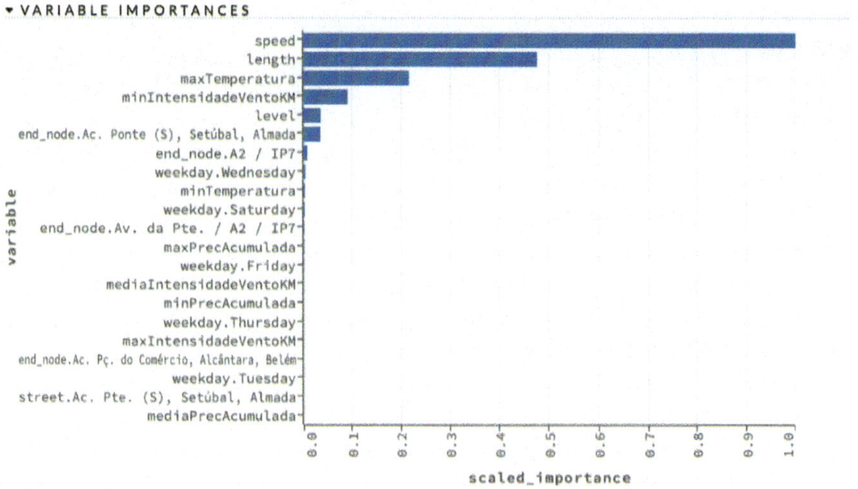

Fig. 1 Based on the original data set, XGBoost algorithm reported the feature (variable) importance

Table 4 Models description

Model	Description
Model A	Waze data set with original features, including `delay` = −1, augmented with contextual data;
Model B	Same as *Model A* but but replacing the `delay` using (1), $n = 1$;
Model C	Same as *Model B*, but removing `speed`, `level` and `length` features
Model D	same as *Model C*, but including short-term historical features

week of 2020 with a total of 97074 records. The split into test and train data could be taken differently, for example, using random sampling. However, we want to build a model that is as close to what really occurs as possible, in which the ML model can be updated regularly, e.g. every week, based on the most recent available data.

We conducted four experiments, described in Table 4. Model A and B are useless for forecasting, since the most important features that predict the `delay` are `length`, `speed` and `level` or a combination of all, which are unknown at the time of forecasting. These models show us the "asymptotically" calculated prediction error. We can see the impact of removing the three aforementioned features on the predictive quality of Model C. It suffers from removing the most significant features to predict the delay. With Model D, some of the information was recovered by including short-term historical features that are known at the time of forecasting. Table 5 presents the performance for each model, using R^2, MAE and RMSE.

Table 5 Model comparison using original and transformed data set. Model D was chosen one, since is the most realistic one, given the domain

Model	R2	MAE	RMSE
Model A	0.972	7.78	21.13
Model B	0.971	142.32	201.69
Model C	0.796	331.737	542.711
Model D	0.936	189.428	304.899

3.4 Road Traffic flow Indicator (RTI)

The RTI is calculated using (2), and captures the flow of the road traffic. It works for any spatial region (e.g. a parish) in a given 30m time period (in the past, the present and the future). It depends on the number of traffic jams and on the delay at the Region Of Interest (ROI) being analysed.

$$\texttt{delayMean}_j^k \times \texttt{\#TrafficJams}_j^k \tag{2}$$

where j is a 30m period, k is a ROI, and `#TrafficJams` is the number of traffic jams, normalised between 0 and 1. If j is in the future, the delay is estimated using the predictive model referred in Sect. 3.3. The `#TrafficJams` is calculated with the previous known value. Normalisation is an adaptive procedure, which uses the maximum value received so far.

4 Interactive Visual Analytical Dashboard—iVAD

The dashboard[5] was prototyped using the unfolded.ai[6] tool, as illustrated in Fig. 2. iVAD contains four areas of information/interaction: Area identified by Ⓐ contains layers' control. There, users can show/hide street-level information and region-level RTI. In Ⓑ, there is the map that supplies contextual spatial information to the RTI. Area Ⓒ shows a temporal distribution of the number of jams for a period of time. It supports time animation, enabling a representation of the evolution of jams in a spatio-temporal context [5]. And finally, Area Ⓓ displays the legend of each represented layer. Next, we will explain in more detail regions Ⓑ and Ⓓ, as they are the most significant in terms of visual representation of the information being displayed.

[5] The dashboard is available at https://studio.unfolded.ai/public/56a543da-063f-4cc2-b558-7cd1287f33aa, in English.

[6] https://studio.unfolded.ai.

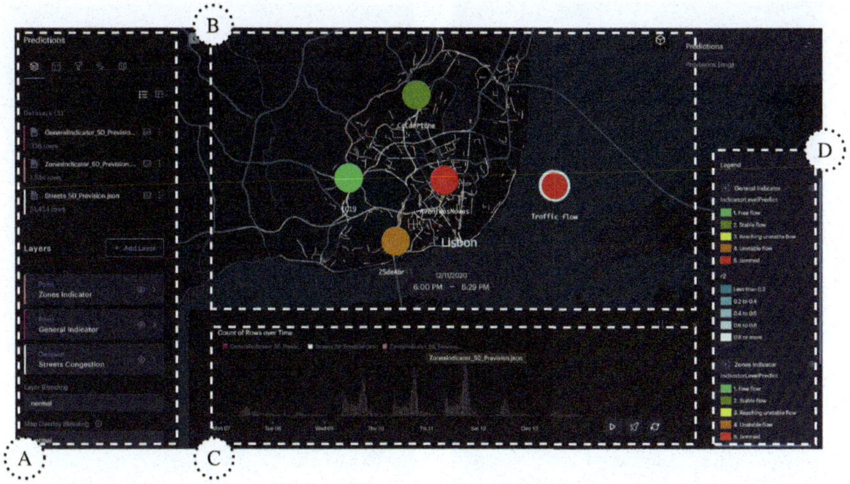

Fig. 2 Snapshot of iVAD, showing the forecast value of global RTI (labelled *Traffic Flow*) and local RTI (at the centre) , where the four main areas are identified: A —Layer control; B —Map context and RTI display; C—Temporal control and animation; and D —Legend

4.1 Map Area

For illustration purposes, four relevant ROIs for the city were selected, three of them at entrances/exits (green and orange, in Fig. 2. B) and one in the city centre (red in the middle of the map). They represent local RTI for a given ROI k. ROIs were calculated using road segments that start or end on the main access road—the city entrance/exit/main road. Domain knowledge is used to select roads that integrate RTI and are city-dependent. It is worth noting that segments of the city are part of the source data, and they do not represent circular areas. Other ROIs are possible, however, they should be reduced in number to leave the dashboard clutter free. In this prototype dashboard, the global RTI is placed outside of the city limits. It was decided to place it on top of the river (a place without road traffic) to distinguish it from the locals RTI. Since we want to make a short-term prediction of 30 minute for the traffic flow of the city, we want to show the analyst the "quality" of the prediction. Two problems must be addressed: (i) What performance measure should be used to communicate the quality of the model (R^2, MAE or RMSE); and (ii) How to visually inform the analyst of the model's performance, in a simple and effective way.

To tackle the first problem, it was decided to use the R^2, which is a statistical measure that represents the proportion of variance for a dependent variable that is explained by the independent variables. While no single measure can be used to assess the quality of the prediction, R^2 has some nice qualities for showing up in a dashboard: (i) it is bounded, with a maximum value of 1; (ii) can be assigned to a colour scale, indicating a strong or weak model. R^2 being bounded is key for creating

(a) A global RTI, showing a good prediction ($R^2 \approx 0.95$) of free flow traffic.

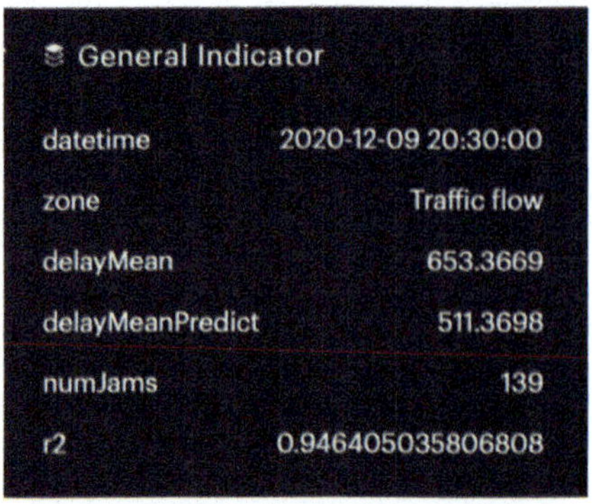

(b) Tooltip with detailed information for Fig. 3a.

(c) A global RTI, showing a poor prediction ($R^2 \approx 0.0.4$) of free flow traffic.

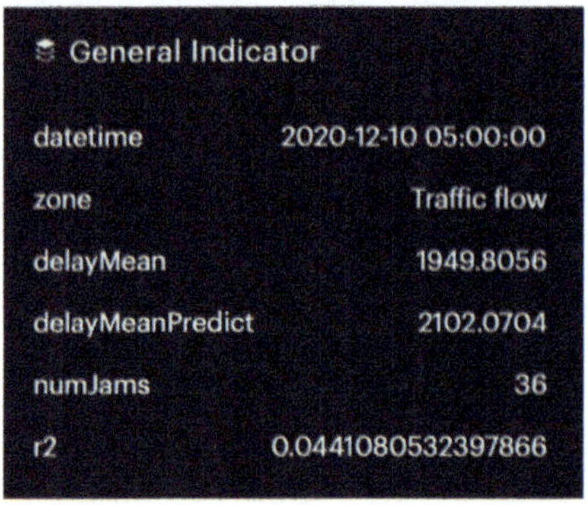

(d) Tooltip with detailed information for Fig. 3c.

Fig. 3 Examples of global RTI prediction for free flow with different values for the goodness of fit (R^2). It shows the analysts how good the prediction is, supporting a better judgement of what is depicted

a visual representation that scales and adapts to different situations. Considering that Waze data is less representative during timeframes where fewer people contribute, this measure seems to incorporate this situation. The results, illustrated in Figs. 3 and 4, confirm the R^2 as a proper "goodness of fit" measure. To address the second problem, "how to visually communicate to analysts the quality of the model", it was decided to use a large coloured border to visually inform the prediction quality of the model. Figure 3 shows how the global RTI is displayed in the iVAD predicting a free flow traffic situation. In one situation, Fig. 3a, the prediction has a good performance, with a white (pristine) coloured border. RTIs are implemented as affordances, i.e., the colours change as a hint that the user can interact with them to gain access to more information. Thus, a single click unveils a tooltip with a configurable number of parameters that exists in the data set used. For the examples shown in Fig. 3b and d

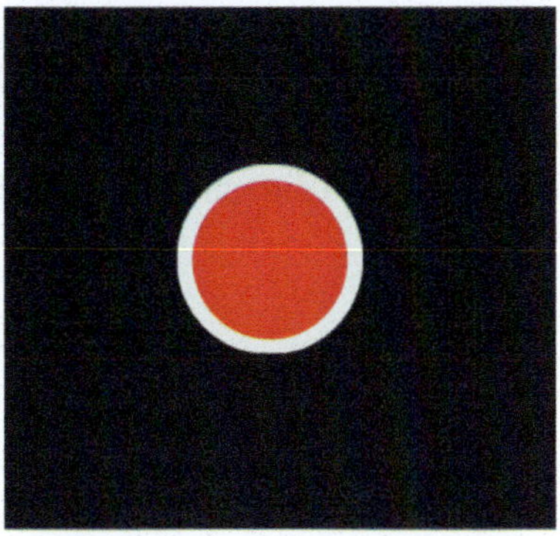

(a) A global RTI, showing a good prediction (R^2 ≈ 0.95) of jammed traffic inside the city.

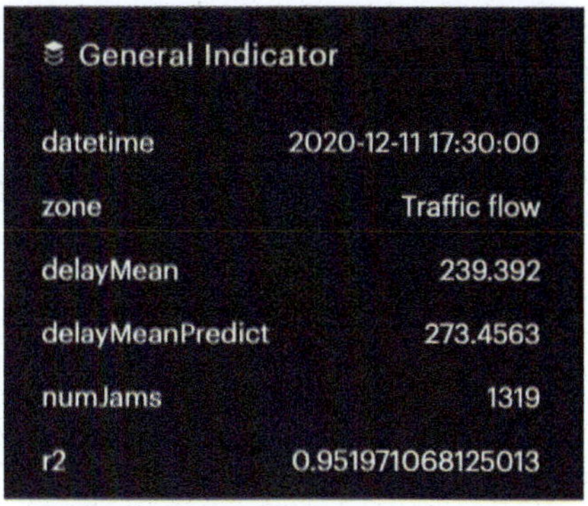

(b) Tooltip with detailed information for Fig. 4a.

(c) A global RTI, showing a poor prediction (R^2 ≈ 0.23) of jammed traffic inside the city.

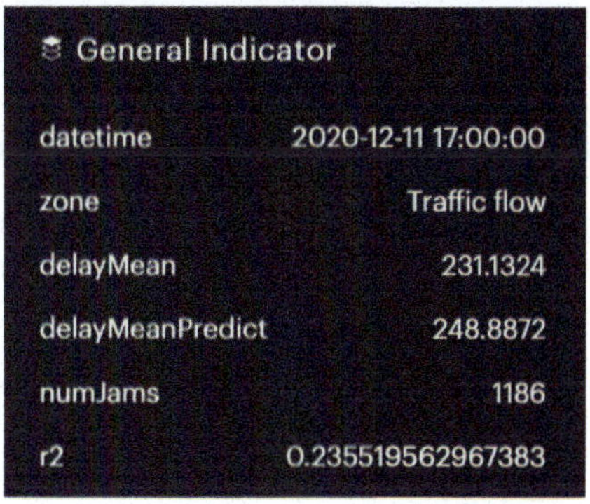

(d) Tooltip with detailed information for Fig. 4c.

Fig. 4 Examples of global RTI prediction for jams with different values for the goodness of fit (R^2). It shows the analysts how good the prediction is, supporting a better judgement of what is depicted

it was decided to include for the demonstration purposes, the original and predicted values, the number of jams, the date and time and the R^2. Do notice the time where the poor prediction occurred—5 AM, an hour with few Waze contributors—as opposed to the time where the good prediction occurred—8.30 PM, an hour with more Waze contributors. For sake of comparison, Fig. 4 shows the global RTI for jam situations. Do notice that, in this case two consecutive periods show different R^2, that cannot be inputted into the number of Waze contributors.

4.2 Legend Area

The dashboard presents all information on a common scale; all the values on the map are normalised between 0 and 1. One can separate two different situations that should have different colouring:

1. the traffic flow; and
2. the R^2 representing the quality of the prediction.

To represent the perception of traffic flow, both at the street level and for the RTIs, it was decided to use a 5 level scale (inspired by the level-of-service standard [15]) to reduce clutter and increase readability of the indicator [25]:

1. Free flow $[0, 0.2[$;
2. Stable flow $[0.2, 0, 4[$;
3. Reaching unstable flow $[0.4, 0.6[$;
4. Unstable flow $[0.6, 0.8[$; and
5. Jammed $[0.8, 1]$.

The colour palette used is inspired by the traffic lights, from green (Free flow) to red (Jammed). For each level, the scale is uniform. Taking into account R^2 the colour scale is taken from a cold colour palette that simultaneously has:

1. a good contrast with the map colour;
2. is different from the colours used to represent the RTI;
3. has a good contrast in most of the RTI colours used;
4. high values, which represent good predictions, is represented with a pristine colour (white); and
5. low values, which represent worse predictions, is represented with a more impure colour (teal).

Both colour palettes are shown in Fig. 2. Ⓓ.

5 User Based Evaluation

A user-based evaluation was carried out to assess the quality of the RTI. The question under evaluation is whether the global RTI can successfully summarise the flow of traffic. A Google form was used for remote testing. There was no interaction between the participants and us during the testing protocol.

5.1 Characterisation of the Participants

A total of 29 users participate in the test, whose age distribution is illustrated in Fig. 5 51.7% of the participants claim they are acquainted with data visualisation tools data

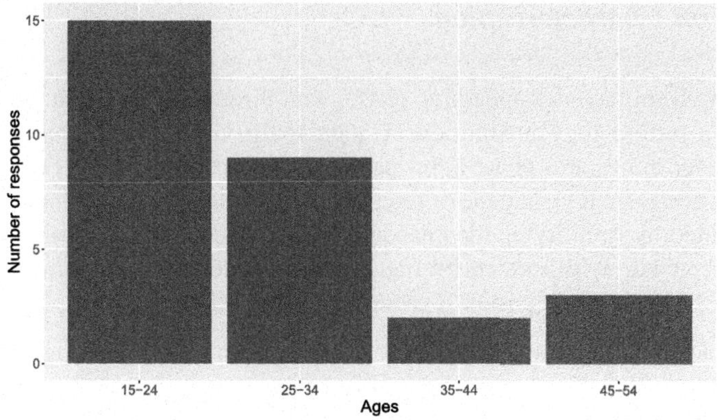

Fig. 5 Age distribution of the participants in the user testing

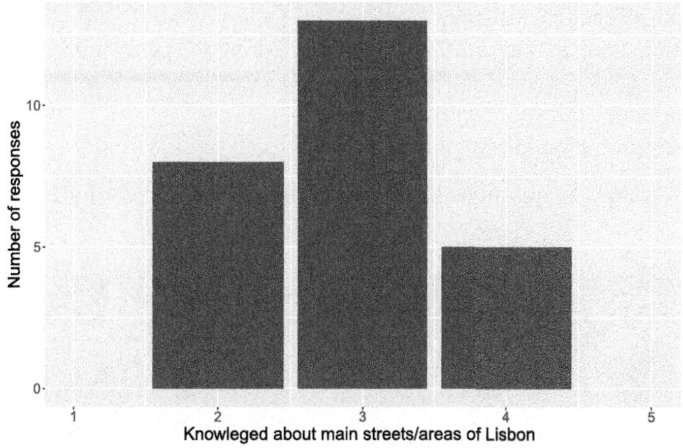

Fig. 6 Distributions of the answers to the question "How well do you know the main streets/areas of Lisbon city where the indicators were presented in the previous images". a 5-level likert scale was used (1—I don't; 5—I know perfectly)

visualisation tools. The majority of the participants (79.3%) claim to drive regularly, however only 37.9% drive regularly in the city of Lisbon. These values might indicate that while users are aware of traffic problems, since they are regular drivers, they have average knowledge about daily traffic problems inside Lisbon. This is supported by the answers to the question "How well do you know the main streets/areas of Lisbon city where the indicators were presented in the previous images", illustrated on Fig. 6.

5.2 Form Characterisation

The testing form includes a learning phase, with three questions, that guideD users through the terminology and introduceD them to the visual representation of iVAD. Among other things, this phase allow participants to know what a RTI is, that there are global and local RTI, and the different LoD the dashboard has. That information is transmitted by simplifying the image being displayed and to describing textually what is depicted, as illustrated by Fig. 7. After the learning phase, users need to answer to 15 evaluation questions, presented in the following order:

A. four questions to guess the global RTI based on the information given by local RTI and street level information (see Fig. 8a and b), with two different difficulties for each variation;
B. four questions to tell if the global RTI is a good resume of the traffic flow given the local RTI and street level information (see Fig. 8c and a), with two different difficulties for each variation;
C. a question about a video with an animation, asking the performance of the global RTI;
D. an open-ended question; and
E. five questions about driving habits and age.

The first 9 evaluation questions has a 5-level likert scale to record the answer.

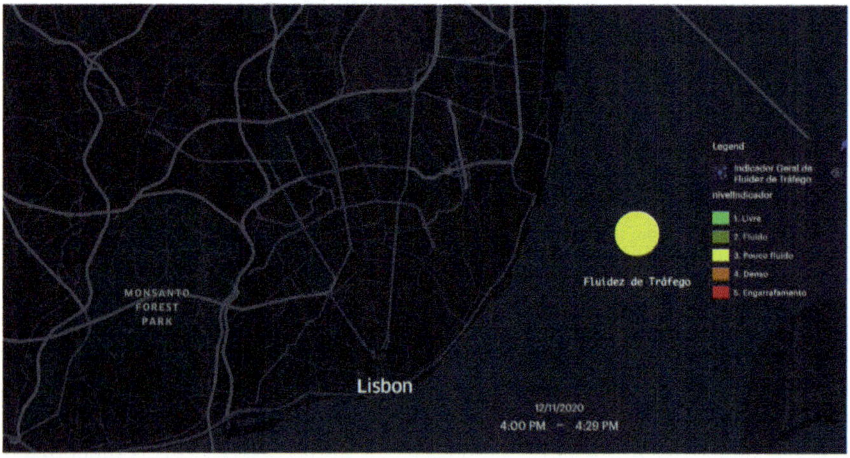

Fig. 7 Example of an image representing global RTI during the learning phase. The image was complemented with by the text " *Global Road Traffic flow Indicator (RTI) for the city of Lisbon on the day 12/11/2020, between 4PM and 4.30PM*"

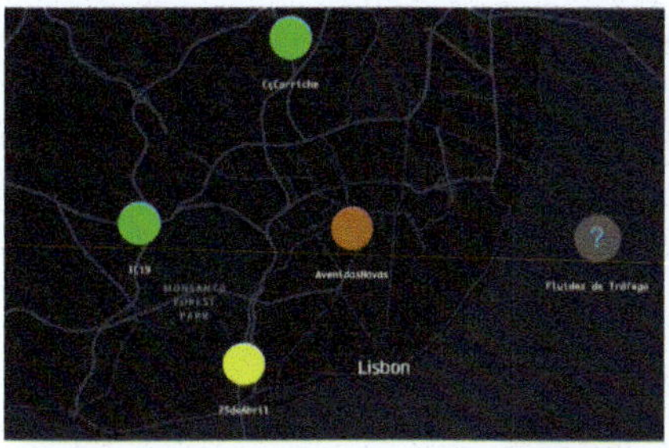

(a) Image used in the question "*What traffic flow should be reported from the global RTI, based on local RTI?*" (two different consecutive LoDs).

(b) Image used in the question: "*What traffic flow should be reported from the global RTI, based on street level info?*" (two different LoDs).

(c) Image used in the question "*How good the consistency is between the global and local RTI?*" (level of agreement between two LoDs).

(d) Image used in the question "*How good the consistency is between the global and street level information?*" (level of agreement between the lowest and the highest LoD).

Fig. 8 A representative set of images used in the user-based evaluation form. (original in Portuguese)

5.3 Discussion

After user evaluation, a power analysis was conducted to determine the strength of the test, given the sample size and the effect size achieved (from medium to large, depending on the questions).The number of users were enough to support the conclusions to get an $\alpha = 0.05$ and $\beta = 0.8$. Figure 8 illustrates some of the images used in the form, to support the discussion. Figure 8c and d illustrate the questions where users need to say if they consider the global RTI to be consistent in reporting traffic flow conditions in a 5-level likert scale (1: not at all consistent to 5: consistent). The answers were positive, i.e., greater then 2.5 ($p-$value$<$ 0.001 and 0.05 respectively). Apparently, it was more difficult for users to compare the consistency between street-level and global RTI. When users were asked to tell which RTI value to use, the result shows that users often got the correct answer, 60% to 80%, when looking at the local RTI (as illustrated by Fig. 8a). However, when looking at the street level (as illustrated by (Fig. 8b), the errors increase. Here, the correct answers vary from 0% to 49%. Figures 9,10, 11 and 12 illustrates the results describe previously. Therefore, it can be concluded that the RTI can provide people with information regarding traffic fluidity.

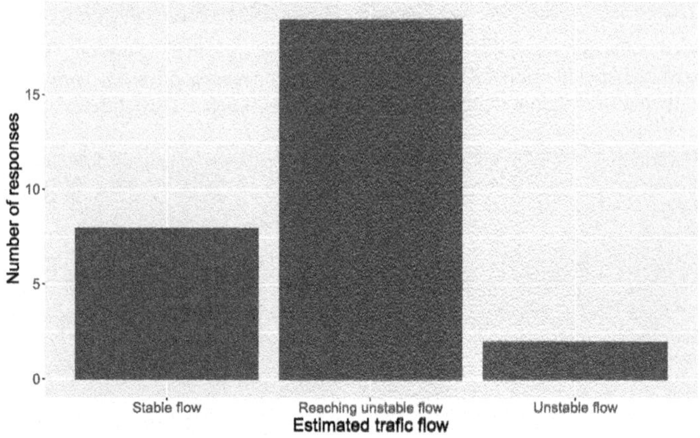

Fig. 9 Results of user' answers for the question supported by Fig. 8a. The correct answer is "Reaching unstable flow". Around 80% of the answers were correct

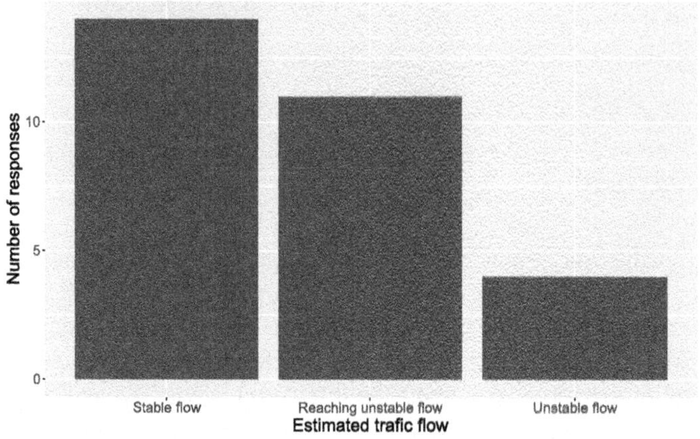

Fig. 10 Results of user' answers for the question supported by Fig. 8b. The correct answer is "Unstable flow". No one answered this correctly

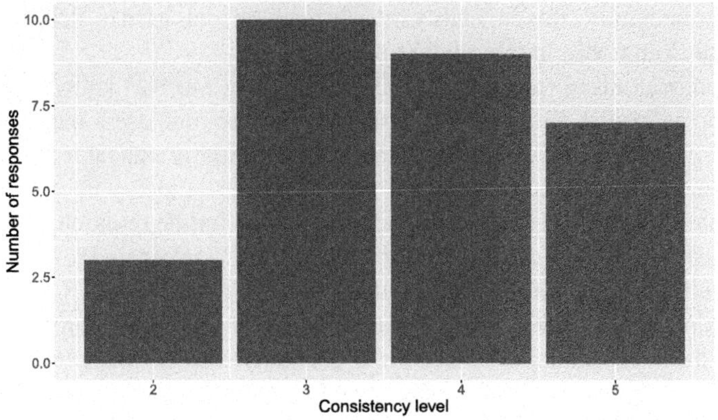

Fig. 11 Results of user' answers for the question supported by Fig. 8c. The responses show a positive result, with $p-$value< 0.001

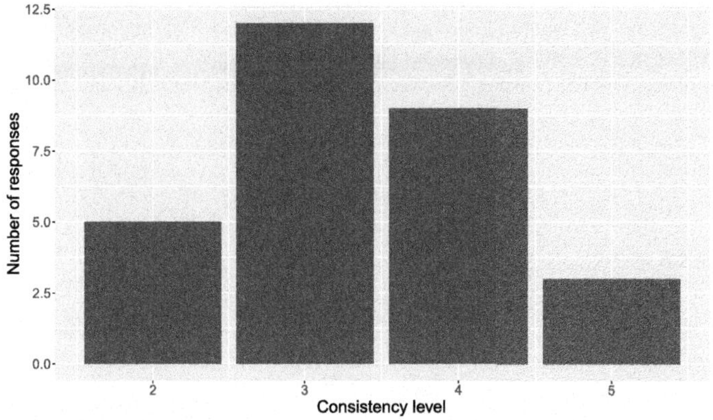

Fig. 12 Results of user' answers for the question supported by Fig. 8d. The responses show a positive result, with $p-$value< 0.001

6 Conclusion

The goal of Visual Analytics and Visual Knowledge Discovery is about discovering new messages, which extends beyond visualising input data. Essentially it's a symbiotic relationship between Visual Tools and Machine Learning; using visual tools to enhance AI/ML, and using AI/ML to enhance visual exploration and representation [18]. In this chapter we presented an iVAD displays both real and predicted data, based on a multilevel Road Traffic flow Indicator (RTI) that can use real or predicted data. The seamlessly integration with real and predictive data is aligned with the visual analytics path, allowing users to explore spatio-temporal patterns and discover new insights unveiled by one of the Levels-of-Detail available. Although iVAD is prototyped in a tool, it is far from providing a finished dashboard. It shows the path for integrating this capability into the analytic tools used by decision-makers, in this case from CML, in their daily operations.

The approach taken to create the ML model is, to our best knowledge, unique. It uses a transformed `delay` feature to estimated the time users will get on roads segments. Thus, it is possible to estimate a traffic fluidity indicator that takes into account one of the most import assets of modern lives—time.

The data set produced in this research is available for the research community to use. The user-based evaluation shows the proposed RTI is capable to summarise the traffic flow behaviour, with statistical support.

For future work, the model should include more contextual data and other sources of traffic data. For example, data on the characteristics of roads, the location of traffic lights and crosswalks, the events, and the importance of roads to overall flow. Additionally, more algorithms should be tested, including LSTM, to achieve long-term prediction. In addition to the grammar of graphics used in the dashboard, the dashboard itself should be reviewed from an interaction perspective.

Acknowledgements We would like to thank the CGIUL team for the support provided under the "Data at the Service of Lisbon" Protocol. This work was supported by Fundação para a Ciência e a Tecnologia (FCT.IP) through NOVA LINCS (UIDB/04516/2020) and LASIGE (UIDB / 00408/2020).

References

1. Albino C, Pires JM, Datia N, Silva RA, Santos MY (2017) Aa-maps-attenuation and accumulation maps for spatio-temporal event visualisation. In: 2017 21st International conference information visualisation (IV). IEEE, pp 292–295
2. Albuquerque V, Oliveira A, Barbosa JL, Rodrigues RS, Andrade F, Dias MS, Ferreira JC (2021) Smart cities: data-driven solutions to understand disruptive problems in transportation-the lisbon case study. Energies 14(11):3044
3. Amin-Naseri M, Chakraborty P, Sharma A, Gilbert SB, Hong M (2018) Evaluating the reliability, coverage, and added value of crowdsourced traffic incident reports from waze. Transp Res Record 2672(43):34–43
4. Bachechi C, Po L, Rollo F (2022) Big data analytics and visualization in traffic monitoring. Big Data Res 27:100292
5. Beck F, Burch M, Diehl S, Weiskopf D (2017) A taxonomy and survey of dynamic graph visualization. In: Computer graphics forum, vol 36. Wiley Online Library, pp 133–159
6. Bratsas C, Koupidis K, Salanova JM, Giannakopoulos K, Kaloudis A, Aifadopoulou G (2020) A comparison of machine learning methods for the prediction of traffic speed in urban places. Sustainability 12(1)
7. Chawla P, Hasurkar R, Bogadi CR, Korlapati NS, Rajendran R, Ravichandran S, Tolem SC, Gao JZ (2022) Real-time traffic congestion prediction using big data and machine learning techniques. World J Eng (ahead-of-print)
8. Chen T, Guestrin C (2016) Xgboost: a scalable tree boosting system. In: Proceedings of the 22nd ACM SIGKDD international conference on knowledge discovery and data mining, pp 785–794
9. Din IU, Guizani M, Rodrigues JJPC, Hassan S, Korotaev VV (2019) Machine learning in the internet of things: designed techniques for smart cities. Futur Gener Comput Syst 100:826–843

10. Dinis D, Teixeira AP, Barbosa-Póvoa A (2020) Foresim-bi: a predictive analytics decision support tool for capacity planning. Decis Support Syst 131:113266
11. Essien A, Petrounias I, Sampaio P, Sampaio S (2021) A deep-learning model for urban traffic flow prediction with traffic events mined from twitter. World Wide Web 24(4):1345–1368
12. Farajiparvar P, Hoseinzadeh N, Han LD, Hedayatipour A (2020) Deep learning techniques for traffic speed forecasting with side information. In: 2020 IEEE green energy and smart systems conference (IGESSC), pp 1–5
13. Fernandes B, Silva F, Alaiz-Moretón H, Novais P, Analide C, Neves J (2019) Traffic flow forecasting on data-scarce environments using ARIMA and LSTM networks. In: World conference on information systems and technologies. Springer, pp 273–282
14. Hoseinzadeh N, Liu Y, Han LD, Brakewood C, Mohammadnazar A (2020) Quality of location-based crowdsourced speed data on surface streets: a case study of waze and bluetooth speed data in Sevierville, TN. Comput Environ Urban Syst 83:101518
15. https://ccag.ca.gov/wp-content/uploads/2020/04/2019-CMP-Final-040920.pdfCMP-Final-040920.pdf (2019) Congestion management agency for San Mateo county
16. Hussain Q, Alhajyaseen WKM, Pirdavani A, Reinolsmann N, Brijs K, Brijs T (2019) Speed perception and actual speed in a driving simulator and real-world: a validation study. Transp Res Part F Traffic Psychol Behav 62:637–650
17. Kovalerchuk B, Andonie R, Datia N, Nazemi K, Banissi E (2022a) Visual knowledge discovery with artificial intelligence: challenges and future directions. In: Integrating artificial intelligence and visualization for visual knowledge discovery. Studies in Computational Intelligence. Springer Nature Switzerland AG
18. Kovalerchuk B, Andonie R, Datia N, Nazemi K, Banissi E (2022b) Visual knowledge discovery with artificial intelligence: challenges and future directions. In: Integrating artificial intelligence and visualization for visual knowledge discovery. Springer, pp 1–27
19. Krajzewicz D, Hertkorn G, Rössel C, Wagner P (2002) Sumo (simulation of urban mobility)-an open-source traffic simulation. In: Proceedings of the 4th middle east symposium on simulation and modelling (MESM20002), pp 183–187
20. Metz D (2018) Tackling urban traffic congestion: the experience of London, Stockholm and Singapore. Case Stud Transp Policy 6(4):494–498
21. Nadrian H, Mahmoodi H, Taghdisi MH, Aghemiri M, Babazadeh T, Ansari B, Fathipour A (2020) Public health impacts of urban traffic jam in Sanandaj, Iran: a case study with mixed-method design. J Transp Health 9:100923
22. Ni M, He Q, Gao J (2014) Using social media to predict traffic flow under special event conditions. In: The 93rd annual meeting of transportation research board
23. Nubert J, Truong NG, Lim A, Tanujaya HI, Lim L, Vu MA (2018) Traffic density estimation using a convolutional neural network. Preprint at arXiv:1809.01564
24. Paiva KM, Cardoso MRA, Zannin PHT (2019) Exposure to road traffic noise: annoyance, perception and associated factors among brazil's adult population. Sci Total Environ 650:978–986
25. Saaty TL, Ozdemir MS (2003) Why the magic number seven plus or minus two.Math Comput Model 38(3–4):233–244
26. Serrador A, Tremoceiro J, Cota N, Cruz N, Datia N (2018) iLX—a success case in public tender methodology. In: ProjMAN 2018—international conference on project management
27. Shahid N, Shah MA, Khan A, Maple C, Jeon G (2021) Towards greener smart cities and road traffic forecasting using air pollution data. Sustain Cities Soc 72:103062
28. Silva RA, Pires JM, Datia N, Santos MY, Martins B, Birrac F (2019) Multimed Tools Appl 78(23):32805–32847
29. Taborda R, Datia N, Pato MPM, Pires JM (2020) Exploring air quality using a multiple spatial resolution dashboard-a case study in Lisbon. In: 2020 24th International conference information visualisation (IV). IEEE, pp 140–145
30. Tempelmeier N, Sander A, Feuerhake U, Löhdefink M, Demidova E (2020) Ta-dash: an interactive dashboard for spatial-temporal traffic analytics. In: Proceedings of the 28th international conference on advances in geographic information systems, pp 409–412

31. Vaz J, Datia N, Pato M, Pires JM (2022) Traffic flow indicator: predicting jams in a city. In 2022 26th International conference information visualisation (IV). IEEE
32. Yuankai W, Tan H, Qin L, Ran B, Jiang Z (2018) A hybrid deep learning based traffic flow prediction method and its understanding. Transp Res Part C Emerging Technol 90:166–180
33. Zerubavel E (1985) Hidden rhythms: schedules and calendars in social life. University of California Press
34. Zhang D, Kabuka MR (2018) Combining weather condition data to predict traffic flow: a GRU-based deep learning approach. IET Intell Transp Syst 12(7):578–585
35. Zhao Z, Chen W, Wu X, Chen PCY, Liu J (2017) LSTM network: a deep learning approach for short-term traffic forecast. IET Intell Transp Syst 11(2):68–75

Guided Visual Analytics—A Visual Analytics Guidance Approach for Systematic Reviews in Research

Lennart B. Sina, Cristian A. Secco, Midhad Blazevic, and Kawa Nazemi

1 Introduction

Visual Analytics systems can be used in various application domains to help users extract meaningful information and knowledge from large and complex data sets. These Visual Analytics systems are often complex systems developed to be primarily used by expert users, such as analysts. The problem is that for many use cases, such as systematic literature review creation through Visual Analytics [38], potential users are not necessarily experts but scientists from different domains with different levels of expertise [7]. Various approaches, models, and systems have already attempted to reduce the complexity of Visual Analytics, e.g., adaptive visualizations [23, 24] and adaptive Visual Analytics [11]. However, the problem is that such systems must offer many functionalities and visual layouts to enable in-depth analyses. This multitude of functionalities can easily lead to a cognitive overload [23, 24, 41]. Guidance approaches are one way to support the user in using complex applications such as Visual Analytics systems. There are several theoretical models on guidance in Visual Analytics [6–8]. Still, no practical applications exist that present a design study or implementation for a guidance design for existing Visual Analytics applications. As we stated in [38] a variety of tools are available that authors can use to manage

L. B. Sina (✉) · C. A. Secco · M. Blazevic · K. Nazemi
Human-Computer Interaction and Visual Analytics, Darmstadt University of Applied Sciences, 64295 Darmstadt, Germany
e-mail: lennart.b.sina@h-da.de

C. A. Secco
e-mail: cristian.a.secco@h-da.de

M. Blazevic
e-mail: midhad.blazevic@h-da.de

K. Nazemi
e-mail: kawa.nazemi@h-da.de

B. Kovalerchuk et al. (eds.), *Artificial Intelligence and Visualization: Advancing Visual Knowledge Discovery*, Studies in Computational Intelligence 1126,
https://doi.org/10.1007/978-3-031-46549-9_11

already extracted sources for their systematic reviews and allow them to collaborate in different ways [16]. These solutions only refer to previously extracted literature or searches in specific databases. Moreover, only a few of these systems use artificial intelligence to find, e.g., similar publications [10]. To the best of our knowledge, no system or approach exists that combines a visual search and exploration by integrating information extraction methods (e.g., through neural networks) with an automatic or semi-automatic reporting functionality. A guided stepwise visual search system that assists users in the whole process of systematic review conduction and creates an automated protocol based on the search terms and interactions with the system to generate a visual and automatic "flow-graph" protocol to meet the requirements of a systematic review [32] does not exist.

We introduce in this work an approach for stepwise visual guidance for systematic literature reviews that we explain based on the PRISMA-2020 statement [32]. The approach provides an interplay of automated steps and manual user validation. Here, steps are performed automatically by the system and only require confirmation or correction by the user. In the different steps, which are visually separated from each other, the user is presented with appropriate visualizations, which he can use, for example, to identify and sort literature. Moreover, we propose an integrated, automatically generated flow graph that is generated and displayed during the stepwise visual guidance and a new visualization that can access different topic models to identify relevant topics. Furthermore, we present the concept for an AI-based assistant we named *"Metadata-driven Artificial Intelligence Assistant"* (**MAIA**). We conducted a design study to plan the various steps of the visual guidance system and to identify the steps that may be automatically conducted by the system or in which the user might need assistance from the AI-based assistant MAIA. Finally, the implementation of the guidance design approach into an existing Visual Analytics solution is elaborated.

In summary, the key research contributions of this work are (1) a stepwise Visual Analytics guidance approach for systematic reviews, (2) a design study used to identify steps that can be automated and assisted in the guidance process of systematic review conduction, (3) a new visualization that can be used to identify topics that may be relevant by accessing different topic models, (4) a concept for the AI-based assistant (MAIA).

2 Related Work

As we stated in [38], to the best of our knowledge, no other Visual Analytics systems for systematic reviews exist [38]. The section on related work in this paper is divided into two parts, as both visual guidance approaches and systematic literature tools are relevant to this work. The first literature review is a systematic review according to the PRISMA statement [32] that focuses on guidance design approaches for Visual Analytics systems. The second review summarizes graphical user interfaces that support researchers in generating systematic reviews.

2.1 Guidance Design in Visual Analytics

Visual Analytics systems are often complex expert systems. Their operation requires high expertise. Guidance approaches aim to support the user and thus simplify operations. There are several approaches to guidance in Visual Analytics systems to reduce complexity for users. There are several approaches to guidance in Visual Analytics systems to reduce complexity for users. The following systematic review will examine and present approaches for guidance in Visual Analytics. The systematic literature review will follow the PRISMA methodology [32].

2.1.1 Methods

To find all relevant literature on guidance in Visual Analytics, a systematic literature review according to the PRISMA statement was carried out across multiple databases with several search queries. Within the framework of this systematic literature review, the databases IEEE, ACM, ERIC, arXiv, ScienceDirect, and DBLP were accessed. Table 1 presents the queries that were applied to the databases.

As shown in Fig. 1, 250 search results were retrieved. Duplicates were removed using the automated tool *DUDE* [14]. Afterward, the search results were manually checked to detect potentially existing duplicates. The resulting 203 results were filtered manually by the authors, reviewing the title and, in some cases, the abstracts to

Table 1 Tabular overview of the queries applied to the different databases

Database	Query
IEEE	• (All Metadata:"guidance") AND (All Metadata:"design") AND (All Metadata:"Visual Analytics") • (All Metadata:"guidance") AND (All Metadata:"Visual Analytics") • (All Metadata:"Visual Analytics") AND (All Metadata:"user support") • (All Metadata:"Visual Analytics") AND (All Metadata:"assistance")
ACM	• (Title:"Visual Analytics") AND (Abstract:"guidance") AND (Abstract:"design") • (Abstract:"Visual Analytics") AND (Abstract:"guidance") AND (Abstract:"design")
ERIC	• (All Metadata:"Visual Analytics")
arXiv	• (All Metadata:"Visual Analytics") AND (All Metadata:"guidance") • (All Metadata:"Visual Analytics") AND (All Metadata:"user support") • (All Metadata:"Visual Analytics") AND (All Metadata:"assistance")
ScienceDirect	• (All Metadata:"Visual Analytics") AND (All Metadata:"guidance") • (All Metadata:"Visual Analytics") AND (All Metadata:"user support") • (All Metadata:"Visual Analytics") AND (All Metadata:"assistance")
DBLP	• (All Metadata:"Visual Analytics") AND (All Metadata:"guidance") • (All Metadata:"Visual Analytics") AND (All Metadata:"user support")

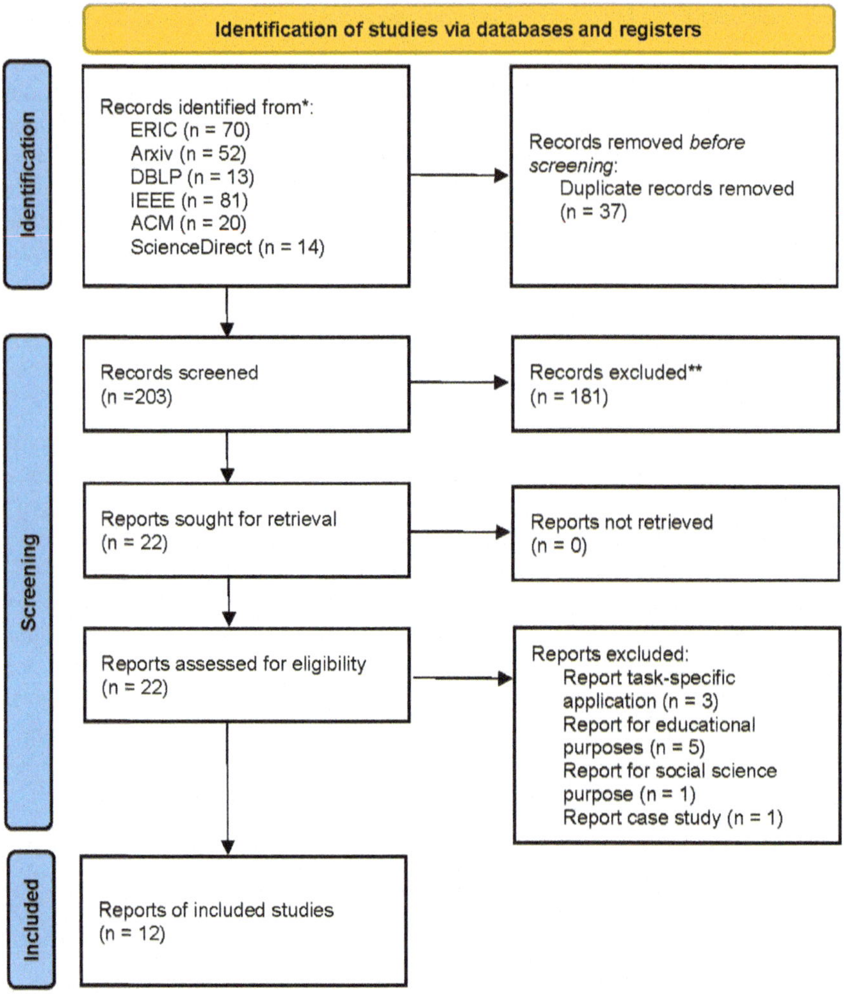

Fig. 1 PRISMA flow chart showing the filtered results for each filter step [32]

filter publications that did not fulfill the research purpose. After the initial screening, the remaining 22 papers were reviewed based on their abstracts, eliminating an additional 10 publications. The publications were filtered according to the following in- and exclusion criteria.

Inclusion Criteria:

1. Libraries, frameworks, and practical guidance design implementations.
2. Reports containing design paradigms and theoretical approaches to guidance design in Visual Analytics systems.

3. Intersectional approaches that combine different methods of analyzing user behavior for better guidance in Visual Analytics systems.
4. Other immediately related works.

Exclusion Criteria:

1. Task-specific implementations of systems without many conclusive insights to generalizable guidance design.
2. Reports that purely emphasize subjective perspectives of educational professionals employing pilot systems with no prior experience in the use of Visual Analytics systems.
3. Reports examining the potential social impacts of automated guidance or its design in data analysis.
4. Reports that are only case studies.

As shown in Fig. 1, 12 papers were examined in the literature review. The main focus of this review lies in finding state-of-the-art guidance design paradigms and practical approaches for Visual Analytics systems, or insights from already implemented examples.

2.1.2 Results

Collins et al. [8] investigated the importance of guidance in visualization and Visual Analytics. They specifically examined the role of guidance in the analytic process. Visualization tasks, implementation, and validation strategies were given an in-depth view. The paper is highly theoretical and does not give any practical specifics regarding implications. Metrics for aspects that should be emphasized in guidance and the (increasingly emerging) importance of guidance in the context of Visual Analytics are clearly and concisely portrayed [8]. Ceneda et al. [7] presented a theoretical definition of guidance and a clear-cut model [7]. The model allows an equally clean thought process in developing a framework or implementation for guidance in Visual Analytics. Despite its theoretical nature, this work can be a cornerstone of the development and design process behind actual implementations [7]. Precisely this model is amended in [6]. This research functioned as an update to previous work on the theoretical definition of guidance characteristics in Visual Analytics. While similar to the previous work, it does not offer any specifics regarding implementation, interface, or framework. Instead, they expressed an even more clearly defined model of what process and shape guidance should take [6]. Stoiber et al. [42] conceptualized the difference between experts and amateurs using Visual Analytics systems, thus showing the necessity of guidance and user support for inexperienced users. They proposed how to model and overcome the knowledge gap in Visual Analytics systems [42]. Pérez-Messina et al. [33] proposed a work that aims to close the gap of understanding the effects of guidance on the analytical discourse, specifically for knowledge generation in Visual Analytics environments. A model from the perspective of user/system dynamics was presented. Additionally, a typology of system guidance tasks was introduced, and several case studies were performed to underline the value of this paper's contribution. Sperrle et al. [40] presented a multi-granular model that allows the Visual Analytics system to progress and learn in accordance with the user. Mutlu et al. [22] presented a Visual Analytics tool that guides users through a given visual task through interactive recommendations. They emphasized identifying patterns in user behavior and user interaction, which can be used to give more accurate predictive guidance and support. The research was based on a fairly small, uniform sample size of 20 participants with prior experience in Visual Analytics. An interface format, as well as timing prompts, were suggested. Additionally, adjustments for user preferences and needs and a larger (online) study were emphasized regarding future work [22]. Rodrigues et al. [35] proposed methods to enhance user guidance in Visual Analytics through content-based data retrieval. They introduced an interface that provided multiple views of data sets, in which results were propagated from one view to the next depending on user interaction. The calculations concerning which data should be emphasized are calculated via different distance metrics. They assessed, in their work, that combining content-based data retrieval with visual data analysis provides users with a more productive and comprehensive tool to fulfill tasks [35]. Monadjemi et al. [21] proposed active search in concatenation with user interaction to uncover analytic goals. Once these goals are determined, the system will suitably guide the user through data exploration tasks [21]. Ceneda et al. [5] proposed a methodology to determine when user guidance is most essential in Visual Analytics tasks. Additionally, a framework was proposed that uses facial recognition software and machine learning to detect these specific moments [5]. Yang [44] proposed an analysis-guided exploration system to tackle frequently encountered problems in Visual Analytics tools. The semantic similarity was used to present data in a more accessible and intuitive manner to the user. Accordingly, guidance was only provided via semantic analysis and an adjusted data visualization [44]. Sperrle et al. proposed with Lotse [39] a practical framework for guidance in Visual Analytics. They introduced a theoretical model and a practical library containing the implementation upon which a framework is set up. They proposed a four-step approach that used different strategy templates converted into guidance specifications during the proposed process and interpreted in this form by a guidance engine. The guidance engine proposed suggestions for the user. The interplay between feedback and the guidance engine is an iterative process aiming to optimize guidance suggestions.

2.1.3 Discussion and Conclusion

The investigated literature mainly focused on theoretical approaches to guidance design in Visual Analytics. These reports focused on theoretical aspects like developing appropriate models and paradigms for guidance in Visual Analytics. Theoretical works like [40] could lead to more desirable guidance in Visual Analytics systems, providing the user with a system that adjusts to the user's behavior and preferences. This model is even shown to be beneficial in other recent works in Visual Analytics. In addition, many intersectional approaches have been found. These approaches focused mostly on using and applying user behavior analysis for more accurate prediction and guidance. Only one publication focused on the practical implementation of a theoretical model. *Lotse* [39] by Sperrle et al. introduced a framework for guidance in Visual Analytics [39]. Their approach proposes a general framework for guidance in Visual Analytics that is based on strategy templates [39].

A practical approach for AI-assisted guidance in Visual Analytics that could be used for conducting systematic literature reviews using Visual Analytics could not be found.

2.2 Systematic Review Tools

In this related work section, we extend the list of tools we presented in [38] that can be used to create systematic literature reviews. Moreover, we want to mention that the online catalog on systematic review tools and guidance (SR-Toolbox [30]) has been updated in 2022 [15].

Systematic review systems typically share a fundamental feature, which is the ability to import and manage literature sources. Some examples of such systems that possess this basic functionality are, for instance, *CADIMA* [16], *Rayyan* [31], and *SLR* proposed by Hinderks et al. [12]. Another crucial task is identifying and removing duplicates. McKeown et al. [17] tested six systems and concluded that *Ovid* and *Covidence* had the highest specificity for identifying duplicate references and that *Rayyan* showed the highest sensitivity [17]. Many solutions can be utilized for systematic reviews across various fields, but some focus on a particular domain or discipline. According to Kohl et al. [16] *SLR*, *REviewER*, and *SESRA* [20] can be used in more than one area, but their focus lies in reviews for software engineering. Kohl et al. [16] found that *SLURP* [4], *StArt* [10], and *PARSIFAL* are limited to the software engineering domain only and that *Covidence*, *RevMan5*, and *RevMan Web* are mainly designed for reviews that are created in the medical field for Cochrane Library [16]. Bustos et al. propose *BUHOS* a web-based systematic literature software for social sciences. *DistillerSR*, *SRDB.PRO*, and *SRDR* allow direct connection to and a real-time search in the *PubMed* database [16]. *DistillerSR* offers the possibility of accessing the *PubMed* database and importing the results. *SRDR* offers the possibility to specify *PubMed* IDs, which are then automatically filled with the corresponding metadata [16]. *SyRF* [1] has a strong focus on pre-clinical studies [16]. Some systems support collaborative work by multiple users, such as *CADIMA* [16], *SESRA* [20] and *EPI-Reviewer 4* [16].

Machine learning and text mining are employed to varying extents in various applications. Among others, methods for document clustering (e.g., in *EPPI-Reviewer 4* and *SLR*) or for calculating similarity (e.g., *SLR* or *StArt*) are used. *Rayyan* also uses similarity graph functions for exploring citation networks [16]. Machine learning approaches are also used to assist in the article screening by the tools Colander, and SyRF [16]. *StArt* calculates scores of likely relevance [10]. *EPPI-Reviewer 4* uses Text mining to assist with identifying relevant studies. *Rayyan* uses a support vector machine classifier that learns from user decisions (in- and excludes docs). Moreover, *Rayyan* scores unclassified studies for likely reference. *SWIFT-Review* proposed by Howard et al. [13] uses Topic Modeling approaches to identify over-represented topics and to rank order titles and abstracts for manual screening [13]. *SyRF* [1] proposed by Bahor et al. [1] also automatically calculates the risk of bias. *DistillerSR* offers an AI-quality check, which was found helpful in validating excluded hits [43].

Our review of related systems and works revealed that most systems allow importing and managing literature sources through import functionalities. Typically, these systems specialize in a specific domain or discipline and do not have a direct literature search feature. Only a few systems incorporate learning methods such as clustering or similarity detection. Only one system [13] uses topic modeling to extract information. Our literature review indicates that a visual system that integrates learning methods, visual interaction, and search does not exist.

3 General Method

Figure 2 illustrates a simplified version of our approach to data transformation adapted from our previous works [25–28, 38]. The iterative transformation process provides bidirectional data flow between the data, in this case, text documents and the Visual Guidance System. In the transformation pipeline, features are extracted from the data and used to create data models, which are then visually depicted through

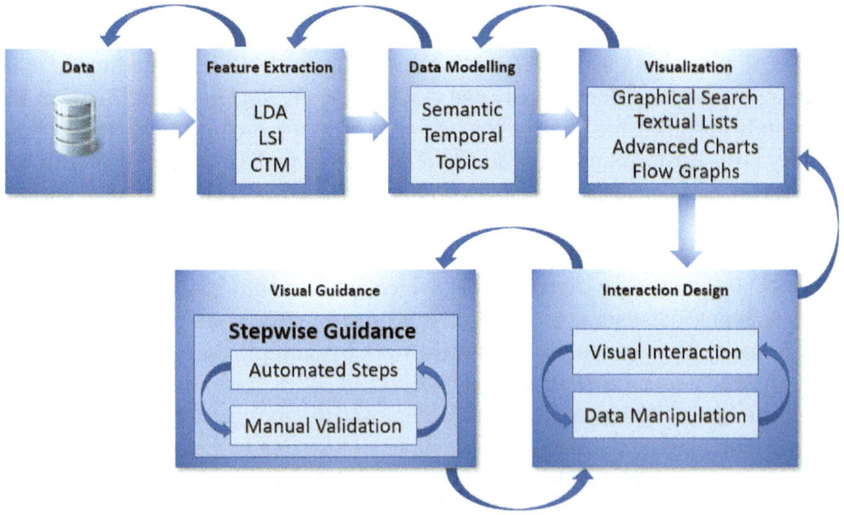

Fig. 2 The conceptual presentation of our abstracted approach, that we adapted from our previous works [25, 27, 28, 38]. It describes an iterative transformation process of heterogeneous data into visualizations that can be interacted with in a stepwise guidance process

various visualizations. As part of the transformation pipeline, features are extracted from the data, and data models are created based on them, which are then visually depicted through various visualizations. The stepwise visual guidance is strongly connected with the visualizations through the interaction design, which enables data manipulations through visual interactions. The following sections explain the individual steps leading to our visual guidance approach.

3.1 Data

In our initial step, we obtain and enhance heterogeneous data by merging it from two databases. To gather the "Document Object identifiers" (DOI), we utilize the DBLP indexing API, which has about six million entries. We resolve the DOIs through databases like Springer, ACM, and IEEE to obtain the abstracts and metadata. For publications without DOI, we match them with their title, author, and publication date. We then combine the gathered metadata with the Springer database, which has about six million full-text articles. The data from both databases are combined, duplicates are eliminated, and any incorrect or incomplete entries are removed. The final dataset is stored in a relational database and includes about ten million entries, including abstracts and about six million full-text articles.

3.2 Feature Extraction

In the second step, we perform feature extraction on the acquired data. In our previous work [29], we conducted a preliminary study that used Latent Dirichlet Allocation (LDA) [3] and Latent Semantic Indexing (LSI) [9] to compare the information contained in abstracts and full-text articles. We improved the topic modeling methods with a Correlated Topic Model (CTM) [2] approach. We evaluated the coherence [34] and u-mass [18] of all three algorithms. The best results were obtained using LDA with 600 topics. The CTM model also had good scores with 720 topics. Each topic includes 30 words and 30 n-grams (bi- and trigrams). We assigned 10 topics to each document with each topic modeling method.

3.3 Data Modeling

As described in our previous works [25, 27, 28, 38], we integrated "aspect-oriented data models" that focus on specific aspects, e.g., semantic-, temporal- or topic aspects that are given in the data. Based on the features of the given data, specific visualizations are available, and different features inside those visualizations are available. Moreover, the topic data models are used further in the process of visual guidance to automate different steps and to train the neural network that assists the users in different tasks. A key feature of the topic data model is that the results of three topic modeling procedures are accessible. This allows accessing the different topics assigned to the documents.

3.4 *Visualization*

In this work, we present four visualizations that we identified as useful in the conduction process of systematic reviews. The visualization dashboard can access other visualizations we presented in previous works [25, 28, 37, 38]. The first visualization to discuss is the graphical search. It allows a search-in-search functionality through Points of Interests (POIs) that are basically the predefined eligibility criteria, as Moher et al. proposed [19]. The PRISMA flow graph visual layout is automatically generated based on the search results of each individual step. Moreover, we propose a textual list that displays a detailed list of publications. The last visualization to discuss is the Treemap Bar Chart, in which the temporal-, topic- and semantic models are merged to visualize information about the temporal development of topics and phrases over the years. The visualization can access the different topics assigned to documents using different topic modeling methods. This enables the identification of further relevant topics.

3.5 *Interaction Design*

Yi et al. [45] proposed seven general categories of interaction techniques that are used in information visualization. They propose the categories *select, explore, reconfigure, encode (abstract/elaborate), filter* and *connect* [45]. As the proposed Visual Analytics system uses most of these visual interaction techniques, we present the visual interaction with the system and the visualizations based on the proposed categories. Since we propose an interplay between visual interaction and data manipulation, the data manipulation steps are explained in the context of the corresponding categories. We propose the selection of concrete entities to highlight specific data entities and mark them as interesting. In some visualizations, such as the proposed Treemap Bar Chart, it is possible to influence the number of elements displayed. Moreover, specific elements may be selected to highlight temporal progressions or display specific details. Moreover, the current step in the PRISMA flow graph is automatically selected and highlighted in a different color. The filtered results during the guided process can be visualized in the Visual Analytics dashboard. This offers the user the possibility to explore the filtered search results directly with the help of different visualizations. The user can interact with the different visualizations to find further relevant topics that were missed by one topic modeling method, e.g. through the Treemap Bar Chart. The Visual Analytics dashboard also allows rearranging the proposed visualizations by removing and adding them to the dashboard. Our approach proposes partial and lazy data loading to reduce the initial loading time of the application. In the first step, a small subset of data entities are loaded to enable a fast first rendering of the web-based user interface. If more entries are available, they will be loaded asynchronously as a background task. As soon as more entries are available, the user interface is updated with the newly available data. Furthermore, we propose that details such as a publication's abstract are only loaded when needed. An example of this is the details in the sorted list visualization. The existing entry in the data model is enhanced with the extended information. We propose data reduction and filtering at different levels with different intentions. Filters to the global semantic data model can be applied through visualizations as the sorted list. Unwanted publications can be excluded from the semantic data model to make them also disappear from the facets. Through facet filters and graphical search [28], the local data models can be manipulated to filter the results displayed. More complex queries can also be applied to the data to filter for relevant data through different filter sets. The proposed AI assistant (MAIA) will show relevant related items that are suitable in the current process step of the stepwise visual guidance process. The assistant proposes, for example, related topics or publications.

3.6 *Visual Guidance*

We propose a visual guidance approach to assist users of the Visual Analytics system in creating systematic literature reviews. The proposed method foresees a guided, stepwise process in which the user receives different supporting visualizations and suggestions from an AI-based assistant (MAIA) for the respective process steps of the review at different points in time. Figure 3 presents an abstract, general approach for AI-based assistance. While the user interacts with the system (1), project-related user data is generated. This includes, e.g., topics, POIs, and other metadata of the search. Data is shared with the AI-based assistant (4) through an application programming interface (2). In (3), available data on Topics (3a) and publication classifications based on them, publication metadata (3b), and user data (3c) that includes topical interests and statistics are stored. The AI-based assistant (4) consists of five subcomponents (4a-e). (4a) Topic similarity determination, (4b) hybrid filtering using collaborative and content-based filters, (4c) user interaction processing and interest detection, (4d) a pre-trained neural network for text similarity detection, and (4e) recommendation detection using similarity metrics or a neural network. The proposed output of the AI-based assistant depends on the available data. If the user creates a new project with too little project-related user data in (3c), more general related topics are suggested. A user interest matrix can be created for topic-like suggestions only through user feedback. Once sufficient project-related user data is available, a synthesis of user interests (4c) and (4d) text similarity can be used to give relevant publication suggestions (4e) based on similarity measurements or through a neural network. The

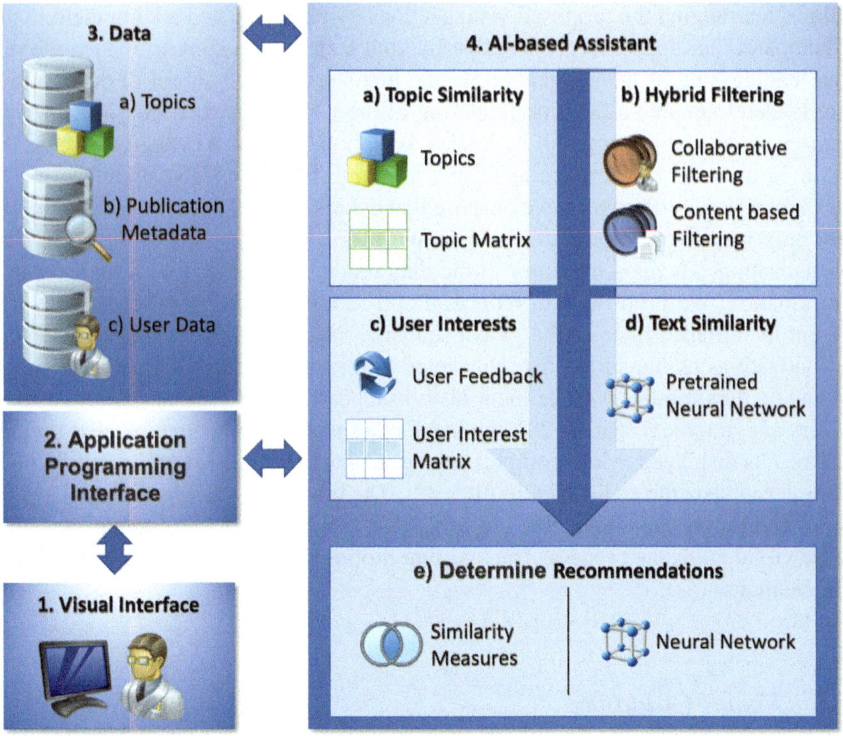

Fig. 3 Our conceptual approach, for AI-based assistance, consists of four parts. Interactions with the visual interface are sent through an API to the AI-based assistant that uses several metrics, algorithms and neural networks to determine recommendations

assistant will be further improved by manual annotation in the form of accepting or rejecting recommendations.

4 Design Study on Guidance for Systematic Reviews

We conducted a design study on a guidance approach that combines artificial intelligence and interactive visualizations to assist users in conducting systematic reviews. The proposed guidance approach is to assist users through the various steps of conducting literature reviews. According to the PRISMA statement, the guidance design approach will be explained in exemplary for systematic reviews. Such reviews follow a strict scheme for *preparation* of the review, the *identification* of relevant literature, and the *screening* procedure. Based on these three phases and a further phase, the stepwise visual guidance approach will be outlined in the following sections based on some exemplary designs.

4.1 Preparation

The preparation phase consists of multiple steps, including checking whether an existing review should be elaborated or a new review should be conducted. For this purpose, it is also retrieved whether a study protocol is already available that contains metadata that could be extracted automatically. In general, the preparation phase is used to gather all metadata necessary for the systematic review's conduction. This metadata includes, on the one hand, information about the authors and their affiliation. On the other hand, depending on the form of systematic review, there is also more specific administrative information, such as the registration number. This is, for example, required for systematic reviews according to the PRISMA Statement and PRISMA review protocols [36].

4.2 Identification

In the first step of the identification phase, relevant search terms can be defined. Here, concatenations of search terms with boolean operators and wildcard symbols (*) are also possible. The system directly returns the number of results found for each search term. The narrowing down of the search results can be done either directly by the concatenation of the individual search terms by the AND operator or in the following step through graphical search as shown in Fig. 5 (1,2) or through facet filters (e.g., years, topics or authors) in the following steps.

The filtered result sets from the identification phase are automatically transferred to the PRISMA flow graph (Fig. 4). As shown in Fig. 6, the user is then shown the current results (2) for his search query (1) as well as any duplicates found, or other documents sorted out (3) in two separate lists. This allows the user to intervene manually and correct potentially undesirable behavior of the system. The results may be visualized through a click on (4), which allows a visualization dashboard to adjust the search filters further or use different visualizations to analyze the data. The AI assistant (5) offers recommendations for the current step. In this example, the AI-wizard offers further publications available by a relevant author concerning a relevant topic.

Before proceeding to the next screening phase, the user is offered the possibility to export the identified search results in BibTex format and download the search steps in XML or JSON format. Downloaded search steps can be used to reproduce the search results.

4.3 Screening

The screening process starts with the preselection step shown in Fig. 7. In the next step, the selected documents can be obtained automatically. Subsequently, the user

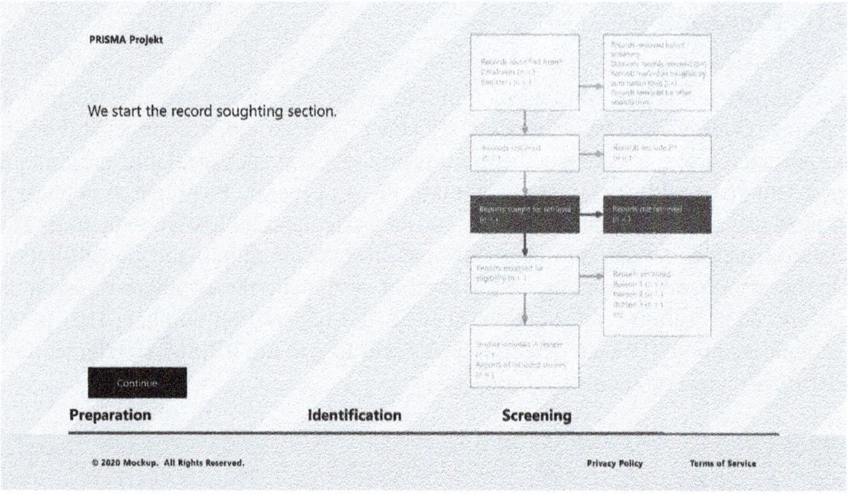

Fig. 4 Design Study—Exemplary representation of an intermediate step in the guided Visual Analytics process. (1) shows the PRISMA flow graph filled in semi-automatically during the process. Depending on the process step, the relevant fields of the flow graph are visually highlighted. (2) Shows us which phase of the review the user is currently in. By clicking on another phase, the user may navigate back and forth. (3) A button can be pressed to move to the next step

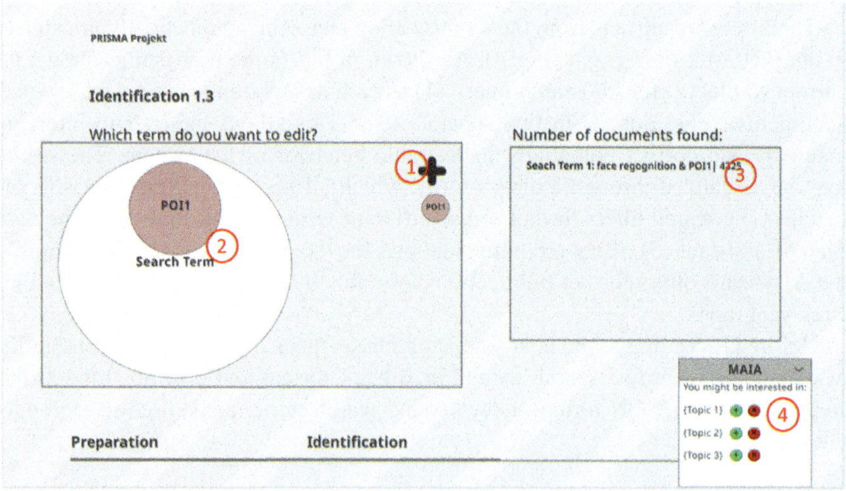

Fig. 5 This screenshot of the Design study shows Identification step 1.3. (1) can be used to create Points of Interest (POI) that can be dragged and dropped into (2) to filter the results. POIs might be nested to create subsets of data. (3) shows the resulting amount for concatenated search strings. In (4), the Metadata-driven Artificial Intelligence Assistant (MAIA) proposes other topics useful as POIs. They can be added using the (+) button or might be discarded. MAIA will learn from the user feedback

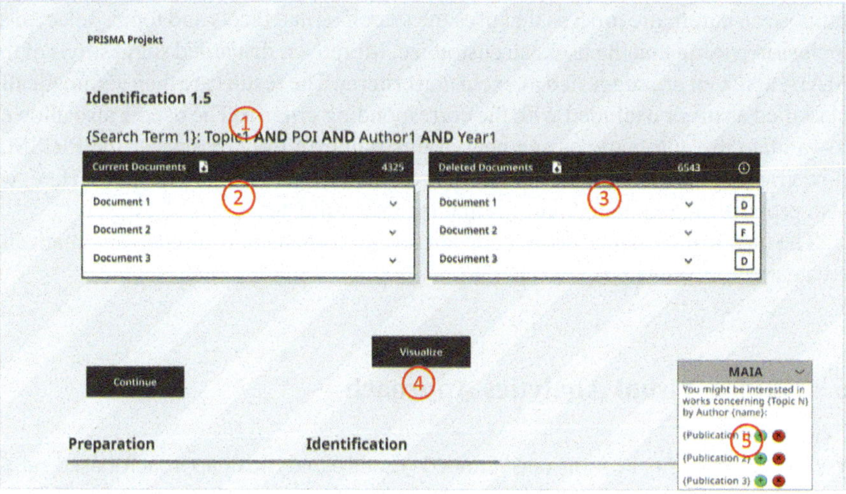

Fig. 6 This screenshot of the Design study shows Identification step 1.5. In (1), the current search term is displayed. (2) displays the current results set, and (3) the out-sorted documents with the corresponding sort-out criteria. (4) allows the visualization of the current result set to identify relevant latent topics and further filter the results. In (5), MAIA displays publication recommendations

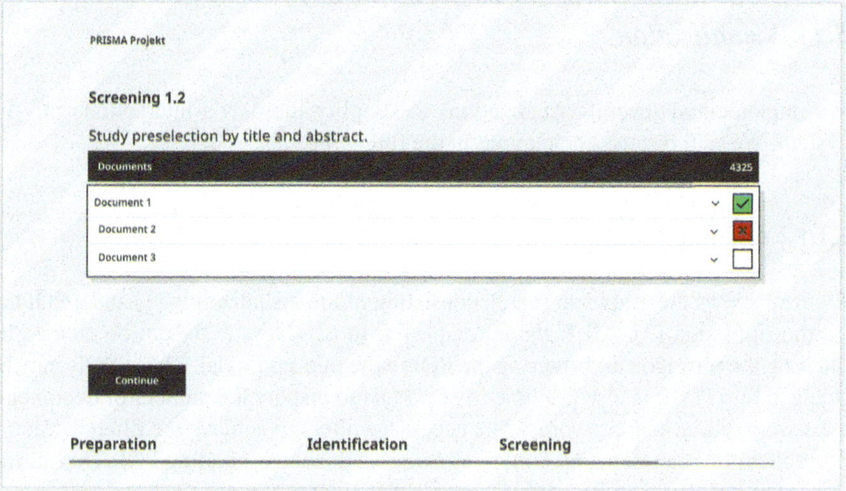

Fig. 7 The user can check documents by their title and summary. By clicking once on the box next to a title, the document is marked as matching (displayed in green). Another click marks the document as unsuitable (displayed in red)

is shown a sorted list with all results. This shows for which documents a full text could be obtained automatically and which documents require manual intervention by the user. According to PRISMA, the results must be assessed for eligibility in the next steps. This can be conducted by defining in- and exclusion criteria. We propose an automatic process that uses the search terms, facets and topics selected as inclusion criteria that the user can customize. Moreover, discarded suggestions of the MAIA assistant are suggested as exclusion criteria. The results are then automatically classified as in- or excluded with the corresponding criteria. The user is also allowed to validate the automatic assignment at this point. In the last step of the PRISMA flow graph, the documents still have to be divided into reviews and studies. Here, we also propose an automated process that the user validates.

The final results can be downloaded in several formats or be used to automatically create a report as we proposed in [38].

5 Guided Visual Analytics Approach

We describe in this section our "Guided Visual Analytics" approach based on an exemplary implementation of the approach as a Visual Analytics system. Thereby the step-by-step guidance process will be described. Since the interaction design takes place in the context of the visualizations and the step-by-step guidance process, we describe the process according to the usage scenario.

5.1 Visualization

We implemented several visualizations to simplify the user's interaction with the system. We will discuss each layout in the following.

5.1.1 Graphical Search

Figure 8 shows the graphical search consisting of the visualization (1) and a POI bar on the right side (2). All POIs are displayed in (2). New POIs can be created by clicking the plus icon and entering the POI in the pop-up modal. Those POIs may be dragged into (1). It is also possible to nest POIs to display the number of documents containing all nested keywords. The necessary filters to achieve the filtered amount of publications can be set by double-clicking on the corresponding POI. This allows the search results to be filtered through graphical interactions.

5.1.2 Textual List

Figure 9 shows the textual list, which provides an overview of the filtered publications. It initially displays the title of the publication and the publication year (1).

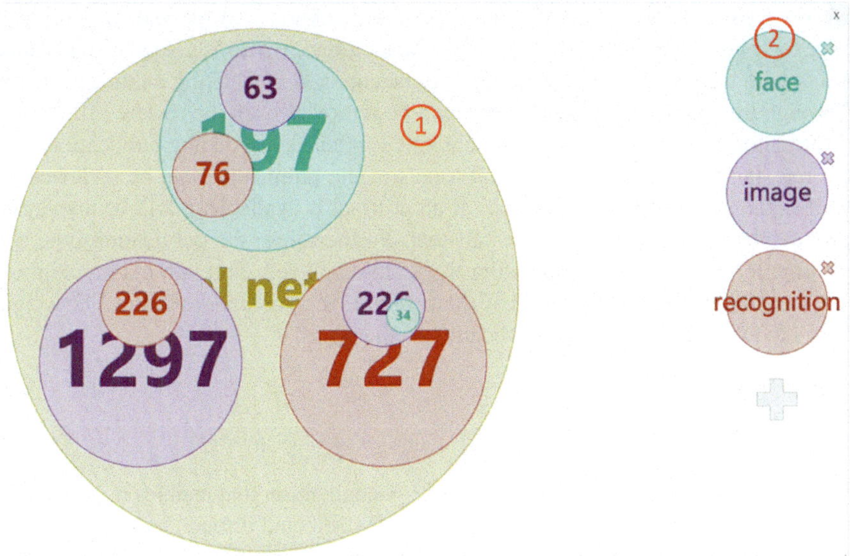

Fig. 8 The graphical search consists of two components, (1) the main chart and (2) the POI-bar. The visualization enables search within the search results to refine data. To do this, the individual POIs from (2) must be dragged into (1). It is possible to layer POIs in several levels

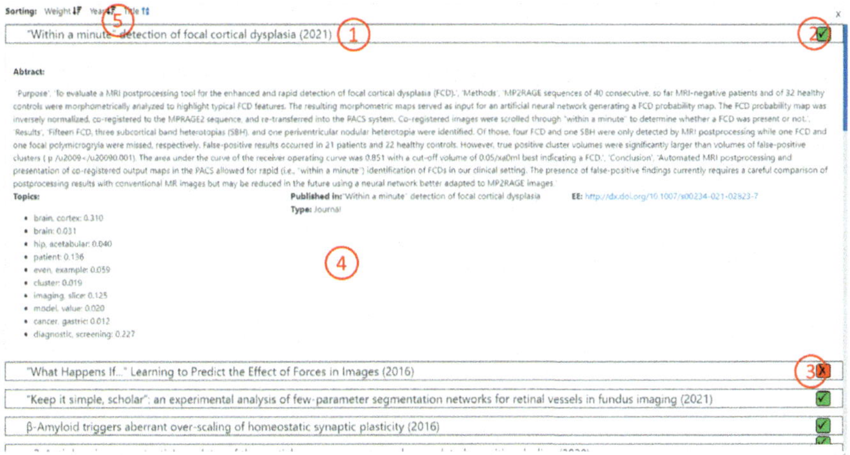

Fig. 9 The textual list that displays initially title and year (1) for each publication. On the right side of each Publication item, a checkbox is displayed. It is initially checked (2), and its state can be changed by clicking on the box which marks it red (3). The publications are sorted by title descendant (5). One publication is toggled and provides additional information

On the right side of each Publication item, a checkbox is displayed. It is initially checked for each publication that should be included, which is visually highlighted green with a checkmark (2). The state can be changed by clicking on the checkbox. Through this interaction, publications can be removed from the review (3). Further information (4) can be displayed by clicking on the corresponding publications. On the one hand, the 10 most important topics of the publication can be reviewed, as well as the percentage of each topic. If an abstract is available, it will be displayed. The displayed link can reach the full text. Furthermore, the publication type and information about the corresponding work are displayed. Through the top-controls (5), the publications can be sorted by their weight, year, or alphabetically. The weight refers to the weighting of the publication to the query.

5.1.3 Treemap Bar Chart

Figure 10 (1) shows the Treemap Bar chart visualization that enables a view of the temporal development of topics. One element is selected through a click, indicated by a higher opacity. Labels are only visible if there is enough space. Zooming in is available through the mouse wheel. The theme's name is also displayed as a tooltip when hovering the mouse over it. This also highlights the temporal progression in orange color. Through the top-controls (2), the topics of different topic modeling methods may be accessed. This allows the user to get insights into different topic models. The right controls (3) influence the number of topics to be displayed. Initially, the top 30 occurring topics are displayed in the chart.

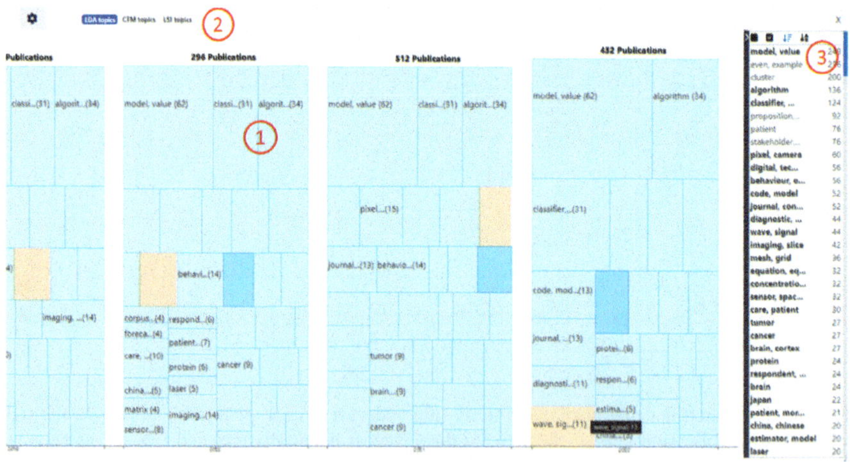

Fig. 10 We propose the Treemap Bar Chart visualization in this work as an advanced visualization. (1) shows the visualization. The top-controls (2) enable switching of the topic models. The right-controls (3) enable influencing the number of topics displayed

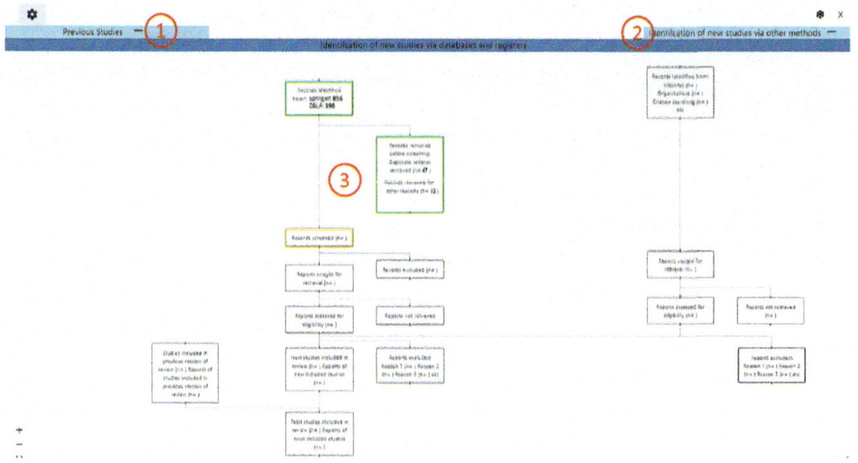

Fig. 11 Flow graph visualization according to the PRISMA 2020 statement [32]. Through (1,2) previous studies and additional web resources may be uploaded by the user. Completed steps are outlined in green, and the current step is indicated by a yellow border (3)

5.1.4 PRISMA Flow Graph

Figure 11 shows our flow graph implementation. The layout of the flow graph follows the PRISMA statement. The number of filtered elements is automatically generated in the stepwise guidance process. The flow graph is filled with the required information, e.g., the number of literature and refinements based on the eligibility criteria. The eligibility criteria are identified through the topic modeling and the researcher's predefined list that leads to in- and exclusion criteria. The user performs the final check. In addition, previous studies can be selected in (1) and then uploaded to the system in BibTex or CSV format. This is another way to specify previous studies in the system if one has not already done so in the guided process. The uploaded data is merged, and duplicates are removed from the result set, as described in the previous section. By clicking (2), additional web resources can be added to the flow graph. The system accepts BibTex and CSV files for previous studies and web resources in the current version. Fulfilled steps are shown as green-bordered boxes (3). A yellow border around the box displays the current step.

5.2 Stepwise Guidance

As mentioned before, Visual Analytics dashboards are complex expert systems. We implemented the proposed stepwise visual guidance that eases our Visual Analytics system usage, as shown in Fig. 12. In the stepwise guidance process, users can create

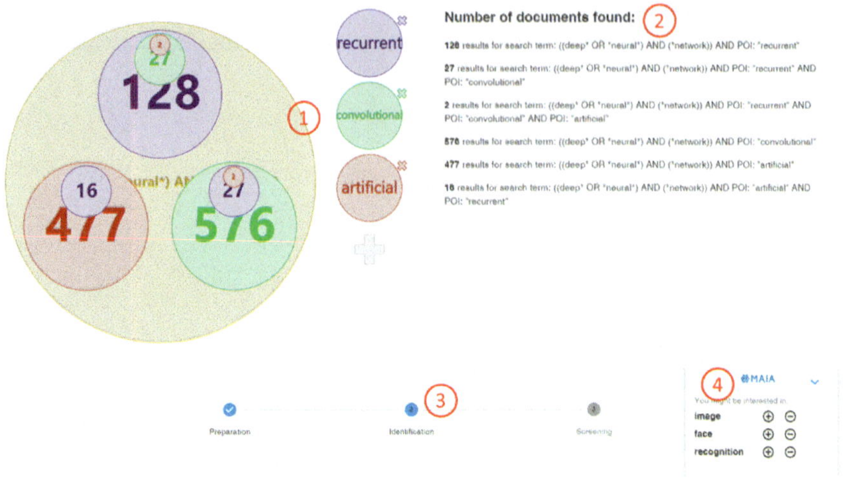

Fig. 12 Implementation of the first step in the Identification phase of the visual guidance system. In (1) we displayed our graphical search. (2) shows the result amount for each unique query. (3) highlights the current phase. (4) displays the MAIA assistant

projects and follow simplified steps to generate the metadata and use different filters during the guided process.

Figure 12 shows one step of the identification phase. We display the graphical search in the area (1). The number of search results for queries generated through the graphical search and the queries are displayed in (2). In (3), the current phase is displayed. The finished preparation phase is marked as checked as the user provided all necessary information. In (4), MAIA proposes the usage of different topics. MAIA is currently only displaying mock-up data. We finished the MAIA design that is not yet coupled with the API.

5.2.1 Visual Analytics Dashboard

The Visual Analytics dashboard is accessible at different points in time in the stepwise guidance process. The POI and facet filters set during the stepwise process are also applied to the dashboard.

The Visual Analytics dashboard consists of four areas, as illustrated in Fig. 13. All areas are dynamic, particularly in data and data entity selection. The top area (1) provides search functionalities. Moreover, it provides color schemes for the entire user interface and visualizations. The lower area of (1) displays all active filters. In the left navigation (2), the facets of the underlying data are generated and illustrated automatically. This area also includes the number of automatically adapted results to the selected facets. Visualizations can be added to the dashboard by dragging them from area (3) into the dashboard and filling the center area (4,5,6,7). The

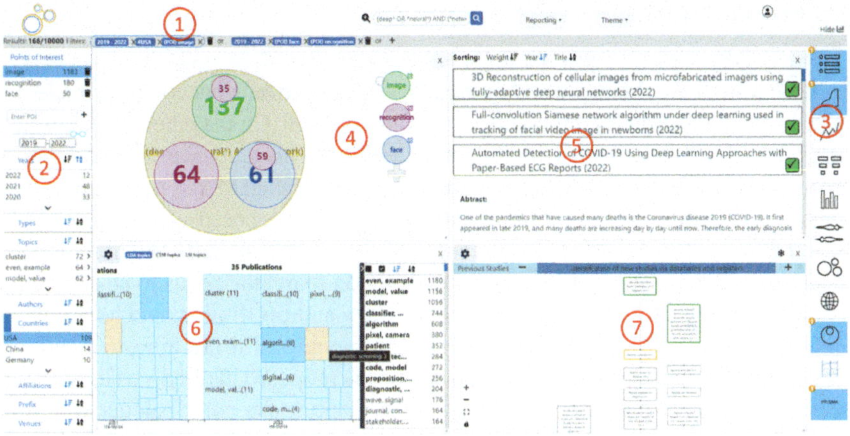

Fig. 13 A screenshot of the web-based interactive visualization dashboard after a search for the term "(deep* OR *neural*) AND (*network)" was performed. Several filters, as well as Points of Interest (POIs), are applied to the data: on the top bar (1), the search term and the chosen facets are illustrated on the left bar (2), enabling users to facet the search results. On the right bar (3), icons for visualizations are displayed that can be chosen. These are based on the underlying data and might change if other data are used. The visual dashboard includes in this example four visualizations: (4) the graphical search that enables a search refinement according to PRISMA, (5) the sorted list presentation, (6) the Treemap Bar Chart, and (7) the PRISMA flow graph

Visual Analytics dashboard can be used to search and filter publications in several ways.

6 Conclusions

We proposed in this work a guided Visual Analytics approach to assist users in conducting systematic literature reviews. Based on a design study, we outlined a stepwise visual guidance process that can be assisted by an AI-based assistant we named MAIA. MAIA aims to suggest relevant information in several steps of the SLR creation process according to the PRISMA-2020 statement [32]. MAIA is supposed to use knowledge derived from the document collection, topic assignments, and user interactions. The general goal of the proposed stepwise visual guidance approach is to ease the usage of a complex Visual Analytics system by simplifying the steps through automation, while allowing the user to validate the automated steps at various points manually. Moreover, we proposed an integrated, automatically generated flow graph that is generated and displayed during the stepwise visual guidance. We also proposed the Treemap Bar Chart visualization that displays the occurrence of topics we extracted with different topic modeling methods over the years. By accessing different topic models, the Treemap Bar Chart can help to identify relevant topics.

7 Future Work

To further improve the presented system, several improvements could be made. On the one hand, more data sources and entries are required. On the other hand, notifications should be implemented when data sources have been updated and new entries are available (monitoring). Moreover, the AI-based assistant MAIA must still be fully implemented and tested. Full implementation of the proposed concept for MAIA would be very beneficial for the guidance system as it could assist the users in various steps of review creation. Furthermore, citation searching needs to be supported, as well as collaborative projects with other review authors. Both would be beneficial and lead to a more reliable system for systematic reviews. The last thing to mention is that user tests need to be performed to evaluate the proposed method's usability more accurately. This also includes evaluating and adapting the color palettes to ensure accessibility for people with visual impairments, such as color blindness.

Acknowledgements We thank Michael Löhn from our course Visual Trend Analytics at the Technische Universität Darmstadt, Susanne Clara, and Pascal Kisker from our course Information Visualization at Darmstadt University of Applied Sciences for their contributions to guidance research. This work was conducted within the research group on Human-Computer Interaction and Visual Analytics at the Darmstadt University of Applied Sciences (https://vis.h-da.de).

References

1. Bahor Z, Liao J, Currie G, Ayder C, Macleod M, McCann SK, Bannach-Brown A, Wever K, Soliman N, Wang Q, Doran-Constant L, Young L, Sena ES, Sena C (2021) Development and uptake of an online systematic review platform: the early years of the CAMARADES systematic review facility (SyRF). BMJ Open Sci 5(1). https://doi.org/10.1136/bmjos-2020-100103
2. Blei DM, Lafferty JD (2005) Correlated topic models. In: Proceedings of the 18th international conference on neural information processing systems, NIPS'05. MIT Press, Cambridge, MA, USA, pp 147–154
3. Blei DM, Ng AY, Jordan MI (2003) Latent dirichlet allocation. J Mach Learn Res 3:993–1022
4. Bowes D, Hall T, Beecham S (2012) SLuRp: a tool to help large complex systematic literature reviews deliver valid and rigorous results. In: Proceedings of the 2nd international workshop on evidential assessment of software technologies, EAST '12. Association for Computing Machinery, New York, NY, USA, pp 33–36. https://doi.org/10.1145/2372233.2372243
5. Ceneda D, Arleo A, Gschwandtner T, Miksch S (2022) Show me your face: towards an automated method to provide timely guidance in visual analytics. IEEE Trans Visual Comput Graph 28(12):4570–4581. https://doi.org/10.1109/TVCG.2021.3094870
6. Ceneda D, Gschwandtner T, May T, Miksch S, Schulz HJ, Streit M, Tominski C (2017) Amending the characterization of guidance in visual analytics
7. Ceneda D, Gschwandtner T, May T, Miksch S, Schulz HJ, Streit M, Tominski C (2017) Characterizing guidance in visual analytics. IEEE Trans Visual Comput Graph 23(1):111–120. https://doi.org/10.1109/TVCG.2016.2598468
8. Collins C, Andrienko N, Schreck T, Yang J, Choo J, Engelke U, Jena A, Dwyer T (2018) Guidance in the human-machine analytics process. Vis Inf 2(3):166–180. https://doi.org/10.1016/j.visinf.2018.09.003

9. Deerwester S, Dumais ST, Furnas GW, Landauer TK, Harshman R (1990) Indexing by latent semantic analysis. J Am Soc Inf Sci 41(6):391–407. 10.1002/(SICI)1097-4571(199009)41:6<391::AID-ASI1>3.0.CO;2-9
10. Fabbri S, Silva C, Hernandes E, Octaviano F, Di Thommazo A, Belgamo A (2016)Improvements in the start tool to better support the systematic review process. In: Proceedings of the 20th international conference on evaluation and assessment in software engineering, EASE '16. Association for Computing Machinery, New York, NY, USA. https://doi.org/10.1145/2915970.2916013
11. Gotz D, Zhou MX (2008) An empirical study of user interaction behavior during visual analysis. IBM Research RC24525 (W0803-127)
12. Hinderks A, Mayo FJD, Thomaschewski J, Escalona MJ (2020) An SLR-tool: search process in practice: a tool to conduct and manage systematic literature review (SLR). Association for Computing Machinery, New York, NY, USA, pp 81–84. https://doi.org/10.1145/3377812.3382137
13. Howard BE, Phillips J, Miller K, Tandon A, Mav D, Shah MR, Holmgren S, Pelch KE, Walker V, Rooney AA, Macleod M, Shah RR, Thayer K (2016) SWIFT-Review: a text-mining workbench for systematic review. Syst Control Found Appl 5(1):87. https://doi.org/10.1186/s13643-016-0263-z
14. Institut HP: Bibtex deduplication using dude. https://hpi.de/naumann/sites/dude/bibtex/
15. Johnson E, O'Keefe H, Sutton A, Marshall C (2022) The systematic review toolbox: keeping up to date with tools to support evidence synthesis. Syst Rev 11. https://doi.org/10.1186/s13643-022-02122-z
16. Kohl C, McIntosh EJ, Unger S, Haddaway NR, Kecke S, Schiemann J, Wilhelm R (2018) Online tools supporting the conduct and reporting of systematic reviews and systematic maps: a case study on CADIMA and review of existing tools. Environ Evid 7(1). https://doi.org/10.1186/s13750-018-0115-5
17. McKeown S, Mir ZM (2021) Considerations for conducting systematic reviews: evaluating the performance of different methods for de-duplicating references. Syst Control Found Appl 10(1):38. https://doi.org/10.1186/s13643-021-01583-y
18. Mimno D, Wallach HM, Talley E, Leenders M, McCallum A (2011) Optimizing semantic coherence in topic models. In: Proceedings of the conference on empirical methods in natural language processing, EMNLP '11. Association for Computational Linguistics, USA, pp 262–272
19. Moher D, Shamseer L, Clarke M, Ghersi D, Liberati A, Petticrew M, Shekelle P, Stewart LA (2015) Preferred reporting items for systematic review and meta-analysis protocols (PRISMA-p) 2015 statement. Syst Rev 4(1). https://doi.org/10.1186/2046-4053-4-1
20. Molléri JS, Benitti FBV (2015) SESRA: a web-based automated tool to support the systematic literature review process. In: Proceedings of the 19th international conference on evaluation and assessment in software engineering, EASE '15. Association for Computing Machinery, New York, NY, USA. https://doi.org/10.1145/2745802.2745825
21. Monadjemi S, Ha S, Nguyen Q, Chai H, Garnett R, Ottley A (2020) Guided data discovery in interactive visualizations via active search
22. Mutlu B, Gashi M, Sabol V (2021) Towards a task-based guidance in exploratory visual analytics. In: 54th Hawaii international conference on system sciences, HICSS 2021, Kauai, Hawaii, USA, vol 5, pp 1–9. ScholarSpace. https://hdl.handle.net/10125/70789
23. Nazemi K (2016) Adaptive semantics visualization. Studies in Computational Intelligence. Springer International Publishing
24. Nazemi K, Breyer M, Forster J, Burkhardt D, Kuijper A (2011) Interacting with semantics: A user-centered visualization adaptation based on semantics data. In: Smith MJ, Salvendy G (eds) Human interface and the management of information. Interacting with Information. Springer, Berlin, Heidelberg, pp 239–248
25. Nazemi K, Burkhardt D (2019) Visual analytics for analyzing technological trends from text. In: 2019 23rd international conference information visualisation (IV), pp 191–200. https://doi.org/10.1109/IV.2019.00041

26. Nazemi K, Burkhardt D, Kock A (2021) Visual analytics for technology and innovation management. Multimedia Tools Appl 81(11):14803–14830. https://doi.org/10.1007/s11042-021-10972-3
27. Nazemi K, Burkhardt D, Kock A (2021) Visual analytics for technology and innovation management: an interaction approach for strategic decision making. Multimedia Tools Appl. https://doi.org/10.1007/s11042-021-10972-3
28. Nazemi K, Feiter T, Sina LB, Burkhardt D, Kock A (2022) Visual analytics for strategic decision making in technology management, 1 edn. Springer International Publishing, Cham, pp 31–61. https://doi.org/10.1007/978-3-030-93119-3_2
29. Nazemi K, Klepsch MJ, Burkhardt D, Kaupp L (2020) Comparison of full-text articles and abstracts for visual trend analytics through natural language processing. In: 2020 24th International conference information visualisation (IV), pp 360–367. https://doi.org/10.1109/IV51561.2020.00065
30. O'Keefe H (2023) The systematic review toolbox. http://systematicreviewtools.com/. Accessed 30 Jan 2023
31. Ouzzani M, Hammady H, Fedorowicz Z, Elmagarmid A (2016) Rayyan-a web and mobile app for systematic reviews. Syst Control Found Appl 5(1):210. https://doi.org/10.1186/s13643-016-0384-4
32. Page MJ, Moher D, Bossuyt PM (2021) Prisma 2020 explanation and elaboration: updated guidance and exemplars for reporting systematic reviews. BMJ 372. https://doi.org/10.1136/bmj.n160
33. Pérez-Messina I, Ceneda D, El-Assady M, Miksch S, Sperrle F (2022) A typology of guidance tasks in mixed-initiative visual analytics environments. Comput Graph Forum 41(3):465–476. https://doi.org/10.1111/cgf.14555
34. Röder M, Both A, Hinneburg A (2015) Exploring the space of topic coherence measures. In: Proceedings of the eighth ACM international conference on web search and data mining, WSDM '15. Association for Computing Machinery, New York, NY, USA, pp 399–408. https://doi.org/10.1145/2684822.2685324
35. Rodrigues J, Romani L, Traina A, Traina C (2010) Combining visual analytics and content based data retrieval technology for efficient data analysis. In: 14th International conference on information visualisation IEEE Press, pp 61–67 (2015). https://doi.org/10.1109/IV.2010.101
36. Shamseer L, Moher D, Clarke M, Ghersi D, Liberati A, Petticrew M, Shekelle P, Stewart LA (2015) Preferred reporting items for systematic review and meta-analysis protocols (PRISMA-P) 2015: elaboration and explanation. BMJ (Clin Res ed.) 350. https://doi.org/10.1136/bmj.g7647
37. Sina L, Burkhardt D, Nazemi K (2020) Visual dashboards in trend analytics to observe competitors and leading domain experts. In: Proceedings of CERC 2020, CEUR workshop proceedings, vol 2815. CEUR-WS.org, Aachen, Germany, pp 222–235
38. Sina LB, Nazemi K (2022) Visual analytics for systematic reviews according to PRISMA. In: 2022 26th International conference information visualisation (IV). IEEE. https://doi.org/10.1109/IV56949.2022.00059
39. Sperrle F, Ceneda D, El-Assady M (2022) Lotse: a practical framework for guidance in visual analytics. IEEE Trans Vis Comput Graph:11. https://doi.org/10.1109/TVCG.2022.3209393
40. Sperrle F, Jeitler A, Bernard J, Keim D, El-Assady M (2021) Co-adaptive visual data analysis and guidance processes. Comput Graph 100:93–105. https://doi.org/10.1016/j.cag.2021.06.016
41. Stab C, Nazemi K, Breyer M, Burkhardt D, Kohlhammer J (2012) Semantics visualization for fostering search result comprehension. In: Simperl E, Cimiano P, Polleres A, Corcho O, Presutti V (eds) The semantic web: research and applications. Springer, Berlin, Heidelberg, pp 633–646. https://doi.org/10.1007/978-3-642-30284-8_49
42. Stoiber C, Ceneda D, Wagner M, Schetinger V, Gschwandtner T, Streit M, Miksch S, Aigner W (2022) Perspectives of visualization onboarding and guidance in VA. Vis Inf 6(1):68–83. https://doi.org/10.1016/j.visinf.2022.02.005

43. Taieb V, Smela-Lipińska B, O'Blenis P, François C (2018) Prm 181—use of artificial intelligence with DistillerSR software for a systematic literature review of utilities in infectious disease. Value Health 21:S387. https://doi.org/10.1016/j.jval.2018.09.2299
44. Yang D, Xie Z, Rundensteiner EA, Ward MO (2007) Managing discoveries in the visual analytics process. SIGKDD Explor Newsl 9(2):22–29. https://doi.org/10.1145/1345448.1345453
45. Yi JS, Kang YA, Stasko J, Jacko J (2007) Toward a deeper understanding of the role of interaction in information visualization. IEEE Trans Vis Comput Graph 13(6):1224–1231. https://doi.org/10.1109/TVCG.2007.70515

Integrating Machine Learning in Visual Analytics for Supporting Collaboration in Science

Midhad Blazevic, Lennart B. Sina, Cristian A. Secco, and Kawa Nazemi

1 Introduction

Collaboration in the scientific community has been on the rise for decades, as described in studies [2], reports [47], and in articles [26] in which they illustrate an analysis of the *Dimensions* dataset. The illustration shows international collaborations have risen in the chosen timeframe from slightly below 15% in 2002 to below 25% in 2020 [26]. Although studies show the figures are higher or lower based on the domain, it is clear that the increase is affecting the community in general. Another relevant factor that has begun to affect the numbers and tendency is the geopolitical impact of US and EU—Russia, and US-China relations, as initial observations show a decline in collaboration between researchers in the arguing nations [26].

Nevertheless, the number of scientific collaborations will keep rising as the benefits are too great to ignore, and the complexity of current challenges is too difficult to overcome alone. The benefits, for example, are the ability to carry out work on complex tasks, save resources [4, 16], improve the work's visibility [1, 4], and improve research quality and efficiency [25, 45], all allow for innovation to be generated faster, and more efficiently [17, 37]. In turn, societies prosper from innovation as it leads to new products that create jobs and opportunities that generate wealth for the

M. Blazevic (✉) · L. B. Sina · C. A. Secco · K. Nazemi
Human-Computer Interaction and Visual Analytics, Darmstadt University of Applied Sciences, 64295 Darmstadt, Germany
e-mail: midhad.blazevic@h-da.de

L. B. Sina
e-mail: lennart.b.sina@h-da.de

C. A. Secco
e-mail: cristian.a.secco@h-da.de

K. Nazemi
e-mail: kawa.nazemi@h-da.de

B. Kovalerchuk et al. (eds.), *Artificial Intelligence and Visualization: Advancing Visual Knowledge Discovery*, Studies in Computational Intelligence 1126,
https://doi.org/10.1007/978-3-031-46549-9_12

people and communities and allows communities to develop solutions for societal, economic, and ecological challenges, The impact of innovation in specific domains, such as medicine, is even broader, as it can lead to cures that help the entire world. However, even if it does not present an immediate cure, it can be used later for new research that builds upon it, leading to a cure or therapy. Both scenarios underline the need for an optimal environment for the scientific community and, thus, for an approach that utilizes appropriate technologies and approaches. Scientists might have domain-specific tools or applications commonly used for specific tasks based on their experience, showing a lack of consensus on a single system [46] for newly formed research teams. This lack of consensus not only hampers the previously mentioned benefits but also causes an organizational hurdle that requires more work and time from the research team to come to a consensus and be able to collaborate at all.

The challenges impeding the creation of such a state-of-the-art methodology and approach have been discussed in various publications throughout the last decades, both technological and human-based. From a technological perspective, the advancement of communication and cloud technologies has improved the situation of real-time communication and data storage issues which was a challenge up until the late 2000s. However, a single source platform integrating technologies such as recommendation systems, natural language, visual analytics, and similarity algorithms is not existing to the best of our knowledge. Some platforms have introduced recommendation systems, but these mainstream platforms have to go beyond recommending publications and researchers that one "might know". As our approach describes, the recommendation system and text analysis technologies are not being utilized to the fullest of their potential. The same can be said for visual analytics and interaction design approaches designed to improve exploratory search capabilities. Studies have validated the benefits of visual analytics systems in this context [41]. However, this is not commonly utilized by academic search platforms, with only a few exceptions, such as Microsoft Academic, which was retired. The human-based challenges are psychological aspects, trust issues, team-building constraints, reluctance to share information, and cognitive distraction through multitasking groupware applications [37].

Our approach demonstrates how a single-source collaborative research platform can utilize approaches to modernize the collaborative research process. It also analyzes the challenges of collaborative research and proposes solutions to overcome them or, at the very least, minimizes their negative impact. More precisely, we define the main contributions based on our previous work [8] as six-fold: (1) a novel approach for visual collaboration combining visual analytics with on-the-fly similarity measurements, (2) a visual and fluid User Interface design for research and exploration based on our previous work on annotation in context, (3) a novel approach to the paper creation and research process based on a literature review over the past decade, (4) a novel approach to assisting ideation with extracted information about research teams and approach recommendations, and (5) and a novel approach to research project profiling that recommendation systems can use to recommend potential team members. Furthermore, our sixth contribution is a usability examination examining young academics' expectations and needs toward collaboration platforms.

2 Related Work

The increased importance of collaboration within the scientific community [20, 35] is due in part to the attention on it by industry, academia [52], and policymakers [24, 37], which has motivated scientists to accept the collaborative mentality and think of innovative means of collaboration. To outline the challenges our work counters, we describe the current landscape of collaborative research, beginning with the various research processes published thus far, proposed improvements to specific parts of the research process, the utilization of recommenders in science, and ideation. We then present our work describing the utilization of recommendation systems in science that assist researchers and describe the ideation processes used in academia to generate innovation.

2.1 Research Process Variations

Kumar states that research methodologies, although varying in detail, have a similarly broad approach and are generally considered supportive in various academic disciplines [22]. A detail that can vary is, for example, the sequence of the steps that can change based on experience and knowledge about research methodologies, as some methods and procedures have been presented [22] and internalized by researchers. Kumar also presents an eight-step process that categorizes the steps into three phases; in phase one, the researcher must define what is to be researched by formulating the research problem, in which the researcher defines what the research should accomplish [22]. Afterward, the second phase describes the planning of a research study, which begins with step two; the conceptualization of the research design [22]. In step two, the goal is to explain how the previously defined research problem will be tackled in terms of logistical arrangements, measurement and sampling methods, and time planning [22]. Step three describes the construction of data collection instruments and is considered the first practical step [22]. In step three, the researcher has to decide how data will be collected [22]. In step four, the researcher must select a sample that adequately reflects their focus group or population [22]. Step five concludes phase two and describes writing a research proposal [22]. Information and results from previous steps are used to explain to a reader what, why, and how is being researched in the research proposal [22]. Phase three focuses on conducting the research study, and the first step of this phase, step six, begins with the data collection task, which was previously defined in the previous phase [22]. The collected data is then processed and displayed accordingly and is used in step eight to write a research report. Writing the report is considered the most challenging and final step of the research process [22].

The business research process presented by Sreejesh et al. consists of seven steps and begins with identifying and defining a problem or opportunity, which consists of two tasks: identifying a problem and defining the problem [43]. The authors associate this first step with exploratory research, allowing for a better definition of the research problem [43]. Completing both tasks allows for a statement of research objectives to be created in which the problem is clearly defined along with the objectives that need to be accomplished in a brief and specific manner [43]. Thereby the statement of research leads to the development of hypotheses [43]. The hypotheses further clarify the research questions from the previous step and allow guesses to be made to push work further in a particular direction [43]. The following step is planning the research design that defines how research should be conducted [43]. The previous steps provide details about what is allocated to this project, what measurements should be used, and how sampling should be conducted [43]. The selection of an appropriate research method is step three and defines which research methods should be used [43]. After defining the research method, the sampling procedure used will be defined in step four [43]. Step five, data collection, uses the previously defined sampling procedure and other steps to collect data, which is examined and evaluated in step six [43]. The final step is the research report preparation and presentation, in which the entire research process, significant findings, and conclusions are written or simply presented [43]. The research report allows for better decision-making, as the report might deliver critical information that can impact the decision of business makers [43].

Cohen et al. published a book that, among other information related to research, describes two models and a definition of the scientific method, which they state can be misleading as varying numbers of methods can be incorporated into a single method.[12]. These methods are sets of procedures describing how researchers come to certain conclusions and should be reproducible with the same or other materials [12]. The first method, known as "Stages in the development of science", is a six-step method based on the work of Kerlinger from 1970 [12]. The method begins by identifying the "phenomena" and the underlying science that is to be examined in order to tackle the "phenomena" [12]. The second step is the observational stage, in which researchers attempt to find, label, and categorize relevant aspects [12]. In the third stage, these aspects are used in correlational research to create theories to be implemented or developed [12]. The fourth step is the experimental phase, where variables are manipulated to reach expected optimal results [12]. The fourth step allows researchers to keep making their hypotheses more and more precise as inchoate ideas get improved as the experiments and results keep coming in [12]. The improved and finalized hypotheses and observations are used in step five and describe the accumulation of these results [12]. In step six, the accumulated results are used to resolve the "phenomena" and can be used for further hypotheses [12]. Cohen et al. also present the eight-stage model that Hitchcock and Hughes suggested in 1995, which begins with the definition of hypotheses, hunches, and guesses before designing the experiment in the second stage along with tasks that gather samples and identify relevant items [12]. In Stage three, the researchers conduct observations to identify correlations and patterns [12]. The observations allow researchers to form hypotheses in stage four, which attempt to explain the observations and irregularities [12]. Stage five is the experimental stage in which the researchers test hypotheses, guesses, and predictions [12]. Depending on stage five results, stage six describes the developed laws resulting from the confirmation of the hypotheses or the rejection of them [12]. Stage seven then generalizes the results, and stage eight creates new theories from the generalization of said results [12].

The "Closed-Loop Conceptualization of the Research Process" by Drew et al. describes eight steps or phases through a closed loop from the investigation of an idea to the development of research questions [15]. Although it is a closed loop, the description begins with the investigation of an idea, followed by the development of research questions [15]. After a research question is developed, it leads to guesses that define how the investigation or study is to be carried out [15]. To answer the research questions, data is commonly gathered and analyzed [15]. The authors point out that the analysis of gathered data should close the loop; however, as the process takes place, new questions or topics of interest can come into existence [15]. These new questions or topics are the reasoning behind the closed-loop perception, as they are used for future work, which in turn begins a new iteration of the process [15].

The "11 Steps" iterative research method by Sonmez is categorized into Manifestation, Description, Prediction, Investigation, Determination, Diagnosis, Verification, Treatment, Reinforcement, Progress, and Tracking [42]. Sonmez reasons the increase in the number of steps with the assumption that this allows for a more systematic and methodological approach that is more meaningful and understandable than others [42]. Manifestation describes the problem that has arisen, and that leads to research that must be conducted [42]. Sonmez points out that there is a difference between problems that have a necessity and those that are imaginary and do not, stating that these problems are not relevant to this model [42]. The description stage is where the problem is defined in detail, going into the problem's respective aspects and prediction containing the task of finding the cause of the problem and attempting to push further work in a particular direction [42]. The investigation stage builds upon the previous stages and is the search for materials of interest, which is the most time-consuming, and thus might take months or years to complete [42]. The determination stage summarizes the found material allowing for better investigations [42]. These summaries are then summarized once again in the following stage, diagnosis [42]. The verification stage utilizes the summarized summarizations to verify or deny their correctness [42]. Afterward, the next stage, treatment, allows for a treatment of said problem to be conceptualized and carried out [42]. The reinforcement stage considers steps for sustainability that should allow for the conceptualized treatment of the problem to be lasting [42]. The eleventh stage is named "progress" and should describe how risks should be removed [42]; however, it is sparsely described and without details about possible approaches or ideas. The final stage, tracking, describes the process of following up on the developments of the problem after the process in order to learn from it [42].

Our research on research processes shows the various understandings published in the last two decades, each with various steps or phases and each with a specific domain or research task focused on. The range of steps and details vary, confirming the work of Kumar, as the supportive nature of the methods is seen in all presented methods. It is also clear that there are steps that, although named differently, are found throughout all methods, such as defining problems and hypotheses, experimentation, and result reporting. These commonly found methods confirm the definition of Cohen et al., stating that methods can be incorporated into others. The varying number of steps and granularity underlines the challenge of finding the correct level of granularity for a general method.

2.2 Collaboration Research Platform Improvements

Zeng et al. [51] describe the relevance of collaboration in general and propose their TSEP platform to model team-based research behavior based on social computer theories and empirical analyses [51]. Through empirical approaches, they attempt to identify aspects that impact the successfulness of collaboration training, such as characteristics of trainers, institutional setups, training structures, and processes [51]. Furthermore, their approach also describes assessment methods and measurements that can be used to monitor and evaluate collaboration [51]. Their work ultimately concludes that collaboration is challenging, especially when considering that collaboration involves combining not simply humans but instead different cultures, domains, and even disciplines; however, with appropriate modeling and analysis, it can be positively influenced. [51].

There are also proposed systems that focus on the collaboration of researchers and practitioners, such as the MEDIATION (Metrics Driven Research Collaboration) proposed by Schreiber et al. [40]. MEDIATION focuses on evaluating and visualizing metrics that can help keep the focus on defined goals of collaboration projects and teams [40]. Their proposal showcases metrics that can be used to keep the focus on goals and is an example of interactive visualizations' relevance in collaboration management. Their work also describes potential conflict points that collaboration can bring, such as missing common understanding or working without structure [40].

Menacer and Arbaoui [28] describe a web-based collaboration infrastructure for research publications, focusing on sharing documents, searching for publications, and feedback posting [28]. The features of their system are the submission of metadata and content, categorization of publications, presentation of relevant conferences and important dates, along with information about conferences such as rank and status [28]. Their presented system has publications and venues at the core of the design, allowing, on the one hand, to assist researchers by using information gathered from publications and venue data and, on the other hand, provide assistance in research by utilizing the gathered data.

Vcelak et al. [46] describe the challenges of research collaboration in the medical domain, with a focus on data and software uniformity to assist researchers in reproducing experiments, and resulting results [46]. Their approach attempts to store and standardize source data, methods used, and results, all of which should be shared via their proposed MRE (Medical, Research, and Education) system [46]. Their related work section, in which they describe the CARMEN virtual laboratory and the GRP (Global Research Platform) [46], explains the importance of standardized data and central storage of methods and tools. This results in the best case in reproducibility of experiments and new or improved ideas or consequent research.

A shift in focus in the field of research collaboration platforms is also described in Wolff et al. [49], where the proposed PIMES (Process Improvement for Mechatronic and Embedded Systems) platform focuses on attempting to improve the transfer of innovative ideas to project funding [49]. Thereby reducing administrative task load so that researchers can spend more time following their ideas and innovate [49]. Kumar et al. [23] focus on creating a collaborative editing tool named LiteDoc, focusing on collaborative writing while countering collaborative writing challenges, such as overwriting or interleaved texts that result in gibberish [23].

As previously mentioned, the proposals in this sections are examples of improvements to collaboration platforms. Our investigation identified examples from different scientific disciplines to underline that creating a collaboration platform is complex due to the various challenges. However, the presented works show that platforms can improve the chances of successful collaboration by analyzing and evaluating collaboration, as described by Zeng et al. and Menacer and Arbaoui. These evaluation and monitoring-focused elements can also be further improved by utilizing interactive visualizations as proposed by Schreiber et al. Furthermore, the varying works show the need to examine every aspect of the research process, including writing, as described by Kumar et al.. The impact that a standardized approach can have on the scientific community is presented to a smaller extent by Vcelak et al., who, although focused on a single domain and aspect, were able to show the importance of standardization. Furthermore, the diversity of domains described in this section shows the relevance of collaboration methodology across all fields of science.

2.3 Recommendation in Science

Zawali and Boukhris proposed cross-domain collaborative filtering with cosine similarity and person correlation approaches to recommend academic venues based on past publications and reference lists, which they tested using a DBLP dataset [50]. Their decision to use a cross-domain approach rather than the in academia commonly used single-domain approach was based on the assumption that cross-domain approaches are more efficient [50]. Another benefit of their approach is that it counters cold start issues as it gathers information from various sources, making it unlikely to have no information about a researcher [50]. Using 0/1 accuracy and the Hamming loss they evaluated their recommendations [50]. They found that the cross-domain recommendations performed better than single-domain ones [50]. Furthermore, cosine similarity resulted in better results than the use of the Pearson correlation [50].

Rodrigues et al. proposed to use recommenders to recommend research collaborations of researchers with similar interests [38]. They found these researchers by using collaborative filtering based on co-authorship and papers read, thereby those researchers who previously had contact in some form with each other [38]. They crawled ResearchGate to gather about 3000 profiles with which they evaluated their approach [38]. Their evaluation examined the precision, recall, and f1 measurements, proving that user-based content-based filtering outperforms collaborative filtering and user-based approaches outperform item-based ones [38]. Their work also identifies challenges, such as a missing standardized approach with which collaboration commonly develops, and that recommendation systems can utilize. Some difficulties exist when working with vocabularies because researchers use different vocabularies to describe their interests, thus increasing the difficulty of ranking based on vocabularies [38].

Ng proposed using recommenders to recommend publications of high quality based on content similarity, peer reviewers, authority, and popularity, using deep learning approaches, Doc2Vec and PageRank [33]. To evaluate the proposed approach, publications from ACM and DBLP were used, 300000 in total, and the results showed an accuracy of 72% in terms of predicting subject areas and the performance evaluation that compared the proposed approach to other systems showed that the proposal outperformed [33].

To summarize our findings about recommendation systems in science, our examples show that recommendation systems that focus on venues, publications, and even collaboration have been proposed. However, most of the publications we found are focused on publications. The three examples we presented in this section are selected from the last five years that show the latest approaches, with which we describe the existing methods and challenges. Although Zawali et al. describe the benefits of cross-domain approaches, Rodrigues et al. point out important issues that also effect cross-domain approaches, the difficulties of varying vocabularies and a lack of unified approaches or behavior throughout the research process. Instead, recommendation systems are based on varying observed real-world approaches or behavior. The common aspects seen throughout various works and our examples are that collaborative filtering approaches and cosine similarity are popular and yield good results. Furthermore, Ng shows with his work that deep learning approaches have also become viable in the field of scientific recommenders. However, there still remains research to be conducted to determine which natural language processing approaches are optimal. A common understanding of the research process would improve the integration of these machine learning-based recommenders throughout the entire research process.

2.4 Ideation

Okudan et al. attempted to improve the ideation performance of undergraduates using the TRIZ method [36]. The TRIZ method is an approach that generates concepts based on knowledge gained from past innovations, from which solution patterns are extracted, while at the same time attempting to push people away from familiar and comfortable solutions [36]. By pushing people away from familiar and comfortable solutions, the TRIZ method attempts to counter the urge in people to stay in their comfort zone rather than explore new and thus innovative ideas [36]. The ideation tools they used within their experiment were brainstorming and sketching [36].

Deo et al. also examined how various tools and methods such as brainstorming, brain sketching, gallery method, and TRIZ impact the ideation of 100 students, but they focused on cognitive aspects to identify the stimuli that impact ideation processes [14]. They describe issues arising when the ideation process is conducted in teams, which can have adverse effects that prevent ideas from being discussed or generated. They describe that larger teams do not necessarily mean more or better ideas [14]; thus, ideation systems should not create large teams and assume better results. Furthermore, the fear of being judged by others on the team can make people keep their thoughts to themselves and "freeride", which describes a lack of participation in the ideation process [14]. This adverse effect has implications that cannot be underestimated, as it dramatically reduces the chances of generating innovative ideas. Deo et al. describe the sheer number of tools and methods also causes challenges, as new researchers do not know about all of the various tools and methods, and others do not want to change their habits or try new ones [14]. Habits also eventually lead to what the authors describe as "fixation", which means that eventually, new ideas generated will have similarities to the previous ideas, thus reducing variety and novelty until some external stimuli or intervention break the cognitive fixation [14]. Their results show that the groups of students that had positive stimuli due to their tasks being aligned with their personal interests and backgrounds had considerably better ideation results when compared to the other group who did not have these positive stimuli [14], further underlining the importance of positive cognition and stimuli.

Zimmerling et al. examined how gamification elements such as quests, points, and virtual awards can be used as stimuli to improve the ideation process [53]. Their results show positive effects such as a higher willingness to share ideas when using, for example, appropriate point systems for ideas [53]. Their work also describes challenges when attempting to improve stimuli with gamification, such as competitiveness [53]. A competitive environment can lead to a decrease in the quality of ideas and over-justified reward systems that give rewards simply due to the number of points gathered for specific tasks [53].

Commonly three ideation metrics are used to measure effectiveness; quantity, variety, and novelty [14, 19, 36]. Novelty measures how unusual an idea is and can be calculated when taking the functions that are needed into consideration and counting the number of occurrences based on the functions for every idea [19, 36]. Novelty is classified into three types; personal, meaning that the idea is new for that person; societal, meaning the idea is new to people in a community; and historical, meaning an idea is the first of its kind in history [19]. The variety describes the uniqueness of an idea, or the dissimilarity of ideas, and is estimated by the variety index, which examines their similarity based on construction levels, that are perceived as a genealogy tree with branches [19, 36]. Quantity is measured by the number of ideas created within a time frame [36]. Okudan et al. expanded upon the commonly used metrics to include quality, which they do not explain in their work [36].

To summarize our research on ideation, various studies have proven that ideation can be improved by using the previously mentioned methods and tools. However, the sheer number of tools and methods, of which the majority are unknown to the public, is overwhelming. Due to the majority being unknown, researchers will commonly choose to use already known tools rather than other tools that might be better suited for the group or project. Furthermore, the described cognitive challenges make it more difficult to generate ideas, as the researchers' creativity seems to eventually reach a ceiling where ideas become similar unless appropriate stimuli or intervention is found. Our research points to a research gap in how the sheer number of tools and methods could be optimally and autonomously presented to teams based on their group constellation, tasks, and their lack of knowledge about the various tools and methods.

3 General Method

Based on Sect. 2.1, we have identified the following steps as key elements of the research process that we use as a foundation for our approach; (1) research idea generation, (2) research planning, (3) research execution, and (4) research result presentation. These steps are generalized and can be adapted to any domain. As described in Sect. 2.1, specific tasks exist for specific domains; however, our goal is to create a general approach. This leads us to conclude that successful research collaboration is built upon four pillars; inspiration, team assembly, ideation, and teamwork, as illustrated in Fig. 1. These four pillars categorize challenges and tasks that are ordered as consecutive steps in which research is commonly conducted. These four pillars require a stable foundation to increase the chances of success. The technologies we have identified as having the highest potential in the context of collaborative research platforms are recommendation systems, natural language processing, visual analytics, and interaction design. As the fusion of machine learning-based technologies and interactive visualizations is visual analytics, the argument can be made that fusing visual analytics into the research process is sufficient to modernize and thus improve collaborative research and the resulting results in the form of innovations. These technologies can be used in all pillars to counter the current challenges and improve the researchers' experience within each respective task.

The research process begins with the inspiration pillar, as a problem, challenge, or idea for a project is born from an inspiring moment described in this pillar. Although external sources, such as the environment or the private life of a researcher, can cause inspiration, our focus is on manipulating internal inspiration sources, with our goal being to increase inspiration leading to innovation. In this context, our most acute challenge is information overload, as researchers will likely not be inspired by information that does not interest them or that they cannot find or see. Once a rough idea has been defined, a team can be assembled. The team assembly is crucial as, based on each team member, the expertise, the team's overall potential, and its chances of success are impacted. The challenge of finding the optimal team member is often underestimated or taken too lightly. Studies have shown that the approach to finding team members is regrettably not always based on expertise but

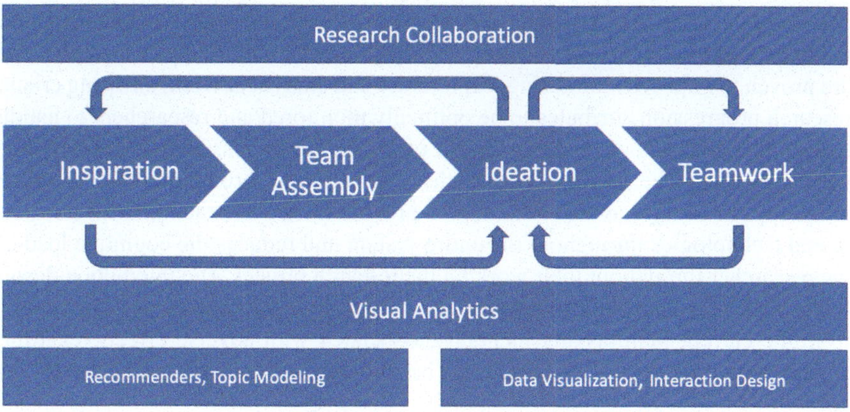

Fig. 1 Pillars of our research collaboration approach: with the main steps of inspiration, team assembly, ideation, and teamwork built on the applied technologies to support each process step

instead on social and geographical proximity [4], which can negatively impact the research quality and chances of successful innovation. An assembled team then begins discussing and further defining the research idea, problem, questions, and hypotheses based on preliminary research. Such discussions are part of the ideation phase, where the team discusses various ideas and approaches. Ideation can happen throughout the project, as new challenges can arise that require new ideas to find working solutions or observation can lead to discussions, resulting in new ideas leading to innovation. Research collaboration is a team-based approach to producing knowledge [20]. After fine-tuning the idea and creating a research plan in the ideation process, the work begins in the teamwork pillar. The teamwork pillar can be further broken down between team management tasks and routine research tasks required to answer research questions or develop a scientific product such as an engineered prototype, medical therapy, or cure. Both task types impact teamwork, as the team management monitors and manages the team, ensuring it works optimally and defuses risks and challenges, while the team works on the project, collaborating to ensure work results are of the highest quality and the results can be published optimally.

Figure 1 shows arrows that connect ideation with inspiration and teamwork. These visualize the interconnectivity of ideation throughout the research process. Inspiration is needed for ideation, but ideation can also lead to inspiration during, for example, discussions during brainstorming. Furthermore, as ideation is commonly a team-based process, it is connected to the teamwork pillar. As teams work together, new ideas are generated, and ideation is also needed when new challenges confront teams. Ideation is a crucial part of the research process.

This leads us to the foundational technologies and their meaning briefly outlined. All four pillars require natural language processing techniques to analyze data generated during the process, which is then used by the other technologies. Artificial intelligence approaches such as recommenders use this information to provide accurate recommendations of publication, venue, dataset, and potential team member recommendations, which cover all four pillars. Interactive visualization approaches are proven to improve monitoring, exploratory and analytical work, allowing crucial research progression attributes to be optimally monitored and researchers to handle better large amounts of data that would otherwise be impossible to overview during exploration. The final technology that can be found throughout all four pillars is interaction design, as it is required to provide a fluid interaction with all mentioned technologies throughout the entire system and reduces the cognitive load on the researcher throughout their work on the research project. The interaction design focus is the reasoning behind a single-source web-based approach, as it shifts away from the current multitasking mentality associated with adverse effects regarding cognition and distraction. In Sect. 4, we briefly outline the subsystems for each pillar.

3.1 Inspiration

Let us assume that the story of Newton, who saw an apple fall from a tree, is true. As he witnessed this and asked himself why it fell straight down, leading him to formulate his gravitational theory in the 17th century, which helped develop the space travel we know today [32]. This story represents the interesting and, at times, unexpected way inspiration works. For our work, we categorized inspiration sources into those that can be manipulated and those that cannot. As the story points out, the inspiration that arises from the randomness of life can come from the environment, life experiences, private social connections, and private media consumption, such as listening to music or watching a movie. Our work focuses on the sources that can be manipulated to increase initial research idea generation and improve the ideation process during work on the research project later on in the research process. The sources that can be manipulated are calls for papers, conferences [44], academic or scientific news, publications, projects, collaborations, and discussions [31]. The crucial part of manipulating these sources is figuring out the interests of the researchers and then using appropriate methods to find relevant resources and present them to researchers, overcoming the current information overload observed throughout all domains. Using the story to describe the problem, if Newton's apple tree were hidden deep inside a forest, he would not have witnessed the apple fall from it. We have to eliminate the non-apple trees to increase the chance of witnessing an apple fall from an apple tree or go even further and also present trees similar to apple trees in the hope that they will be as inspiring. The aspect of inspiration is even more complex when we consider that every human is different and their thought process and perception can work differently [18, 21, 37, 45]. This difference in thought processes and perception means that different researchers might gain different thoughts, inspirations, or insights when examining the same resource. Using the Newton story again, if Newton had witnessed the apple falling together with a colleague from another domain, two ideas might have come into existence. Where Newton had a eureka moment about gravity, his colleague from engineering might have thought of an innovative idea to make the impact on the falling apple less harmful. The same observation, but different outcomes in the form of ideas and innovations resulting from the same inspiring moment. What exactly forms the thought process of individuals is still being researched, but a few factors have been agreed upon, such as life experiences, cultural background, and social networks. This understanding can be fruitful in terms of inspiration, as Moser et al. [31] point out that social interactions, discussions, and resulting consensus leads to higher chances of ideation between parties with different views. These different perspectives clash due to researchers coming from different cultures [39] or having different experiences, but these differences hold different insights that, when discussed appropriately, can lead to inspiration. This contradicts the current approaches that is seen in social media, where like-minded bubbles are created, and users are not thrown into bubbles where people of different opinions are; instead, people are thrown into bubbles of like-mindedness, leading to a loss of inspiration potential. This is another point that has yet to be adequately addressed in studies. The inspiration aspect of any modern research project requires assistance via recommendation systems that are built upon the optimal extraction of researcher information to provide relevant sources of inspiration while at the same time utilizing varying perspectives and thought processes to improve the chances of inspiration.

3.2 *Team Assembly*

The commonly practiced approach to research team assembly is often based on the social proximity of the researchers [4], which is counterproductive to a certain extent. The core belief of collaborative research is that complex tasks can be better handled by gathering expertise in a team [31]; thus, the better the expertise, the more capable the team is. By only collaborating with researchers from the workplace or only those whom the researcher knows, the best expertise required for a project is not utilized. However, it is vital to bear in mind that not every expert is a positive influence on the team. Many attributes must be considered when assembling a team, of which expertise is only one of many. Studies have shown that the following attributes are also of interest; language, availability, interests, compatibility with project owners and the team [27], and ethical and cultural aspects [39]. To modernize and improve the research team assembly, two challenges must be addressed for cutting-edge approaches. The first is optimally analyzing researchers to identify the previously mentioned attributes, and the second is identifying the project's needs to recommend the correct researcher. We mention visual analytics at this point as the number of possible candidates whose attributes correspond with the needs of the project might be large, and it would be helpful if the attributes could be visualized in a form that would allow for easier comparisons to be made by the project management. Although recommendation systems could find potential team members, thus narrowing the list of potential candidates and recommending them, it is the decision of the project manager who is chosen from the candidates to be sent an invitation. The previously mentioned second challenge is more complex as it asks how a project's needs can be identified. As team assembly is a task carried out relatively early on in collaborative research projects, identifying the needs of a project requires an early analysis of the project, which we discuss in the following section, as ideation proves to be a treasure trove of information about the project. The integration of machine learning approaches such as recommenders also requires thought to be placed into the optimal weighting of the various attributes of researchers and the project's needs.

3.3 *Ideation*

When examining the simplified pillar model, one might ask, where is the difference between the pillars inspiration and ideation? The main goal of the ideation pillar is to facilitate and motivate the collaborative creation of ideas, which can be for research questions or even for challenges arising during project work. Commonly used methods are brainstorming, whiteboarding, mind mapping, innovation jams, or discussions or social interactions within the workgroup [3, 10, 31]. These methods and tools encourage creating ideas that should lead to solutions that are discussed within the team and lead to a consensus. Ideation is also possible by examining existing knowledge and ideas, improving them, or generating new ideas based on them. As previously stated, observations have underlined that sharing knowledge and resources within a group can also lead to innovative ideas based on differences in perspective, perception, and general thought processes; however, this requires appropriate stimuli that motivate researchers to share their ideas. The recognition of ideation as a strategically important activity [11] in combination with the increased presence of technology in the ideation process leads us to examine what added value can be extracted from the combination. As the related work points out, the challenges are a large number of tools and methods, lack of stimuli, and other cognitive challenges that cause a reduction in idea quantity, variety, and novelty.

Examining the history of specific innovative projects shows that there have been similar projects that were simultaneously worked on by autonomous groups without them ever having any collaboration or contact with each other. Two projects with similar goals needlessly used double the amount of resources to accomplish similar goals because they did not collaborate, likely due to the groups not knowing about each other or due to personal reasons. If the teams had collaborated, the resources and expertise could have been combined, and results would have likely been better and achieved quicker. This raises the question of what ideation has to do with workgroups collaborating. The answer is that by examining the results of the ideation process, we can accurately represent the project, which can be used for any recommendation systems requiring project information, including recommendations that support the further iterations of the ideation process. Furthermore, we can use similarity measurements to find and recommend similar projects. By doing so, we facilitate collaborations and fusions between autonomous teams working on similar projects, improving the results by showing research teams that they are not alone with their work on a particular subject.

3.4 Teamwork

The teamwork pillar is the generalized step of the research process in which collaborative work is carried out. In contrast to the examples described in the related work section, our teamwork pillar contains tasks such as data gathering, data analysis, and writing the results, and if required, domain-specific tasks such as the creation of prototypes or the creation and testing of therapies are all part of the pillar. This pillar also contains both administrative or management tasks as well as operative tasks. In terms of administrative or management tasks, we have to improve research project monitoring which requires the research project to have evaluation metrics with which interactive visualizations and dashboards can be created. This allows the project management to monitor the project's progression, task completion rates, and deadlines and identify potential setbacks or risks on time. Identifying risks on time can be countered, improving the chances for success. As described, our approach is not domain focused, and thus the operative work is focused on research and the creation of publications in general. We propose to combine the writing task with the explorative research task. Commonly scientific papers are written in Latex. Studies have also proven the value of visual analytics in their powerful ability to disseminate literature [5, 30], ability to create overviews and filter results, and annotate in the context of exploration [7, 10]. Examples of this are SurVis [5], TreeVis [41], etc. Although visual analytical approaches are rarely provided to researchers within modern search platforms, these are of the utmost importance and can be further enhanced by understanding what the researcher is working on.

As we examined how to combine both tasks into a single approach and prototype application, we identified a fundamental challenge with the currently practiced research process, the multitasking mentality. Collaborative research requires that researchers have many applications running as they work. When researching, they will likely have search engines in browsers, literature management software, and a note-taking or latex editor application running and will be switching through the applications as need be. At the same time, collaboration requires communication; thus, Outlook and chat applications likely further divert attention with every message, call, or email notification. This application-switching mentality has been studied and proven to affect productivity and attention issues, among others, adversely [6, 29, 48]. This challenge leads us to believe that human cognition in visual analytics [5] and user interface and interaction design [13] should be more considered when designing collaborative research platforms. A single-source web-based approach is an optimal solution, as it allows researchers to work in a single environment that provides uniformity through all applications or features, reducing application switching and multitasking, thus reducing their adverse effects. Furthermore, this would eliminate software deviations that might have otherwise arisen if researchers had worked with different applications or versions of applications.

4 Collaborative Research System

As described in the previous section, the broad spectrum of tasks in the research process holds a broad spectrum of challenges that must be counter to create an appropriate environment for researchers in terms of complexity and pragmatism. This section presents our system in brief detail, based on our described pillars of research collaboration, with the goal of integrating machine learning-based approaches throughout the research process to increase the overall quality of research while reducing hurdles for collaboration.

4.1 Inspiration

As described in the previous sections, information overload is challenging in the current research environment, so we propose two approaches to increase the chances of inspiration; a personalized inspiration hub and an intelligent Latex editor with an integrated publications recommender and visual analytics system. Our approach utilizes natural language processing and recommender systems to counter this challenge and improve the inspiration pillar. The intelligent editor analyzes written text in real-time and, based on discovered latent topics, provides recommendations for publications. We previously evaluated various combinations of topic models and similarity algorithms, which showed that the best combination of compared combinations for recommending publications is LDA and cosine similarity [9]. As researchers write their text in the editor, the top N publications ranked by their similarity score are presented. An alternative form of using the real-time analysis is shown in Fig. 2. The user can choose to see results via a visual analytics system that presents all found similar publications. The results are visualized in various forms so that the researcher can select the visualizations that appeal to them. Furthermore, filter options allow for data exploration during writing and searching.

Recommendations based on researcher profiles should also be used in academic and social network feeds, presenting relevant publications, conferences, calls, and news. Information about these researcher profiles is found in Sect. 4.2. Using a similar recommendation approach as previously described, we analyze the data that the researchers input about themselves (languages, interests, research description) and further enhance this via a topic model analysis of authored publications gathered via OrcID connection. This combination allows us to find latent relevant information about the researcher's interests and place interesting and new publications, conferences, calls, and news in their feed, thus providing suitable material that should inspire the researcher while being independent of user-generated vocabularies. How-

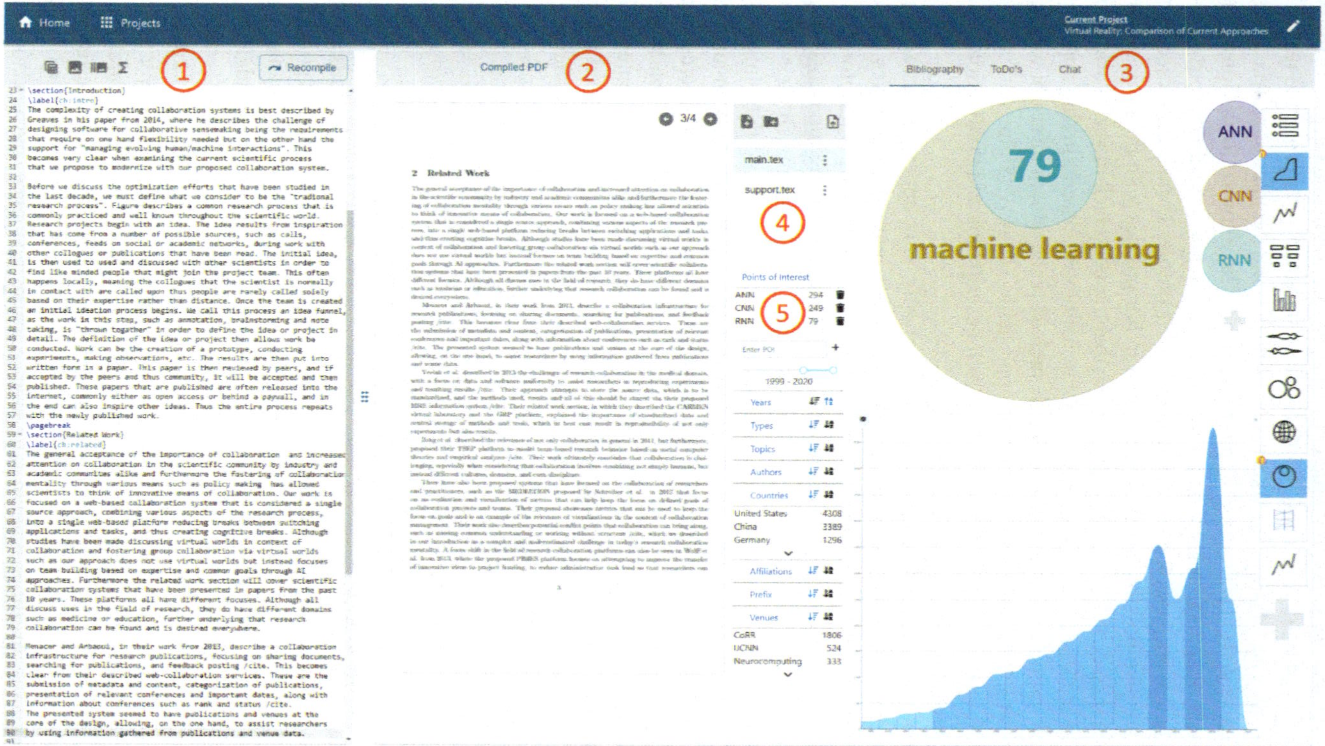

Fig. 2 UI Overview featuring a Latex Editor (1), the PDF preview (2), the automated search results based on topic modeling visualizations (3). (4) shows tex files of that project, and allows users to manage the files and create new files and folders. (5) shows filter options that users can use for the visualizations to explore the data

ever, the requirement of data for this approach can cause cold start problems for new researchers that have yet to publish any papers. A recommendation can still occur as the profile should yield enough information; however, to improve the results and minimize the risk of cold start issues, we propose using the interest matrix approach that we presented [7] and a cross-domain approach. This allows information about the researchers' search interests to be gathered based on annotated publications collected during their work, which is associated with their profile. Project-based recommendations are also possible by combining researcher profiles with the project profiles described in Sect. 4.3. It is important to note that there is a lack of real-world explicit ratings in academia that can be utilized for collaborative-based filtering approaches. Although implicit ratings can be utilized, such as reads and citations, however, our work focuses on content-based filtering approaches. These examine and utilize the content, in our case, textual information from descriptions and publication content, to generate topically relevant recommendations.

An optimal presentation is crucial, as commonly, sites appear to spread recommendations throughout various pages, which is not optimal as it makes elements fight for the researcher's attention. Instead, we propose having a hub for the user that solely consists of elements meant to inspire them, including top N lists of publications, upcoming venues, and news and research feeds that match their interests. This personalized site uses the interests gathered from extracting their interests via manually inputted data describing themselves and their interests and an analysis of their publications. Furthermore, this is enhanced with a personalized interest matrix, which analyzes the lists of annotations that have been accumulated via topic modeling and stores the identified topics. This allows the system to recommend content appropriately based on content rather than keyword queries. Based on observed behavior, we assume that researchers who learn to check the hub daily will have higher chances of being inspired by the various recommended elements.

4.2 *Team Assembly*

To create teams efficiently, two profiles are required. A project profile describes the research projects, and a researcher profile describes the researchers. The project profile contains goals, a description of the project, information about language, team members, notes, associated annotations, and with information from the ideation process that we later describe. Using topic modeling, the system analyzes the gathered information to identify the project topic, which will be compared later to the researcher's profiles to identify researchers that work on similar topics. We vectorize researchers based on their profiles to compare project profiles with researchers. The profile consists of user input and the published work of the respective researcher. The researcher creates the researcher profile that contains attributes and information about, for example, known languages, current residence, current and former workplaces, interests, preferred domains, and a Boolean value for the desire to be recommended to collaborative projects. The previously named LDA topic modeling approach further enhances the researcher's input. Their published work is analyzed to determine the domains they traverse and their expertise in the respected found domains. The topic modeling approach shows the topics the researcher has worked on thus far and the input of the other aspects previously described works to make sure they meet the other requirements. This information allows the system to identify persons of interest and compare them to the project profile. Identifying persons of interest does not guarantee that the proposed researchers will desire to work on the project nor that the project owner will request the proposed researchers. A list of potential candidates is presented to the project owner, ranked by expertise and compatibility with the project. If a candidate appeals to the project owner, that candidate is sent a request, which the candidate must accept or deny. A hybrid filtering approach based on collaborative and content-based filtering approaches built upon the proposed topic modeling approach allows the recommender system to identify relevant persons and projects while being independent of manually created vocabularies. This approach counters the previously described difficulties with finding partners based on expertise, but adding team members manually is also possible, making the researcher recommender optional.

4.3 Ideation

Our novel approach to ideation has two focus points; the utilization of information gathered from ideation tools to describe projects and an automated approach that encourages the utilization of various ideation methods while using positive stimuli to encourage quantity, variety, and novelty. We first gather information about a research project by analyzing initial data from notes, manual project descriptions, and annotated and gathered publications. Using topic modeling, which we described in the previous sections, we can determine what topics best describe the project. This is used for the team assembly approach, described in Sect. 4.2, but we assume that the team is assembled at the advanced ideation phase, and thus the gathered information about the project is enhanced with information about the team members, more precisely their roles in the team, expertise and their experiences with ideation tools and methods. The researchers are categorized into team roles based on a questionnaire. The questionnaire is required until the system has gathered enough information that can be used later on by a neural network to determine the roles automatically based on the researcher's respective attributes. The combination allows the system to recommend a workflow consisting of several appropriate tools and methods in iterations until the result of the ideation process is completed, meaning until the group or the leader of the group is satisfied. Each iteration consists of one tool or method being used until either a set amount of time has passed or until the group choices to end the iteration, after which the system asks if the ideation process is complete. If the group chooses yes, then the ideation process ends, but if not, the system will recommend a group of adequate tools and methods that can be used for the next iteration. The various iteration of tools and methods present a larger number of possibilities to think of and present ideas, overcome the fixation problem described in Sect. 2.4, and assist leadership in finding the best approach to ideation for each project and team. Throughout the ideation process, the leadership can activate additional optional gamification features, such as a quest master, virtual awards in the form of badges, and a point system, to motivate ideation. These features are project-based and optional because it is unclear if researchers have a positive attitude or stimuli towards this feature.

The ability to describe projects based on an analysis of the initial research work and ideation results allows the system to accomplish various tasks, such as recommendations for researcher collaboration, venues, publications, and team members. Comparisons between projects are also possible and desired, as previously described. These comparisons require creating appropriate similarity algorithms and vectorizing the information gathered. The comparison is based on the similarity of projects and allows the system to present identified similar systems if the research group leader activates the option to search for similar projects. Giving the leaders this choice creates a sense of privacy, which we consider to be of the utmost importance.

4.4 Teamwork

Figure 2 shows the initial version of the proposed system in a typical use case, where a researcher is writing a paper in the Latex editor shown in (1) and (2) shows the PDF preview. In (3), the researcher can see visualizations based on the results of the analysis of their written text. The system uses topic modeling to analyze the text written in real-time. The writer's text is analyzed and compared to publications stored in a database using similarity measurements. The results are then sent to the visual analytics system, where they are presented to the researcher. The researcher can also decide to see the results in an annotation in the context of the exploration approach, which we described in our previous work [7, 8]. This allows the researcher to quickly and fluidly examine potential relevant publications and annotate them. The annotated publications are accessible to all members of the project. (3) is dynamic based on the user's cognitive desires and tasks and the results of the written text analysis. Figure 3 shows a use case where the researcher chooses to use an interactive list view (3) to examine results. The researcher is presented with a list view of the results that they can open by clicking on the title, to reveal the abstract and a DOI link.The researcher has the ability to annotate publications that they find interesting by clicking the blue bookmark button, that then turns to yellow. Furthermore if the researcher finds a publication that they to consider to be very interesting they can commence a search query by clicking on the "find similar" button.

As shown in Fig. 3, the identified publications are visualized in a visual research analytics system which the researcher can further filter or change visualizations as desired. This visual analytics system is stored in one tab. It can easily be switched on the fly for other components such as chat, Trello board, annotation, and note-taking on gathered publications. All team members can access gathered publications

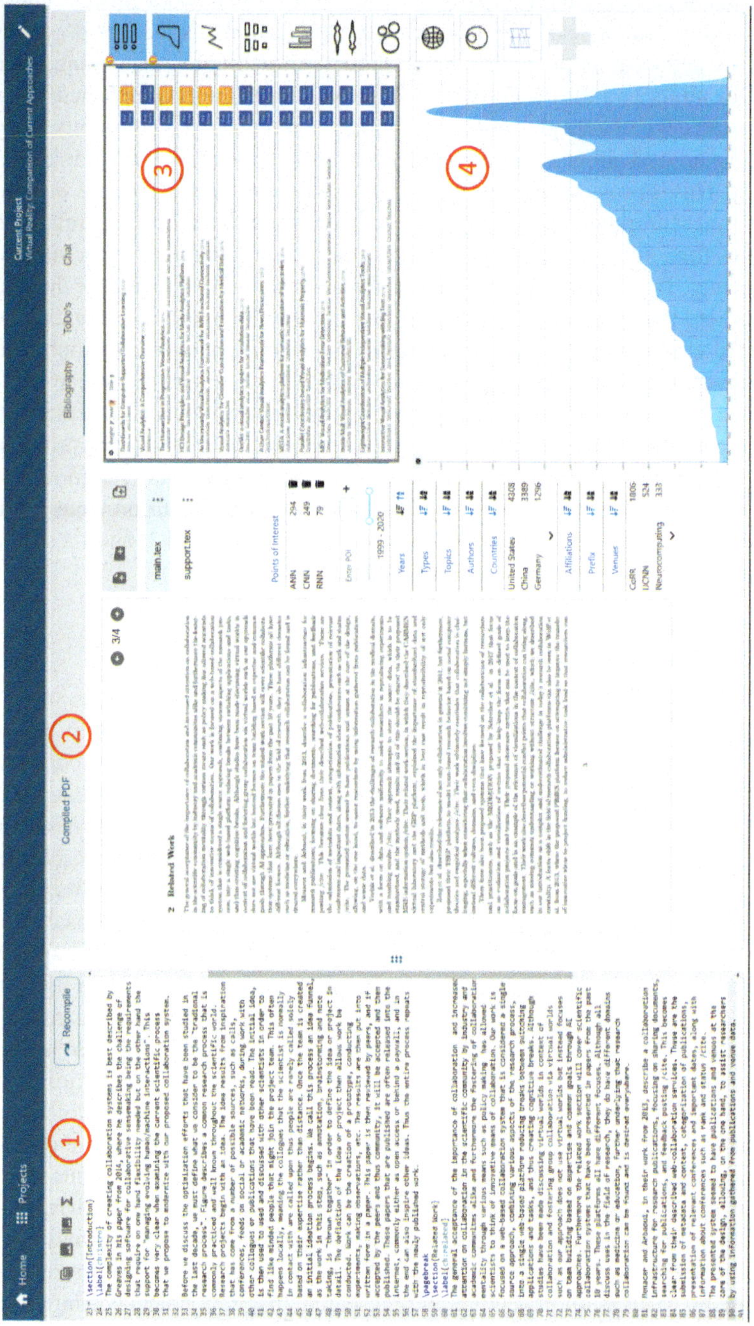

Fig. 3 UI overview featuring a Latex Editor (1), the PDF preview (2), the automated search results as a interactive listed view visualization (3) and (4) a temporal overview. Both (3) and (4) are interchangeable with any visualization that is seen on the right side

that are stored online, and users can highlight text, add notes to marked areas, and place comments. This allows the user to be fluid in their work process and keep focus on the task. By not having to change applications as they work, cognitive breaks are reduced, as the collaborative process of gathering, annotating, sharing, communicating, and ultimately writing a paper can all be accomplished from the single-source web-based platform. In addition, the written text is constantly analyzed in the background allowing for recommendations to be made that adapt to the user's progress. This is project-based, as the user can switch the project they are working on via the project dropdown. Changing a project means tabs that are project-related, such as publication, storage, annotation, Trello boards, and communication, all change dynamically. This dynamic change means that users only need to change their active projects, and the system will adapt. Furthermore, our approach to annotation in the context of exploration [7] is also integrated to quickly gather relevant publications to improve the overall research process for both the project and the researcher.

Figure 4 shows the initial testing approach, which allows advanced users to choose which topic modeling approaches, combined with specific similarity algorithms, the recommendations should be created with. A summary of similarity algorithms that are used in the field of document similarity can be found in our previous work [7, 8]. The advanced approach currently consists of four topic models and six similarity algorithms from which an advanced user can choose.

5 Heuristic Evaluation

To understand young academics' expectations, needs, and desires, we conducted a qualitative evaluation based on Nielsen's approach [34]. Nielsen states that the most errors can be identified through a qualitative evaluation with five people based on his heuristics per iteration, as findings tend to reach a ceiling and repeat themselves [34]. Thereby not necessarily leading to more insight [34]. We conducted a first iteration based on Nielsen's approach. We expanded upon this and focused on usability and user interface but also asked about application features to expand on what we discovered in our related work section. The examination was conducted with five people from the domains of electrical engineering, logistics management, biotechnology, architecture, and psychology, who were between the ages of 24 to 29. When asked what scientific, social collaboration platforms they know of, if any, a majority of the testers commonly associate ResearchGate followed by GitHub, with collaborative or social research platforms. This influences what users expect in terms of user interface and features. When asked what features of the platforms they commonly use, they stated that they mostly use the platform to assist them with their writing, as a source of inspiration, and for literature research tasks. When asked what improvements can be made, they stated that exactly these commonly used features should be improved, such as improved literature research, inspiration, and writing features. In terms of collaborative writing, we began by asking which applications they are familiar with for writing. The majority of them were familiar with Google Docs, and

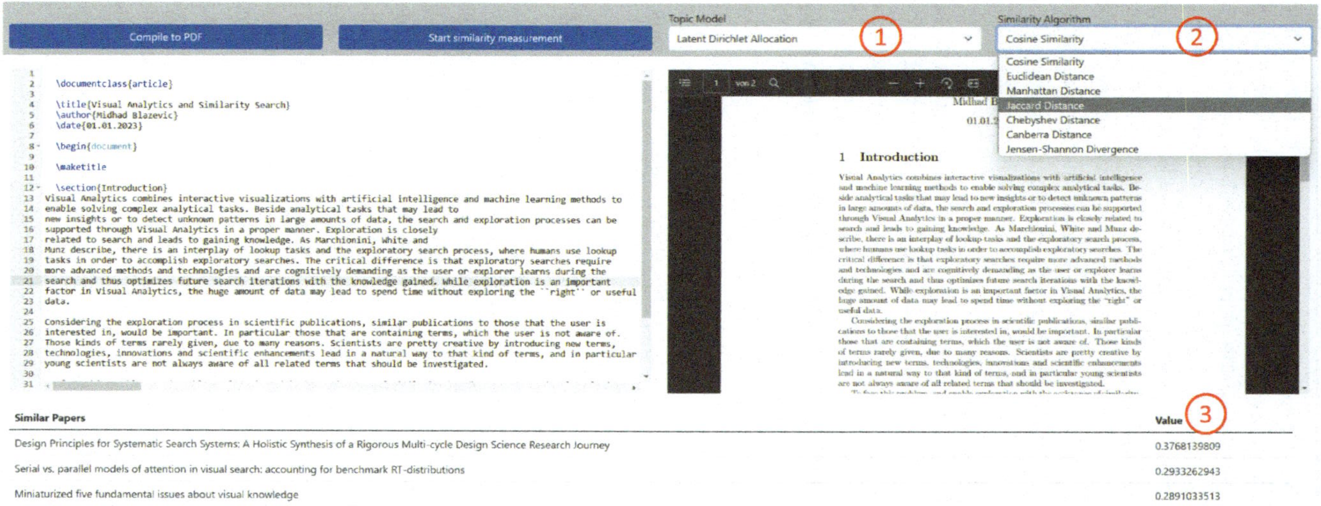

Fig. 4 Advanced mode of the Latex Editor for users who want to create their own combinations of topic modeling and similarity algorithm. (1) shows a dropdown where the user can choose a topic modeling approach for example LDA, LSA or NNMF. (2) shows a dropdown where users can choose a similarity algorithm for example Cosine, Jaccard, or Manhatten. Advanced users can also see the calculated value in (3) to examine the impact of their choice

Microsoft Word, both being well known by four out of five people, and two out of five stated Latex. However, they stated that they commonly collaborate with others via Google Docs and Microsoft Word, with only one of them commonly collaborating via Overleaf. Examining our system's usability, testers were given tasks to test four parts of the proposed platform: paper creation, profile updating, workgroup task monitoring, and collaborative literature management. 100% of the testers had a very positive experience while testing the collaborative literature management part. 60% of testers rated the other three parts as very positive, and the remaining 40% as a positive experience. None of the testers rated the system negatively, and all of the testers completed their tasks. When asked which features were the most important for them, they stated that collaboration management features were the most important, followed by features for intelligent collaborative editing, recommendation systems, and networking features, in that order.

The user study shows that users have expectations based on the popular ResearchGate, but also recognize issues or missing features on that platform. This observation confirms our assumptions based on our research, which also points to an overall lack of research in machine learning-supported collaborative research platforms and methodologies. The fact that ResearchGate is often used for research tasks but lacks visual analytical features also underlines the importance of utilizing visual analytics in our approach. However, even though our approach deviates from ResearchGate, it has received very positive feedback due to our applications that utilize machine learning and visual analytical approaches that, in comparison, improve the experience when working on research, writing publications, and collaboration.

6 Conclusions

Our work describes a generalized approach of the commonly practiced research process with a heavy emphasis on integrating machine learning-based approaches throughout every process step, as illustrated in Fig. 1. The conducted examination outlined the expectations and needs of young academics, which showed that the expectations are based on ResearchGate, which is well known among them. However, their needs reflect collaboration and research features that are currently missing. The primary approaches we utilized to improve the steps that yield great potential are recommendation systems, natural language processing, visual analytics, and interaction design, which are utilized in our approach. By extracting and analyzing data during the writing step of the process in real-time and creating an on-the-fly visual analytics tool, researchers are presented with relevant publications that are similar to their work, which they can optimally visually explore and annotate. Advanced users can create their own combination of topic modeling and similarity algorithms based on topic models and similar algorithms presented as options. The proposed real-time analysis of written text during the publication writing process shows an interesting on-the-fly-based approach to creating a foundation for recommendations and visual analytics systems independent of manually created vocabularies. Our work also covered the ideation process in science, which studies have not as extensively covered as in education or product design. Based on studies from other domains, we identified ideation tools and methods that have been proven to improve ideas' variety, quantity, and novelty. We built upon that knowledge by integrating machine learning-based approaches to counter the fixation and stimuli challenges by recommending viable tools and approaches in multiple iterations. These recommenders require information about the researchers, publications, and research projects. We use topic modeling to analyze as much information about the projects as possible; for example, by analyzing project descriptions, the team working on the project, annotations, and information gathered during the ideation process. This approach allows for projects to be described automatically and makes projects comparable. Furthermore, the adverse cognitive effects on researchers based on the mentality of multitasking and multi-app workflows during research collaborations that are common at this time, are countered by using the proposed single-source. The single source approach is enhanced with a fluid UI, allowing researchers to concentrate on tasks at hand while creating a workspace that suits their tasks.

7 Future Work

Our approach presents a broad range of modernization throughout the research process by integrating machine learning approaches and improving interaction design. The broad spectrum of modernization also means that various aspects should be further studied to ensure that researchers truly have the best possible environment and tools to create sustainable research and innovation. The core of our approach is the utilization of topic modeling. Although we have investigated the various combinations of topic modeling methods and similarity algorithms, we will also investigate autoencoders and other vectorization approaches, such as word2vec and doc2vec, to examine if these approaches yield better results than we currently have. The results impact all of the presented pillars. We also are increasing the number of publications that the system analyzes and will be able to recommend and base recommendations on, which will improve the overall quality of recommendations and thus support. Evaluation metrics should also be further examined in the future, as such a complex methodology requires that the implementation be tested and the effect on the project and researchers be observed and analyzed in detail. Evaluation metrics that should be further examined are task completion time, perceived workload strain, number of publications produced, number and quality of teams created with and without completed projects, and attention tracking. These evaluations require a baseline; thus, a survey and observations of a large pool of researchers from various domains must also be undertaken for comparison. This extensive evaluation could confirm previous studies and observations and help improve our approach by showing benefits and shortcomings that we have yet to cover in detail, such as the impact on the exploratory search process. The lack of studies focusing on the ideation of scientists rather than students or designers is a crucial point of improvement to understand the specific challenges that can be encountered when conducting ideation as scientists, which could also improve the identification of viable ideation and tools that can be used in research projects. Although we focus on content-based filtering approaches due to an abundance of academic content within papers and descriptions, it is important to also evaluate hybrid-based approaches and examine how collaborative-based approaches can be utilized with the data available. The various tasks and respective goals make it challenging to find a general approach, but as we outlined in this chapter, the potential for such an approach and the positive impact it would have on research quality, innovation, and the research process, in general, should not be underestimated.

Acknowledgements We thank Kjell Kunz, Viet Anh Ly, and Sascha Haas from the Darmstadt University of Applied Sciences and Jessica Bersch and Adrian Lumpe from our course Visual Trend Analytics at the Technische Universität Darmstadt, who contributed to this research. This work was conducted within the research group on Human-Computer Interaction and Visual Analytics at the Darmstadt University of Applied Sciences (https://vis.h-da.de).

References

1. Abbasi A, Hossain L, Owen C (2012) Exploring the relationship between research impact and collaborations for information science. In: 2012 45th Hawaii international conference on system sciences, pp 774–780. https://doi.org/10.1109/HICSS.2012.664
2. Abramo G, D'Angelo CA, Di Costa F (2009) Research collaboration and productivity: is there correlation? High Educ 57(2):155–171. https://doi.org/10.1007/s10734-008-9139-z
3. Adikari S, Keighran H (2016) Ideation governance for human-centered innovation in information systems. In: 3rd International conference on computing for sustainable global development (INDIACom), pp 1327–1332
4. Alves TPV, Borges MR, Vivacqua AS (2013) An environment to support the discovery of potential partners in a research group. In: Proceedings of the 2013 IEEE 17th international conference on computer supported cooperative work in design (CSCWD), pp 344–349. https://doi.org/10.1109/CSCWD.2013.6580986
5. Beck F, Koch S, Weiskopf D (2016) Visual analysis and dissemination of scientific literature collections with survis. IEEE Trans Visual Comput Graphics 22(1):180–189. https://doi.org/10.1109/TVCG.2015.2467757
6. Bellur S, Nowak KL, Hull KS (2015) Make it our time: in class multitaskers have lower academic performance. Comput Hum Behav 53:63–70. https://doi.org/10.1016/j.chb.2015.06.027 www.sciencedirect.com/science/article/pii/S0747563215004677
7. Blazevic M, Sina L, Burkhardt D, Siegel M, Nazemi K (2021) Visual analytics and similarity search-interest-based similarity search in scientific data. In: 25th International conference information visualisation (IV), pp 211–217. https://doi.org/10.1109/IV53921.2021.00041
8. Blazevic M, Sina LB, Nazemi K (2022) Visual collaboration—an approach for visual analytical collaborative research. In: 26th International conference information visualisation (IV), pp 293–299. https://doi.org/10.1109/IV56949.2022.00057
9. Blazevic M, Sina LB, Secco CA, Nazemi K (2023) Recommendation of scientific publications—a real-time text analysis and publication recommendation system. Electronics 12(7). https://doi.org/10.3390/electronics12071699. https://www.mdpi.com/2079-9292/12/7/1699
10. Boeckle M, Novak J (2015) Explorative analysis of applying collaborative visual annotations in online discussions to support the ideation of products or services. In: IEEE 19th international

conference on computer supported cooperative work in design (CSCWD), pp 159–164. https://doi.org/10.1109/CSCWD.2015.7230951
11. de Carvalho MA (2009) Effective new product ideation: ideatriz methodology. In: Tan R, Cao G, León N (eds) Growth and development of computer-aided innovation. Springer, Berlin, Heidelberg, pp 127–140
12. Cohen L, Manion L, Morrison K (2009) Research methods in education, 6, ed. Routledge, London
13. Crescenzi A, Ward AR, Li Y, Capra R (2021) Supporting metacognition during exploratory search with the orgbox. Association for Computing Machinery, New York, NY, USA, pp 1197–1207. https://doi.org/10.1145/3404835.3462955
14. Deo SR, Hölttä-Otto K (2017) Application of external stimuli during ideation: Impact of background and inclination based stimuli on novice minds. In: International conference on transforming engineering education (ICTEE), pp 1–12. https://doi.org/10.1109/ICTEED.2017.8585668
15. Drew CJ, Hosp JL, Hardman ML (2008) Designing and conducting research in education. SAGE Publications, Los Angeles. https://doi.org/10.4135/9781483385648
16. Fraser S, Brodeur SA, Katz R, Liu XS, Molthagen-Schnöring S, Pao SYA (2021) 2021 strategies for "socially distant" university-company collaborations. In: 2021 IEEE/ACM 8th international workshop on software engineering research and industrial practice (SER IP), pp 26–27. https://doi.org/10.1109/SER-IP52554.2021.00011
17. Fraser S, Mancl D (2021) Exploring the dimensions of university-company collaborations: Research, talent, and beyond. In: IEEE/ACM 8th International workshop on software engineering research and industrial practice (SER IP), pp 57–64. https://doi.org/10.1109/SER-IP52554.2021.00017
18. González-Mendívil JA, Rodríguez-Paz MX, Caballero-Montes E, Garay-Rondero CL, Zamora-Hernández I (2019) Measuring the developing of competences with collaborative interdisciplinary work. In: IEEE global engineering education conference (EDUCON), pp 419–423. https://doi.org/10.1109/EDUCON.2019.8725163
19. Jagtap S, Larsson A, Hiort V, Olander E, Warell A (2015) Interdependency between average novelty, individual average novelty, and variety. Int J Des Creativity Innov 3(1):43–60. https://doi.org/10.1080/21650349.2014.887987
20. Katz J, Martin BR (1997) What is research collaboration? Res Policy 26(1):1–18. https://doi.org/10.1016/S0048-7333(96)00917-1 www.sciencedirect.com/science/article/pii/S0048733396009171
21. Knoll SW, Horton G (2010) Changing the perspective: improving generate thinklets for ideation. In: 43rd Hawaii international conference on system sciences, pp 1–10. https://doi.org/10.1109/HICSS.2010.103
22. Kumar R (2010) Research methodology: a step-by-step guide for beginners, 3rd ed. edn. SAGE, London
23. Kumar S, Pan H, Wang R, Tseng L (2020) Litedoc: make collaborative editing fast, scalable, and robust. In: IEEE international conference on pervasive computing and communications workshops (PerCom Workshops), pp 1–6. https://doi.org/10.1109/PerComWorkshops48775.2020.9156221
24. Leckel A, Veilleux S, Dana LP (2020) Local open innovation: a means for public policy to increase collaboration for innovation in SMEs. Technol Forecast Soc Chang 153:119891. https://doi.org/10.1016/j.techfore.2019.119891 www.sciencedirect.com/science/article/pii/S0040162518315403
25. Li P (2020) Exploring the scientific research collaboration network in sustainability science in china. In: 2020 39th Chinese control conference (CCC), pp 6682–6689. https://doi.org/10.23919/CCC50068.2020.9188926
26. Maher B, van Noorden R (2021) How the covid pandemic is changing global science collaborations. Nature 594(7863):316–319. https://doi.org/10.1038/d41586-021-01570-2 www.nature.com/articles/d41586-021-01570-2

27. Melin G (2000) Pragmatism and self-organization: Research collaboration on the individual level. Res Policy 29(1):31–40. https://doi.org/10.1016/S0048-7333(99)00031-1 www.sciencedirect.com/science/article/pii/S0048733399000311
28. Menacer MA, Arbaoui A (2013) Design principles for a dedicated web-based collaboration infrastructure for research publications and resources. In: World congress on computer and information technology (WCCIT), pp 1–3. https://doi.org/10.1109/WCCIT.2013.6618749
29. Moisala M, Salmela V, Hietajärvi L, Salo E, Carlson S, Salonen O, Lonka K, Hakkarainen K, Salmela-Aro K, Alho K (2016) Media multitasking is associated with distractibility and increased prefrontal activity in adolescents and young adults. Neuroimage 134:113–121. https://doi.org/10.1016/j.neuroimage.2016.04.011 www.sciencedirect.com/science/article/pii/S1053811916300441
30. Mooranian M, Dillon T, Chang E (2011) Overview of cognitive visualisation. In: 5th IEEE international conference on digital ecosystems and technologies (IEEE DEST 2011), pp 138–142. https://doi.org/10.1109/DEST.2011.5936613
31. Moser C, Birkholz JM, Deichmann D, Hellsten I, Wang S (2013) Exploring ideation: knowledge development in science through the lens of semantic and social networks. In: 46th Hawaii international conference on system sciences, pp 235–243. https://doi.org/10.1109/HICSS.2013.218
32. National Geographic (2022) Isaac newton: who he was, why apples are falling|national geographic society (17.10.2022). https://education.nationalgeographic.org/resource/isaac-newton-who-he-was-why-apples-are-falling
33. Ng YK (2020) Research paper recommendation based on content similarity, peer reviews, authority, and popularity. In: IEEE 32nd international conference on tools with artificial intelligence (ICTAI), pp 47–52. https://doi.org/10.1109/ICTAI50040.2020.00018
34. Nielsen J, Landauer TK (1993) A mathematical model of the finding of usability problems. In: Proceedings of the INTERACT '93 and CHI '93 conference on human factors in computing systems, CHI '93, pp 206–213. Association for Computing Machinery, New York, NY, USA. https://doi.org/10.1145/169059.169166
35. Oh JS, Jeng W (2011) Groups in academic social networking services–an exploration of their potential as a platform for multi-disciplinary collaboration. In: IEEE third international conference on privacy, security, risk and trust and 2011 IEEE third international conference on social computing, pp 545–548. https://doi.org/10.1109/PASSAT/SocialCom.2011.202
36. Okudan GE, Chiu MC, Lin CY, Hernandez NV, Schmidt LC, Linsey J (2010) A pilot exploration of systematic ideation methods and tools on design learning. In: 9th International conference on information technology based higher education and training (ITHET), pp 102–107. https://doi.org/10.1109/ITHET.2010.5480052
37. Ordóñez-Matamoros G, Cozzens SE, García-Luque M (2011) North-south and south-south research collaboration: What differences does it make for developing countries?—the case of colombia. In: 2011 Atlanta conference on science and innovation policy, pp 1–10. https://doi.org/10.1109/ACSIP.2011.6064479
38. Rodrigues MW, Brandão WC, Zárate LE (2018) Recommending scientific collaboration from research gate. In: 2018 7th Brazilian conference on intelligent systems (BRACIS), pp 336–341. https://doi.org/10.1109/BRACIS.2018.00065
39. Sasaki H, Kajikawa Y, Fujisue K, Sakata I (2010) Detecting the valley of international academic collaboration in renewable energy. In: IEEE international conference on industrial engineering and engineering management, pp 99–103. https://doi.org/10.1109/IEEM.2010.5674434
40. Schreiber M, Kraft B, Zündorf A (2017) Metrics driven research collaboration: Focusing on common project goals continuously. In: 2017 IEEE/ACM 4th international workshop on software engineering research and industrial practice (SER IP), pp 41–47. https://doi.org/10.1109/SER-IP.2017.6
41. Schulz HJ (2011) Treevis.net: a tree visualization reference. IEEE Comput Graph Appl 31(6):11–15. https://doi.org/10.1109/MCG.2011.103
42. Sönmez S (2018) 11 steps process as a research method. Univ J Educ Res 6(11):2597–2603. https://doi.org/10.13189/ujer.2018.061125

43. Sreejesh S, Mohapatra S, Anusree MR (2014) Business research process. In: Sreejesh S, Mohapatra S, Anusree MR (eds) Business research methods. Springer International Publishing, Cham, pp 13–22. https://doi.org/10.1007/978-3-319-00539-3_2
44. Su X, Wang W, Yu S, Zhang C, Bekele TM, Xia F (2016) Can academic conferences promote research collaboration? In: Proceedings of the 16th ACM/IEEE-CS on joint conference on digital libraries, JCDL '16, pp 231–232. Association for Computing Machinery, New York, NY, USA. https://doi.org/10.1145/2910896.2925446
45. Thavorn J, Gowanit C, Muangsin V, Muangsin N (2021) Collaboration network and trends of global coronavirus disease research: a scientometric analysis. IEEE Access 9:45001–45016. https://doi.org/10.1109/ACCESS.2021.3066450
46. Vcelak P, Kratochvil M, Kleckova J (2013) Collaboration and research information system used for sustainable long-term research. In: Fourth world congress on software engineering, pp 307–310. https://doi.org/10.1109/WCSE.2013.56
47. White K (2023) Publications output: U.S. trends and international comparisons|nsf - national science foundation (17.01.2023). https://ncses.nsf.gov/pubs/nsb20206/international-collaboration
48. Wiradhany W, Koerts J (2021) Everyday functioning-related cognitive correlates of media multitasking: a mini meta-analysis. Media Psychol 24(2):276–303. https://doi.org/10.1080/15213269.2019.1685393
49. Wolff C, Igel B, Lauschner U (2013) Innovation management processes for academic research. In: IEEE 7th international conference on intelligent data acquisition and advanced computing systems (IDAACS), vol 02, pp 526–529. https://doi.org/10.1109/IDAACS.2013.6662980
50. Zawali A, Boukhris I (2020) Cross domain collaborative filtering recommender system for academic venue personalization based on references. In: IEEE symposium series on computational intelligence (SSCI), pp 2829–2835. https://doi.org/10.1109/SSCI47803.2020.9308377
51. Zheng X, Ke G, Zeng DD, Ram S, Lu H (2011) Next-generation team-science platform for scientific collaboration. IEEE Intell Syst 26(6):72–76. https://doi.org/10.1109/MIS.2011.104
52. Zhou X, Liang W, Wang KIK, Huang R, Jin Q (2021) Academic influence aware and multidimensional network analysis for research collaboration navigation based on scholarly big data. IEEE Trans Emerg Top Comput 9(1):246–257. https://doi.org/10.1109/TETC.2018.2860051
53. Zimmerling E, Höflinger PJ, Sandner P, Welpe IM (2016) Increasing the creative output at the fuzzy front end of innovation—a concept for a gamified internal enterprise ideation platform. In: 49th Hawaii international conference on system sciences (HICSS), pp 837–846. https://doi.org/10.1109/HICSS.2016.108

Visual Knowledge Discovery and Criminal Justice. Insights from a Computational Crime Analysis Research

Nicola Lettieri, Alfonso Guarino, Delfina Malandrino, and Rocco Zaccagnino

1 Introduction

Extracting actionable knowledge from criminal proceedings is one of the primary challenges in the research field coming up at the frontier between computer science, law, and computational crime analysis. Public prosecutors, on the other hand, generally have at their direct disposal information systems that do not provide advanced information extraction functionalities: databases containing complaints, criminal records, or police reports accessible via traditional textual interfaces that lack advanced and visually-intuitive information extraction functionalities.

As a matter of fact, regardless of the legal system considered, the administration of justice has dawdle in taking the advantages and opportunities provided by data and computation. Innovative approaches to make sense of information concealed in legal proceedings stay outside the horizons of legal experts and scholars. It is no coincidence that the EU Commission (see the recent Communication "*Digitalisation of Justice in the European Union. A toolbox of opportunities*"—2.12.2020

N. Lettieri
National Institute for Public Policy Analysis, 00198 Rome, Italy
e-mail: n.lettieri@inapp.org

A. Guarino (✉)
Department of Law, Economics, Management and Quantitative Methods, University of Sannio, Piazza Arechi II, 82100 Benevento, BN, Italy
e-mail: alguarino@unisannio.it

D. Malandrino · R. Zaccagnino
Department of Computer Science, University of Salerno, Via Giovanni Paolo II, 132, 84084 Fisciano, SA, Italy
e-mail: dmalandrino@unisa.it

R. Zaccagnino
e-mail: rzaccagnino@unisa.it

B. Kovalerchuk et al. (eds.), *Artificial Intelligence and Visualization: Advancing Visual Knowledge Discovery*, Studies in Computational Intelligence 1126,
https://doi.org/10.1007/978-3-031-46549-9_13

COM(2020)710 final[1]) has set the digital evolution of justice as a crucial objective for the future.

In such a scenario, conceiving advanced tools for knowledge mining—here understood as the "extitderivation of high-level concepts and descriptions from data through symbolic reasoning involving both data and relevant background knowledge [1]"-ecomes ever more a worthy challenge.

This paper focuses on how visualization can be composed with other computational heuristics—artificial intelligence in the first place—to develop solutions enhancing judges' ability to make sense of data and information available (and often concealed) within investigative documents.

In our perspective, resonating with the visual knowledge discovery perspective [2], visual metaphors represent not merely a way to show information but also an essential component of knowledge formation processes: the visual interaction with visual metaphors has an intrinsic heuristic value by reasons of which it makes sense to discuss "visual knowledge mining". The analysis will start from *CrimeMiner*—an ongoing research project exploring innovative computational approaches to the study of criminal networks, and then we present several visualizations designed to help Public Prosecutors make sense of the information at their hand and to facilitate a deeper understanding of criminal organizations under inquiry.

We present three visualizations designed to tackle different tasks within the project: (a) highlighting the structural and qualitative features of both criminal organizations and their components (or members); (b) depicting the progress and transformation of criminal networks over time; (c) enhancing the interaction between the legal scholar/expert and computational heuristics (social network analysis and machine learning) in the knowledge formation process. Beyond addressing specific demand for information, the visualizations have been put at the disposal of a unique knowledge-mining strategy intended for the legal world, which is still a little used to computation and visualization.

The rest of the paper is organized as follows: Sect. 2 presents the key literature neighboring this paper, while Sect. 3 depicts the application scenario of the visualizations conceived, i.e., *CrimeMiner*. In Sect. 4, we describe in detail the three visualizations proposed (contributions of this paper), and in Sect. 4.4, we illustrate the preliminary study conducted to evaluate the visualizations implemented in *CrimeMiner*. Section 5 concludes the works with final remarks and future research directions.

2 Related Work

In this section, we frame the related work of this research. It is primarily split into what concerns the computational crime analysis (Sect. 2.1) and the visual analytics initiatives in the legal context (Sect. 2.2).

---

[1] https://bit.ly/2QbZEKE.

2.1 Computational Crime Analysis

The "visual knowledge-mining" strategy presented in this paper lays in between the multidimensional scenario connecting computer science, computational legal studies [3] and law, specifically the areas of positive/procedure law dealing with criminal trials and inquiries.

Over the last twenty years, the analysis of crime with computational methods has attracted a lot of interest. An ever-growing horizon of approaches and methods ranging from data mining to social network analysis, from geographic information systems to agent-based modeling has progressively infiltrate the scientific and investigative study of criminal phenomena taking place both at the individual and social levels.

Since the second half of 80s, research in this field has tackled a large variety of topics [4, 5]: quantitative simulations of jury opinion emergence; assessment of eyewitness credibility; construction of argument tree containing claims, reasons, and objections; forensic geology for investigation; facial reconstruction. Recently, Big Data and the advancements in Artificial Intelligence nourished an extensive interest towards the integration of different data-driven heuristics for crime analysis. The main aim is coming up with models of crime that allows to predict both (a) where and when crimes will happen and (b) who will commit them or become a victim [6].

In general terms, the literature in the field illustrates a multifaceted scenario regarding goals and methods (e.g., [7]). We can identify different categories of works we can taxonomize as follows:

- *Studies based on offenders' profiling data.* For instance, [8], proposed a machine learning method to find "crime patterns" understood as a series of crimes committed by the same offender or group of offenders. The authors presented the method and tested it with real data from the Crime Analysis Unit of the Cambridge Police Department.
- *Articles focusing on the analysis of spatio-temporal data.* These works primarily handle such data in the form of a grid to predict crime hotspots/location [9–12], adopting supervised or unsupervised learning techniques, e.g., classifiers based on deep neural networks and K-means, respectively. Mohler and Porter [13] proposed an approach—tested using geographic data provided by the Portland Police Department—aiming to optimize the spatial discretizations of data used for predictions. The authors developed a procedure allowing (1) to select the optimal grid size and orientation, (2) to learn a scoring function that maximizes crime prediction accuracy. On this page, other works use spatio-temporal data combined with other kinds of data. For instance, [14] adopted criminal records to predict the typology of future crime events in a specific area. Bogomolov et al. [15] exploited mobile phone and demographic data to predict crime hotspots. The critical point of their work is using aggregated and anonymized human behavioral data captured from mobile network activity to face the crime prediction problem.
- *Works focused on predicting the number of specific offences.* In this group, we find initiatives to predict homicides in Brazilian cities [16] or Colombia [17], or cell

phone theft [18, 19]. In particular, [16] found that the education and unemployment levels of the area are among the most significant features for predicting homicides.

These works have not implemented tools usable and thought for end-users, but they mainly present proofs-of-concept of methods and approaches. Therefore, the solutions presented in the literature are not available for judges or Public Prosecutors to carry out inquiries and crime analysis and, however attractive, remain at a low technology readiness level.

2.2 Visual (Legal) Analytics

A highly cited paper by Herbert A. Simon and Jill Larkin [20] underlines the *"dramatic effects that alternative representations"* (textual and visual) may produce both on *"search and recognition processes"* and on human *"inference capabilities"*. Visualization advantages, according to the authors, *"are computational"*: diagrams can be better representations *"not because they contain more information, but because the indexing of this information can support extremely useful and efficient computational processes"*.

Today, Simon's and Larkin's intuitions materialize and take new forms into computational tools that provide individuals with innovative ways to turn data into knowledge while enabling them to act on their findings in real time. Particularly interesting, in this regard, are the methodological and technological achievements from Visual Analytics (VA), a fledgling research field [21] that explores new ways to: (i) synthesize information and derive insights from massive, dynamic, ambiguous, and often conflicting data; (ii) detect the expected and discover the unexpected; (iii) provide timely, defensible, and understandable assessments; (iv) communicate these assessments effectively for action.

In the legal field, the idea of exploiting visual metaphors to ease the management and understanding of documents and data has repeatedly appeared throughout history. The use of diagrams and maps dates back to the Middle Ages with the so-called *"arbor"* diagrams: maps graphically exemplifying legal concepts like the impediments to marriage or the various stages of procedures in Roman Law.

In much more recent times, the heuristic power of visualization has been put at the core of a new type of diagrams—the so-called "*Wigmore charts*"—conceived by Henry Wigmore to support lawyers dealing with complex amounts of ambiguous evidence so to facilitate reasoning processes required to confirm or rebut hypotheses presented in court. With the advent of digital technologies, visualization came back into the limelight, inspiring many works leveraging graphic representations—e.g., Bayesian networks—to make more intuitive and insightful intellectually challenging tasks like organizing complex sets of evidence [22–24] or supporting argumentative tasks for lawyers [25, 26]. More recently, the availability of ever more powerful technologies and insightful computational heuristics has boosted the interest in tools allowing to explore the law and its data visually. Based on most recent research experiences, three main ways of using visualization in the legal field to make some clarity and conceptual order.

- *Enhanced information retrieval.* The first group of visual solutions is designed to make data recovery and interaction more intuitive and insightful faced with the complexity of a legal universe made up of heterogeneous documents (statutory norms, case law, papers, administrative documents) produced by authorities of different levels (local, national, international) and often interrelated among them [27, 28].
- *Legal corpora analysis.* A second group of applications—resulting from the integration of graphic representation and computational heuristics ranging from NA to data mining techniques—is oriented to discover intrinsic properties (e.g., relevance, structural features, evolution dynamics) of documentary corpora and their components [29–32].
- *Analysis of legally relevant empirical phenomena.* Visualization techniques can also be exploited to support the representation and the understanding of facts playing a role for the purposes of applying legal rules. For instance, it can be used to ease the interaction with large amounts of public sector data and offer users crosscutting reading of heterogeneous information [33].

3 Background

For the sake of clarity, we split the background into two parts: the *CrimeMiner* project (Sect. 3.1), and the knowledge mining architecture underlying visualizations proposed in this work (Sect. 3.2).

3.1 CrimeMiner: A Computational Crime Analysis Project

The starting point of this paper is *CrimeMiner*, an ongoing project moving in the perspectives presented in Sect. 2 and exploring the application of computational social science [34] methods to study crime for investigative and scientific aims. The project draws on a collaboration with Public Prosecutors from the Italian Anti-Mafia Investigation Directorate. It aims to provide the judiciary with a deeper and more comprehensive understanding of criminal organizations' structural and functional features.

In this respect, we highlight that if we ignore more "traditional" information systems (databases containing complaints, criminal records, police reports, and so on), criminal investigation primarily stays, for Public Prosecutors, a "pen and paper" task. This is why, they would be handy tools allowing to collect, analyze, visualize and mine information from investigative documents produced by the Public Prosecutor's office and police. Beyond everything, knowledge about criminal networks' structural and functional characteristics (and individuals therein) is vital for inquiry and enforcement goals.

In such a context, in addition to facing theoretical and methodological issues, the *CrimeMiner* project geared towards the development of an online analytical platform [35]. Such a platform aims to explore several computational heuristics (also combined together) for a a deeper understanding of structural and functional characteristics of criminal organizations based on the analysis of simple relational and investigative data. Within the project, we have experimented with data mining, social network analysis, machine learning and data visualization, as well as agent-based modeling. Over the years, the project has tackled a multitude of problems: (i) automated extraction, from unstructured case files, of information (e.g., social entities and relationships) to be used in subsequent study; (ii) creation of criminal network graphs[2] and the application of network analysis to learn the characteristics of criminal organizations and persons therein [36]; (iii) visual browsing and visual analytics [37] of investigative data; (iv) integration of social network analysis with Agent-Based Models [38] preparing the ground to "*in silico*" what-if analysis.

Among the project's main achievements, there is a series of findings derived with the sole application of social network analysis metrics to the telephone tapping graph. Different centrality metrics have offered as much information. *PageRank* [39], for instance, has proven to be a good proxy of criminal leadership, while betweenness centrality [39] has shown to be a valuable support for the identification of intermediaries and drug mules [36].

Despite the outcomes achieved, the project has highlighted the need to enhance the quality of knowledge extracted from trial documents. Due to the complexity of the criminal phenomenon, the assessment of essential issues for the inquiries can hardly rely on centrality measures and network analysis only. A particularly challenging topic, in this respect, is the evaluation of criminal dangerousness (see Sect. 4.3). A better evaluation of this characteristic in individuals and groups is fundamental on different levels: for the continuation of investigations, more empirically grounded crime prevention strategies, and the overall trial outcome. The point is that dangerousness is an inherently hybrid concept emerging from the combination of legal and empirical (social, criminological) factors. The reflection on this point led us to figure out a knowledge mining approach aiming to "empower" Public Prosecutor's ability to interpret evidence.

[2] The platform handles data about social relations that are represented as a graph $G = (V, E)$, where V = persons included in the case files, and E = relation, such as telephone or environmental tappings.

3.2 A Three-Tier Knowledge-Mining Architecture Supporting the Evaluation of Criminal Dangerousness

In light of the above, a particularly challenging topic is the evaluation of criminal dangerousness here understood as the propensity to execute and organize illegal activities and/or the possibility to further commit crimes. A better assessment of this characteristic in both individuals and groups is essential on several levels: for the survival of the investigations; for more empirically-grounded crime prevention tactics; for the overall trial outcome. The point is that dangerousness is an inherently hybrid concept which results from the analysis of legal, and empirical (social, criminological) elements.

To tackle this problem, we conceived a *visual knowledge-mining* approach aiming to "empower" Public Prosecutor's ability to understand evidence rather than just forecasting. The approach has two guidelines: (a) select and combine computational heuristics suitable to integrate and jointly analyze "*social* facts", for instance, interactions captured by social network analysis, and "*legal* data", for instance, criminal records defining the profile of organization's members; (b) examine innovative ways to embed the Public Prosecutor's know-how into the knowledge-mining process by means of human-machine cooperation. The result is a multi-tier visual knowledge-mining architecture in which the formation of new knowledge is conceived as an iterative process developing on *three* tiers (see Fig. 1).

1. *Data:* extraction of information from trial documents containing the legal and relational information requested to study both the criminal network and the dangerousness of persons under inquiry.

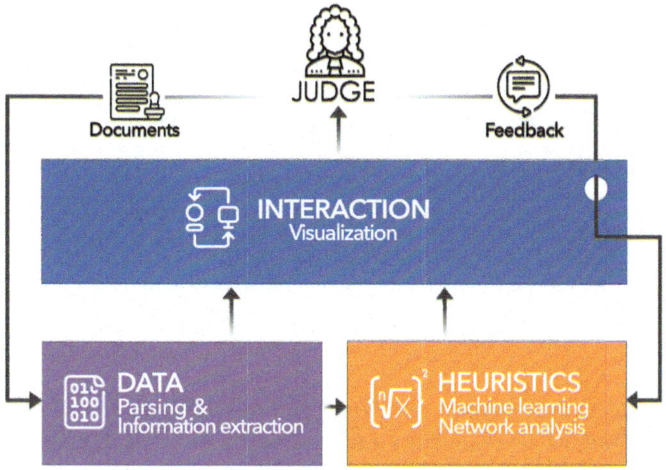

Fig. 1 *CrimeMiner*: the three-tier knowledge mining architecture

2. *Heuristics:* study of data extracted from documents through network analysis approaches useful to assess the degree and type of the social activity of groups and persons therein.
3. *Interaction:* domain experts interaction with data and the results of their analysis (e.g., network-based inference regarding the role and activities of persons under inquiry) and with the models at the lower tier (e.g., domain expert's feedback can improve the performance of online learning methods aimed at supporting the assessment of criminal dangerousness).

Visualization has a key role in the architecture above sketched. We have ascribed a vital function to visual metaphors (in line with [20]): to explore/retrieve/manipulate information and to intuitively interact with the computational heuristics developed and deployed into the *CrimeMiner* analytical platform.

4 Integrating Visualization in a Knowledge Mining Strategy

This section is devoted to present the primary contribution of this paper. It schematically depitcs how visualization can support Public Prosecutors in highlighting the structural and qualitative features of both criminal organizations and their members (details in Sect. 4.1)—also over time (see Sect. 4.2), and how it makes possible the "visual knowledge-mining" process by acting as the interaction layer between Public Prosecutors and *CrimeMiner* (details in Sect. 4.3). For each visualization proposed, first, we highlight its aim within the knowledge-mining strategy; second, we offer details concerning how it works and how it has been developed and deployed in *CrimeMiner*.

4.1 Similarities Between Individuals Under Inquiry at a Glance

Highlight: *The study of similarity between persons in the criminal network can assist the dangerousness assessment. Virtually dangerous persons may have common characteristics; hence they are similar.*

We have developed within *CrimeMiner* a *Similarity module* implementing *SimRank* [40]. *SimRank* is a similarity measure based on a graph-theoretical model. The module provides the user with information concerning similarities (in terms of person's features and social relationships) that may exist between persons within the criminal network at hand. To increase the understanding of similarity metrics

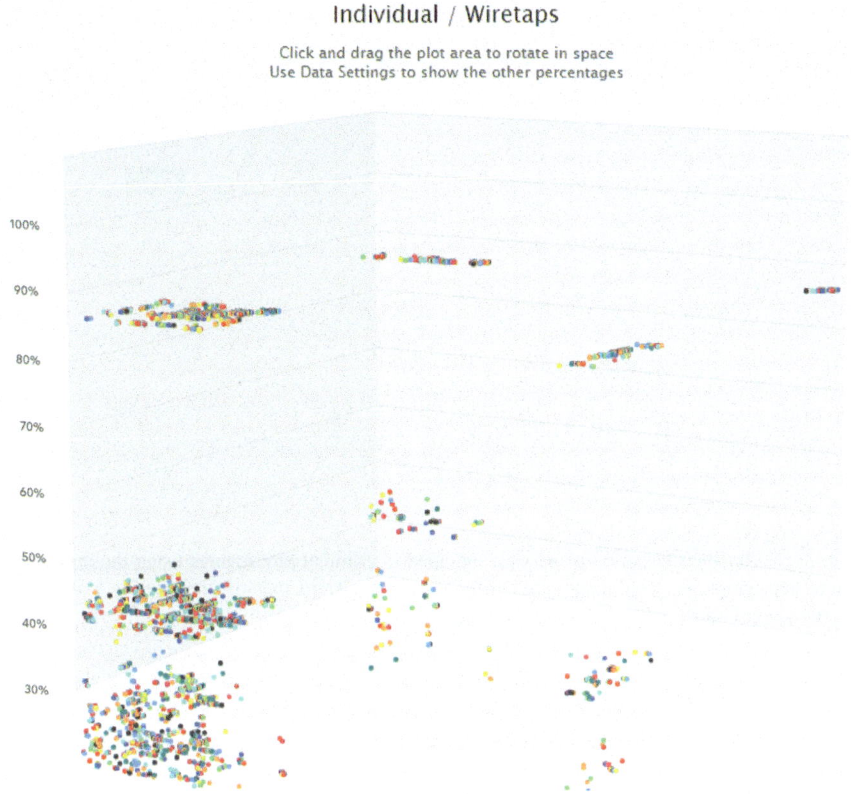

Fig. 2 *SimRank* visualization of persons included in criminal proceedings using the individual-phone calls graph. Persons' names are blurred for privacy. Here, in particular, *SimRank* between each person under inquiry

applied on several graphs' types, we create 2D and 3D graphics, through the use of Highcharts[3] JavaScript Library. Figure 2 shows a 3D graphic in which x–z axes are for individuals' names and the y-axis represents the similarity percentages between each pair of persons, respectively. The pair is represented as a point in the Cartesian graph. Through this arrangement on a Cartesian graph, we can intuitively highlight that the top points represent the more similar pairs. With a mouse-over, the *Similarity module* unveils names and similarity percentage of each pair of persons under inquiry. By means of the "Data Settings" option, the user can change the *SimRank* threshold to better visualize the degrees of similarity he/she is interested in (by default the threshold is set to [30%, 100%]). For example, by setting the threshold at 70%, the module will show only the pairs of persons with a similarity higher or equal than 70%, thus excluding the ones exhibiting lower similarity. Moreover, the user can select just one person to get his/her *SimRank* with all the others (see Fig. 3). In this

[3] https://www.highcharts.com.

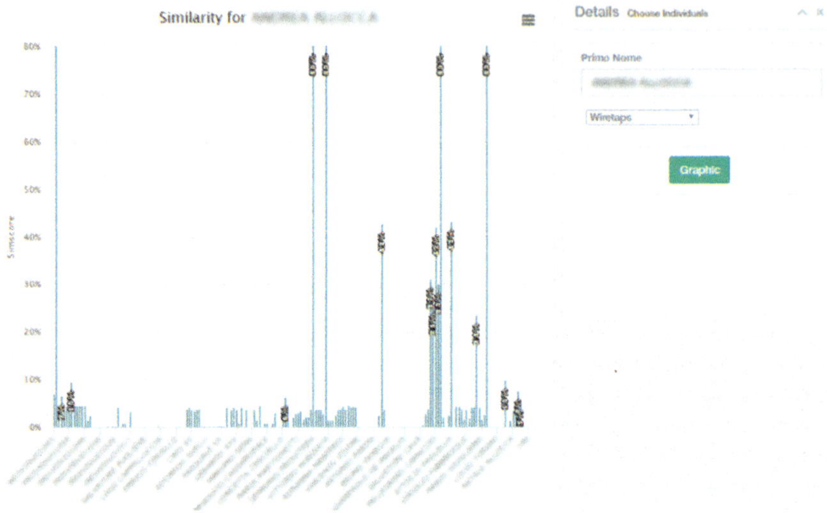

Fig. 3 *SimRank* visualization of persons included in criminal proceedings using the individual-phone calls graph. Persons' names are blurred for privacy. Here, in particular, *SimRank* of a single person under inquiry against all others

case, the similarity module makes use of a bar-chart where the bars represent the similarity (the higher, the more similar) and on the x-axis there are the persons under inquiry.

4.2 *A Question of Time: Visualizing the Evolution of Criminal Networks*

Highlight: *Analyzing the historical evolution over time of the criminal network can better support the assessment of dangerous individuals, highlighting suspicious social patterns.*

We have developed within *CrimeMiner* a *timed graph-based visualization module*, providing the user the possibility to browse the social relations between individuals in the criminal network over the investigation period. This aid is thought to support grasping social patterns' evolution, highlighting suspicious ones. An abstracted version of the visualization is available in Figs. 4 and 5.

The timed-graph-based visualization module works with every kind of graph (social relations) *CrimeMiner* handles (e.g., phone or environmental tappings); here, we present the visualization's details using phone calls as a testbed. Figures 4, 5, 6 and 7 depict the

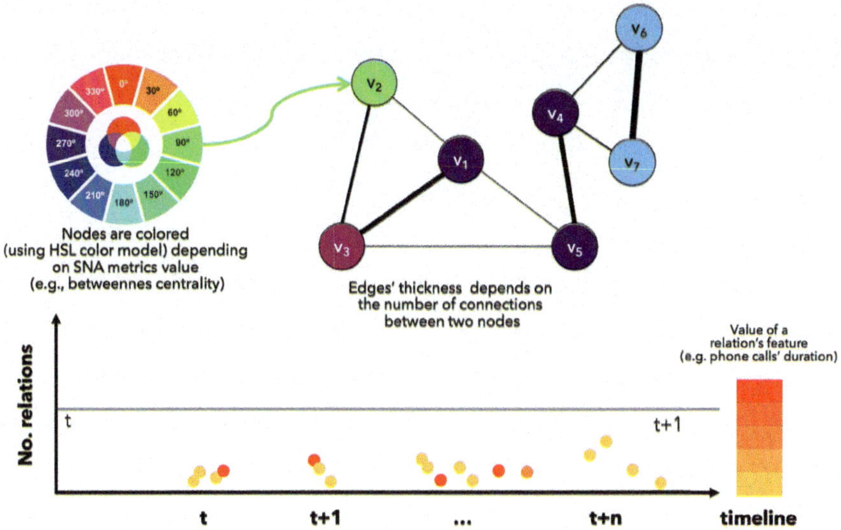

Fig. 4 timed graph visualization module: abstracted version. The visualization includes two main areas: **a** the timed graph (a multi-graph), depicting the criminal network during the inquiry over time; **b** the timeline, allowing users to select a time period of interest and visualize, accordingly, the number and "intensity" (e.g., duration of phone calls) of the relation among individuals under analysis

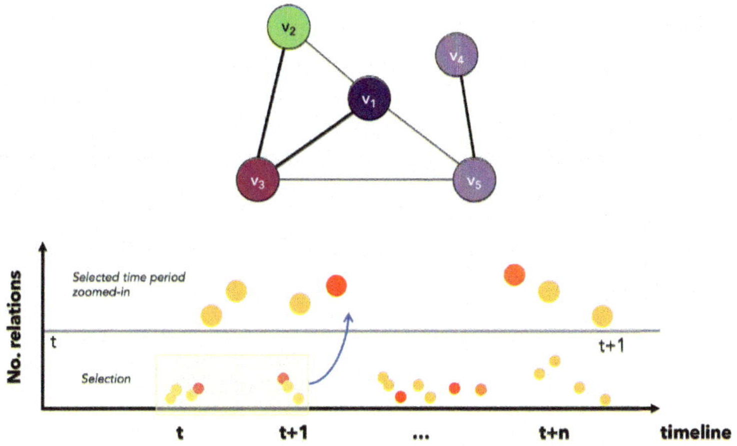

Fig. 5 timed graph visualization module: abstracted version. The timed graph depicts the criminal network in the selected time period (i.e., $(t, t{+}1)$. Nodes' color and edges' thickness changes accordingly. Observe that two individuals, v_6 and v_7, have not been investigated in the time period $(t, t{+}1)$, hence they do not appear

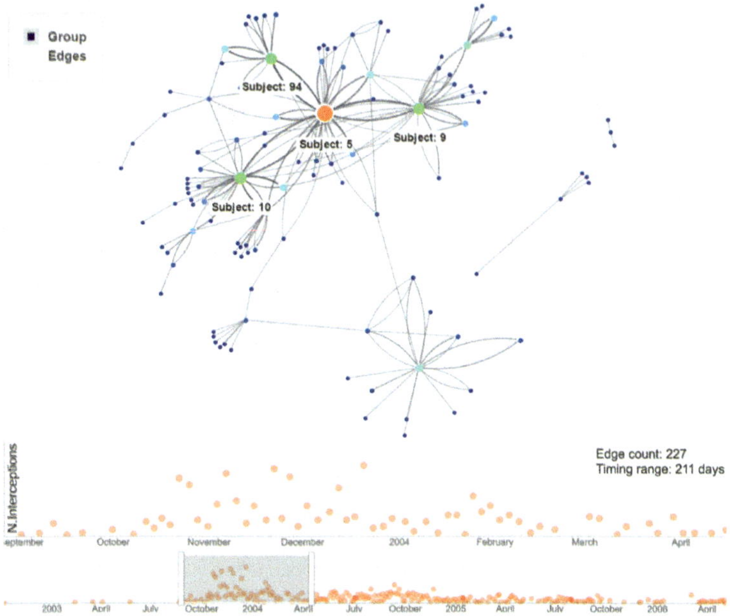

Fig. 6 timed graph visualization module: instantiated version in *CrimeMiner*. *Individual-Phone calls multi-graph*. Here, the timed graph depicts inquiry materials from October 2003 to April 2004

timed multi-graph $G = (V, E)$. Formally, let $V = \{v_1, v_2, \ldots, v_n\}$ be a set of *individuals*, and E be a set of *tapped phone calls* such that $(v_i, v_j)_t \in E$ if there exists a *tapped phone call* from v_i to v_j, with $1 \leq i, j \leq n$ and $i \neq j$ at time t. We define $C = \{0, 1, \ldots, 360\}$ as the set of colors according to the *HSL* metrics (Hue, Saturation, Lightness)–which goes from blue to red–, and a color function $color : V \rightarrow C$, assigning to every node $v_i \in V$ a color $c \in C$. The color of v_i depends on the sum of outgoing and ingoing edges from v_i. We define $\forall v_i \in V, E_{v_i} = \{(vj, vk)_t | v_j = v_i \vee v_k = v_i\}$, with $E_{v_i} \subseteq E$, as the set of outgoing and ingoing edges of v_i at time t. Therefore, the shade of v_i is given computing

$$color(v_i) = \left\lfloor \frac{(bc_{v_i,G_t} - m_t)}{(M_t - m_t)} \times 360 \right\rfloor$$

where m_t and M_t are the minimum and maximum centrality value at the time t respectively.

We define a thickness function $thick : E \rightarrow \mathbb{N}$ that sets a thickness to the edges of the timed graph, acting as a grouping function to reduce the visual clutter. The thickness of a visualized edge at time t is proportional to the number of edges between two nodes at time t. Formally, let $v_i, v_j \in V$ two nodes, we compute the thickness

$$thick((v_i, v_j)_t) = |E_{v_i} \cap E_{v_j}|$$

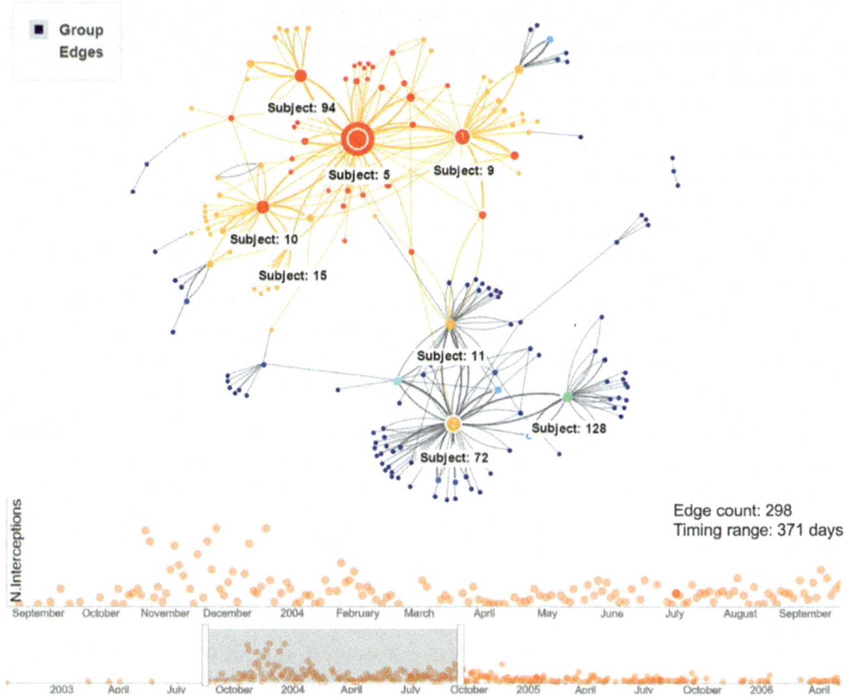

Fig. 7 timed graph visualization module: instantiated version in *CrimeMiner*. *Individual-Phone calls multi-graph.* Here the timed graph depicts inquiry materials from October 2003 to October 2004. Subject 5 and its connections have been highlighted

Through a double-click on a node of interest, the domain expert highlights all node's social relationships, belonging to the first (persons called or that have called, informally "friends") and second level (persons called or been called by individuals of the first level, informally "friends of friends"). The timeline at the bottom includes all the inquiry period, in our case, from October 2002 to 2006. By default, the timed graph displayed—in turn with the timeline—includes nodes and relations regarding the whole inquiry period. However, the timeline allows to display the criminal network at a time t or in a time range $[t_s, t_e]$ (see Figs. 6 and 7), and, therefore, reduce the number of nodes and relations displayed to only those belonging to the range $[t_s, t_e]$. The timeline is composed of two graphics sharing the x-axis as the time axis. The y-axis represents the number of tapped phone calls. Therefore, a depicted point $P(t_p, pc_p)$ in the timeline represents the number of tapped phone calls $pc_p = |E|$ in the timed graph at a specific time t_p. Furthermore, these points are colored up according to the total duration of tapped phone calls; the more the duration, the darker the color.

To develop this module, we exploited two JavaScript libraries: Ogma.js[4] e D3.js.[5]

[4] https://doc.linkurio.us/ogma/latest/.

[5] https://d3js.org.

4.3 Visual Interaction with Computational Heuristics Supporting the Evaluation of Criminal Dangerousness

Highlight: *Visually interacting with the knowledge base and the computational heuristics can help Public Prosecutors in the process of assessing the criminal dangerousness.*

We have implemented in *CrimeMiner* a *dangerousness module*, which is an online learning module supporting the Public Prosecutor (and users in general) to assess the criminal dangerousness of persons under inquiry. We think that systems that can explain the *how* and *why* of their predictions (thus more transparent), in our case through a visualization layer, are more helpful for users who will have a better perception of control and therefore will trust such systems more, and can correct their shortcomings. The basic idea is to provide a "graspable Artificial Intelligence" module within *CrimeMiner*, in line with the visual knowledge discovery perspective discussed in [2].

The assessment of persons' dangerousness and, on an aggregate scale, organizations plays a central role in crucial stages of criminal trial and, even before, in the inquiry phase. Whatever the legal system considered, it represents a vital factor not only for sentencing purposes [41]—to determine the scale of penalty—but also for the adoption of interim and preventive measures[6] the judicial authority can enact both during the preliminary investigation and the hearing. Such relevance comes, however, with a high level of intricacy. Dangerousness is a multifaceted phenomenon in which different aspects converge. If, in general terms, it can be conceived as the likelihood that an individual commits crimes, seen more closely, it coincides with the propensity of the subject to engage in criminal activities, the so-called "criminal tendency". The latter is usually identified based on a variety of indexes: the reasons underlying illegal activity; the personality of the offender[7]; its conduct of life and living (individual, family and social) conditions; the presence of any judicial precedents.

The module relies on a *KStar* [42] model implemented with WEKA library in Java. *KStar* is an instance-based classifier, that is the class of a test instance is based upon the class of those training instances similar to it, as determined by a similarity function. It differs from other instance-based learners in WEKA in that it uses an entropy-based distance function.

[6] The expression refers to immediately executive measures of coercion resulting in limitations of personal freedom or the availability of goods. Such measures against the accused aim: (a) to prevent inappropriate behaviors during the course of the criminal proceeding (e.g., attempts to conceal evidence or to commit other crimes); (b) to ensure the enforcement of the judgment.

[7] It can only be assessed with the contribution of specific categories of domain experts like psychiatrists or psychologists.

KStar model has been fit on a dataset of 3,100 persons under inquiry. Such persons have been manually labeled—by five experts in criminal justice—into two classes, i.e., "*virtually dangerous*" and "*virtually not dangerous*". Such domain experts have obtained, in annotation task, a Fleiss' Kappa (adaptation of Cohen's Kappa [43] for three or more raters) of $k = 0.87$, thus an almost perfect agreement. Each person under inquiry has been represented by using a combination of handcrafted features (reported in Table 1) obtained from:

- social network analysis metrics (e.g., betweenness centrality, degree centrality),
- criminal proceedings (e.g., criminal records).

To effectively fit the *KStar* model, we followed the widely adopted steps in machine learning research [44]:

1. we have divided the dataset into training and testing. Here, we opted for a 80–20 ratio, i.e., 80% for the training set and the remaining 20% for the testing set;
2. we have performed a validation phase using k-fold cross-validation method with $k = 5$. This phase has been conducted exploiting only the training set;
3. we have performed a testing phase by employing the test set after fitting the model on the training set. This phase allows us to evaluate the performance of the KStar on new (unseen) samples.

We obtained promising performance: 0.95 accuracy ($SD = 0.06$) in validation, and 0.93 accuracy in the testing phase. The *KStar* model acts as an online learning model, waiting for the Public Persecutor's feedback to adapt its internal weights. The *dangerousness module* makes use of interactive graph visualization allowing the Public Prosecutor to input such feedback to the underlying *KStar* model (depicted in Fig. 8). The user must click (single-click) on the node he/she wants to study $v_i \in V$, and the module displays a callout where it can be expressed the new label for v_i through a radio button (between "*dangerous*" and "*not dangerous*"). Once submitted the label (using "Submit" button), the module computes the features for v_i (the ones in Table 1) and forwards them to the *KStar* along with the new label. *KStar* updates its internal weights with the `updateClassifier()` function.

Table 1 Type of features used to train the *KStar* model

Criminal profile features
Crime records
Crime charges
Criminal profile features
Betweenness centrality
Degree centrality
Outdegree centrality
Indegree centrality
Page Rank

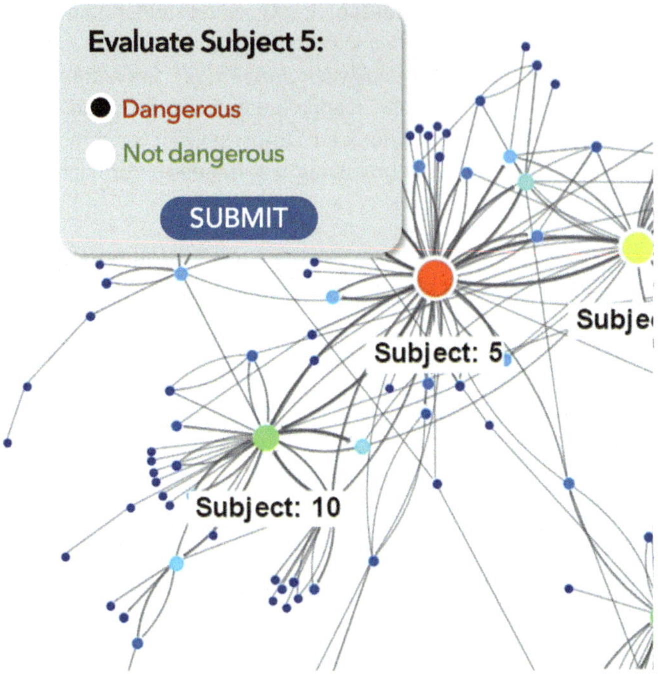

Fig. 8 *CrimeMiner*: the user can interact with the online learning module through the timed graph visualization

We empirically observed that the model correctly learns the user feedback, updating the assessment of criminal dangerousness for specific persons; however, the use of this kind of system shall undergo severe tests to study: (a) the degree of its adaptation on the Public Prosecutor's feedback, (b) the "speed" of the model to change its internal weights, and (c) the spillovers of the feedback on all the other persons throughout the inquiry.

However, we propose here, loosely based on [45], an "in-silico" evaluation of the performance of this model as the number of "*Dangerous*" feedback increases. All the actions have been automated through Selenium.[8] We automated the click on a node of the criminal network and the click on feedback button, in particular the choice of the "*Dangerous*" option. Specifically, randomly selecting a node, we send a negative feedback and then repeat the classification of the entire dataset. This process has been repeated for 10, 20, 30, 40, and 50 nodes in the network. The results are available in Fig. 9. We can observe how *KStar* behaves after receiving a certain number of feedback about persons under investigation. The accuracy decreases as feedback are sent. After 50 nodes/individuals randomly chosen and suggested to be "*Dangerous*", the model exhibits accuracy close to guessing.

[8] https://www.selenium.dev/documentation/webdriver/.

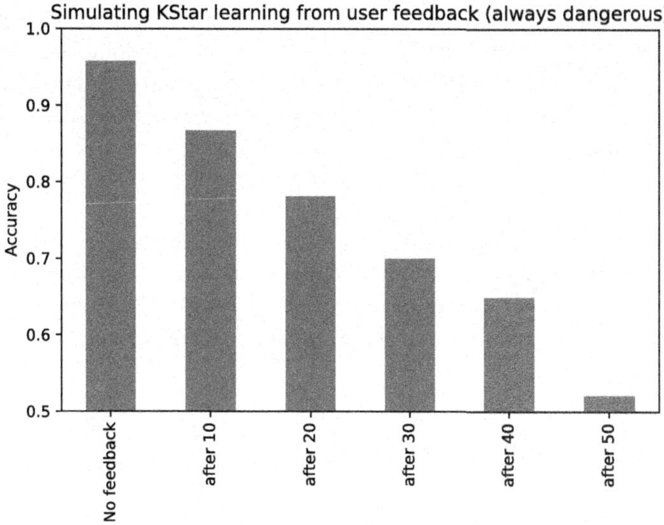

Fig. 9 How *dangerousness module* learns from users' feedback. Accuracy results after sending 10, 20, 30, 40, and 50 feedback

On a more general level, the ability of the system to learn also according to user feedback is the feature that aroused the most significant interest in the deputy prosecutors interviewed. In their view, the prospect of man-machine heuristic cooperation ensuring both a better control over the system and the exploitation of the domain expert know-how takes on priority compared to the—also highly appreciated—"insightfulness" of computational heuristics and the intuitive forms of interaction with data enabled by visual analytics.

4.4 Evaluate the Visualizations Proposed: A User Study

In this section, we offer a preliminary usability study with end users on the proposed new visualizations. As done in earlier studies [28, 46], we enrolled seven (7) participants among legal scholars (29%) and public prosecutors (71%). Prior research concerning usability testing indicates that five users are sufficient to uncover approximately 80% of usability problems [47]. By following a methodology also applied in [48], at the first stage, we gave to participants information about *CrimeMiner* and its main functionalities, then we asked them to use *CrimeMiner* for a 15-minutes session and then accomplish two tasks requiring to go through both the similarity and the timed graph modules. Specifically, "*Task (1): Given that Subject 5 is dangerous, find another person with the same criminal profile within the network through the Similarity module*" and "*Task (2): Using the timed graph module, analyze the evolution of a given sub-community of the network in terms of both activity and structure*".

At the end of this testing phase, we asked users to spend another 10 min to fill in the standard QUIS questionnaire [49]. As a result of the quantitative evaluation, all participants rated (on a 5-point Likert scale) as easy to perform the assigned tasks (Mean = 4.3 for both tasks), while about the qualitative evaluation, the most positive results were about the overall reaction to the software (Mean=7.2, St. Dev = 0.9), and how easy was learning to operate the proposed system (Mean = 7.5, St. Dev = 0.5).

Through this study with end users, we also got the chance to gather suggestions about future developments of the tool. Among the most interesting ones, we find *"It would be interesting and beneficial to drill-in the phonecalls' contents, the actual text, to understand who had said what, to whom, and when he/she did it"*.

5 Conclusions and Future Works

In this paper, we have presented new visualizations developed and deployed into *CrimeMiner*, an experimental crime analysis platform. In particular, such visualizations play three goals: (a) highlighting the structural and qualitative features of both criminal organizations and their components; (b) depicting the progress and transformation of criminal networks over time; (c) enhancing the interaction between the legal scholar/expert and computational heuristics in the knowledge formation process (in particular for the dangerousness assessment of persons within the criminal network at hand). The visualizations have been positively assessed by 7 domain experts.

Taken as a whole, the experience made throughout the project has clearly shown the role that visualization can play in supporting the extraction of actionable knowledge from data in the legal scenario, where domain experts would otherwise struggle to interact with computational methods lying far beyond their skills.

Visualizations developed and deployed within *CrimeMiner* have shown they can offer unprecedented and intuitive ways of grasping phenomena and circumstances that would be materially inaccessible without computational backing. Reading thousands of pages of telephone/environmental tappings and procedural documents clearly does not prompt in the mind of a judge the picture of a social graph with all its features. Meanwhile, the experience made in assessing social dangerousness has shown how visualization can represent a notably suitable "interface" for implementing new forms of human-machine interaction and cooperation in computational crime analysis.

The efficacy of the "visual knowledge-mining" strategy in supporting the discovery of dangerous persons and groups will be evaluated in the next future with Public Prosecutors and several criminal organizations as a testbed. However, it is worth highlighting that, although conceived for the legal domain, our proposal is definitely amenable to applications in other fields of Humanities and Social Science, where the use of computational instruments and techniques still has a long way to go.

As future works for this project, we have already planned experiments with visual text mining approaches to extract knowledge also from the contents of the phone calls as well as the tapped discussions held during in-person meetings. The idea is that of providing through *CrimeMiner* an intuitive way to delve into the individuals' communications content and check for critical topics, significant or suspect keywords.

References

1. Kaufman KA, Michalski RS (2005) From data mining to knowledge mining. Handb Statist 24:47–75
2. Kovalerchuk B (2018) Visual knowledge discovery and machine learning, vol 144. Springer
3. Lettieri N (2020) Law in the turing's cathedral: notes on the algorithmic turn of the legal universe. In: Barfield W (ed) The Cambridge handbook of the law of algorithms. Cambridge University Press, pp 691–721
4. Chan JB (2001) The technological game: how information technology is transforming police practice. Crim Justice 1(2):139–159
5. Nissan E (2009) Legal evidence, police intelligence, crime analysis or detection, forensic testing, and argumentation: an overview of computer tools or techniques. Int J Law Inf Technol 17(1):1–82
6. Hvistendahl M (2016) Crime forecasters. Science 353(6307):1484–1487
7. Tashea J (2017) Calculating crime. ABAJ 103:54
8. Wang T, Rudin C, Wagner D, Sevieri R (2013) Learning to detect patterns of crime. In: Joint European conference on machine learning and knowledge discovery in databases. Springer, pp 515–530
9. Kim S, Joshi P, Kalsi PS, Taheri P (2018) Crime analysis through machine learning. In: 2018 IEEE 9th annual information technology, electronics and mobile communication conference (IEMCON). IEEE, pp 415–420
10. Lin YL, Chen TY, Yu LC (2017) Using machine learning to assist crime prevention. In: 2017 6th IIAI international congress on advanced applied informatics (IIAI-AAI). IEEE, pp 1029–1030
11. Rummens A, Hardyns W, Pauwels L (2017) The use of predictive analysis in spatiotemporal crime forecasting: building and testing a model in an urban context. Appl Geogr 86:255–261
12. Wheeler AP, Steenbeek W (2020) Mapping the risk terrain for crime using machine learning. J Quant Criminol:1–36
13. Mohler G, Porter MD (2018) Rotational grid, PAI-maximizing crime forecasts. Stat Anal Data Min ASA Data Sci J 11(5):227–236
14. Esquivel N, Nicolis O, Peralta B, Mateu J (2020) Spatio-temporal prediction of Baltimore crime events using CLSTM neural networks. IEEE Access 8:209101–209112
15. Bogomolov A, Lepri B, Staiano J, Oliver N, Pianesi F, Pentland A (2014) Once upon a crime: towards crime prediction from demographics and mobile data. In: Proceedings of the 16th international conference on multimodal interaction, pp 427–434
16. Alves LG, Ribeiro HV, Rodrigues FA (2018) Crime prediction through urban metrics and statistical learning. Phys A 505:435–443
17. Ordoñez-Eraso HA, Pardo-Calvache CJ, Cobos-Lozada CA (2020) Detección de tendencias de homicidios en colombia usando machine learning. Revista Facultad de Ingeniería 29(54):e11740–e11740
18. Delahoz-Dominguez EJ, Fontalvo-Herrera TJ, Mendoza-Mendoza AA (2020) Definición de perfiles geográficos de hurto de automóviles. caso aplicado en cartagena. Justicia 25(37):99–108

19. Meneses-Escobar CA, Castillo-Rodríguez CM, Rodas-Vásquez A (2019) Análisis espacial y temporal del hurto de celulares, pereira, risaralda, año 2018. Revista Logos Ciencia Tecnología 11(2):167–175
20. Larkin JH, Simon HA (1987) Why a diagram is (sometimes) worth ten thousand words. Cogn Sci 11(1):65–100
21. Keim D, Kohlhammer J, Ellis G, Mansmann F (2010) Mastering the information age: solving problems with visual analytics
22. Hepler AB, Dawid AP, Leucari V (2007) Object-oriented graphical representations of complex patterns of evidence. Law Probab Risk
23. Taroni F, Biedermann A, Bozza S, Garbolino P, Aitken C (2014) Bayesian networks for probabilistic inference and decision analysis in forensic science. John Wiley & Sons
24. Tillers P (2005) Picturing factual inference in legal settings
25. Gordon TF (2007) Visualizing carneades argument graphs. Law Probab Risk
26. Verheij B (2007) Argumentation support software: boxes-and-arrows and beyond. Law Probab Risk
27. Lettieri N, Altamura A, Malandrino D (2017) The legal macroscope: experimenting with visual legal analytics. Inf Vis 16(4):332–345
28. Lettieri N, Guarino A, Malandrino D, Zaccagnino R (2020) The affordance of law sliding treemaps browsing hierarchically structured data on touch devices. In: 2020 24th International conference information visualisation (IV). IEEE, pp 16–21
29. André O, Peter F, Nellen S (2016) A visual approach to the history of swiss federal law. In: DHd 2016: modelling-networking-visualization
30. Lettieri N, Altamura A, Faggiano A, Malandrino D (2016) A computational approach for the experimental study of EU case law: analysis and implementation. Soc Netw Anal Min 6(1):56
31. Kuppevelt D, Dijck G (2017) Answering legal research questions about dutch case law with network analysis and visualization. In: Legal knowledge and information systems: JURIX 2017: the thirtieth annual conference, vol 302. IOS Press, p 95
32. du Toit N (2019) Network visualisation as a citator user interface. J Open Access L 7:1
33. Guarino A, Lettieri N, Malandrino D, Russo P, Zaccagnino R (2019) Visual analytics to make sense of large-scale administrative and normative data. In: 2019 23rd International conference information visualisation (IV). IEEE, pp 133–138
34. Cioffi-Revilla C (2014) Introduction to computational social science. Springer, London
35. Lettieri N, Altamura A, Giugno R, Guarino A, Malandrino D, Pulvirenti A, Vicidomini F, Zaccagnino R (2018) Ex machina: analytical platforms, law and the challenges of computational legal science. Future Internet 10(5):37
36. Lettieri N, Malandrino D, Vicidomini L (2017) By investigation, I mean computation. Trends Organ Crime 20(1–2):31–54
37. Lettieri N, Guarino A, Malandrino D (2018) E-science and the law three experimental platforms for legal analytics. In: Legal knowledge and information systems—JURIX 2018: The thirty-first annual conference, Groningen, The Netherlands, , pp 71–80, 12–14 Dec 2018
38. Lettieri N, Altamura A, Malandrino D, Punzo V (2017) Agents shaping networks shaping agents: integrating social network analysis and agent-based modeling in computational crime research. In: EPIA conference on artificial intelligence. Springer, pp 15–27
39. Koschützki D, Lehmann KA, Peeters L, Richter S, Tenfelde-Podehl D, Zlotowski O (2005) Centrality indices. In: Network analysis. Springer, pp 16–61
40. Jeh G, Widom J (2002) Simrank: a measure of structural-context similarity. In: Proceedings of the eighth ACM SIGKDD international conference on Knowledge discovery and data mining, pp 538–543
41. Floud J (1982) Dangerousness and criminal justice. Br J Criminol 22(3):213–228
42. Cleary JG, Trigg LE (1995) K*: an instance-based learner using an entropic distance measure. In: Machine learning proceedings 1995. Elsevier, pp 108–114
43. McHugh ML (2012) Interrater reliability: the kappa statistic. Biochemia Medica 22(3):276–282
44. Zaccagnino R, Capo C, Guarino A, Lettieri N, Malandrino D (2021) Techno-regulation and intelligent safeguards. In: Multimedia tools and applications, pp 1–22

45. Guarino A, Malandrino D, Zaccagnino R (2022) An automatic mechanism to provide privacy awareness and control over unwittingly dissemination of online private information. Comput Netw 202:108614
46. Guarino A, Malandrino D, Marzullo F, Torre A, Zaccagnino R (2022) Adaptive talent journey: optimization of talents growth path within a company via deep q-learning. Expert Syst Appl 209:118302
47. Virzi RA (1992) Refining the test phase of usability evaluation: how many subjects is enough? Hum Factors 34(4):457–468
48. Malandrino D, Guarino A, Lettieri N, Zaccagnino R (2019) On the visualization of logic: a diagrammatic language based on spatial, graphical and symbolic notations. In: 2019 23rd international conference information visualisation (IV). IEEE, pp 7–12
49. Chin JP, Diehl VA, Norman KL (1988) Development of an instrument measuring user satisfaction of the human-computer interface. In: Proceedings of the SIGCHI conference on Human factors in computing systems, pp 213–218

Illuminating Text: Unleashing Knowledge with Visual Discovery in Text Mining and Natural Language Processing

Designing and Evaluating Context-Sensitive Visualization Models for Deep Learning Text Classifiers

Andrew Dunn, Diana Inkpen, and Răzvan Andonie

1 Introduction

The language models of the day need to be able to process or generate text, as well as predict missing words, sentences or relations depending on the task. The latest Deep Learning (DL) models applied to text data are extremely complex, and can contain hundreds of millions of different trainable parameters. This high level of complexity makes it very difficult for humans to understand their inner workings. Visual analytics can be applied to text data as a means to bridge the complementary strengths of people and computers.

Visualization methods can give us an insight into these black-box models, and there is still a clear role for human analysis and decision-making. Visualization can be considered a preliminary step for interpreting or explaining a trained ML model.

The terms interpretability and explainability are often used interchangeably, even if it is generally accepted that there is a difference between them. Interpretability in an ML model means that the model behaves in a consistent and predictable manner, so that changes to model parameters result in expected changes in the model's output. Explainability refers to the process of providing *post hoc* explanations for the behavior of black box models [24]. In this work, we are primarily concerned with the explainability of DL models.

Currently, there is a huge interest in visualizing ML for NLP models. In contrast to the visualization methods applied to image processing ML tasks, visualization is less intuitive when it is applied to text. Saliency maps are a popular tool for gaining insight into DL models. In this case, saliency maps are typically depicted as heatmaps

A. Dunn · R. Andonie
Central Washington University, 400 E University Way, Ellensburg, WA 98926, USA
e-mail: andrew.dunn@cwu.edu

D. Inkpen
University of Ottawa, Ottawa, ON, Canada
e-mail: diana.inkpen@uOttawa.ca

R. Andonie (✉)
Transilvania University, Bulevardul Eroilor 29, Braşov 500036, Romania
e-mail: razvan.andonie@cwu.edu

B. Kovalerchuk et al. (eds.), *Artificial Intelligence and Visualization: Advancing Visual Knowledge Discovery*, Studies in Computational Intelligence 1126,
https://doi.org/10.1007/978-3-031-46549-9_14

of neural layers, where "hotness" corresponds to regions that have a big impact on the model's final decision. Grad-CAM is a state-of-the art technique for saliency map visualization [31] using the gradient information obtained from backpropagating the error signal from the loss function with respect to a specific feature map, at any layer of the network.

Current solutions to visualizing NLP models involve heatmaps similar to Grad-CAM, and visualizing connections between tokens in models that utilize attention mechanisms [3, 4]. These solutions typically look at the importance of individual tokens on the model output. Before being passed to an NLP model, the text must be broken up into a series of tokens that are then transformed into numerical vectors. Each token typically represents a single word or character in the input text. The goal of these NLP visualization methods is to highlight the most significant parts of the text that have the greatest impact on the model output. For instance, which tokens contribute most to the decision that a text is classified as "fake news" [3]. While effective, these methods look at each separate token independently, thus lacking context from the entire text. We believe the combination and order in which the words appear is significant and that they influence a modern NLP model's interpretation of the text.

Typically, for tasks such as machine translation, the output is in textual form, and an encoder-decoder mechanism is used. The weakness of this approach is that the encoder produces a finite-length vector, without regards to whether or not any of the inputs are more important (or less important). A solution to this problem is the *attention* mechanism. Several variants of this mechanism have been introduced: convolutional, intra-temporal, gated, and self-attention [26]. Self-attention has recently become very popular in the state-of-the art model called *Transformer*, and provides attention to words in the same sentence. Transformer models can extract complex features from the input data and effectively solve NLP problems.

Visualization is often the bridge that language model designers use to explain their work. Coloring of the salient words or sentences, and clustering of neuron activations can be used to understand the model. In case of Transformers, explaining by visualization the information processing flow is difficult because of the complexity of these models.

In May 2022, during the Dagstuhl Seminar *Visual Text Analytics*, interdisciplinary working groups of experts from visualization, NLP, and ML examined several topic areas to reflect on the state of the field, identify gaps in knowledge, and create an agenda for future cross-disciplinary research [7]. According to their conclusions, visualizing textual information is a particularly challenging area of information visualization research. The visualization techniques for text have specific constraints to consider, including the provision of context for text fragments of similar or different size and structure, depicting embeddings and high dimensional representations, and ensuring legibility of text incorporated into visualizations. In addition, there are difficulties in inferring the semantic meaning of ambiguous terms and determining the referencing between subsequent statements.

One of the seminar themes was about Toolkits and Standards ("What success stories regarding existing text visualization approaches and systems can we learn from?

What is needed?"). In the Tool Demos section of the report, several text visualization tools were presented [7]: Cartolabe, SPIKE, Storifier, LingVis.io, ALVA, t-viSNE, etc. The report mentions among the remaining challenges in visual text analytics "a proper and comprehensive analysis of these tools". It would be beneficial for users and developers of visual text analytics systems able to compare such tools.

A primary objective of visual explanations in Natural Language Processing (NLP) is to identify the text segments that exert the most significant influence on the model's output. In NLP classifiers, this is typically achieved by breaking down the text into tokens, such as words or character sequences, which are then converted into numerical vectors, often referred to as word embeddings [25]. More sophisticated representations may also consider syntactic and semantic structures. For instance, Yun et al. utilized dictionary learning techniques to offer detailed visualizations of Transformer representations, shedding light on the underlying semantic structures [43].

One common method for measuring the impact of individual elements of an input data on a model's decision-making process is called leave-one-out, and can be used for visualization. The leave-one-out (LOO) process is a feature erasure technique that systematically identifies and removes a single feature from the input data to measure its effect. In an NLP model, a feature is a vector representing a single token in the input text, typically a word or character, although character n-grams and subwords may also be selected as tokens. The modified data is then passed through the model and the difference in output strength from the full input is recorded. This process is repeated for every feature. The token influence is calculated by subtracting the output strength of the modified inputs from the output strength of the original unmodified input. The output strengths of the removed features are then ranked in order of their influence score, and all tokens that had a positive difference in output value when removed are considered as the set of most influential features on the model's output [11].

The LOO approach examines each input token individually, which inherently has its limitations. On the other hand, advanced language models, such as BERT [10], possess the capability to grasp intricate phrases and contextual relationships between words. In our prior research [11], we introduced a novel context-sensitive visualization method for NLP Transformers. This technique harnesses existing NLP tools to identify crucial token groups that profoundly impact the model's output. By visualizing these groups, explanations are generated while still preserving context from the original text. Termed as the leave-n-out (LNO) method, it computes the significance of token groups by iteratively performing classification without them and evaluating the resulting loss in accuracy.

In the following step, we aimed to evaluate our model from [11] by comparing it with contemporary state-of-the-art explanation/visualization methods presented in [12]. However, we encountered challenges as conventional classification accuracy measures proved insufficient for assessing the quality of visualizations. We required more sophisticated criteria to measure the usefulness of the extracted knowledge in explaining the classifier's decision-making process. Additionally, we sought to visualize differences between the knowledge extracted by various models to facilitate ranking. Our investigation revolved around evaluating explanations/visualizations generated by ML models for NLP, specifically text classification. Unfortunately, there is limited research on assessing the quality of visualizations in the NLP domain [9]. Our primary focus in [11] was not to improve the text classifier's accuracy, but rather to evaluate the visualizations that shed light on its decision-making process. As a result, we emphasized knowledge extraction visualizations rather than prioritizing prediction accuracy.

Our current work represents a substantial expansion of the concepts discussed in our previous paper [12]. Here, we present various methodologies for evaluating the quality of NLP visualizations, encompassing both automated techniques employing quantifiable measures and subjective techniques reliant on human judgments. Additionally, we examine the performance of our LNO method on a multi-classification task by testing it against a dataset containing more than two classes. In Sect. 3, we offer multiple case studies showcasing visualizations generated by both the LOO and LNO methods. Subsequently, in Sect. 4, we explore diverse techniques to assess the quality of NLP visualizations. Initially, we investigate similarity scores by calculating the Rank-biased Overlap [41] between the top-ranked words selected through various visualization techniques, aiming to identify any correlations between their word choices. Next, we delve into an automated evaluation technique for NLP visualizations, which involves assessing the classification of a minimal text generated by selecting the top-n words from the visualization and comparing it with the classification based on the original text. Lastly, we present the outcomes of a small-scale survey in which human participants are asked to choose the visualization they believe best supports the model's decision.

Our proposed method highlights the important/significant set of tokens in a text that are strongly associated with the predicted class, but not with the other classes. This helps a user understand why a class was predicted. Explanations like this are becoming of primary importance, according to the 2020 European Commission's White Paper on AI [8]. They are also crucial for the field of human-centered AI. Users in many application domains cannot use AI-based systems that are not explainable (for example, in the medical domain [2] or in the legal domain [20]).

The subsequent sections of this chapter are organized as follows: Sect. 2 provides an overview of related research in the field. In Sect. 3, we present a synopsis of our context-sensitive visualization method, as introduced in [11], along with the incorporation of a new model trained on emotion detection. Section 4 outlines the evaluation techniques we propose and presents the corresponding results. Lastly, Sect. 5 offers conclusions drawn from our findings, along with highlighting open problems for future exploration.

2 Related Work

The early introduction of visualizations for neural NLP models can be attributed to the work by Li [21]. In this paper, four strategies for visualizing neural NLP models were presented, drawing inspiration from analogous techniques employed in computer vision. However, it is important to note that NLP visualization poses greater complexity compared to image recognition models. Recently, specialized techniques have been devised to address this complexity, with Transformers-based deep networks emerging as the most prominent approach in NLP visualization [3].

The Transformer model integrates an encoder, decoder, and self-attention mechanism, enabling it to establish connections between inputs in sequential data, akin to recurrent neural networks like LSTM models [37]. This unique architecture empowers the Transformer to extract intricate features from the input data, which has been instrumental in its success in addressing NLP challenges.

In 2018, Google introduced BERT [10], a cutting-edge NLP model renowned for its numerous achievements in various common NLP tasks, including classification, sentence prediction, and named entity recognition. BERT, an acronym for Bidirectional Encoder Representations from Transformers, is constructed using a stack of Transformer encoders. With a vast network comprising 110 million trainable parameters (BERT-base) or 345 million parameters (BERT-large), training BERT from scratch is very time-consuming. The model's intricate structure, due to the sheer number of parameters, poses challenges when attempting to explain the behavior or individual components of BERT.

The widespread success of BERT has spurred researchers to develop several similar language models based on Transformers, each with distinct goals. For instance, RoBERTa [22], DistilBERT [29], and XLNet [42] have been introduced to either enhance model accuracy or reduce training time and complexity.

In 2019, OpenAI introduced GPT-2 [27], another large-scale language model that, like BERT, relies on Transformers and is designed in multiple sizes. GPT-2 Large, with 762 million trainable parameters, surpasses BERT-large in scale. While BERT utilizes a stack of Transformer encoders, GPT-2 is built on a stack of Transformer decoders. Additionally, GPT-2 employs masked self-attention, restricting attention weights to the current and preceding tokens in the sequence. In contrast, BERT uses unmasked self-attention, allowing attention weights to be generated between all tokens in the text, irrespective of their position.

In 2020, OpenAI released GPT-3 [5], an exceptionally large language model with 175 billion trainable parameters, standing as the largest non-sparse language model at that time. Among the latest versions from the GPT-3.5 series is ChatGPT, which leverages reinforcement learning to implement intent detection capabilities.[1]

Understanding the intricate information processing flow and outcomes within a Transformer model can be quite challenging due to its inherent complexity. One effective and current approach to address this complexity is visualization, which offers a certain level of explainability. An initial survey on visualization techniques tailored for Transformers can be found in [3, 4].

In certain instances, visualizing complex models, including those built with Transformers, can be achieved through the use of model-agnostic tools specifically designed for benchmarking or hyperparameter scoring. An example of such a tool

[1] https://openai.com/blog/chatgpt/.

is *Weights and Biases*,[2] which provides an extensive array of visualization and customization capabilities. However, more intriguing are the visualizations specifically developed for Transformers, serving either to explain the models themselves, like ExBERT [15], or to elucidate certain model-specific attributes, such as embeddings or attention maps [38]. Detailed information can be found in [3, 4], which also includes a list of recent articles focusing on explaining Transformer-related topics through visualizations.

A general goal would be to visualize the entire lifecycle of a Transformer model. However, none of the current visualization systems (considered in [3, 4]) has yet managed to examine all the facets of the Transformers. This is due to the fact that this area is relatively new and there is no consensus on what needs to be visualized. We can conclude that the visualization of Transformer models is still under development.

In addition to Transformer visualizations, there are several specific BERT visualization approaches [3, 4] focused on the role of embeddings and relational dependencies within the Transformer learning phase [28], the role of attention during pretraining or training [35, 38], and the importance of various linguistic phenomena encoded in its language model [6].

Han et al. [14] mapped saliency over NLP inputs to explain predictions via influential training examples using a BERT-based model. They found that in a lexicon-driven sentiment analysis task, saliency maps and influence functions are largely consistent with each other. However, these results were not consistent for the more complex task of Natural Language Inference, a classification problem that concerns the relationship between a premise sentence and a hypothesis sentence. An explanation could be that saliency maps focus on the input, and may neglect to explain how the model makes decisions. Some saliency methods are incapable of supporting tasks that require explanations that are faithful to the model or the data-generating process [1].

Recently, a team led by Yann LeCun from Facebook AI Research, UC Berkeley, and New York University conducted research involving dictionary learning techniques to produce in-depth visualizations of transformer representations, thus providing valuable insights into the semantic structures [43]. They have developed an interactive website,[3] enabling users to gain further understanding of transformer models through latent space visualization. Their evaluation experiments were based on a pretrained BERT model. The transformer factors were categorized into three levels: low level (word level), mid level (sentence level), and high level (long-range patterns). As far as we are aware, the closest existing work to our contribution is presented in [43], although it is essential to note that our research was conducted before their publication.

AllenNLP [13] is a free open-source Python library built on top of PyTorch and it provides a simplified interface for building and training deep learning NLP models. In addition to providing many popular models and frameworks, AllenNLP includes a few visualization techniques called AllenNLP Interpret [39]. Using this

[2] https://www.wandb.com/.

[3] https://transformervis.github.io/transformervis/.

library, we are able to directly compare visualization results between the various methods. Specifically, the three methods used in AllenNLP are saliency maps based on Guided Back-propagation [32], Integrated Gradients [36], and SmoothGrad [33]. These methods produce weight vectors that can easily be normalized and visualized as heatmaps, for a direct comparison with our Leave-one-out and Leave-n-out visualizations [11].

Integrated Gradients [36] is a visualization technique created by Sundararajan et al. that attributes a prediction of a deep network to its input features by calculating the path integral using all gradients on the straight line path through the model, from a baseline input to the given input.

Smooth Gradients [33] is a visualization technique created by Smilkov et al. that aims to improve the quality of saliency maps by removing noise. This is achieved by averaging the outputs of several noisy saliency maps.

While these visualization methods can be utilized with NLP models, the primary focus of the authors was on image-based networks. As a result, there are limited available techniques that specifically address visualizing text classification models. One such method is the approach proposed in [11], which entails a context-sensitive visualization technique designed to identify the most significant token groups influencing the model's output.

Explanations are important for humans, in order to accept and start using a model. Human-based evaluation is a necessity, [17] calling it domain knowledge constituency of the ML model. In terms of evaluating the quality of the visualizations, the only method used in related work is human-based evaluation [9].

LMdiff [34] is a visual tool to compare language models. It allows observing qualitative differences between language models of similar type and of the same tokenization method. It visually compares probability distributions of two models that differ, e.g., through fine-tuning, distillation, or simply training with different parameter sizes. LMdiff can assist in choosing interesting text instances by identifying the most important phrases from large corpora.

Domain knowledge-based automatic scoring proves to be more effective in NLP when it is not solely derived from training data. In the NLP domain, individuals possess the potential to be "domain" experts, facilitating human-based evaluation. While there exists work on automatic scoring, such as [23], it primarily evaluates extracted key phrases based on their contributions to the classifier's accuracy, without generating visualizations. Our approach in this work involves a similar automatic scoring technique, but we apply it to visualizations, relying on extracted syntactic dependencies. Additionally, we introduce a novel rank-based evaluation method.

3 Context-Sensitive Visualization

To make this work self-contained, we present here a summary of our results on context-sensitive visualization of text from [11], also adding new experimental results.

Based on LOO, the leave-n-out (LNO) process systematically removes adjacent and non-adjacent tuples of n tokens from the input text, producing several new texts with those tokens missing. Where LOO removes a single token at a time, LNO removes multiple tokens at each step. The resulting texts are then passed to the model and their classification output is compared with that of the full text, and the difference in the activation strength is recorded. The modified texts that produce the largest difference in the target classification output neuron are selected, and the combination of removed words is then considered to be the most influential on the model's output. In our experiments, we elected to use word pairs, and thus use tuples of size two. Tuples of more than two tokens could be also used and may provide different results. However, the computational overhead may increase in this case exponentially and the benefits are questionable.

Our approach looks at combinations of words and phrases to discover their effect on the model's output, producing a visualization that is more context-sensitive to the original text. Our method leverages existing NLP tools for dependency parsing combined with a model-agnostic LNO token pre-processing approach to identify and highlight contextual groups of tokens that have the largest perceived effect on the model's classification output. The model itself can be any binary or multi-label NLP neural network classification model as long as the output layer is a fully-connected layer with the same number of neurons as classes.

Since the LNO processing step is computationally expensive and the number of possible word combinations explodes exponentially, we use a sentence-level dependency parser from the spaCy NLP library[4] to highlight promising word groups. The dependency parser creates a tree of relationships between the words in the sentence. By only selecting groups of words that are connected via a branch in the dependency tree, we are able to greatly reduce the search space for the algorithm.

Using the previously described processes, the steps required to visualize a text using our method are as follows. First, an input text is passed through the classification model and the class prediction and output strength are recorded as the baseline. Next, each sentence in the text is passed through a dependency parser which produces a tree structure highlighting dependency relations between the various words in each sentence. Using these dependency trees, new texts are generated with word pairs removed. Word pairs are limited to those that are connected via a branch in the dependency tree. Each generated text has only one-word pair removed at a time, and words are unique to their position in the sentence. After all new texts are generated, we proceed to the LNO process. All generated texts are passed through the model and their class prediction and output strengths are recorded. Once all texts have been processed, the output strength of the predicted class neuron for each generated text is then compared to output strength of the original baseline text. The heatmap is then generated by first sorting all word pairs by their output strength difference to that of the original text. Word pairs that did not contribute to the class prediction are culled. If an individual word appears in more than one pair, the maximum output strength for that word out of all pairs is selected. Finally, the output strengths for all remaining

[4] https://spacy.io/.

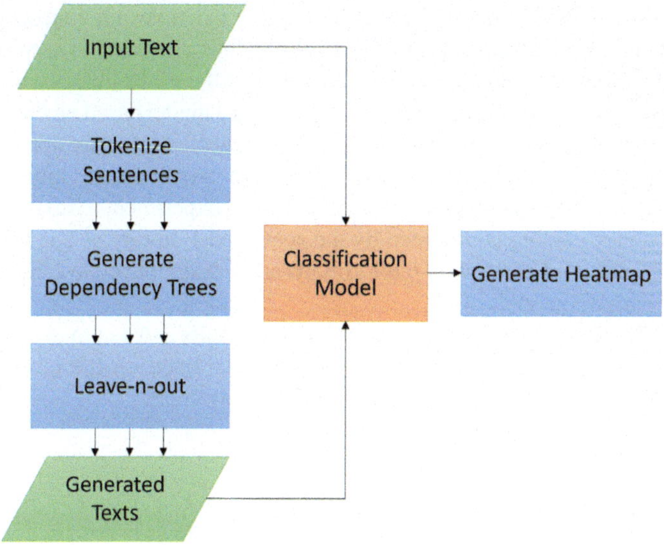

Fig. 1 LNO process flowchart

words are normalized using a linear scale. Figure 1 contains a simplified flowchart detailing the order of operations for our method.

For our experiments, we used three datasets. The first one was the IMDB Large Movie Review sentiment analysis dataset[5] from Stanford University, which contains 25,000 training and 25,000 testing reviews labelled with two categories: positive reviews and negative reviews. We used 20% of the training set as a separate validation dataset. As mentioned, the test set is separate from the training and validation sets.

Second, we used the Amazon Reviews sentiment analysis dataset from Xiang Zhang,[6] which was randomly sampled to contain 600,000 training and 260,000 testing reviews evenly split between two classes: one-star and five-star reviews, representing negative and positive reviews respectively. Again, 20% of the training set was used as a separate validation dataset.

Third, we tested a multi-classification model using the emotion dataset[7] from Elvis Saravia et al. [30], which contains 416,809 short texts that were taken from Twitter messages and have been labeled with one of six different emotions: anger, fear, joy, love, sadness, and surprise. This dataset was randomly sampled to contain 333,448 training and 83,361 testing texts. As with the other datasets, 20% of the training set was used as a separate validation dataset. This particular dataset was trained over the course of 20 epochs and achieved an overall testing accuracy of 53%. The lower accuracy of this model is expected due to the overlap between different emotions,

[5] https://ai.stanford.edu/amaas/data/sentiment/.

[6] https://www.kaggle.com/bittlingmayer/amazonreviews.

[7] https://huggingface.co/datasets/emotion.

making it difficult to distinguish between them. For example, a loving sentence may be classified as a joyful sentence, or a fearful sentence classified as surprising.

All code is written for Python 3.8 and utilizes the AllenNLP[8] library which is based on PyTorch.[9] All experiments were conducted using a model based on the pre-trained BERT-base model. A dropout layer and a fully connected layer are appended to the pre-trained BERT-base model to create our classification model. Only the weights in the fully connected layer were trained for these experiments. Text tokenization and dependency parsing is provided by the spaCy NLP library.[10] The visualization images are generated using the Matplotlib library.[11] Example code can be found on GitHub.[12]

In the following, we will present experimental results and heatmap visualizations for both the LOO and LNO methods for models trained on the aforementioned datasets. All the models are based on the same architecture.

3.1 IMDB Large Movie Reviews Dataset Model

Figures 2 and 3 show a short example text that represents a positively-classified movie review. In Fig. 2, the LOO process identified the word "best" as the most influential to the classification. We can see in Fig. 3 that more of the sentence context is highlighted, with the word pair "best movies" most prominently highlighted, showing that LNO picked out more of the context.

Figures 4 and 5 show a mixed positively-classified review. Again, Fig. 4 shows the resulting visualization from the LOO process. Note that the word "mindless" is not highlighted, but the word "but" is. Interestingly, Fig. 5 shows that heatmap result from LNO has reversed for these two words, with "mindless" highlighted with a stronger influence on the classification and "but" being not highlighted. Also note that the LOO result gives the word "with" a stronger influence than the words "cool effects", which is contrary to what we might expect. In contrast, the LNO heatmap gives more importance to the "cool effects", which is more in line with our expectations.

3.2 Amazon Reviews Dataset Model

Figure 6 contains a positive Amazon product review visualized using the LOO process, while Fig. 7 shows the same text visualized using the LNO process. Here we

[8] https://allenai.org/allennlp.

[9] https://pytorch.org/.

[10] https://spacy.io/.

[11] https://matplotlib.org/.

[12] https://github.com/dunn-cwu/context-sensitive-viz.

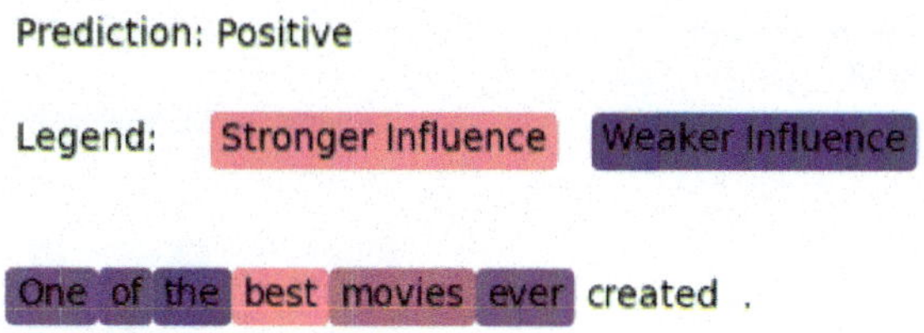

Fig. 2 Example LOO heatmap output—positive movie review

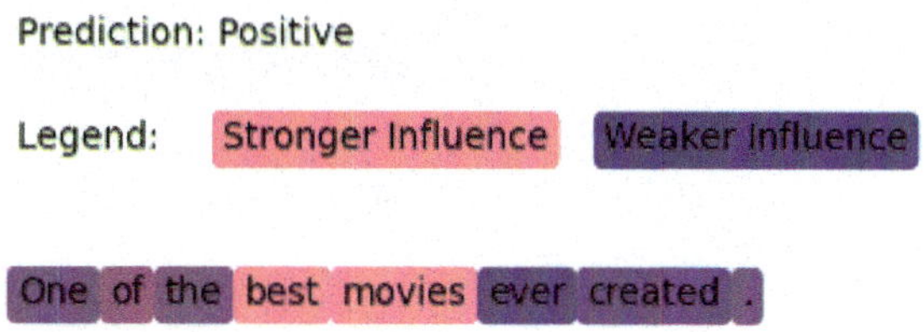

Fig. 3 Example LNO heatmap output—positive movie review

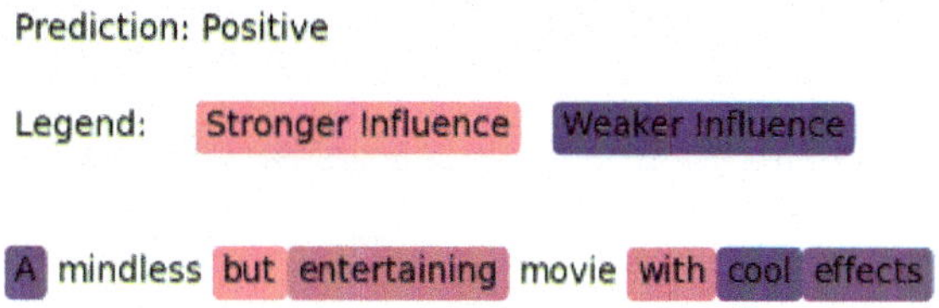

Fig. 4 Example LOO heatmap output—mixed positive movie review

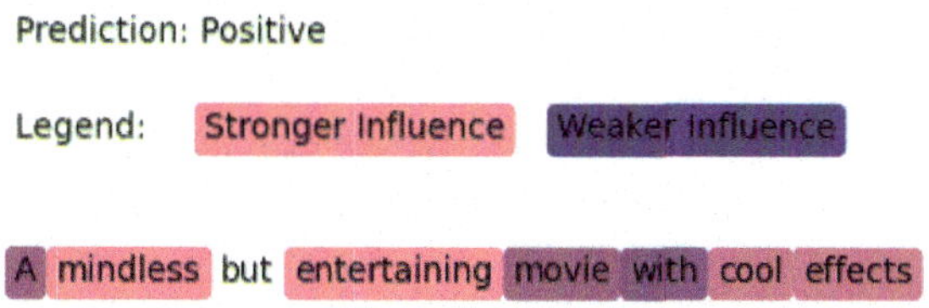

Fig. 5 Example LNO heatmap output—mixed positive movie review

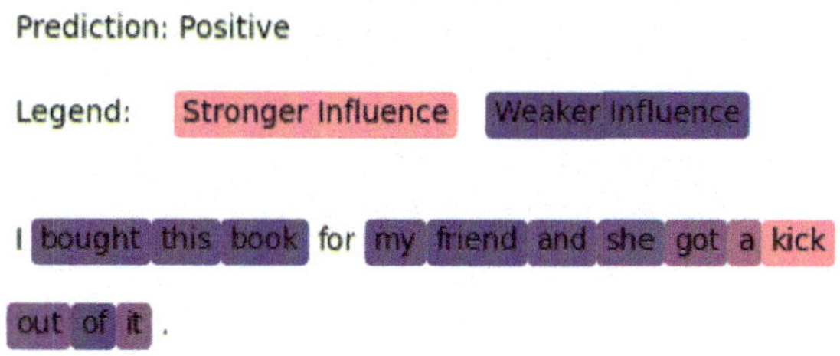

Fig. 6 Example LOO heatmap output—positive product review

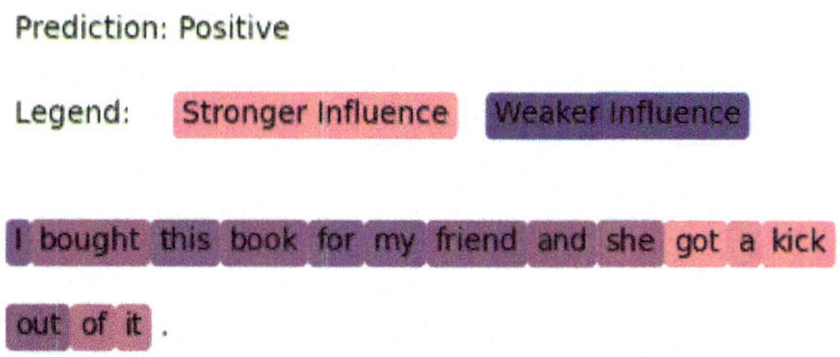

Fig. 7 Example LNO heatmap output—positive product review

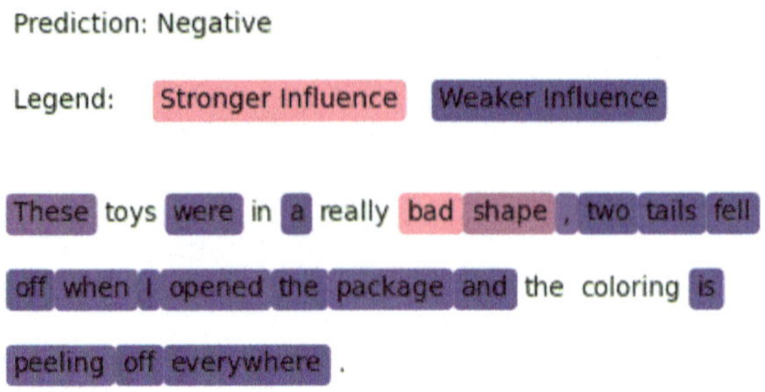

Fig. 8 Example LOO heatmap output—negative product review

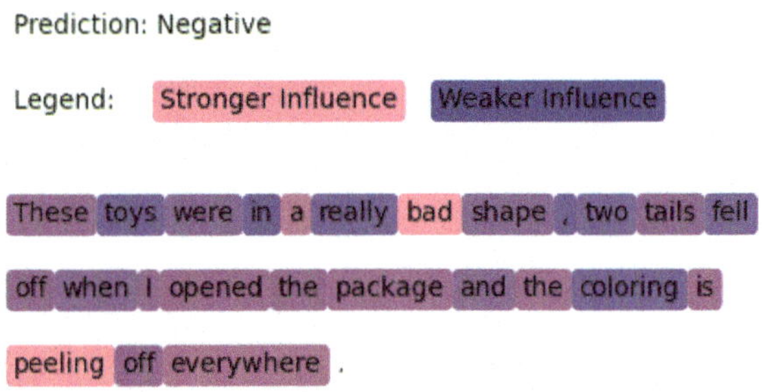

Fig. 9 Example LNO heatmap output—negative product review

observe that the phrase "got a kick" shows a stronger influence with the LNO process over LOO, highlighting more of the context around the word "kick".

Figures 8 and 9 display a negative product review. In Fig. 8, the LOO process visualization highlighted the words "bad shape" most strongly, while the LNO process highlighted more of the text. Notably, the LNO process selected the word "peeling" as strongly influencing the classification, which was shown as having a weaker influence in the LOO process. Finally, we observe that, in general, the LNO process utilizes more of the heatmap color range than the LOO process.

3.3 *Emotion Dataset Model*

To see how the LNO method may perform in a multi-classification task, we trained the same BERT based model on the emotions dataset[13] of which has six different classes. Figures 10 and 11 show a short text that was classified to contain a sad emotion. In Fig. 10, we see that the visualization for LOO primarily highlighted the words "left" and "distraught", while not considering most of the text as influencing the classification. Figure 11 in contrast once again shows that the LNO process is highlighting more of the text, and primarily focuses on the phrase "I was feeling a bit distraught".

[13] https://huggingface.co/datasets/emotion.

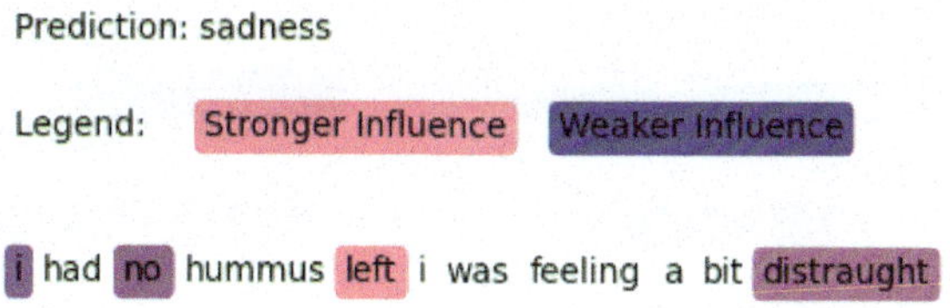

Fig. 10 Example LOO heatmap output—sadness emotion prediction

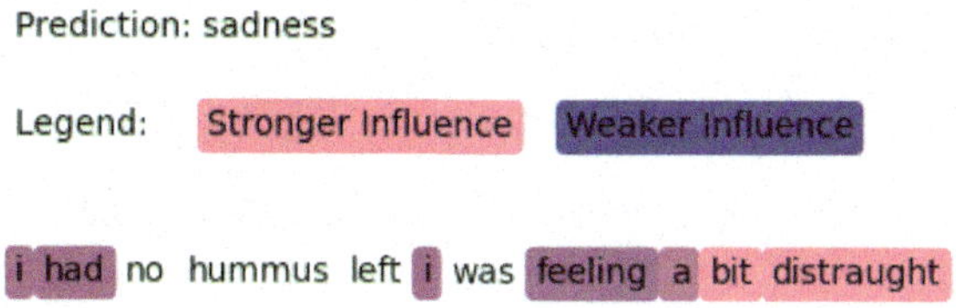

Fig. 11 Example LNO heatmap output—sadness emotion prediction

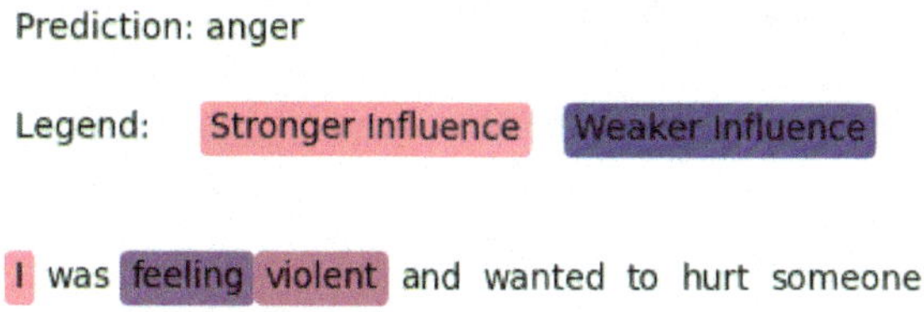

Fig. 12 Example LOO heatmap output—anger emotion prediction

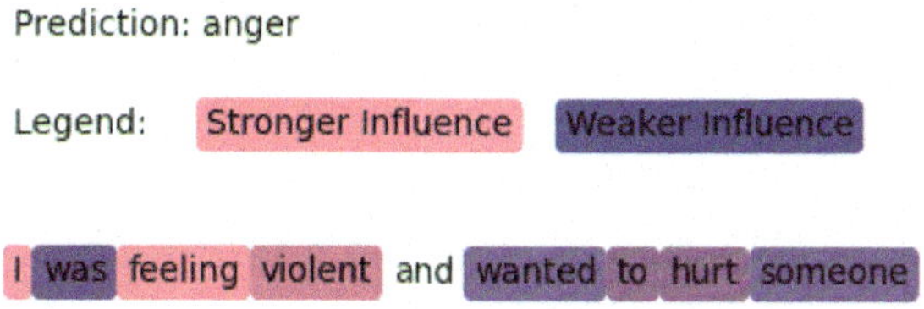

Fig. 13 Example LNO heatmap output—anger emotion prediction

In the next pair of examples found in Figs. 12 and 13, we see a short text that was classified as representing the anger emotion. As with the previous example pair, the LNO process has selected more of the text as contextually relevant to the classification. Looking at Fig. 12 we see that the second half of the sentence was not considered at all relevant to the classification, while LNO highlighted the entire phrase "wanted to hurt someone" as contributing in Fig. 13.

In Figs. 14 and 15 we see a longer text was classified as containing a joyful emotion. In Fig. 14, we see that LOO highlighted much of the text as relevent to the classification, and primarily focused on the word "awe". Looking at the results from LNO in Fig. 15, we see similar results to LOO but with more of the central phrase "tablecloths from my grandmother and feel envy" highlighted. LOO appears to have completely ignored the word "envy" while LNO considers it of high importance to the classification. Interestingly, we see an increasing gradient of influence for the central phrase in LNO, while in LOO it is more random.

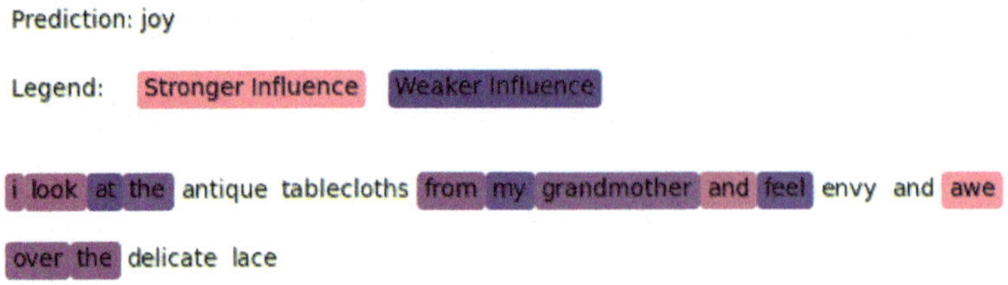

Fig. 14 Example LOO heatmap output—joy emotion prediction

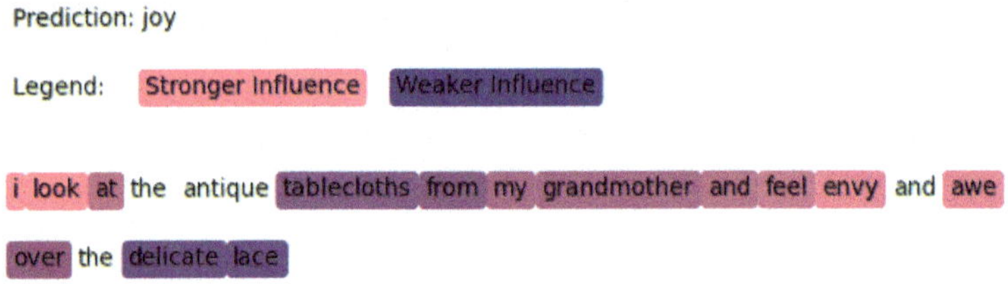

Fig. 15 Example LNO heatmap output—joy emotion prediction

Looking at these emotion dataset examples, we see that the LNO process can perform just as well on multi-classification tasks as binary classification tasks in visualizing the importance of tokens on the overall text classification. Despite the low overall testing accuracy of this model, the visualization results overall were of relatively high quality in our testing.

4 Evaluation Techniques and Results

In Sect. 3 we presented a context-sensitive visualization method for text classification. The problem we address now is how to assess the efficiency of the method. In other words, we want to develop techniques for evaluating the quality of text visual explanation performances. The ultimate goal is to be able to compare and rank visual explanation models.

4.1 Ranked-Biased Overlap

For assessing the similarity between different visualization techniques, we employed Ranked-biased Overlap (RBO) [41] as a measure. RBO is a recently developed statistical method that handles comparisons between disjointed ranked lists. Its key features include monotonicity and a weighting mechanism that prioritizes higher-ranking items, influencing the overall score accordingly. By utilizing RBO, we can

Table 1 Average RBO score (p-value = 0.6, 5,000 samples)

	LNO	LOO	Saliency	Integrated grad	Smooth grad
LNO	1.0	0.364	0.134	0.144	0.170
LOO		1.0	0.151	0.179	0.175
Sal			1.0	0.159	0.198
IntGr				1.0	0.138
SmGr					1.0

Table 2 Standard deviation RBO score (p-value = 0.6, 5,000 samples)

	LNO	LOO	Saliency	Integrated grad	Smooth grad
LNO	1.0	0.235	0.175	0.186	0.200
LOO		1.0	0.189	0.214	0.208
Sal			1.0	0.198	0.211
IntGr				1.0	0.181
SmGr					1.0

compute a similarity score between the top-ranking words for diverse visualization methods, even if they rarely share all the same words. In addition, RBO includes a hyperparameter called the p-value, which changes the weighting of the top-ranked items contribution to the overall score. When calculating the RBO ranking, a p-value of 0.6 was used, which weighs the top 5 tokens such that they contribute to 97.67% of the total score. This value was chosen because the sample texts used for this experiment were relatively short, ranging in length from two to three sentences, and were randomly sampled from the Amazon Reviews dataset.[14] We computed the RBO score for each pair of visualization techniques, specifically LNO, LOO, saliency maps, Integrated Gradients, and Smooth Gradients over 5,000 text samples.

Looking at Table 1, we can see that LNO and LOO shared the highest average RBO score at 0.364, but also the largest variance according to Table 2, with a standard deviation of 0.235. Although there does not seem to be a strong correlation for the top-ranking words between any two visualization techniques, the similarity between LNO and the gradient-based techniques is lower on average than for LOO.

4.2 Automatic Evaluation

In pursuit an automated evaluation technique to compare visualization methods, we employed the following approach: for various values of n ranging from 5 to 35, we

[14] https://www.kaggle.com/bittlingmayer/amazonreviews.

Table 3 Average minimum classification difference

Top n	LNO	LOO	Saliency	Integrated grad	Smooth grad
5	−0.232	−0.105	−0.274	−0.219	−0.244
10	−0.165	−0.027	−0.220	−0.162	−0.186
15	−0.132	0.013	−0.183	−0.131	−0.144
20	−0.107	0.038	−0.143	−0.096	−0.111
25	−0.088	0.045	−0.106	−0.066	−0.103
30	−0.059	0.042	−0.085	−0.053	−0.076
35	−0.040	0.037	−0.062	−0.050	−0.068

extracted the top-n words from the text using each technique. Subsequently, we constructed minimal texts by keeping only these top-n words and removing all others from the original text. We then fed these minimal texts into the classifier and calculated the difference in classification probability. A negative value signifies reduced confidence in the top-n word classification compared to the full-text classification, whereas a positive value indicates higher confidence. It is important to note that, during the construction of the top-n minimal text, we preserved the word order from the original text, irrespective of the individual word rankings. We obtained the results presented in Table 3 by conducting this evaluation on 5,000 randomly sampled texts from the Amazon Reviews dataset. All sampled texts were ensured to have a minimum length of 50 words.

In Table 3, we can see that as the top-n words increase, so does the classification probability of the minimal texts for each technique. In this experiment, LOO produced the best average classification accuracy and is the only technique that produced on average a better than baseline classification probability for top-n minimal texts where $n >= 15$. We suspect this is due to LOO producing input texts that are more similar to the full text than LNO due to removing fewer words, but we leave further investigation for future work. LNO produced minimal texts that gave classification probabilities that are in line with the gradient-based techniques, while performing better for each value of n than both the standard saliency maps based on Guided Backpropagation and SmoothGrad.

4.3 Evaluation with Human Judges

To ensure our method yields valuable visualizations for NLP models, we conducted two small-scale surveys with human judges to compare the perceived quality and usefulness of our visualizations. As interpretation tools primarily benefit humans, we consider a technique useful if it imparts knowledge on the model's effectiveness and offers insights into the data. For both surveys, three participants were presented with

Table 4 LNO versus LOO visualization survey

	LNO	LOO	Diff
Participant 1	73	27	+46
Participant 2	62	38	+24
Participant 3	59	41	+18

Table 5 LNO versus SmoothGrad visualization survey

	LNO	Smooth Grad	Diff
Participant 1	56	44	+12
Participant 2	65	35	+30
Participant 3	58	42	+16

automatically-generated heatmap visualizations, equally distributed between positive and negative reviews sourced from the Amazon Reviews dataset. Each example included the classification prediction indicating whether it was classified as positive or negative. The participants viewed the examples in a randomized order for the outputs of two methods and were asked to choose the visualization that best represented the predicted classification between the two methods. In case the participants encountered incomprehensible text or nonsensical information, they were permitted to skip the example, and the process continued until 100 examples were evaluated. It is crucial to note that the participants were unaware of the underlying methods employed to produce the visualizations, and they were not computer scientists, though they were native English speakers.

The results of both surveys are presented in Table 4 and Table 5, respectively. These surveys offer valuable insights into the effectiveness and user-perceived quality of our visualization technique for NLP models.

Table 4 contains the survey results comparing visualizations generated from our LNO method versus the standard LOO method. On average, each participant chose our method more frequently, with the common reasoning being that words relevant to the classification were highlighted more frequently and prominently than the LOO method. The feedback from the participants also shows that LNO tends to include more contextual information from the text over LOO.

Table 4 contains results from the second survey we conducted, to compare our method LNO to SmoothGrad [33], the technique based on Guided Backpropagation Saliency Maps that uses random noise to sharpen the output sensitivity maps. We chose the this method from the three included in AllenNLP because it seems to provide better heatmap visualizations in our tests. In addition, SmoothGrad was developed to be an improvement over other saliency map-based methods. Again, the participants' selections indicated that LNO produced better visualization of the text over the alternative method.

4.4 Execution Time

To demonstrate the advantages of employing the dependency parser for input reduction, we conducted benchmark experiments using our LNO method. Table 6 presents the benchmark results, comparing the usage of the dependency parser for input reduction against the exhaustive search method, which generates all word pairs for the given text. Due to the exponential time complexity associated with computing all word pairs, we restricted these experiments to shorter texts. The text samples were randomly selected from the Amazon Reviews dataset, comprising 5,000 positive and negative reviews evenly split, with lengths ranging from two to three sentences.

Reviewing the outcomes in Table 6, it becomes evident that utilizing the dependency parser significantly reduces the overall execution time required to process a text when compared to exhaustive search. Even for these short texts, using the dependency parser decreased the execution time by an average factor of 14.62 and reduced the number of generated word pairs by an average factor of 16.88. The Pair-Text Length Ratio data entries in Table 6 represent the ratio of the number of generated pairs to the number of words in the input text. With a Max Pair-Text Length Ratio of 0.977, we can observe that, for this dataset, the number of word pairs generated during the LNO process never exceeded the number of words in the text when using the dependency parser. In contrast, the exhaustive search generated up to 20.023 times more word pairs than there were words in the text, and this exponential time complexity would worsen as the text length increases. These results clearly highlight the efficiency of using the dependency parser for input reduction in our LNO method.

In addition to recording time execution metrics, the top 5 words were extracted from both methods for each sample and passed through the classifier again using the

Table 6 LNO dependency parser versus exhaustive search benchmarks

	Dependency parser	Exhaustive search
# Samples	5000	5000
Minimum time (sec)	0.076	0.200
Maximum time (sec)	0.628	10.482
Average time (sec)	0.289	1.853
Std. dev.	0.076	1.111
Min # pairs	7	18
Max # pairs	51	861
Avg. # pairs	26.885	182.497
Min pair-text length ratio	0.741	1.158
Max pair-text length ratio	0.977	20.023
Avg pair-text length ratio	0.912	6.097
Std. dev.	0.0326	3.011
Avg. top 5 min. class. diff.	−0.102	−0.112

Table 7 LNO, LOO, and AllenNLP benchmarks (5,000 samples)

	Min tme (sec)	Max time (sec)	Avg. time (sec)	Std. dev.
LNO	0.181	7.076	1.365	0.511
LOO	0.230	7.294	1.528	0.551
Saliency	0.244	0.606	0.343	0.033
Integrated grad	2.016	4.153	2.567	0.186
Smoothgrad	2.027	4.265	2.576	0.185

same word order as the original text. On average, the top 5 words had an average classification probability difference of -0.102 for the dependency parser, and a -0.112 difference when using exhaustive search. This result implies that not only does usage of the dependency parser improve the time complexity, but also improves the classification performance of the minimal text over exhaustive search. This may be due to the dependency parser filtering out word combinations that are less relevant to the sentiment analysis task.

Table 7 contains benchmark results comparing execution times for generating the heatmap data from texts that range between 1 and 4 sentences in length. All five visualization techniques were tested on the same set of 5,000 randomly sampled texts from the Amazon Reviews dataset. Looking at the results, our method LNO performed marginally better than LOO in execution time, and required less time on average to process the texts. Given that LOO is of linear time complexity, these results suggest a similar time complexity for LNO in this limited experiment. The standard saliency maps using guided back-propagation performed the best by a wide margin, requiring on average just 0.343 s to process a text, an improvement factor of 3.98 times over LNO in second place. While both Integrated Gradients and SmoothGrad have a higher initial overhead, they have a lower time complexity than LNO and LOO in relation to the text length. Therefore, we would expect the average execution time for Integrated Gradients and SmoothGrad to become less than LNO and LOO with longer input texts. However, for shorter texts, LNO performs better on average than the other methods tested in this experiment.

5 Conclusion and Future Work

We presented our proposed visualization method leave-n-out which can be leveraged to produce context-sensitive visualizations, as well as some case studies showing the types of heat map visualizations that can be produced. We also directly compared the LOO and LNO methods, and highlighted some perceived differences in the output visualizations We investigated how to evaluate explanations/visualizations resulting from machine learning models for text classification. This is an area of research that very few available results. We presented several methods for evaluating the quality of NLP visualizations, including both automated techniques based on quantifiable measures and subjective techniques based on human judgements. Our goal was not to improve the accuracy of a particular text classifier but to assess the quality of the visualizations that explain its decisions.

Based on our experimental findings, the LNO method demonstrates comparable performance to the other tested techniques in terms of visualization quality and execution time. However, a significant open problem remains, namely, the automated evaluation of visualizations aligning with human perceptions regarding their usefulness. This challenge proves to be highly complex to address. Recognizing the importance of visualizations being intended for human interpretation, it becomes evident that a model's superior performance in certain statistical measures does not automatically equate to generating more useful visualizations for humans. This issue underscores the need for further research in developing evaluation techniques that truly reflect the value of visualizations from a human perspective.

Going beyond current visualizations that are model-agnostic, future frameworks will have to provide visualization components that focus on the important Transformer components like corpora, embeddings, attention heads, or additional neural network layers that might be problem-specific [18]. Visualization of context-sensitive syntactic information and semantic structures is one of the hottest applications of ML in NLP.

We make a distinction between attention and explanation. These terms are frequently used in ML visualization, for instance, when using saliency maps in deep learning visualization. Such saliency maps generally visualize our attention but do not "explain" the deep learning model [16]. A challenging and open problem is to use visualization as an explanation tool for ML models. In other words, visualizing the combination of words according to which the text is classified as "fake news" highlights our attention or even may explain the classification decision.

Combining advances in artificial intelligence, ML, DL, and visualization to explain models to humans, especially for live-critical applications like health care, is an emerging field. Model explanation is fundamentally human activity, not only an algorithmic one [18]. This synthetic research field recently emerged as Visual Knowledge Discovery [19]. In the future, we hope for more robust and capable tools for generating explanations of ML models, as the field and complexity of models are advancing at a rapid pace.

References

1. Adebayo J, Gilmer J, Muelly M, Goodfellow I, Hardt M, Kim B (2018) Sanity checks for saliency maps. In: Advances in neural information processing systems, pp 9505–9515
2. Amann J, Blasimme A, Vayena E, Frey D, Madai V (2020) Explainability for artificial intelligence in healthcare: a multidisciplinary perspective. BMC Med Inf Dec Making 20. https://doi.org/10.1186/s12911-020-01332-6

3. Braşoveanu AMP, Andonie R (2020) Visualizing transformers for NLP: a brief survey. In: 2020 24th international conference information visualisation (IV), pp 270–279. https://doi.org/10.1109/IV51561.2020.00051
4. Braşoveanu AMP, Andonie R (2022) Visualizing and explaining language models. In: Kovalerchuk B, Nazemi K, Andonie R, Datia N, Banissi E (eds) Integrating artificial intelligence and visualization for visual knowledge discovery. Springer International Publishing, Cham, pp 213–237. https://doi.org/10.1007/978-3-030-93119-3_8
5. Brown TB, Mann B, Ryder N, Subbiah M, Kaplan J, Dhariwal P, Neelakantan A, Shyam P, Sastry G, Askell A, Agarwal S, Herbert-Voss A, Krueger G, Henighan T, Child R, Ramesh A, Ziegler DM, Wu J, Winter C, Hesse C, Chen M, Sigler E, Litwin M, Gray S, Chess B, Clark J, Berner C, McCandlish S, Radford A, Sutskever I, Amodei D (2020) Language models are few-shot learners. https://github.com/openai/gpt-3
6. Clark K, Khandelwal U, Levy O, Manning CD (2019) What does BERT look at? an analysis of BERT's attention. CoRR abs/1906.04341. http://arxiv.org/abs/1906.04341
7. Collins C, Fokkens A, Kerren A, Weaver C, Chatzimparmpas A (2022) Visual Text Analytics (Dagstuhl Seminar 22191). Dagstuhl Reports 12(5):37–91. 10.4230/DagRep.12.5.37. drops.dagstuhl.de/opus/volltexte/2022/17443
8. Comission E (2023) White paper on artificial intelligence: a european approach to excellence and trust. https://commission.europa.eu/publications/white-paper-artificial-intelligence-european-approach-excellence-and-trust_en
9. Danilevsky M, Qian K, Aharonov R, Katsis Y, Kawas B, Sen P (2020) A survey of the state of explainable AI for natural language processing. In: Proceedings of the 1st conference of the Asia-Pacific chapter of the association for computational linguistics and the 10th international joint conference on natural language processing. Association for Computational Linguistics, Suzhou, China, pp 447–459. https://aclanthology.org/2020.aacl-main.46
10. Devlin J, Chang M, Lee K, Toutanova K (2019) BERT: pre-training of deep bidirectional transformers for language understanding. In: Burstein J, Doran C, Solorio T (eds) Proceedings of the 2019 conference of the North American chapter of the association for computational linguistics: human anguage technologies, NAACL-HLT 2019, Minneapolis, MN, USA, June 2-7, 2019, volume 1 (Long and Short Papers). Association for Computational Linguistics, pp 4171–4186. https://doi.org/10.18653/v1/n19-1423
11. Dunn A, Inkpen D, Andonie R (2021) Context-sensitive visualization of deep learning natural language processing models. In: 2021 25th international conference information visualisation (IV), pp 170–175. https://doi.org/10.1109/IV53921.2021.00035
12. Dunn A, Inkpen D, Andonie R (2022) Evaluation of deep learning context-sensitive visualization models. In: 2022 26th international conference information visualisation (IV), pp 359–365. https://doi.org/10.1109/IV56949.2022.00066
13. Gardner M, Grus J, Neumann M, Tafjord O, Dasigi P, Liu NF, Peters M, Schmitz M, Zettlemoyer LS (2017) Allennlp: a deep semantic natural language processing platform
14. Han X, Wallace BC, Tsvetkov Y (2020) Explaining black box predictions and unveiling data artifacts through influence functions
15. Hoover B, Strobelt H, Gehrmann S (2019) exBERT: a visual analysis tool to explore learned representations in transformers models. CoRR abs/1910.05276. http://arxiv.org/abs/1910.05276
16. Jain S, Wallace BC (2019) Attention is not explanation. CoRR abs/1902.10186. http://arxiv.org/abs/1902.10186
17. Kovalerchuk B, Ahmad MA, Teredesai A (2021) Survey of explainable machine learning with visual and granular methods beyond quasi-explanations. Interpretable artificial intelligence: A perspective of granular computing pp 217–267
18. Kovalerchuk B, Andonie R, Datia N, Nazemi K, Banissi E (2022) Visual knowledge discovery with artificial intelligence: challenges and future directions. In: Kovalerchuk B, Nazemi K, Andonie R, Datia N, Banissi E (eds) Integrating artificial intelligence and visualization for visual knowledge discovery. Springer International Publishing, Cham, pp 1–27. https://doi.org/10.1007/978-3-030-93119-3_1

19. Kovalerchuk B, Nazemi K, Andonie R, Datia N, Banissi E (2022) Integrating artificial intelligence and visualization for visual knowledge discovery. Springer
20. Lettieri N, Guarino A, Malandrino D, Zaccagnini R (2020) Knowledge mining and social dangerousness assessment in criminal justice: metaheuristic integration of machine learning and graph-based inference. Artif Intell Law. https://doi.org/10.1007/s10506-022-09334-7
21. Li J, Chen X, Hovy E, Jurafsky D (2015) Visualizing and understanding neural models in NLP. arXiv preprint arXiv:1506.01066
22. Liu Y, Ott M, Goyal N, Du J, Joshi M, Chen D, Levy O, Lewis M, Zettlemoyer L, Stoyanov V (2019) RoBERTa: a robustly optimized BERT pretraining approach. CoRR abs/1907.11692. http://arxiv.org/abs/1907.11692
23. Lucaci D, Inkpen D (2021) Towards unifying the explainability evaluation methods for NLP. In: Wang L, Feng Y, Hong Y, He R (eds) Natural language processing and Chinese computing—10th CCF international conference, NLPCC 2021, Qingdao, China, October 13-17, 2021, Proceedings, Part II, Lecture Notes in Computer Science, vol 13029. Springer, pp 303–314. https://doi.org/10.1007/978-3-030-88483-3_23
24. Marcinkevics R, Vogt JE (2020) Interpretability and explainability: a machine learning zoo mini-tour. ArXiv abs/2012.01805
25. Mikolov T, Yih Wt, Zweig G (2013) Linguistic regularities in continuous space word representations. In: Proceedings of the 2013 conference of the North American chapter of the association for computational linguistics: human language technologies. Association for Computational Linguistics, Atlanta, Georgia, pp 746–751. https://aclanthology.org/N13-1090
26. Otter DW, Medina JR, Kalita JK (2021) A survey of the usages of deep learning for natural language processing. IEEE Trans Neural Netwo Learning Syst 32(2):604–624. https://doi.org/10.1109/TNNLS.2020.2979670
27. Radford A, Wu J. Child R, Luan D, Amodei D, Sutskever I (2019) Language models are unsupervised multitask learners. https://github.com/openai/gpt-2
28. Reif E, Yuan A, Wattenberg M, Viégas FB, Coenen A, Pearce A, Kim B, Visualizing and measuring the geometry of BERT. In: Wallach et al. [40], pp 8592–8600. http://papers.nips.cc/paper/9065-visualizing-and-measuring-the-geometry-of-bert
29. Sanh V, Debut L, Chaumond J, Wolf T (2019) DistilBERT, a distilled version of BERT: smaller, faster, cheaper and lighter. CoRR abs/1910.01108. http://arxiv.org/abs/1910.01108
30. Saravia E, Liu HCT, Huang YH, Wu J, Chen YS (2018) CARER: contextualized affect representations for emotion recognition. In: Proceedings of the 2018 conference on empirical methods in natural language processing. Association for Computational Linguistics, Brussels, Belgium, pp 3687–3697. https://doi.org/10.18653/v1/D18-1404. https://www.aclweb.org/anthology/D18-1404
31. Selvaraju RR, Cogswell M, Das A, Vedantam R, Parikh D, Batra D (2019) Grad-CAM: visual explanations from deep networks via gradient-based localization. Int J Comput Vis 128(2):336–359. https://doi.org/10.1007/s11263-019-01228-7
32. Simonyan K, Vedaldi A, Zisserman A (2014) Deep inside convolutional networks: Visualising image classification models and saliency maps. In: Bengio Y, LeCun Y (eds) 2nd international conference on learning representations, ICLR 2014, Banff, AB, Canada, April 14-16, 2014, Workshop Track Proceedings. http://arxiv.org/abs/1312.6034
33. Smilkov D, Thorat N, Kim B, Viégas FB, Wattenberg M (2017) SmoothGrad: removing noise by adding noise. CoRR abs/1706.03825. http://arxiv.org/abs/1706.03825
34. Strobelt H, Hoover B, Satyanaryan A, Gehrmann S (2021) LMdiff: a visual diff tool to compare language models. In: Proceedings of the 2021 conference on empirical methods in natural language processing: system demonstrations, pp 96–105. Association for Computational Linguistics, Online and Punta Cana, Dominican Republic. https://doi.org/10.18653/v1/2021.emnlp-demo.12. https://aclanthology.org/2021.emnlp-demo.12
35. Su W, Zhu X, Cao Y, Li B, Lu L, Wei F, Dai J (2020) VL-BERT: pre-training of generic visual-linguistic representations. In: 8th international conference on learning representations, ICLR 2020, Addis Ababa, Ethiopia, April 26-30, 2020. OpenReview.net. https://openreview.net/forum?id=SygXPaEYvH

36. Sundararajan M, Taly A, Yan Q (2017) Axiomatic attribution for deep networks
37. Vaswani A, Shazeer N, Parmar N, Uszkoreit J, Jones L, Gomez AN, Kaiser L, Polosukhin I (2017) Attention is all you need. In: Guyon I, von Luxburg U, Bengio S, Wallach HM, Fergus R, Vishwanathan SVN, Garnett R (eds) Advances in neural information processing systems 30: annual conference on neural information processing systems 2017, 4-9 December 2017, Long Beach, CA, USA, pp 5998–6008. http://papers.nips.cc/paper/7181-attention-is-all-you-need
38. Vig J (2019) A multiscale visualization of attention in the transformer model. In: Costa-jussà MR, Alfonseca E (eds) Proceedings of the 57th conference of the association for computational linguistics, ACL 2019, Florence, Italy, July 28–August 2, 2019, Volume 3: System Demonstrations, pp 37–42. Association for Computational Linguistics. https://doi.org/10.18653/v1/p19-3007. https://doi.org/10.18653/v1/p19-3007
39. Wallace E, Tuyls J, Wang J, Subramanian S, Gardner M, Singh S (2019) Allennlp interpret: a framework for explaining predictions of NLP models. CoRR abs/1909.09251
40. Wallach HM, Larochelle H, Beygelzimer A, d'Alché-Buc F, Fox EB, Garnett R (eds) (2019) Advances in neural information processing systems 32: annual conference on neural information processing systems 2019, NeurIPS 2019, 8-14 December 2019, Vancouver, BC, Canada. http://papers.nips.cc/book/advances-in-neural-information-processing-systems-32-2019
41. Webber W, Moffat A, Zobel J (2010) A similarity measure for indefinite rankings. ACM Trans Inf Syst 28(4). https://doi.org/10.1145/1852102.1852106
42. Yang Z, Dai Z, Yang Y, Carbonell JG, Salakhutdinov R, Le QV, Xlnet: Generalized autoregressive pretraining for language understanding. In: Wallach et al. [40], pp 5754–5764. http://papers.nips.cc/paper/8812-xlnet-generalized-autoregressive-pretraining-for-language-understanding
43. Yun Z, Chen Y, Olshausen BA, LeCun Y (2021) Transformer visualization via dictionary learning: contextualized embedding as a linear superposition of transformer factors. arXiv preprint arXiv:2103.15949

How NLP and Visual Analytics Can Improve Asset Management

Pedro Santos, Matilde P. M. Pato, Nuno Datia, and José Sobral

1 Introduction

Asset management is a complex process involving the coordination of several disciplines to ensure physical assets are maintained, upgraded, and operated correctly. In an attempt to deal with such complexity, Vanier [39] describes asset management as the successful implementation of data collection related to six questions: (i) What do you own? (ii) What is it worth? (iii) What is deferred maintenance? (iv) What is its condition? (v) What is the remaining service life? (vi) What do you fix first? The data source needed to answer most of the above questions are documents called Work Order (WO). WOs provide all the relevant information regarding a situation with an asset and outline the steps involved in completing them. Creating and managing WOs enables rigorous control of assets during their life cycle. Analysing WOs is crucial to any business. It helps to ensure that all tasks are completed efficiently and on time while also providing valuable insights into the current state of operations. By reviewing WOs, businesses can identify fields needing improvement or potential opportunities to increase productivity. Furthermore, analysing WOs can

P. Santos (✉)
ISEL - Lisbon School of Engineering, Politécnico de Lisboa, Lisbon, Portugal
e-mail: pedro.santos@deetc.isel.ipl.pt

M. P. M. Pato
NOVA LINCS & LAGIGE & Future Internet Technologies, ISEL - Lisbon School of Engineering, Politaécnico de Lisboa, Lisbon, Portugal
e-mail: matilde.pato@isel.pt

N. Datia
NOVA LINCS & Future Internet Technologies, ISEL - Lisbon School of Engineering, Politécnico de Lisboa, Lisbon, Portugal
e-mail: datia@isel.ipl.pt

J. Sobral
ISEL - Lisbon School of Engineering, Politécnico de Lisboa, Portugal & Center for Marine Technology and Ocean Engineering (CENTEC), Instituto Superior Técnico, Universidade de Lisboa, Lisbon, Portugal
e-mail: jose.sobral@isel.pt

B. Kovalerchuk et al. (eds.), *Artificial Intelligence and Visualization: Advancing Visual Knowledge Discovery*, Studies in Computational Intelligence 1126,
https://doi.org/10.1007/978-3-031-46549-9_15

help organisations better understand customer needs and preferences to provide a higher level of service. Although many fields in a WO may have a restricted set of values, such as identifying an asset, the fields relating to the problem and the steps taken to resolve it are free text. Free text is a powerful tool for communication, but it can also create issues that need to be addressed. For instance, the text used to describe the problem/solution is left to the technician's discretion. This is open to spelling errors, non-existent words, or inconsistent terminology usage. For example, we can instinctively use the terms "valve replacement" or "valve substitution" or "valve changing". Strategies must be developed to overcome text quality and consistency issues, especially when such information is needed to run a decision-making process. Addressing these inconsistencies or variations can be challenging [9], but several approaches and techniques can be used, including (i) Spell checking and correction, (ii) Domain-specific dictionaries and terminologies, (iii) Named Entity Recognition (NER), (iv) Curation by domain experts, (v) Natural Language Processing (NLP) processing tools. By combining these techniques and adapting them to the specific domain and context, textual inconsistencies can be greatly reduced [9]. It is, however, necessary to know the existing corpus in the WO text to perform a more detailed problem analysis and decide what the best solution is towards the usage of a normalised set of terms. This analysis is done partially using NLP.

One of the most popular techniques used in NLP [31] is Word Embeddings (WE), which involves mapping a set of words to vector representations to capture their semantic meanings and linguistic relationships between words using deep neural networks [7, 25]. WE assigns numerical values to each word based on its context within a given sentence or document. The vectors created are then compared against other vectors representing similar concepts or topics within the corpus being analysed; this comparison allows for a more accurate representation of relationships between words than traditional methods like bag-of-word models [44] alone could provide. Moreover, since these vectorised representations can be mathematically manipulated, they can also be used for clustering and dimensionality reduction tasks. These tasks further enhance the accuracy and robustness of NLP systems built with them. The WEs into Machine Learning (ML) models can enable raw text data contextualisation using ML techniques. Several studies have applied WEs to common NLP tasks, such as information extraction [42], information retrieval [10], machine translation [29], question answering [12, 30], sentiment analysis [37, 43], text classification [35] and summarisation [32].

The asset management process can be improved by ML according to several studies [13]. However, to our knowledge, no one has found a way to implement an algorithm for generating specialised word suggestions based on WOs data. The motivation of this work is the filling of this research gap. This chapter explores the potential of ML to improve descriptions of WOs in a commonly used language. It uses a visual analytics approach [17] to support an interactive exploratory visualisation of words using their relationships within a vector space, with the intent of discovering latent knowledge present in the WE that allow us to understand better how the words are used in this specific domain of asset management. In this case, data visualisation is used for knowledge discovery in formulating hypotheses about the relationships between words and in the observation of underlying semantic properties since the exploratory analysis of the data helps contextualise problems and understand the intricacies of the domain. The dataset is created from real WO data from asset management operations in a Portuguese health facility.

This work extends and updates our previous work [33] and presents the application of modern NLP techniques to process the WOs description while presenting a comparison between two WE models—Word2Vec [24] and Fasttext [4], through semantic similarity tests between the encoded words, as well as a visualisation of the vector space through dimensionality reduction of the encoded vectors.

The main contributions of this chapter are as follows:

1. The application of the studied NLP techniques to the asset management domain;
2. The discovery of a visual pattern, formed by the words in the WOs of a specific domain—asset management in a Portuguese health facility.

The chapter is organised as follows. After the introduction, Sect. 2 presents the research background before this research, Sect. 3 describes the data used, Sect. 4 presents the modelling of the problem and the steps taken towards the objective. The last section concludes the work and points out future work directions.

2 Research Background

Word embeddings have been successfully used in a variety of NLP applications, so the concept is not new. In recent years, several WE models have been developed, such as Word2Vec [24], Fasttext [4] and Global Vectors for Word Representation (GloVe) [27], among others [19]. We have compiled a list of relevant references, divided into three main topics, concerning our work: (1) asset management, (2) visualisation of the models, and (3) the performance of the models.

2.1 NLP With Potential to Revolutionise Asset Management

A few studies use NLP techniques to improve asset management. Bouabdallaoui et al. [5] use natural language processing to classify maintenance requests. Due to WE, the problem raised in the cited paper is quite similar to the one presented in our study. The authors used pre-trained WE from Fasttext, which enabled transfer learning. Transfer learning involves fine-tuning the words in a WE model for a particular scenario to make it more appropriate for the problem domain. Convolutional Neural Networks (CNN), CNN with filters, Long Short-Term Memory (LSTM), and another form of ML model were used to generate four separate models [1, 8]. Accuracy was chosen because the dataset was not imbalanced, and the CNN with filters performed better than any other mode.

The article by Stenstrom et al. [36] describes how NLP techniques can be applied to WO descriptions and discusses the benefits of doing so. The authors discuss their case study, the Swedish railway infrastructure, and address problems related to data quality in WOs. These problems were solved using NLP methods. The subset of words that remained from the original 11000 word list after applying an NLP pipeline was 8442. From this group of words, it was possible to perform analyses that helped the decision-making process. The paper's conclusion confirms that NLP techniques improve error detection and thus increase the overall data quality.

Brundage et al. [6] discuss how current NLP tools are inefficient when applied to industrial use cases, meaning that "out of the box" NLP is not suitable for more technical domains. The current tools and implementations are insufficient to deal with the complexity of more specialised fields, so the authors propose a new paradigm they call Technical Language Processing (TLP)—emphasising that industrial asset management companies should redouble their efforts to produce domain-specific resources. TLP is a methodology that refines the NLP paradigm by obtaining agreement over linguistic resources (e.g., dictionaries, datasets) pertaining to technical domains across all industry applications.

2.2 *Visualisation of Word Embeddings Models*

A vector representation of words is generated using neural language models, and then the resulting vector spaces can be used for visual analysis. This has become a powerful tool for understanding the underlying semantic patterns in words and how they are used in the domain. Visual analysis of high-dimensional vector spaces allows insight into the relationships between words.

Liu et al. [20] argue that researchers often use t-distributed stochastic neighbour embeddings (t-SNE) [38] and Principal Component Analysis (PCA) techniques [14] to create lower dimensional embeddings (perform dimensionality reduction) that allow us to understand the overall structure and explore linear relationships. However, dimensionality reduction techniques can be misleading if not applied correctly. To explore WE spaces, a unique visualisation tool is introduced. Furthermore, linear projection techniques are developed to highlight analogy relationships, and a novel hypothesis-testing approach is designed for emerging structures.

Katricheva et al. [15] present a Python solution to visualise vector representations (for example, WE) as dynamic and interactive graphs. Vertices of the graph correspond to words, while edges indicate semantic similarity between them. The proximity of the vertices can be observed in the graph, as well as the deeper structure of the connections between words, allowing for the exploration of clusters of semantically related words. Visually, the closer the words are in the vector model, the shorter the edges between their vertices in the graph. A word's nearest neighbours are calculated through cosine similarity. The ability to represent the words present in the WE model as a graph can be helpful to filter noise relations, which in turn can lead to important insights about the structure of human language semantic space.

Furthermore, it becomes possible to employ operations of graph theory like random walks, community detection or graph statistics.

In Oubenali et al. [26], the intention is to conduct a scoping review describing the methods and strategies used to visualise medical concepts represented by WE methods. WE are created to visually analyse unstructured textual data in Electronic Health Record (EHR) so that the implicit semantics of words can be captured and the relationships between words and their usage can be understood. This intention to perform visual analysis in a specific domain is similar to the research conducted in this article, as the aim is also to understand the underlying semantic patterns present in words and how they are applied in the domain; the only difference is that this research focuses on asset management concepts rather than medical concepts. The aim is to create structured representations of unstructured data, making it more understandable and easier to use in subsequent modelling tasks. Word embedding methods have become an essential reference for solving this problem. Visualisation methods and techniques make it possible to analyse the resulting vector space visually. The visualisation methods/techniques t-SNE [38], PCA [14], Uniform Manifold Approximation and Projection (UMAP) [23], and K-Means [11] were considered from a set of 7 articles selected for analysis (more details about the selection criteria are presented in [26]). The scoped review aims to broaden the understanding of the usefulness of visualising WE, so the papers are described one by one and conclusions are drawn about how the visual analysis of WE allowed us to understand the given problem in each paper better. It turns out that visualisation helps to explore the embedding results, but it is not an autonomous method for evaluating the relevance of WE.

2.3 *Performance of Word Embeddings*

Understanding which WE model fits better for the given problem is complex. This is because each model has its intricacies, and each issue has its complexity. To determine which WE model is appropriate for a particular task, it is necessary to evaluate its performance in different situations.

In Schnabel et al. [34], a study on evaluation methods for unsupervised embedding techniques, it is pointed out that there is only sometimes one more efficient solution to go for. Different scores rank embedding methods in different ways. Wang et al. [41] empirically evaluated WE trained on four different corpora, namely (i) clinical notes, (ii) biomedical publications, (iii) Wikipedia, and (iv) news. For clinical notes and biomedical publications, WE was trained using unstructured EHR data available at Mayo Clinic, and articles (MedLit) from PubMed Central,[1] respectively. Pre-trained WE GloVe [27] and Google News were used for Wikipedia and the news. It was concluded that WE trained in EHR and MedLit can better capture the semantics of medical terms. It also finds semantically relevant medical terms closer to human judgements than those trained in GloVe and Google News. It was also concluded

[1] Available at https://www.ncbi.nlm.nih.gov/pmc/.

that there is no consistent global ranking of WE for all downstream biomedical NLP applications. As a last conclusion, WE trained in the biomedical domain corpora does not necessarily perform better than those trained in the general domain corpora for any given downstream biomedical NLP task.

Bhoir et al. [2] present a comparative analysis of different WE models: (i) Continuous Bag of Words (CBoW) [24], (ii) Skip gram [24], (iii) GloVe [27] and (iv) Hellinger—Principal Component Analysis (PCA) [18]. Different models are compared based on several other parameters, including how well they perform concerning the size of their training data set, basic overviews showing overall model performance and rates at which words in the context window can be used as potential targets - or not - by each classifier within that input layer. Memory consumption is also taken into account. After extensively comparing each model using the above metrics and weighing all the pros and cons, the authors conclude that GloVe [27] is superior to other approaches. The decision is supported by how the model scales to a large corpus but works well with a small corpus. The improved quality of learned representations by normalising counts and log smoothing makes this work appropriate. The improved quality of the learned representations through normalising counts and logarithmic smoothing make this work right.

3 Data Description

The data used were extracted from Maintenance records collected from a healthcare facility in Portugal. Each word describes an event related to the health facility (e.g. equipment failure, routine inspection, component replacement). Therefore, it is possible to identify trends and relationships between the words describing the situation/problem. The collected maintenance records include WOs documents, which consist of several fields (75 fields to be precise) where the technician fills in information that most accurately describes what needs to be done. Table 1 shows an example of real values for some of the main fields extracted.

From the information in these fields, it is possible to analyse the lexicon used by technicians and study the relationships between the words used to describe day-to-day maintenance situations in this health facility. All WOs present in the healthcare facility's Computerised Maintenance Management System (CMMS) were gathered into a single *JSON* file, from which it was feasible to read and parse the WOs and their most significant fields into memory. No preprocessing of the WOs was done, except for selecting the most relevant fields. There are no standard terms in the WO descriptions written by technicians. Each describes a given situation, resulting in various possible explanations of the same problem. Despite the word differences, the asset management situation is the same. The aim is to create consistency in the language used when the same maintenance situation arises. Consistency is maintained by suggesting words as the technician fills in a description. This suggestion is based on the registered words and knowledge of maintenance situations, their definitions, and the words used.

Table 1 Description of the most significant fields from a Work Order and a sample with real values

Field name	Sample example
Word order ID	16715
Job number (*WOs may have may jobs*)	1
Asset complete identifier	"06001MCTEEE"
Asset ID	6001
WO description	"Periodic review of leaks"
Date when the WO began	"2020-06-04T09:00:00"
Date when the WO ended	"2020-06-04T11:00:00"
Description of the work performed	"Periodic review of leaks"
Asset designation	"Chiller"
Asset Family	"CT"
Asset Subfamily	"CH"

To accomplish this objective, a NLP pipeline is created that takes WOs as input. This pipeline gathers linguistic information about the words used in the descriptions and the relationships established between them. The following section describes it step-by-step. For this study, only the WO Description (the *corpus*) attribute was used.

4 Modelling

Once the WOs have been parsed into memory, they can enter the NLP pipeline depicted in Fig. 1. This pipeline analyses the fields (namely *WO Description*—corpus) present in each WO and enables us to conclude the words used and the relationships between them. NLP pipelines are needed to perform complex tasks on natural language data. For this particular domain, apart from the classical steps (e.g. tokenisation, lowercase and stop words removal), breaking down sentences and performing morphological analysis to keep only the relevant words are crucial to reducing word variability and enabling automation of WOs analysis. For good results, it is essential to include steps focused on the specific problem (e.g. multi-word token, post-tagging, and lemma) [6].

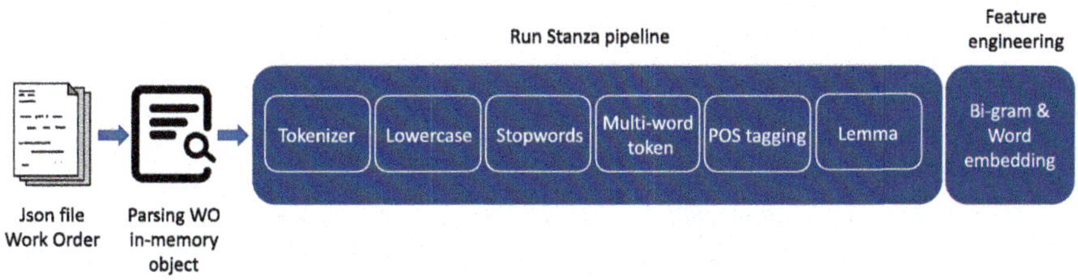

Fig. 1 Workflow of the word embeddings process for the Work Order

4.1 Natural Language Processing Pipeline

There are several NLP libraries providing sets of linguistic annotations (tokenisation, sentence splitting, paragraph boundary detection, spell checking, lemmatisation, etc.). Some are Java-based, like OpenNLP[2] and CoreNLP[3] [22], while others are Python-based, like NLTK,[4] SpaCy[5] and Stanza[6] [28]. The NLP pipeline is created using the Stanza framework, which has pre-trained models in Portuguese and allows for custom processing. Stanza's pipeline consists of modules that range from tokenising raw text to performing syntactic analysis on an entire sentence. Below is a brief description of these modules.

Tokenization Given a sentence and a defined document unit, tokenisation is the task of slicing it up into pieces, called tokens. Certain characters, such as punctuation, are filtered out during the process. Stanza combines tokenisation and sentence segmentation from the corpus into a single module [28]. Multi-word tokens (MWT) (e.g. "life-cycle"), are dealt with separately;

Lowercase Module designed to convert every description to lowercase letters;

Stop word removal After applying lowercase, it is removed words that do not have relevance to the problem (also known as stop words [16]). Since only relevant words are considered, removing stop words will increase the speed and accuracy of subsequent algorithms. Here, we have a domain expert indicating which words should be discarded;

Multi-word token expansion The MWT expansion module treats compound tokens not processed during the tokenisation module. Stanza identified MWT in the tokeniser module. Let's use the "life-cycle" example again. In this case, "life cycle" is considered a token composed of two separate words "life" and "cycle";

Part-of-speech (POS) tagging POS tagging [40] provides important clues that allow the computer to identify relationships between words and sentences, analyse each word in context, and determine how each word affects the meaning of a sentence. The tool categorises words in a corpus according to the POS defined for each word by its definition and uses in the context. Stanza assigns a POS for each word in a sentence and analyses its universal morphological features (e.g. singular/plural, 1st/2nd/3rd person, etc.);

Lemmatisation The knowledge gained from POS tagging is then used to perform lemmatisation, a process in which the inflected forms of a word are grouped and identified as a lemma. Lemmatisation converts words to their canonical form (e.g. "leaks" → "leak", "fixed" → "fix"). Reducing the number of words to a finite set is commonly used as an initial step in ML algorithms. This greatly reduces

[2] Available at https://opennlp.apache.org/.

[3] Available at https://stanfordnlp.github.io/CoreNLP/.

[4] Available at https://www.nltk.org/.

[5] Available at https://spacy.io/.

[6] Available at https://stanfordnlp.github.io/stanza/.

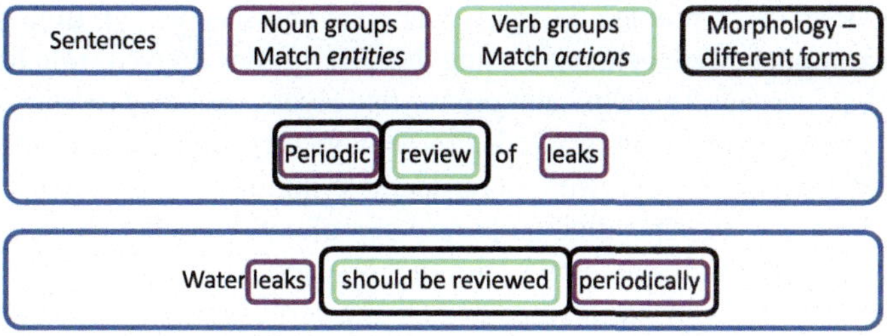

Fig. 2 Linguistic processing with NLP. Sentences are broken down into their constituent parts, with emphasis on verbs and nouns

complexity and variance from data sets by limiting possible solutions, allowing for more effective analysis later on. Figure 2 depicts an example of a lemmatisation module for better understanding, considering the *corpus* in Table 1.

4.2 *Feature Engineering*

After the NLP pipeline has been completed; it is essential to create n-grams and WE as feature engineering techniques to improve the accuracy of ML models.

Creation of n-grams An *n-gram* is a group of *n* words that often occur together. Keeping common words together is the objective. In this particular module, bi-grams and tri-grams are generated. For example, considering the example "Periodic review of leaks" (see Table 1), the tri-gram is rewritten as "periodic review leak" after the lemmatisation module. Then, in the n-gram module, all its words are concatenated using underscores since the three words often appear together. So, the final result is "periodic_review_leak";

Word embeddings Computers operate mostly using numerical representations, so they do not understand textual representations like humans do. A way to overcome this problem is to convert each word (including the *n-gram*) into a numeric vector that tries to capture the word's meaning through numbers, taking into account its context. For words with similar meanings, we have an equal numerical representation.

4.3 *Word Embeddings*

Multiple libraries are available for generating word embeddings from a text corpus. In this study, Word2Vec [24] and Fasttext [4] were selected for creating WE from scratch. The decision to avoid pre-trained WE was deliberate, given that the domain-specific lexicon in the problem is highly technical and specific [3]. Both Word2Vec and Fasttext models were generated using the descriptions derived from the *n-gram*.

Word2Vec is a neural network comprising two layers, designed to learn and reconstruct the linguistic contexts of words. To achieve this, it takes a corpus of words and generates a vector space with a considerable number of dimensions (typically between 100 and 300, depending on corpus complexity). Each distinct word in the corpus is assigned a corresponding vector in this space. These word vectors are positioned so that words with similar meanings in the corpus are placed closely together within the vector space. This arrangement allows Word2Vec to capture semantic relationships between words that appear in related contexts and share similar meanings will have vectors that exhibit proximity in the vector space. The vector space representation enables the model to understand the relationships between words. It can be used to perform various natural language processing tasks, such as word similarity, analogy detection, and even powering larger language model. It is possible to generate the Word2Vec model from two different architectures [24]: (i) Continuous Bag of Words (CBoW), and (ii) Skipgram. CBoW predicts target words from surrounding context words, while Skipgram predicts surrounding context words from target words (inverse of CBoW). In the Word2Vec neural network architecture, there is only one hidden layer. Throughout the training process, the objective is to adapt the weights of this hidden layer to minimize a specified loss function. Surprisingly, the outputs of the neural network are disregarded, and instead, the hidden layer weights themselves serve as WE. Fasttext follows similar principles as Word2Vec but with a minor distinction: it functions at the character level, whereas Word2Vec operates at the word level. As a consequence, the word vector in Fasttext is constructed by summing up the vectors of its individual constituent characters. For example, the word vector “matter” is a sum of the vectors of the constituents <“ma”, “mat”, “att”, “tte”, “ter”, “r”>, where ‘<’ means *Start of Word* and ‘>’ *End of Word*.

Both Word2Vec and Fasttext are designed as neural network models and offer the CBoW and *Skipgram* architectures. However, Fasttext takes a step further by enhancing WE through the incorporation of sub-word information [4].

5 Exploration of the Word Embeddings

Both libraries underwent testing, and WE were generated using WO descriptions. Both the Fasttext and Word2Vec models were created using standard parameter values. From a total of 38445 descriptions, both models learned 2435 words. As each word embedding contains 100 dimensions, it was necessary to perform dimensional reduction to visualize the vector space in 3D. Principal Component Analysis (PCA) [14] was applied first, followed by t-distributed Stochastic Neighbour Embedding (t-SNE) [38]. The reason for using PCA before t-SNE was to reduce noise and speed up pairwise distance computation. It is important to note that t-SNE operates in a stochastic manner, meaning that different representations can be generated each time a plot is created. This variation occurs because t-SNE randomly selects a start-

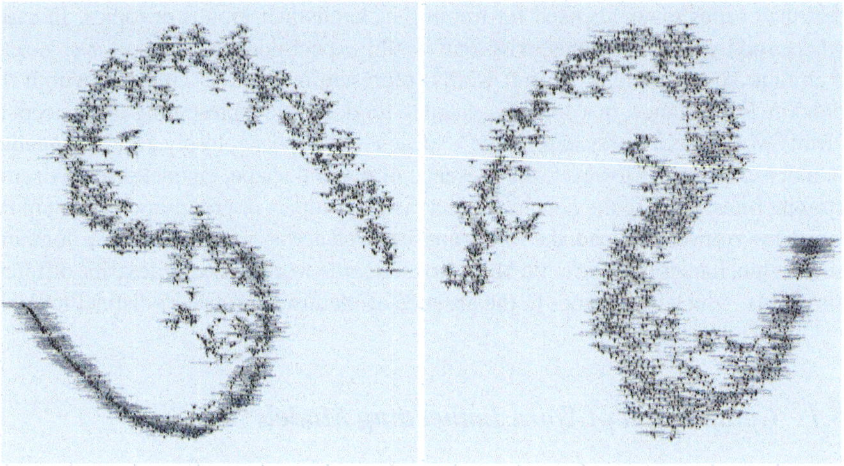

(a) Fasttext (b) Word2Vec

Fig. 3 Examples of word embeddings, represented in 2D, learned by Fasttext and Word2Vec models, respectively

ing point in space, which influences how it reduces the dimensions. After computing several plots, a consistent trend emerged: both the Fasttext and Word2Vec models displayed words in a helical shape in 3D space. However, due to the density and complexity of 3D representations, an interactive interface was required for thorough analysis. Therefore, to facilitate display purposes, further dimensional reduction was conducted to visualize the data in two dimensions, as depicted in Fig. 3. Nevertheless, the same conclusions can be drawn from these reduced 2D representations.

Upon observing both images, the initial impression is that the words are arranged in a helical shape. However, there are subtle variations between the two models. In the plot of the Fasttext model, the spiral appears meticulously drawn. This is particularly noticeable in the lower left corner where the words are nearly aligned in a straight line. Similarly, in the Word2Vec plot, a similar spiral pattern is evident. However, compared to Fasttext the Word2Vec spiral appears less precise, with words not forming as straight a line. The reason for this discrepancy in the representations being mirror images of each other is attributed to the stochastic nature of the dimensional reduction technique, t-SNE, utilised in the process. As for the less precise arrangement of words in the Word2Vec model, it could be linked to the algorithm's lower accuracy in precisely capturing the meaning of each word. This results in less accurate word embeddings than FasttextThis discrepancy may be due to the fact that Fasttext incorporates sub-word information in generating the final word embeddings, whereas Word2Vec solely relies on whole words. Further analysis and discussion of this hypothesis are presented in this document.

Additionally, it is crucial to analyse the significance of the helical shape observed in the word embedding model visualisation. The presence of a spiral shape indicates that the corpus of words used for training lacks distinct groups or topics. In cases where such groups or topics exist, one would expect dense clusters of words (e.g. a circular shape made of several words) representing abstract concepts within the domain. For instance, in a domain related to food, a topic representing the concept of "fruit" would have words like "apple", "peach", and "pear" closely grouped together in a dense manner. However, the presence of a spiral shape, characterised by semi-straight lines, leads to the conclusion that no clear topics or groups are evident in the word descriptions. Instead, the words are scattered across various expertise domains, such as mechanics or electricity. Moreover, the same word is used to describe different situations, which contributes to the absence of identified groups or distinct topics.

5.1 Comparison of Word Embedding Models

In this study, we compare WE performance generated by different libraries. This is done while considering the domain where these techniques are applied and what conclusions can be drawn. The evaluation and comparison of both word embedding models involves querying words that exhibit semantic similarity to a given word. The assessment is based on each model's response and the coherence of the returned words. This evaluation takes place after creating the word embedding models, utilising specific methods in each library to find the closest words in the vector space to a given input word. Although it is not possible to objectively compare the semantics of the word and the domain problem, a qualitative comparison is necessary to understand their respective performance. The first example of the words searched is the word "2020". On the one hand, the Fasttext model returned the words "2019", "2017", "2018" and "2016" as the closest words. On the other hand, the Word2Vec model returned the words "alarm", "cable_passage", "anomaly", and "collage". The second example is "break_room", and the words returned by Fasttext were "delivery_room", "work_room", "waiting_room" and "room". The words returned by Word2Vec were "floor_six", "dressing_room", "fall" (verb) and "fix". The third example is "employee"; the words returned by Fasttext are "schedule", "malfunction", "employee_entrance" and "necessary". The words returned by Word2Vec are "sug" (not a word, probably it is an acronym from the domain), "badly", "together" and "key". The last example presents an intriguing case to analyze because the Fasttext model returns words with a weak semantic link. However, it is still possible to perceive a relationship between the queried word "employee" and the returned words, such as "schedule", "malfunction", "employee entrance" and "necessary". Regarding the words "malfunction" and "employee entrance", it can be deduced that the Fasttext algorithm successfully captured the existing relationship between the queried word and these responses. This is evident because a "malfunction" typically requires an employee to fix it, establishing a connection between the two words. Similarly, the concept of an "employee entrance" inherently incorporates the idea of an employee, further illustrating the captured relationship. Even though the semantic similarity might not be very strong, the returned words still demonstrate relevant associations with the query word. This indicates the model's ability to uncover meaningful connections.

For the other two cases, it is not possible to assume whether the Fasttext model captured the relationship between an employee and a schedule (an employee has a schedule) and between an employee and the concept of necessity (an employee is needed to perform some operation), or if there was a matching of sub-word information, since in Portuguese the three words (employee, schedule and necessary) end with the same four letters, that is, "funcionário", "horário" and "necessário", respectively.

After examining numerous examples, of which the three provided are just a representative sample, it was evident that the semantic accuracy of the WE generated by the Word2Vec model was unsatisfactory. This became apparent as the returned words for most of the examples had little or no relevance to the queried word. Conversely, the Fasttext model produced responses to the queried examples that were generally meaningful, showing semantic similarity or relevance to the queried words. Although the Fasttext's WE yielded sensible results in most cases, there were instances where the returned words for more technical domain-specific terms did not align well, indicating room for improvement. Fine-tuning the parameters of the Fasttext's training algorithm could be explored to enhance the WE in such cases.

5.2 *Exploration of Fasttext Model*

The next step consisted of using the Fasttext model to understand better how the usage of the words might differ when comparing subfamilies. Some assets cover one hundred and eight different subfamilies in the case study context. A subfamily is a domain-specific from second-level classification, where assets with similar characteristics (e.g. similar components) are grouped to form a subfamily. The idea is to understand if the exact words are used across the subfamilies and, more importantly, how they are used. The first analysis consisted of plotting a Fasttext model with all of the words from the entirety of the WO and identifying the twenty most used words in a given subfamily. Figure 4 presents the Fasttext model, this time in three dimensions, and the twenty most used words for the subfamily "Rede Águas Residuais Doméstica" (Domestic Residual Water Network). It is possible to observe the dimensionality reduction of the model to 3D maintains the same spiral-like shape as in 2D.

What stands out immediately is how the words are scattered across the spiral-like shape. Word embedding models are generated based on the meaning of the words and through the contexts where they are found. It would be interesting to see the identification of words that appear in a common context (the subfamilies), so their position would be close in the model. This is not verified, so it is assumed that the Fasttext model could not capture the fact that some sets of words are correlated with a given subfamily. Now, the question is whether that correlation exists and whether it is possible to link a group of words with a specific subfamily or whether the words

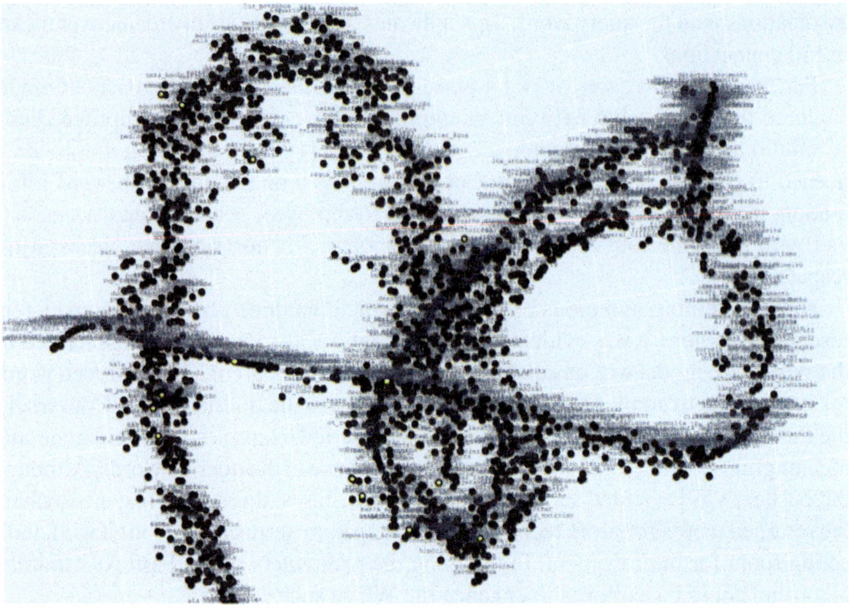

Fig. 4 Fasttext model with the 20 most used words for subfamily "Rede Águas Residuais Doméstica"

are used so sparsely that all of them are found in most of the subfamilies. Additional explorations were made with different subfamilies. However, the same was verified: the essential words in a given subfamily are scattered in the Fasttext model.

Another analysis was made to explore further the Fasttext model's capability to identify subfamily-specific information. The idea is to find the most critical word in a given subfamily (the most frequently used) and observe the closest words to it. This analysis is made in the first place with the subfamily "Rede Águas Residuais Doméstica" (Domestic Residual Water Network), where the most frequent word is "sewer". Then, it was possible to analyse the closest words to "sewer" by verifying the position that each vector encodes and finding the nearest places to the word "sewer is". It is also essential that the most immediate words are confirmed in the same subfamily, so it is dismissed if a word is closely related but not found in the subfamily. The algorithm to find the closest neighbours to the given the word consists in a k-dimensional tree [21], which is a space-partitioning data structure for organising points in a k-dimensional space (where k equals the length of the vector's dimension, in this case); the cited article [21] describes the details of its implementation, which is available in the *scipy* Python module. The conclusions are that the words "photo" (photography), "pantry", "exterior" and "zone" are the closest neighbours. Reminding of the hypothesis that words that appear in similar contexts should be placed closely in the WE model, it would be expected that the words that are the closest to "sewer" would be used in the same contexts, or at least in the exact phrases that appear in the WOs from the subfamily "Rede Águas Residuais Doméstica". However, they practically don't appear alongside "sewer", with only one evidence of the word "pantry".

Other tests in other subfamilies were conducted, such as in the "Proteção Contra Intrusão" (Intrusion Protection) subfamily that belongs to the *Information Systems* family. The most frequent word in the WO s that belongs to this family is "tag". Applying the same algorithm, the closest words to "tag" are "several", "detector" and "according_to". Again, it was verified that these words do not appear alongside "tag" or in similar contexts.

From these findings, it is possible to conclude: Each word's position in the WE model is not influenced by subfamily-specific information but rather by the information in the entire corpus. To understand whether the WE model lacked accuracy and could not learn specific subfamily patterns or if there weren't subfamily-specific patterns in words and how they are used, tag clouds are created.

5.3 Visualisation of Tag Clouds

Within the case study context, assets cover eighty different subfamilies.[7] The 38445 WOs are filtered and grouped based on their respective subfamilies. Using each description, tag clouds are generated for each asset subfamily (Fig. 5b to d). *Tag clouds* offer a visual representation of words, with each word appearing based on its frequency in the corpus. The font size and/or color highlight the most frequently occurring words. Additionally, a tag cloud is created using all WOs (Fig. 5a) regardless of their subfamily, providing a general overview of the lexicon used. It is worth noting that the words in the tag clouds are in Portuguese, as the primary objective is to analyze the frequency of each word in the subfamilies "Piping" (Fig. 5b), "Chiller" (Fig. 5c), and "Gas Centrals" (Fig. 5d), along with the general view of the lexicon (Fig. 5a). Despite the fact the words are in Portuguese, the patterns of occurrence for the most common words in the three subfamilies of assets are visible, (i) "Piping" (Fig. 5b), (ii) "Chiller" (Fig. 5c), (iii) "Gas Centrals" (Fig. 5d), , along with the tag cloud that provides a general lexicon view (Fig. 5a). The words that stand out instantly in all of the tag clouds, thus for any arrangement of WOs, are the same, namely, "substituição", which means **substitution**, "inspeção" which means **inspection**, "reparação" which means **repair** and others less noticeable like "retina", meaning **routine**, and "anual", meaning **annual**. It is interesting to observe that in three vastly different asset subfamilies, the most frequent words are virtually identical, just like in the global tag cloud (Fig. 5a). One potential explanation for this observation is that these words may have generic applicability within the domain of the mentioned activity. For instance, a word like "substitution" can be used in various scenarios where component replacement is required. This wide range of usage leads to the proliferation of such words across all subfamilies. This finding also aligns

[7] A subfamily is a domain-specific second-level classification.

(a) *Tag Cloud* computed for all the WOs

(b) *Tag Cloud* for subfamily "Piping"

(c) *Tag Cloud* for subfamily "Chiller"

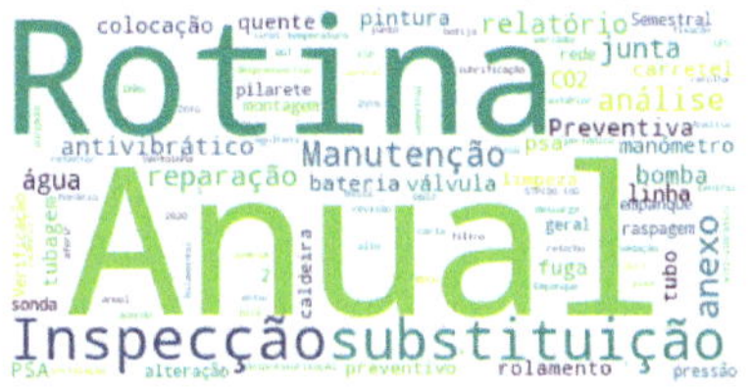

(d) *Tag Cloud* for subfamily "Gas Centrals"

Fig. 5 Illustration of some of the generated *Tag Clouds*. The words are in the original language of the WO—Portuguese

with the insights discussed in the previous subsection concerning word embedding models visualisation. The corpus words are dispersed and utilised in diverse contexts, without specific word groups tailored to particular situations. Consequently, the same words are used to describe distinct asset maintenance scenarios, diminishing their informativeness. Hence, to detect differences between the lexicon employed in these three subfamilies, one must examine the less frequent words.

Based on the analysis of the tag clouds, it can be deduced that a general machine learning model trained on the entire set of WO descriptions is unlikely to yield satisfactory results. The global tag cloud shares almost the same important words as the three subfamilies, indicating that the model might struggle to identify subfamily-specific data essential for accurate learning. Consequently, alternative solutions must be considered, such as employing a separate machine learning model for each subfamily. Although this approach requires managing multiple models, it offers the advantage of preserving specific details unique to each subfamily. Also, word frequency is just one of many indicators of importance. Relying solely on word frequency might overlook the significance of words that appear infrequently (e.g., once or twice). In the domain context, rare words can hold equal importance as frequently occurring ones. Critical failures in assets, fortunately, are not common occurrences, resulting in infrequent usage of specific words related to these situations. However, when such crucial failures do happen, they provide vital information essential for asset management decision-making processes. As a result, these rare words should not be disregarded, as they contribute significantly to the overall understanding of asset-related insights.

6 Conclusions

This chapter presents the application of NLP to automate the processing of WO descriptions. By visualising the words within the corpus, it is possible to understand better how they are used in this specific domain and visually analyse the findings that provide insight into deeper semantic patterns. Therefore, creating WE models, with subsequent dimensionality reduction to 3-D, allowed us to visually explore the semantic and syntactic relationships between words. This visual analysis made it possible to compare two models, Word2Vec and Fasttext. The conclusion was that with the given corpus of words, the performance of Fasttext was superior. This was noticeable in visual representations, where the helical shape was drawn more thoroughly, and in semantic similarity tests, where specific words were queried for the models. Fasttext reported better results by responding to words that made sense in semantic terms, while Word2Vec often returned unrelated words. Although the performance of Fasttext was better than Word2Vec it still has some shortcomings regarding the more technical words. These shortcomings can be tackled with an empirical study on selecting the model parameter values. However, it was not the end goal of this research.

The fact that it is possible to analyse the WE in the corresponding vector space visually enables us to draw conclusions that otherwise would need further computational explorations. For example, the conclusion that there are no groups/topics in the corpus.

Using *tag clouds* gives us a broad perspective of how the lexicon is spread across the several asset subfamilies. It was also concluded that words are distributed dispersedly throughout asset subfsubfamilies. Therefore, finding a direct relationship between a set of words and an asset subfamily is difficult. In other words, it is impossible to identify what kind of asset is being treated only by looking at the words used in a given description. This is because the exact words are used in a variety of contexts.

From the observation of the tag clouds and the conclusions taken from visually analysing the WE, it is also possible to conclude that a general ML model trained on the entirety of the WOs descriptions will not likely yield good results, because the global tag cloud presents roughly the exact most important words as the three subfamilies, so it may have difficulty in identifying subfamily-specific data that is required for it to learn correctly. Therefore, it is necessary to consider other solutions, such as an ML model per subfamily. The downside of this solution is that it requires managing a large number of models. However, it has the advantage of retaining specific details unique to each subfamily.

Acknowledgements This work was supported by the REV@CONSTRUCTION mobiliser project, under the grant LISBOA-01-0247-FEDER-046123 from ANI - National Innovation Agency, and by NOVA LINCS (UIDB/04516/2020) and LASIGE (UIDB/00408/2020) with financial support from FCT— Fundação para a Ciência e a Tecnologia, through national funds. This work contributes to the Strategic Research Plan of the Centre for Marine Technology and Ocean Engineering (CENTEC), which is financed by FCT under contract UIDB/UIDP/00134/2020.

References

1. Albawi S, Mohammed TA, Al-Zawi S (2017) Understanding of a convolutional neural network. In: 2017 international conference on engineering and technology (ICET). IEEE, pp 1–6
2. Bhoir S, Ghorpade T, Mane V (2017) Comparative analysis of different word embedding models. In: 2017 international conference on advances in computing, communication and control (ICAC3). IEEE, pp 1–4
3. Bojanowski P, Celebi O, Mikolov T, Grave E, Joulin A (2019) Updating pre-trained word vectors and text classifiers using monolingual alignment. arXiv:1910.06241
4. Bojanowski P, Grave E, Joulin A, Mikolov T (2017) Enriching word vectors with subword information. Trans Assoc Comput Linguist 5:135–146
5. Bouabdallaoui Y, Lafhaj Z, Yim P, Ducoulombier L, Bennadji B (2020) Natural language processing model for managing maintenance requests in buildings. Buildings 10(9):160
6. Brundage MP, Sexton T, Hodkiewicz M, Dima A, Lukens S (2021) Technical language processing: unlocking maintenance knowledge. Manuf Lett 27:42–46
7. Chandrasekaran D, Mago V (2021) Evolution of semantic similarity—a survey. ACM Comput Surv (CSUR) 54(2):1–37
8. Cheng J, Dong L, Lapata M (2016) Long short-term memory-networks for machine reading. arXiv:1301.3781
9. Dima A, Lukens S, Hodkiewicz M, Sexton T, Brundage MP (2021) Adapting natural language processing for technical text. Appl AI Lett 2(3):e33
10. Ganguly D, Roy D, Mitra M, Jones GJ (2015) Word embedding based generalized language model for information retrieval. In: Proceedings of the 38th international ACM SIGIR conference on research and development in information retrieval, pp 795–798
11. Hartigan JA, Wong MA (1979) A k-means clustering algorithm. Appl Stat 28(1):100–108
12. Huang X, Zhang J, Li D, Li P (2019) Knowledge graph embedding based question answering. In: Proceedings of the twelfth ACM international conference on web search and data mining, pp 105–113
13. Idowu S, Strüber D, Berger T (2021) Asset management in machine learning: a survey. In: 2021 IEEE/ACM 43rd international conference on software engineering: software engineering in practice (ICSE-SEIP), pp 51–60
14. Jolliffe IT, Cadima J (2016) Principal component analysis: a review and recent developments. Philosophical transactions of the royal society a: mathematical, physical and engineering sciences 374(2065):20150202
15. Katricheva N, Yaskevich A, Lisitsina A, Zhordaniya T, Kutuzov A, Kuzmenko E (2020) Vec2graph: a python library for visualizing word embeddings as graphs, vol 02. Springer, Berlin, pp 190–198
16. Kaur J, Buttar PK (2018) A systematic review on stopword removal algorithms. Int J Futur Revolut Comput Sci Commun Eng 4(4):207–210
17. Kovalerchuk B, Andonie R, Datia N, Nazemi K, Banissi E (2022) Visual knowledge discovery with artificial intelligence: challenges and future directions. In: Integrating artificial intelligence and visualization for visual knowledge discovery. Springer, Berlin, pp 1–27
18. Lebret R, Collobert R (2013) Word emdeddings through hellinger PCA. arXiv:1312.5542
19. Li Y, Yang T (2018) Word embedding for understanding natural language: a survey. In: Guide to big data applications. Springer, Berlin, pp 83–104
20. Liu S, Bremer P-T, Thiagarajan JJ, Srikumar V, Wang B, Livnat Y, Pascucci V (2018) Visual exploration of semantic relationships in neural word embeddings. IEEE Trans Visual Comput Graphics 24(1):553–562
21. Maneewongvatana S, Mount DM (1999) Analysis of approximate nearest neighbor searching with clustered point sets. arXiv:cs/9901013
22. Manning CD, Surdeanu M, Bauer J, Finkel JR, Bethard S, McClosky D (2014) The stanford CoreNLP natural language processing toolkit. In: Proceedings of 52nd annual meeting of the association for computational linguistics: system demonstrations, pp 55–60

23. McInnes L, Healy J, Melville J (2018) UMAP: uniform manifold approximation and projection for dimension reduction. arXiv:1802.03426
24. Mikolov T, Chen K, Corrado G, Dean J (2013) Efficient estimation of word representations in vector space. arXiv:1301.3781
25. Mikolov T, Yih WT, Zweig G (2013) linguistic regularities in continuous space word representations. In: Proceedings of the 2013 conference of the north American chapter of the association for computational linguistics: human language technologies, pp 746–751
26. Oubenali N, Messaoud S, Filiot A, Lamer A, Andrey P (2022) Visualization of medical concepts represented using word embeddings: a scoping review. BMC Med Inform Decis Mak 22:03
27. Pennington J, Socher R, Manning CD (2014) GloVe: global vectors for word representation. In: Proceedings of the 2014 conference on empirical methods in natural language processing (EMNLP), pp 1532–1543
28. Qi P, Zhang Y, Zhang Y, Bolton J, Manning CD (2020) Stanza: a python natural language processing toolkit for many human languages. arXiv:2003.07082
29. Qi Y, Sachan DS, Felix M, Padmanabhan SJ, Neubig G (2018) When and why are pre-trained word embeddings useful for neural machine translation? arXiv:1804.06323
30. Ren M, Kiros R, Zemel R (2015) Exploring models and data for image question answering. Adv Neural Inf Process Syst 28
31. Rodriguez PL, Spirling A (2022) Word embeddings: what works, what doesn't, and how to tell the difference for applied research. J Polit 84(1):101–115
32. Rush AM, Chopra S, Weston J (2015) A neural attention model for abstractive sentence summarization. arXiv:1509.00685
33. Santos P, Datia N, Pato M, Sobral J (2022) Comparing word embeddings through visualisation. In: 2022 26th international conference information visualisation (IV). IEEE, pp 91–97
34. Schnabel T, Labutov I, Mimno D, Joachims T (2015) Evaluation methods for unsupervised word embeddings. In: Proceedings of the 2015 conference on empirical methods in natural language processing, pp 298–307
35. Stein RA, Jaques PA, Valiati JF (2019) An analysis of hierarchical text classification using word embeddings. Inf Sci 471:216–232
36. Stenström C, Al-Jumaili M, Parida A (2015) Natural language processing of maintenance records data. Int J COMADEM 18(2):33–37
37. Tang D, Wei F, Yang N, Zhou M, Liu T, Qin B (2014) Learning sentiment-specific word embedding for twitter sentiment classification. In: Proceedings of the 52nd annual meeting of the association for computational linguistics, pp 1555–1565
38. Van der Maaten L, Hinton G (2008) Visualizing data using t-SNE. J Mach Learn Res 9(11)
39. Vanier DJ (2001) Asset management A to Z. In: Innovations in urban infrastructure, pp 1–16
40. Voutilainen A (2003) Part-of-speech tagging. In: The oxford handbook of computational linguistics, pp 219–232
41. Wang Y, Liu S, Afzal N, Rastegar-Mojarad M, Wang L, Shen F, Kingsbury P, Liu H (2018) A comparison of word embeddings for the biomedical natural language processing. J Biomed Inform 87:12–20
42. Wang Y, Wang L, Rastegar-Mojarad M, Moon S, Shen F, Afzal N, Liu S, Zeng Y, Mehrabi S, Sohn S, Liu H (2018) Clinical information extraction applications: a literature review. J Biomed Inform 77:34–49
43. Zhang L, Wang S, Liu B (2018) Deep learning for sentiment analysis: a survey. Wiley Interdiscip Rev: Data Min Knowl Discov 8(4):e1253
44. Zhao R, Mao K (2017) Fuzzy bag-of-words model for document representation. IEEE Trans Fuzzy Syst 26(2):794–804

Similarity in Visual Analytics—A Visual Analytics Approach for Finding Similar Publications

Midhad Blazevic, Lennart B. Sina, Cristian A. Secco, and Kawa Nazemi

1 Introduction

The knowledge, which researchers produce and store in databases, has reached the point where finding relevant publications cannot be adequately accomplished by simple list representation. The information overload situation has led researchers to the point where the risk of exploring data without exploring potentially relevant data may occur. This lack of exploratory search leads to the question of how researchers can be enabled to gather potentially relevant data. The topic of exploration has been discussed by Marchionini [27], White and Roth [46], and Munzner [28], who stated that exploration and lookup tasks are entwined, as can be seen when examining how researchers explore using searches. The essential difference between exploration and lookup tasks is the complexity of tasks and the knowledge of users [27, 28, 46]. Exploration requires advanced tools and methods to balance out the lack of users' knowledge in a certain domain and provide a larger space of search results, requiring the researcher to learn from their searches and become able to optimize future iterations of their search [27] and reduce the uncertainty in the search process [46]. The more relevant information the researcher gathers, the more knowledge they gain and, ultimately, the more precise their exploration becomes.

There is a need for approaches, systems, and tools to enable researchers to find information that might be of great interest to the targeted research even if the terminology differs. Since the exploration process can be supported ideally through

M. Blazevic (✉) · L. B. Sina · C. A. Secco · K. Nazemi
Human-Computer Interaction and Visual Analytics, Darmstadt University of Applied Sciences, 64295 Darmstadt, Germany
e-mail: midhad.blazevic@h-da.de

L. B. Sina
e-mail: lennart.b.sina@h-da.de

C. A. Secco
e-mail: cristian.a.secco@h-da.de

K. Nazemi
e-mail: kawa.nazemi@h-da.de

B. Kovalerchuk et al. (eds.), *Artificial Intelligence and Visualization: Advancing Visual Knowledge Discovery*, Studies in Computational Intelligence 1126,
https://doi.org/10.1007/978-3-031-46549-9_16

visual analytics, we propose a visual analytical approach for exploring undetected terms, knowledge, and patterns. Visual analytics combines interactive visualizations with artificial intelligence and machine learning methods to enable solving complex analytical tasks [21, 29, 30] by optimally presenting large amounts of data that the researcher can use during the search process. An approach that allows "annotation in context" would further improve the ability to conduct exploratory tasks. Our visual analytics system can assist researchers in finding relevant scientific publications. The approach integrates machine learning and natural language methods such as topic modeling, which identify clusters of terms within text that can be used to find similar publications [44]. The resulting clusters are helpful as they contain terms that a machine has learned, but a human might not know of or associate with a cluster. To find the optimal terms that can be used to assist the researcher, the system requires information about relevant publications. This presents an optimal opportunity to utilize a researcher's annotations. Researchers commonly annotate publications that are relevant to their work. In that case, the list consists of solely relevant publications, and the latent topics resulting from the analysis of this list are optimal for finding similar and relevant publications. This complex issue resulting from information overload is a general issue not limited to specific domains. Our approach can be generalized and adapted to various domains of applications.

Furthermore, our approach's interactive visualizations, which researchers use to identify and annotate relevant publications as they explore and discover knowledge, is linked to the emerging field of visual knowledge discovery (VKD) [20, 21]. The approach heavily emphasizes human's ability to interpret visualized representations of publications and their similarity and importance within respective topics, a fundamental yet significant aspect of VKD [21]. The fusion of a flexible visual analytics approach with machine-learning and AI-based approaches allows the researcher to choose which interactive visualizations they can use to interpret the results best, and annotating based on their interpretation also provides the system with more input improving future recommendations and results. This fusion creates flexibility that allows for a more precise similarity analysis based on a researcher's interest rather than a simple publication-based similarity search found in publication search engines.

Our contribution is five-fold and enhances the main contributions of our previous works [3, 4, 29]: (1) an approach to generate and label vectors automatically using natural language processing and machine learning on lists of annotated publications, (2) a novel user interface approach for annotating and visualizing, (3) an approach that utilizes the annotation behavior of researchers to identify topics of interest, (4) an approach that facilitates and encourages annotation during exploration (5) an approach for the scientific community to reduce cold start issues in recommendation systems.

We begin this chapter by introducing the related work that discusses how similar publications can be identified and how interactive visualizations can assist the research process. The related work investigates work that has been published within the past fifteen years. After that, we describe our general approach that describes the steps ranging from data extraction over topic modeling to interactive visualizations. We describe each step and our decisions for using specific algorithms and technologies with which we discover relevant publications before we present how they can be optimally visualized to accommodate researchers' needs. We conclude our work by briefly summarizing our results and outlining possible future developments.

2 Related Work

Our approach includes both similarity algorithms and visual analytics for publication search. For that reason, the related work section is divided into two parts. In the first review, an overview of similarity algorithms is given, and in the second review, different works and systems are presented that introduce visual analytics in publication searches. To the topic of similarity algorithms, we present a brief overview of recently published relevant related works of both areas of interest. Our approach uses topic extraction methods from text and measures the similarity of a set of documents in a "visual analytics" system to enable users to get similar document results visually represented. Similarity can have different "levels of granularity", e.g. "word-to-word similarity", "sentence-to-sentence similarity" or "document-to-document similarity" that can be combined as "word-to-sentence" or "sentence-to-document similarity" [9, 32]. The various possible approaches are the reason why we present work that examines how to calculate similarity with various approaches.

2.1 Similarity Algorithms

Mahmood et al. [25] proposed to examine similarity and dissimilarity measurements to find document relatedness. They extracted keywords or terms from the contents of research papers, such as the title and abstract, compared these between papers, and found and compared synonyms for each keyword [25]. Both similarity and dissimilarity measurements are then normalized and used to calculate the document-relatedness [25]. A disadvantage of this approach is the technique limitation because it is based on the word level, thus not taking sentences into consideration.

Niraula et al. [32] measured the accuracy, "F-measure" and "Kappa" for "semantic similarity and found out that for LDA, the "LDA-Greedy" provides the best results. The "cosine similarity" showed promising results, such as in the work of Sitikhu et al.[39], Thada and Jaglan [41] and in the case study of Soyusiawaty and Zakaria [40]. An approach from Sitiku et al. involved using the "soft cosine measure" [39], which takes, for example, similar word meanings into consideration [34, 38]. As Rahutomo et al. have explained, this approach yields difficulties, as it tends to produce lower similarity results based on semantic relationships on a dimensional level [34], which is confirmed by the results of Sitikhu et al. [39]. These observations lead to prefer the standard "cosine similarity" approach for such projects. Further studies were made into "soft cosine similarity" due to the promising potential of a more semantic-focused approach towards topic similarity.

Mahmood et al. [26], in a later published case study, developed an Onthology named "Content based Onthology for Research Paper Similarity" (COReS) [26]. In this study, they model different document similarity techniques and locate relationships between them to identify their similarities [26]. This approach attempts to examine similarity based on one calculation, which might have a disadvantage in specific scenarios or domains, but instead considers many, and by doing so, might overcome some disadvantages of specific calculations.

Ng [31] examined the problem that users face when starting research. The difficulty that users have when formulating search queries [31]. Using a deep learning model that analyzes a research paper provided by the user, the model attempts to identify the category of the provided paper [31]. The model then presents results ranked with content similarity, peer reviews, the expertise of the authors, and the number of references of documents closely related to the provided paper with which the search was initiated [31]. Doc2Vec generates vectors for the documents, such as the abstract and title [31]. SentiWord scores, h-index, and PageRank values are all considered with this approach [31]; thus, this model attempts to combine content similarity with other important factors to provide the researcher with optimal results. This model's disadvantage lies in its similarity approach that only examines the title and abstract. Thus important content information might get overlooked. This approach to calculating similarity based on title and abstract can also be found in Ristanti et al. [35] work, which used cosine similarity to calculate similarities between two documents for a classification task. This work described that the advantage of cosine lies in its ability not to be affected by the length of the document but instead by a document's terms and low error rates [35].

Lu et al. [24] described the use of "bibliographic coupling" and "co-citations" to calculate document similarity, more precisely, research paper similarity. However, they stated that the disadvantage of this approach lies within the algorithm that would not be able to judge new papers correctly, as they have yet to be cited [24]. This challenge is significant for scientific publications as the landscape changes daily due to new publications published or uploaded daily. To overcome this, Lu et al. used the "common citation x inverse document frequency" (CCIDF) [24] proposed by Lawrence et al. [22], which is considered similar to "tf/idf".

Alschareef et al. [1] also examined citation network similarity and article similarity to create similarity matrices between articles, authors, and venues, which they calculated using cosine similarity [1]. Thus, taking a citation-based approach and further developing it to examine venues that might showcase similar work by creating a "venue-article latent preference matrix".

Heidarien and Dinneen [14] proposed a new hybrid geometric approach to measure the similarity levels among documents and document clustering that they named "TS-SS" [14]. The "TS-SS" algorithm attempts to calculate the similarity between two vectors based on cosine angle and euclidean distance, as well as the difference between their magnitudes [14]. Heirdarien and Dinneen create a "triangle's area similarity" (TS) based on the Euclidean distance of the vectors and then calculate the magnitude difference between the vectors as this component is needed to calculate the "sector's area similarity" (SS), along with the angular difference of the vectors

[14]. TS is then multiplied with SS to create the "TS-SS" [14]. As stated by Heidarien and Dinneen, cosine is not robust enough to distinguish similarity at high levels, and the euclidean distance could have problems when analyzing larger datasets than used in that paper [14].

Huang et al. [16] calculated text similarity using the key phrase vector model and combined two similarity measurements to create a final book similarity weight [16]. Furthermore, intending to use this for book recommendations in a recommendation system, they also applied a customer similarity measurement based on demographic data. They linked both book similarity to customer similarity layers [16].

In this subsection, we introduced related works that use different similarity and scoring approaches for providing similar data entities. We focused our investigation on works with more than two vectors for comparison. Our literature review illustrated that combining vectors as a matrix does not provide any labeled vectors of interest. By combining different vectors as a matrix, labeling such a matrix is essential for gathering the context of search and exploration. We address these limitations in our work.

2.2 *Visual Analytics in Publication Search*

Caillou et al. proposed their Cartolabe system, a web-based system that extracts entities from publications using natural language processing methods and utilizes topic modeling to create representations with which similarity measurements are used to cluster data and visualize the similarity of the entities [8]. The authors state their system is supposed to make larger datasets better understandable [8]. To achieve this, they use a heatmap approach with color-coded regions, clustering, and overlaying topic labels to visualize the HAL dataset [8]. Their visualization approach allows them to zoom in and out and filter results [8]. The evaluation of their proposed system was in the form of a usability study in which seven users from their institution were asked to perform tasks, answer a questionnaire, and were asked to provide feedback on various aspects [8]. The results showed that all tasks were completed, the system was perceived as trustworthy, and the top 10 identified similar publications satisfied the testers based on the questionnaire results [8]. The feedback also showed deficits in the system's user interface, such as suboptimal icon usage to toggle features, missing bookmarking and monitoring capabilities, suboptimal highlighting, and topic cluster labeling during navigation [8].

Zhao et al. also group results based on topic analyses via LDA and biclusters based on a two-mode network that perceives entities, in this case, authors and documents, as nodes and relations, such as authorship, as links [47]. They proposed a heatmap visualization of relevant documents in which the topics' two-mode networks are visualized in rows, the top three topic words are in the column, and the documents and authors are ordered by similarity [47]. The authors evaluated their visualization approach via a user study with twelve participants, comparing the proposed visualization with visualizations without topics and a text-based approach [47]. The study examined task completion and completion time, asked twelve questions in the form of a questionnaire, and examined strategies in which the participants described their approaches for the presented tasks [47]. Their study shows that two-dimensional layouts present 28% more information than one-dimensional [47]. However, their proposed visualization approach slowed the participants more than their simplified approach, which is why they stated that more studies are required to understand the results better [47]. The questionnaire results show that their approach in terms of usability is positive, as the participants state that the interface is easy to use and learn, it is easy to find search results with the proposal, and the interface is attractive [47]. However, due to the interface being new to the participants, the interface is somewhat unclear [47]. When examining the task strategies report results, eight out of twelve participants used the topic keywords to begin their exploration [47]. For the other compared systems, the starting strategy for nine out of twelve participants was to look at the titles first and then at the abstracts [47].

Breitinger and Reiterer propose a user-customizable visualization approach for the STEM community that also considers non-textual elements of publications [7]. They state that although recommendations are commonly used in nearly all digital libraries, the visualization approach of ranked lists could be more optimal [7]. List-based visualization of results is considered intuitive and straightforward, but researchers must work harder to explore publications with this form of visualization due to limitations in the number of presented results and features that only allow for quick identification of relevant publications [7]. The authors calculate and visualize citation, keyword, figure, and mathematical expression similarity within their RecVis frontend [7]. The publications are presented in a graph-based visualization where they are arranged as nodes depending on the closeness of their relevance scores, with shorter and thicker edges the closer related papers are [7]. The citation-based similarity is visualized using interactive circle graphs that show shared citations between a user-selected publication and another that the user chooses from the recommended publications [7]. The keyword-based similarity is visualized by presenting a color-coded comparison of keywords in a source document on one side, in a compared to document recommended by the system and chosen by the user on the other side, and in the middle keywords that are in both colored red [7]. The figure-based and mathematical formula-based similarity visualizations are side-by-side comparisons [7]. Users can adjust a sensitivity threshold that affects how many similar elements from the recommended publications are shown based on their similarity to the source document [7]. The authors evaluated their proposed system with an expert study with ten experts, using a think-aloud protocol, a fourteen-task task-based evaluation, a survey about ease of use and usefulness, and the commonly used standardized PSSUQ questionnaire [7]. The results show the average completion rate of the fourteen questions was 95%, and overall, the visualizations were perceived fondly with average scores for the keyword-based approach 4.0, the mathematical formula-based approach 4.25, and the citation-based approach 4.68 out of 5.0 [7]. The feedback from the participants also shows that the visualizations improved the user experience, and the results of the PSSUQ questionnaire [23] showed that system usability, information quality, and interface quality were also favorable [7]. The authors conclude that by going beyond the similarity of text and dividing similarity visualization between the extracted features, their approach gave researchers a better understanding of the system's recommendations, allowing researchers to find relevant information better [7].

Breitinger et al. describe a research gap in information retrieval and recommendation-based systems and propose an approach that examines how chemical entities can be extracted, utilized, visualized, and explored [6]. Although focused on the chemical domain, their approach shows that the comparison and visualization of entities are possible on textual data and visualizations [6]. Their graphical interface allows users to compare the extracted chemical compounds and browse through similar visualized chemical entities [6].

The related work in this subsection shows that there have been various methods of visualizing publication searches and similarities between publications. The varying visualizations hold a tremendous amount of potential if utilized correctly. In general, these related works confirm previous studies that show the positive effects visualizations yield when researching. However, to utilize visualization to its fullest potential, a customizable user interface is needed, as shown by Breitinger and Reiterer, as it improves the overall usability and cognitive reception of the visualized data. Furthermore, an aspect neglected in studies is the comparison and visualization of non-textual elements described by Breitinger et al. which are also of importance for certain domains more than for others.

3 General Approach

This section describes our approach that counters the challenge of the ever-increasing information overload that researchers encounter when searching for relevant publications. Figure 1 illustrates our approach in an abstract way and describes the steps of our general model. The goal is to provide researchers with relevant publications and visualizations that allow them to explore and annotate the relevant results on the fly, reducing cognitive distractions during exploration. In our case, annotating means bookmarking publications of relevance. The search for similar publications based on annotations utilizes the extracted topics from the bookmarked publications to create an interesting matrix, with which the search attempts to find similar publications. This raises the issue of cold start, more precisely, how is the initial search for publications conducted that leads to an interest matrix based on annotations. The initial publication search for first-time users can be conducted via term or author search, PDF paper, or text upload. The upload approaches analyze the uploaded text via topic modeling to compare the texts to known publications. The respective approach's results are visualized so researchers can optimally examine the search results and quickly annotate relevant publications. Our model begins by extracting data from various data sources and storing the extracted data. Afterward, the extracted data is analyzed using topic modeling and visualized. The initial visualization allows the researcher to annotate during exploration, which we illustrate and describe in the following sections. The

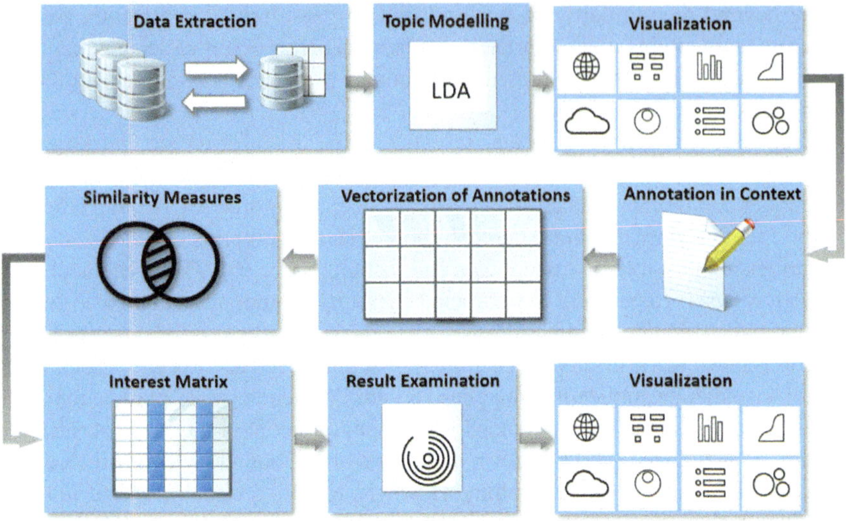

Fig. 1 The general approach with the steps of *Data Extraction*, *Topic Modelling*, *Visualization*, *Annotation in Context*, *Vectorization of Annotations* and *Similarity Measures* based on our previous works [3, 4, 29, 30]

annotated relevant publications and, thus, the underlying sets of topics are gathered and used by the system to calculate the similarity to stored documents. The documents that have a high similarity are considered to fit the researcher's interests and are provided to the researcher for further examination and further visualization to allow a more precise exploration with a focus on only highly relevant publications. Each step is described in more detail in the following sections.

We have an authentication server that allows researchers to login in and store their search information on our server linked to their profiles approach is suited for both researchers with a profile and those without. However, login is optional as our approach works without it by storing results locally as cookies. Researchers with a profile have their results and information stored online and linked to their profiles. This allows for the proposed system to examine much more than a single session, instead presenting the possibility to either store various sessions, for example, for multiple projects, or gather information from entire careers that span over the years, resulting in more precise recommendations. The topics stored online and linked to profiles can be used for various recommenders and a monitoring feature. The monitoring feature visualizes the topics considered to be of interest to the respective researcher. A search for that topic begins by clicking on a topic, as described in Sect. 3.4.

Regarding recommendation systems, our general approach can be expanded to examine the similarity between publications, venues, and researchers, as the extracted latent topics can be used to describe venues and researchers with a similar approach. This requires that the publications be correctly linked to the venues and researchers, which can be challenging when confronted with ambiguity. Linking publications with researchers can be accomplished by examining the OrcID, which counters the ambiguity issue; however, although increasing in popularity, the OrcID is not always given, nor do venues have a commonly used identifier; however, this information is stored as metadata in widely used databases which allows us to link both researchers and venues to publications.

3.1 Data Extraction

Our model starts with gathering scientific publication data from various resources. We use the application programming interfaces of "IEEE" in particular "IEEE Computer Society (Computer.org)", "Springer" and "Crossref" to gather data. The usage of different resources, in particular "Crossref" with "Springer" and "IEEE" leads to many duplicates stored in different databases. Not all duplicates can be identified by applying only the "Document Object Identifier" since many publications do not have a DOI. To identify the entire duplicates, we integrated the following steps that were provided in a first examination of the extracted data no duplicates [3]:

- use the DOI if given, but compare the titles of the publications
- use the title of a publication
- if using the title, compare the authors and years of the publication to confirm that the papers are identical.

Our investigation of about 20 million publications showed that there are, besides duplicates, an unneglectable number of publications with the same authors (sometimes in different order of authors), different titles, and very similar content. We assume that these are reprints or extended versions of the papers. In the case of reprints with precisely the same content, we categorize the according publications as duplicates. Therefore, the entire text is compared. The DOI is, therewith, not the best way to identify duplicates. Two identical papers may have different DOIs since they are published in two books. These reprints are identified as reprints and duplicates, including the title, authors, and years.

In the first step of our model, the entire data is stored in a database, including the metadata that provides information about authors, year of publication, publication type, and further specific information, e.g., authors' countries.

3.2 Topic Modeling

Topic modeling is a machine learning approach with which latent structures, concepts, or topics can be identified [44]. This is based upon the assumption that each document, or in our case, publication, consists of various terms that describe it or its topic. The groups of words that the topic modeling algorithm clusters represent said document or its topic and can be used for comparison purposes [33], as we will describe later on. The challenge when using topic modeling is finding the best approach that suits the tasks and the data to be analyzed. Commonly there is a differentiation between probabilistic and non-probabilistic approaches, in which the two widely used LDA [5] and LSI [10] are divided.

LDA is one of the most popular topic modeling methods currently used. LDA examines a text's words to determine which topics a document belongs to. LDA is built on the assumption that each topic has specific words that belong to it; thus, if a document contains certain words, there is a high probability that the document belongs to the associated topic. Words are sorted based on their probability score or can be filtered using a threshold. Words are considered in a bag of words, ignoring grammatical aspects, stopwords, and word order. Furthermore, the number of topics must be predetermined.

Although divided by category, the underlying natural language processing approaches to tokenization are the same. The ability to fine-tune these approaches allows us to examine various combinations of approaches to identify the most optimal for our task and dataset [3]. We examined using LDA and LSI with and without standard tokenization and lemmatization on full-texts and abstracts for 20 million documents. The results lead us to conclude that the best approach is LDA as it shows that the identified topics are more accurate in two out of three cases with a K of 20 [30]. The best results were found with 500 topics in the 20 million document dataset, with the extraction of unigram topics and n-grams with two or more words that built a topic, resulting in 500 words or phrases [29].

Every document is assigned 20 topics consisting of 20 words and 20 phrases, illustrated in Fig. 2, where the generated topics for a search on "visual analytics" are shown. On the left side, we can see the unigrams, which consist of one word, whereas, on the right side, the phrases consisting of two words are shown. Although the n-grams on the right side have a lower distribution, the accuracy is higher than that of the unigrams [3].

3.3 Weighting

To properly weigh the terms, we examined literature in which comparisons of various weighting methods in the context of topic modeling were conducted to identify the optimal weighting approach. Weighting is used to determine the importance of terms in a text. Some terms appear very often in a text that are of little or no importance, commonly defined as stopwords. They do not help extract the meaning of the text. The most common approaches are examining the number of occurrences of terms in texts and documents, the term frequency, inverse document frequency (TF-IDF), and Okapi BM25, which we previously mentioned [43]. Examining the word frequency or its occurrences is not optimal, as high occurrence does not necessarily mean that representing the document's content is important. TF-IDF corrects this issue by

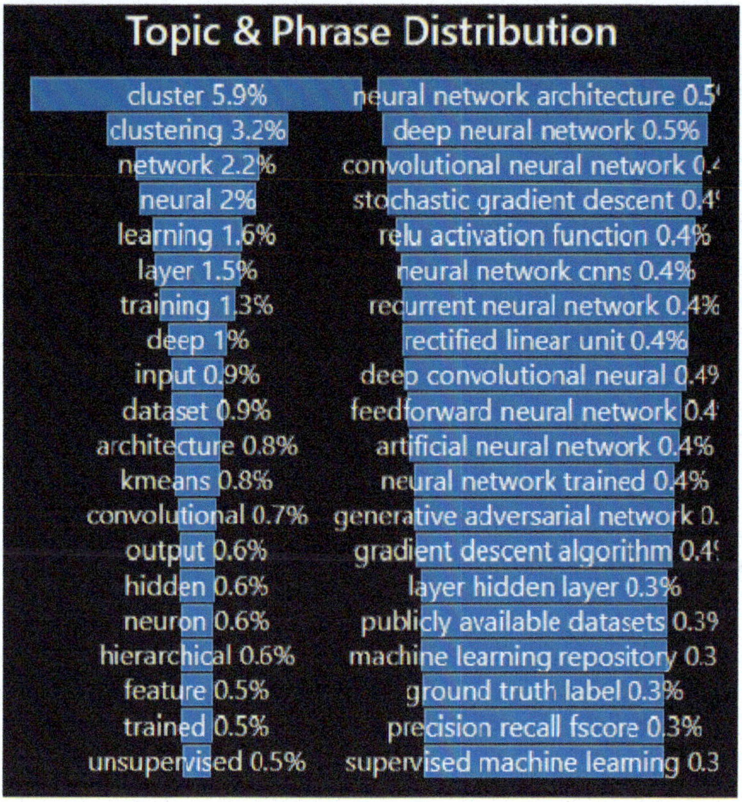

Fig. 2 "Generated single-word topics and n-gram-topics for a search on "visual analytics" with focus on "clustering" [29]

examining the rarity of words. If a word frequently appears in one document but not across multiple documents, then it is considered relevant in describing the content of that one document. Kadhim compared TF-IDF with BM25 using Twitter data and found that TF-IDF outperforms BM25 based on the F1 score and presented more stable results throughout his content categories [18]. This is confirmed by Truica et al., that conclude their work by stating that LDA works better with TF-IDF when having a more extensive vocabulary [43]. Further improving upon this, studies have combined the named approaches with "Naive Bayes" to examine its impact on Naive-Bayes-based approaches. Hairani et al. examined how TF-IDF in combination with a Naive Bayes classifier impacts precision and accuracy when classifying topics of thesis abstracts [12]. Naive Bayes without TF-IDF had an accuracy of 88.69% precision of 89.76% while with TF-IDF 81.74% accuracy, and 86.1% precision [12].

We use the identified topics of interest as features in a Naive Bayes approach. This approach considers the probability of specific topics and defines publications as relevant. The measured probability provides a relevance weight through Naive Bayes for each publication. The correlated feature-weighted (CFW) filter for the Naive Bayes approach described by Jiang et al. shows how features are assigned weights without alleviating the independent-based approach Naive Bayes is known for [17]. Jiang et al. [17] proposed that highly predictive features that correlate with a class but do not correlate to other features provide optimal weighting values [17]. According to the work of Jiang et al., we use the identified topics or term clusters as a highly predictive feature and, as a category, the annotation status of a publication. The CFW filter was tested with 36 UCI machine-learning benchmark datasets, which describe that CFW is at the very least as good as the standard Naive Bayes classification and significantly better in twelfe of the 36 datasets [17]. It is also at least as accurate as KLMFW, GRFW, and DFW and more accurate in three of the 36 datasets [17].

3.4 Visualization

The visualization of the results is performed with two complementary approaches. We used the "visual information seeking mantra" proposed by Shneiderman [37] with "overview first", "zoom and filter", then "details-on-demand" [37, p. 337]. Besides the"visual information seeking mantra", we visualized the results of a search and used the approach of van Ham and Perer [13] of "search", "show context" and "expand on demand". Figure 4 illustrates the start screen, where the most emerging topics in a certain database are visualized [29].

The users can search for queries manually via search bar or, if vectors of interest are available, use the interactive keyword visualization generated based on previous annotations as shown in Fig. 6. The results can be visualized in different ways. As described in the related work section, the sheer number of visualizations that can be used to visualize the search results holds vast potential. Data can be optimally explored by presenting interactive visualizations that suit the data and the researchers' cognition. This is why our approach is customizable and allows researchers to visualize data with various visualization approaches that suit their cognitive needs. The initial presentation of query results is accomplished with temporal spread of the query results as shown in Fig. 3 (1). Users can place further visualizations in the web-based application or replace a single visualization. In Fig. 3, the user has chosen beside the temporal overview that shows the number of publications per year in timeframe (1), an interactive listview of the publications (2), a simple graph-visualization for co-authored relations (3), and the spread of topics in a chosen year (4). This combination of visualizations, which is just one of many possible combinations, is a prime example of how our approach supports researchers.

Combined, these visualizations present an informative overview of a query or topic that allows the researcher to have various paths forward with which they can find relevant publications, annotate them, and commence further search iterations. These iterations can begin with the click of a button from the listview, as explained later in detail with Fig. 5, or by the researcher manually discovering terms or authors of interest via the visualizations. Individually each visualization has its value and

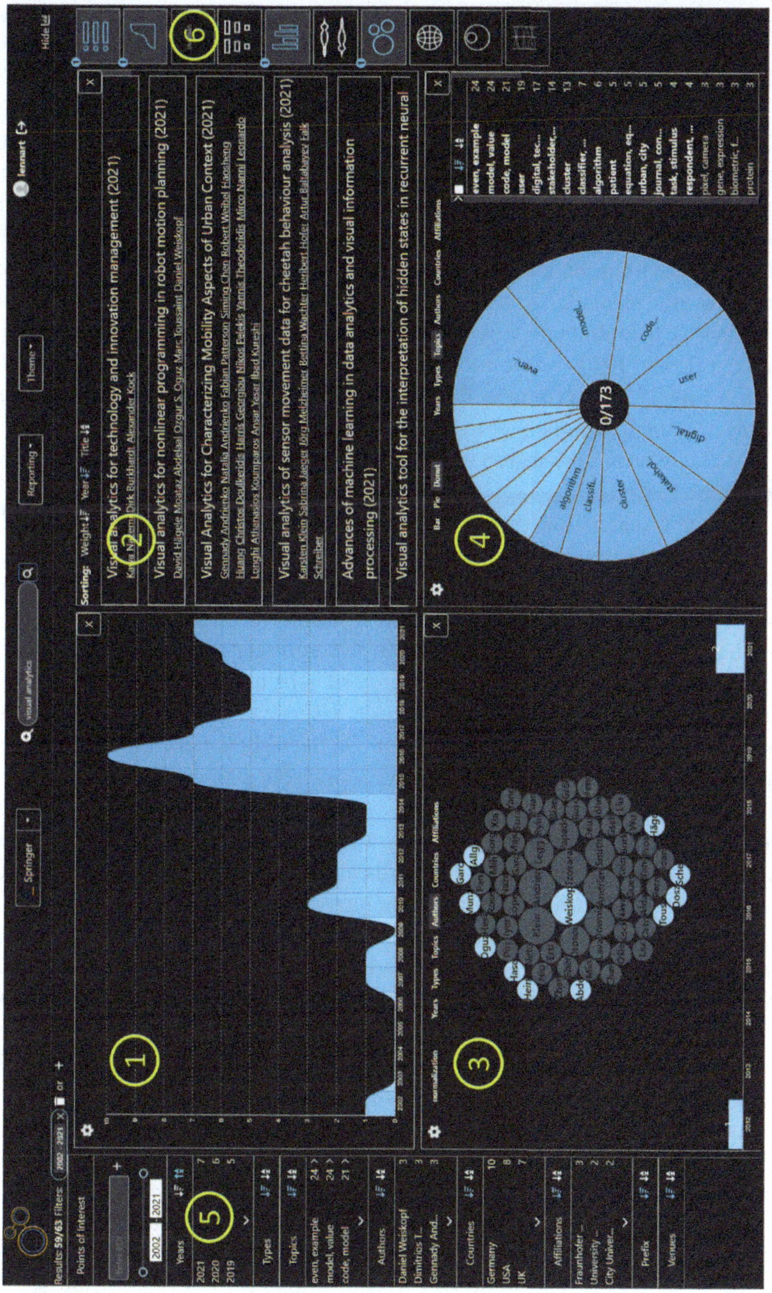

Fig. 3 Visualization of the results on the query "visual analytics". The user has chosen various visualizations for the result-set which are in the areas numbered 1–4. In area 5 the user can filter the results, and in area 6 they can choose different visualizations

purpose for the researcher. The researcher can see when the highest number of publications were published and when the first publication was published about the topic in the temporal overview (1). By presenting this information, the researcher can better understand the matureness of a topic and find the publication that first described the initial idea of something, such as technology. The listview (2) provides researchers with an optimal brief overview of each publication that has been found, allowing for a quick dissection of the most relevant information needed to decide if a publication is relevant. Furthermore, the buttons in that overview allow for on-the-fly annotation, meaning that researchers do not lose cognition or attention by being forced out of the exploration process and onto a separate page to analyze the publication. Their attention stays within the exploration process even as they check the relevance of publications and annotate. Knowing who the experts in a particular field are can support finding potential collaboration partners and finding published work in their respective relevant fields or monitoring them as they will publish relevant work in the future. Finally, by examining the spread of topics, the researcher can discover keywords relevant to the query or topic that the researcher is unaware of. By showing them these keywords, new search iterations and understanding of the query or topic is possible.

Besides the visualization, three more interaction areas allow the refinement of the results, general settings, and the choice of databases or allow choosing visualizations. (5) shows a set of facets that allow refinement of the search, e.g., years, topics, authors, etc. On top, general settings on the user interface can be adjusted, e.g., the color scheme or database choice. In Fig. 3, the Springer database, consisting of 2 million open-access publications at the time of writing, is chosen to visualize an aggregated result set of the Springer database. Users can alternatively choose other databases, such as DBLP, IEEE, or Crossref. On the right bar, the users can choose a variety of visualizations, whereas only those shown are supported by the underlying data models.

The treemap visualization shown in Fig. 4 illustrates the keyword composition of query results. The composition of keywords allows the researcher to discover keywords that are of relevance for their search, thus influencing future search iterations during the explorative research process. This visualization is shown when the user clicks on a year in the temporal visualization. The user can click on any keyword to see the main topic's underlying words. As the user clicks on a topic, the visualization zooms in, as illustrated by the black arrow.

Figure 4 shows the process with the search query "artificial neural network", which a user is interested in, and as they see the topics that are commonly associated with the search query in our database, they find "classifier, learning" as a topic. Clicking on this, the visualization zooms in. It shows an underlying treemap in which the user can see terms such as "classifier" or "learning" with which they further explore the dataset and bookmark relevant publications, which helps the system understand the users' interests.

In Fig. 5, the user has chosen other visualizations for the same result set. The default theme (light mode) was selected that uses another color scheme. On top, two visualizations enable the refinement of search results. (1) shows the "graphical

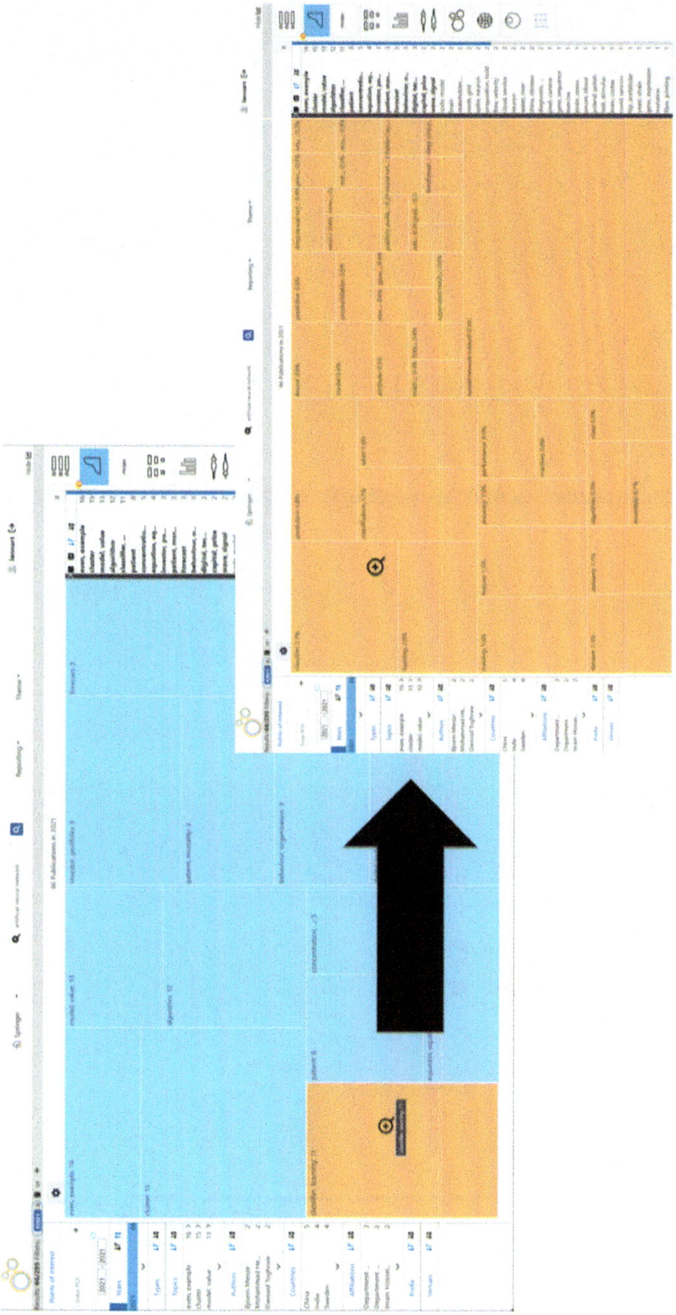

Fig. 4 An example of a "Zoom and Filter" approach that is based on Shneiderman's mantra [37] in which the user has searched for "artificial neural network" and is examining the keyword that his query results are composed of

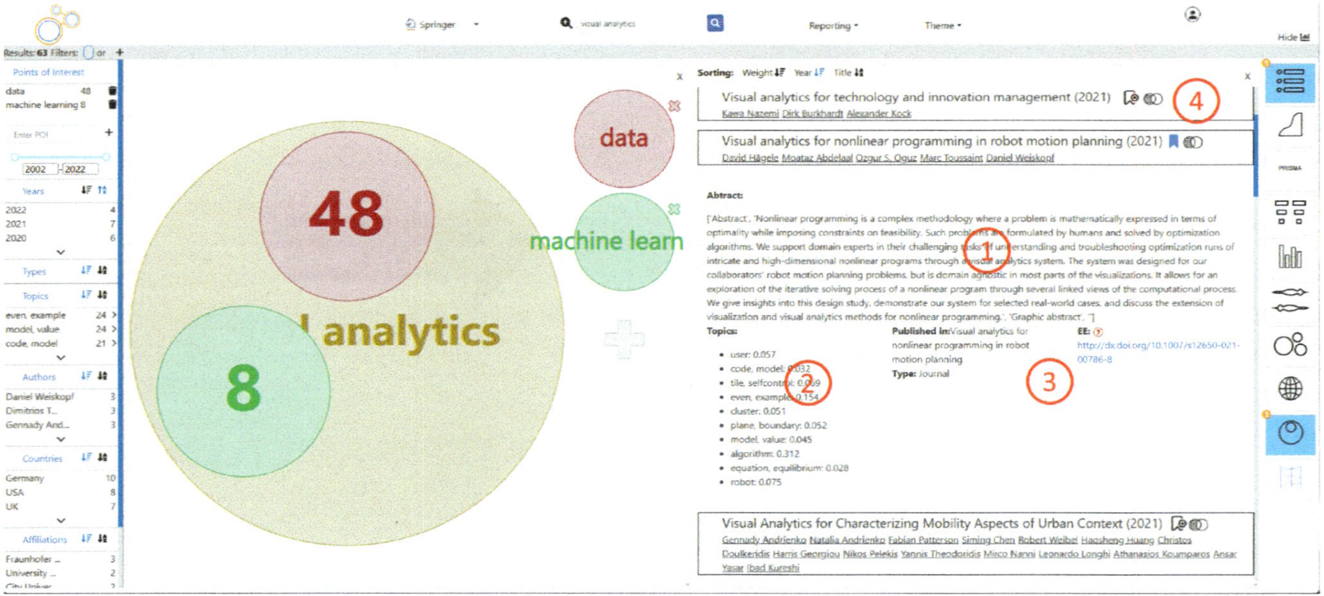

Fig. 5 Visual dashboard with a light theme: illustrating the graphical search on the left, and a list view on the right side. (1) shows an abstract, (2) the identified latent topics, and (3) the metadata. (4) buttons to bookmark or begin a new search for similar publications

Fig. 6 Keyword visualization that shows the topics of interest of a researcher. The blue cloud icon shows that the word cloud visualization is currently active and being visualized

search" visualization that allows the user to search within a search result set. In this case, the user chose the terms "machine learning" and "data". The circle in the center shows the number of results. (2) shows the refined search results as an interactive list. When clicked upon a title, the publication is opened, and more information about it is presented, such as an abstract seen in (1), identified topics (2), and the metadata (3). Furthermore, (4) shows two buttons, one for showing similar publications and one for bookmarking a publication are shown to the right of each title. Once bookmarked, the bookmark symbol turns blue to show the user that the publication is already bookmarked.

Figure 6 visualizes the topics the system has associated the user with based on their annotations. The word cloud visualization is interactive, meaning that the user can click on the terms that commence a search for publications with which the topic modeling system matches them. The colors are randomly generated, and the size is based on the number of elements found in the database. This instance of the word cloud is generated dynamically based on the identified vectors of interest, that is calculated with the annotations made. It is dynamic as it can evolve with newly made annotations.

3.5 Vectorization of Annotations

As researchers only annotate relevant publications, their annotations list is a list consisting of only relevant publications that can be used to identify latent topics that interest the researcher or are relevant to their research. Using the LDA topic modeling method, we identify the topics of interest [3]. We store 20 topics and a unique key per publication either as cookies to those who do not use our authentication server or saved to the user profile on our server. The LDA algorithm is calculated and renewed consistently, and doing so can lead to new topics or clusters. Furthermore, this eases the system's workload, as the LDA calculation takes time and can be RAM and CPU-intensive. Vectorizing each publication's topics allows comparisons and similarities on a topic level to be calculated. In the case of more than ten identical topics, a single interest vector is generated that contains all topics from both papers. A new vector is generated in the case of less than ten identical topics. So a "Vector of Interest" v_{I_j} may contain N topics as illustrated in Eq. 1, whereas topic t_0 is the name of the vector [3].

$$v_{I_j} = (t_1, t_2, t_3, \cdots t_n) \tag{1}$$

3.6 Similarity Measurements

Several commonly used similarity algorithms have been used in the context of publications; however, algorithms such as the Okapi BM25 ranking approach that search for and are word-matching focused [36, 42] are not appropriate as our topic modeling approach is topic focused. Thus, we calculate similarity with approaches that consider the distance between sets or distributions, allowing us to more optimally utilize our knowledge about the latent topics of the publications [3]. Cosine, Jaccard, and Chebyshev have been studied before and showed great potential in working with text [11, 15, 45]. We examined LDA combined with Cosine, Jaccard, and Chebyshev to find the combination with the best results [3]. The "Jaccard index", also known as the "intersection over union" and "Jaccard similarity coefficient", is a statistic used to measure similarity between finite sets of samples and the intersection between these two sets divided by the size of the union of the sets. Equation 2 illustrates the "Jaccard Similarity" according to [2, 3]. In the context of documents, this means that the "Jaccard Distance" of two documents can be measured by dividing the number of unique words that appear in both documents by the total number of unique words that appear in both documents. The result can be a 0 if there are no unique words found in both documents, 1 if both documents have all the same unique words, and results between 0 and 1 are based on the number of unique words found in both documents [2].

$$J(A, B) = \frac{|A \cap B|}{|A \cup B|} = \frac{|A \cap B|}{|A| + |B| - |A \cap B|} \tag{2}$$

The "Chebyshev Distance" is also known as the "Maximum Value Distance", "Chessboard Distance" or the "Maximum Metric", and this attempts to examine the magnitude of differences between the coordinates of a pair of objects based on the maximum difference over any of their axis values [19].

$$DCheb(p, q) = max_i|p_i - q_i| \tag{3}$$

We concluded that Cosine outperformed Chebyshev and Jaccard in that order, so we decided to use Cosine similarity. The decision to use Cosine similarity is described in detail with a human-based evaluation in Sect. 4. The "cosine similarity" approach allows us to examine the angle, thus the orientation, rather than the magnitude, which would be calculated using distance measurements. We calculate the similarity between the vectors of interest v_{I_j} and all topic vectors v_i stored in our database through "cosine similarity" based on the works of Sitikhu et al. [39]. When examining the vectorization of annotations made by the user, we calculate the "cosine similarity" to other topics due to our assumption that the annotated publications, thus topics, are considered interesting or relevant to the user as they bookmarked or annotated them. v_{I_j}, as illustrated in Sect. 3.5, of the documents is compared by examining the "cosine angle" between them. We consider t_0 of a vector v_{I_j} when examining the annotated topic vectors and calculating the similarity to all other topics stored as vectors v_i, each with 20 topics. The more different the compared vectors are, the more the value will tend towards -1, showing a dissimilarity of the compared publications. Publications that tend to go towards -1 are likely uninteresting for the researcher. The more similar the compared vectors are, the more the similarity score goes towards 1. The cosine approach is illustrated in Eq. 4.

$$cos(\theta) = \frac{v_{I_j} \cdot v_i}{||v_{I_j}|| \cdot ||v_i||} \tag{4}$$

3.7 *Interest Matrix*

The result of the similarity calculation utilizing the vectors of interest allows the system to identify the researcher's interests and generate an "interest matrix". Analyzing the accumulated latent topics and finding similar publications based on them will enable the system to go beyond simple similarity results as latent words and topics are taken into consideration that would have otherwise been neglected or not taken into consideration. This makes the interest matrix a more flexible approach to finding relevant publications than matching words, as the comparison takes place on a topic level, and the matrix is filled with topics. Unknown words might be part of a particular topic and, if identified as relevant for the researcher, will be considered without the researcher needing to search for them explicitly. The search results are based on the interest matrix shown in Table 1. Figure 5 shows that the system displays the results in a weighted form in which the highest weighted results are presented on the

Table 1 Interest matrix of annoted content

Interest matrix				
Interest term	Topic 1	Topic 2	...	Topic N
Visual analytics	Interactive visualization	Data visualization	...	Visualization data
Artificial intelligence	Neural network	Machine learning	...	Cognitive psychology

top and are labeled with the first and highest weighted phrase, for example, "Visual Analytics" and "Artificial Intelligence". The figure shows two identified topics of interest, each with four publications. Although the publications do not have the topic in their respective titles, the analysis indicates that the content matches the topic. As more information about the interests of the respective researchers is gathered, the interest matrix scales along increasing the number of interesting terms that are associated with a researcher.

4 Evaluation

To determine the best combination of topic modeling approaches and similarity measurements, we expanded upon our previous work and examined the similarity of resulting publications based on human judgment. We randomly chose one publication from our field of research that the topic modeling approaches LDA and LSA analyzed and then let the system find 50 publications that were the most similar based on each combination of topic modeling and similarity algorithm [30]. This means that we manually examined 300 full-text publications to identify how similar the content was to our publication, and we rated the content similarity on a three-point scale; low/no similarity, medium similarity, and high similarity. In practical terms, both medium and high similarity presented publications similar to the original and could be used to improve the original research. Low/no similarity papers were not of any practical use.

Figure 7 illustrates the results of our comparison and shows the proportion of the approach combinations as blue is cosine similarity, orange is Jaccard, and gray is Chebyshev. (1) shows the results for LDA, and (2) LSA. We can see that Jaccard is the similarity measurement with the most publications with low/no similarity on both sides. At the same time, Jaccard has the most insufficient amounts of publications with medium and high similarity. Jaccard is marginal percentage points better in combination with LDA than LSA, which is shown when comparing the high similarity results. Chebyshev presents interesting results, as although it, in combination with LDA, is responsible for almost 30% of the low/no similarity publications, it is also accountable for 40% of the medium similarity results and 35% of the high similarity publications found. Similar observations are shown in combination with

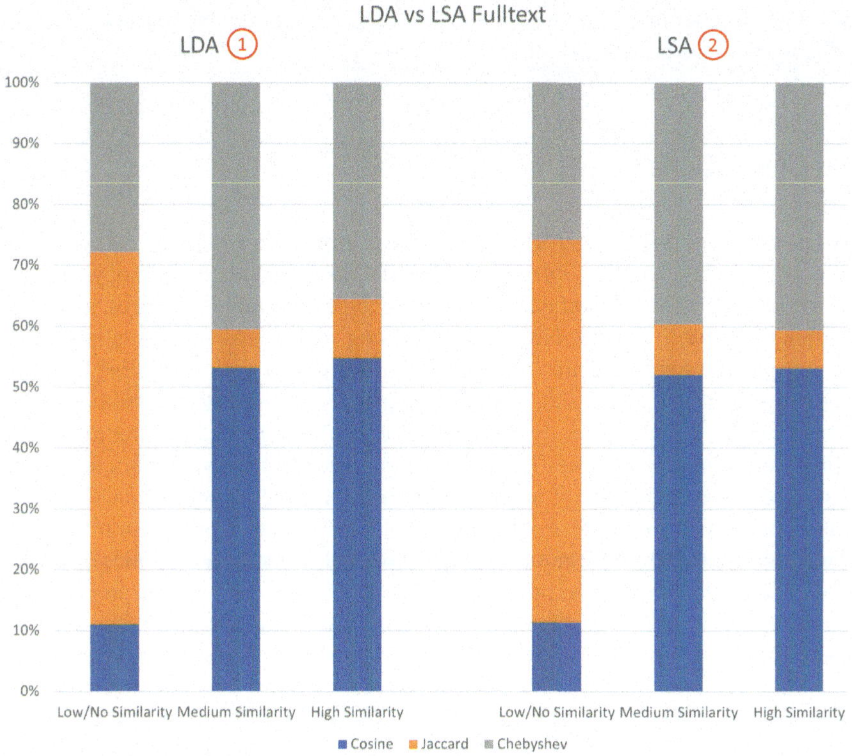

Fig. 7 Comparison of combinations of LDA (1) and LSA (2) with Cosine, Jaccard, and Chebyshev similarity algorithms based on identified content-based similar publications

LSA, where Chebyshev' proportion even increases to over 40% in the high similarity section. These results make Chebyshev better than Jaccard but not as accurate as Cosine. Cosine is the best similarity approach with both LDA and LSA, with being marginally better with LDA. Cosine caused the least amount of low/no similarity publications while having a majority of about 55% of the medium and high similarity publications in combination with LDA. Our evaluation clearly shows that Jaccard presents the most significant number of useless recommendations. We hypothesize that this is due to our approach having a high threshold that filters out many topics that do not have a high probability. The Jaccard method can compare topics more optimally if we lower the probability threshold, thus increasing the number of topics that Jaccard compares. Regarding the positive results from Cosine similarity, this is likely due to Cosine not focusing on the magnitude of our vectors, which word counts commonly influence. Overall, this impacts the decision between LDA and LSA; however, the intensity of the decision depends on the similarity algorithm that it will be combined with. We chose LDA and Cosine in our proposal, and the results are slightly better than if we had selected LSA. If we had decided to use Chebyshev, LSA would have been better suited.

5 Conclusions

We proposed in this chapter a visual analytics approach to support researchers in finding relevant scientific publications, even if terms are used that are not yet known. The exploratory character of our approach combines similarity measures, a machine learning approach, natural language processing and interactive visualizations as an integrated visual analytics model. The utilization of topic modeling, machine learning-based weighting, and similarity algorithms provide relevant publications as search results that significantly improve the exploration process. Using topic modeling, our approach identifies latent topics and leads to labeled vectors of interest. Using user-generated lists of annotated publications, these vectors consist of relevant publications for the researchers and their work. Thus, using them to calculate the similarity to the publications in the database allows relevant publications to be better identified, filtered, and presented. The Naive Bayes approach we propose takes the identified topics of interest and uses them as weighted features. Our approach can furthermore be extended by searching for venues and researchers that can be identified and presented [4]. The topic modeling approach also intercepts the issue of emerging terms that might be unknown to researchers due to the terms being yet used in mainstream literature, the researcher's missing knowledge of the domain, or the rarity of unique approaches. Allowing the system to use these vectors rather than a simple search query, allows for being more flexible than an exact match of search query terms in publications. Our approach of utilizing the annotation behavior of researchers to identify their topics of interest is novel and presents another approach to gathering information that can be used for recommendation systems.

The second main area that our work presents is the utilization of interactive visualizations and the introduced annotation in the context of exploration. This approach is crucial to the exploration of publications as it allows the researcher, on the one hand, to analyze and explore large amounts of publications while annotating on the fly without interrupting the search and exploration process or causing attention or cognitive breaks. Using the pipeline described in our general approach, we minimize non-relevant publications. By focusing on only relevant publications, the impact of visualizations is enhanced, creating a more optimal exploratory search approach. As the figure is a general model of the entire process, researchers can further improve upon or experiment with different technologies for each step to fine-tune our proposal. Our approach of allowing researchers to choose the visualizations that suit them and their needs, zoom and filter, and annotate on the fly, while behind the scenes intelligent approaches analyze and improve the search results with each annotation and search iteration, presents an innovative fusion of interactive visualizations and machine learning that supports researchers on their exploratory journey.

6 Future Work

Our work showcases the potential of utilizing researchers' interests and exploratory behavior, and visual exploration to improve the finding of relevant publications. However, questions remain on the potential of using this approach in research platforms or in combination with recommendation systems. Recommendation systems are commonly confronted with the cold start problem. Although our proposed system also requires information in the form of annotated publications and thus is also affected by the cold start problem, the approach to creating the interest matrix utilizing topic modeling can assist recommendation systems in analyzing researchers more efficiently by examining their published publications. Furthermore, expanding the information gathered and utilized can yield significant improvements and develop use cases, allowing the proposed system to provide more than relevant publications. For example, the co-authorship analysis would allow for networks of authors to be visualized and better analyzed. By examining which venues published each paper, a connection between the venues and the identified topics can be made, which also would allow for determining the credibility of the publications and respective venues. The focus on interactive visualizations to assist and encourage exploration enables researchers to solve complex analytical tasks. Observing and studying researchers is crucial as they use our proposed interactive visualizations for exploratory purposes to determine what effect visualizations have in scientific research.

Acknowledgements We thank Kjell Kunz from the Darmstadt University of Applied Sciences who made an important contribution to the research on this work. This work was conducted within the research group on Human-Computer Interaction and Visual Analytics at the Darmstadt University of Applied Sciences (https://vis.h-da.de).

References

1. Alshareef A, Alhamid M, El Saddik A (2019) Academic venue recommendations based on similarity learning of an extended nearby citation network. IEEE Access 1. https://doi.org/10.1109/ACCESS.2019.2906106
2. Bengfort B (2018) Applied text analysis with python: enabling language-aware data products with machine learning. https://learning.oreilly.com/library/view/-/9781491963036
3. Blazevic M, Sina LB, Burkhardt D, Siegel M, Nazemi K (2021) Visual analytics and similarity search—interest-based similarity search in scientific data. In: 2021 25th international conference information visualisation (IV), pp 211–217. https://doi.org/10.1109/IV53921.2021.00041
4. Blazevic M, Sina LB, Nazemi K (2022) Visual collaboration—an approach for visual analytical collaborative research. In: 2022 26th international conference information visualisation (IV), pp 293–299. https://doi.org/10.1109/IV56949.2022.00057
5. Blei DM, Ng AY, Jordan MI (2003) Latent dirichlet allocation. J Mach Learn Res 3. http://www.jmlr.org/papers/v3/blei03a.html

6. Breitinger C, Herklotz K, Flegelskamp T, Meuschke N (2022) Recommending research papers to chemists: a specialized interface for chemical entity exploration. In: 2022 ACM/IEEE joint conference on digital libraries (JCDL), pp 1–4
7. Breitinger C, Reiterer H (2021) Visualizing feature-based similarity for research paper recommendation. In: 2021 ACM/IEEE joint conference on digital libraries (JCDL), pp 212–221. https://doi.org/10.1109/JCDL52503.2021.00033
8. Caillou P, Renault J, Fekete JD, Letournel AC, Sebag M (2021) Cartolabe: a web-based scalable visualization of large document collections. IEEE Comput Graphics Appl 41(2):76–88. https://doi.org/10.1109/MCG.2020.3033401
9. Dagan I, Lee L, Pereira F (1997) Similarity-based methods for word sense disambiguation. In: 35th annual meeting of the association for computational linguistics and 8th conference of the European chapter of the association for computational linguistics. Association for Computational Linguistics, Madrid, Spain, , pp 56–63. https://doi.org/10.3115/976909.979625. https://www.aclweb.org/anthology/P97-1008
10. Deerwester S, Dumais ST, Furnas GW, Landauer TK, Harshman R (1990) Indexing by latent semantic analysis. J Am Soc Inf Sci 41(6):391–407
11. Gomaa WH, Fahmy AA (2013) A survey of text similarity approaches. Int J Comput Appl 68:13–18
12. Hairani H, Anggrawan A, Wathan AI, Latif KA, Marzuki K, Zulfikri M (2021) The abstract of thesis classifier by using naive Bayes method. In: 2021 international conference on software engineering & computer systems and 4th international conference on computational science and information management (ICSECS-ICOCSIM), pp 312–315. https://doi.org/10.1109/ICSECS52883.2021.00063
13. van Ham F, Perer A (2009) Search, show context, expand on demand: supporting large graph exploration with degree-of-interest. IEEE Trans Vis Comput Graph 15:690–953
14. Heidarian A, Dinneen MJ (2016) A hybrid geometric approach for measuring similarity level among documents and document clustering. In: 2016 IEEE second international conference on big data computing service and applications (BigDataService), pp 142–151. https://doi.org/10.1109/BigDataService.2016.14
15. Huang A (2008) Similarity measures for text document clustering. In: Proceedings of the 6th New Zealand computer science research student conference
16. Huang Z, Chung W, Ong TH, Chen HC (2002) A graph-based recommender system for digital library. In: ACM/IEEE joint conference on digital libraries, pp 65–73. https://doi.org/10.1145/544220.544231
17. Jiang L, Zhang L, Li C, Wu J (2019) A correlation-based feature weighting filter for naive Bayes. IEEE Trans Knowl Data Eng 31(2):201–213. https://doi.org/10.1109/TKDE.2018.2836440
18. Kadhim AI (2019) Term weighting for feature extraction on twitter: a comparison between BM25 and TF-IDF. In: 2019 international conference on advanced science and engineering (ICOASE), pp 124–128. https://doi.org/10.1109/ICOASE.2019.8723825
19. Klove T, Lin T, Tsai S, Tzeng W (2010) Permutation arrays under the Chebyshev distance. IEEE Trans Inf Theory 56(6):2611–2617. https://doi.org/10.1109/TIT.2010.2046212
20. Kovalerchuk B (2018) Visual knowledge discovery and machine learning. In: Intelligent systems reference library
21. Kovalerchuk B, Andonie R, Datia N, Nazemi K, Banissi E (2022) Visual knowledge discovery with artificial intelligence: challenges and future directions. In: Kovalerchuk B, Nazemi K, Andonie R, Datia N, Banissi E (eds) Integrating artificial intelligence and visualization for visual knowledge discovery. Studies in computational intelligence, vol 1014. Springer International Publishing, Cham, pp 1–27. https://doi.org/10.1007/978-3-030-93119-3_1
22. Lawrence S, Lee Giles C, Bollacker K (1999) Digital libraries and autonomous citation indexing. Computer 32(6):67–71. https://doi.org/10.1109/2.769447
23. Lewis JR (1992) Psychometric evaluation of the post-study system usability questionnaire: the PSSUQ. Proc Hum Factors Soc Annu Meet 36(16):1259–1260. https://doi.org/10.1177/154193129203601617

24. Lu W, Janssen J, Milios E, Japkowicz N, Zhang Y (2007) Node similarity in the citation graph. Knowl Inf Syst 11:105–129. https://doi.org/10.1007/s10115-006-0023-9
25. Mahmood Q, Qadir MA, Afzal MT (2014) Finding relatedness between research papers using similarity and dissimilarity scores. In: Li F, Li G, Hwang SW, Yao B, Zhang Z (eds) Web-age information management. Springer International Publishing, Cham, pp 707–710
26. Mahmood Q, Qadir MA, Afzal MT (2017) Application of cores to compute research papers similarity. IEEE Access 5:26124–26134. https://doi.org/10.1109/ACCESS.2017.2771207
27. Marchionini G (2006) Exploratory search: from finding to understanding. Commun ACM 49(4):41–46. https://doi.org/10.1145/1121949.1121979
28. Munzner T (2014) Visualization analysis and design. Taylor & Francis Inc
29. Nazemi K, Burkhardt D (2019) Visual analytics for analyzing technological trends from text. In: 2019 23rd international conference information visualisation (IV), pp 191–200. https://doi.org/10.1109/IV.2019.00041. Best Paper Award
30. Nazemi K, Klepsch MJ, Burkhardt D, Kaupp L (2020) Comparison of full-text articles and abstracts for visual trend analytics through natural language processing. In: 2020 24th international conference information visualisation (IV). IEEE CPS, pp 360–367. https://doi.org/10.1109/IV51561.2020.00065
31. Ng YK (2020) Research paper recommendation based on content similarity, peer reviews, authority, and popularity. In: 2020 IEEE 32nd international conference on tools with artificial intelligence (ICTAI), pp 47–52. https://doi.org/10.1109/ICTAI50040.2020.00018
32. Niraula N, Banjade R, Ştefănescu D, Rus V (2013) Experiments with semantic similarity measures based on LDA and LSA. In: Dediu AH, Martín-Vide C, Mitkov R, Truthe B (eds) Statistical language and speech processing. Springer, Berlin Heidelberg, Berlin, Heidelberg, pp 188–199
33. Prasetya DD, Wibawa AP, Hirashima T (2018) The performance of text similarity algorithms. Int J Adv Intell Inform 4(1):63–69. https://doi.org/10.26555/ijain.v4i1.152. http://ijain.org/index.php/IJAIN/article/view/152
34. Rahutomo F, Kitasuka T, Aritsugi M (2012) Semantic cosine similarity. In: The 7th international student conference on advanced science and technology
35. Ristanti P, Wibawa A, Pujianto U (2019) Cosine similarity for title and abstract of economic journal classification. In: 2019 5th international conference on science in information technology (ICSITech), pp 123–127. https://doi.org/10.1109/ICSITech46713.2019.8987547
36. Russell SJ (2012) Künstliche intelligenz : ein moderner ansatz. http://d-nb.info/1023010321/04
37. Shneiderman B (1996) The eyes have it: a task by data type taxonomy for information visualizations. In: VL, pp 336–343
38. Sidorov G, Gelbukh A, Gomez Adorno H, Pinto D (2014) Soft similarity and soft cosine measure: similarity of features in vector space model. Computación y Sistemas 18. https://doi.org/10.13053/cys-18-3-2043
39. Sitikhu P, Pahi K, Thapa P, Shakya S (2019) A comparison of semantic similarity methods for maximum human interpretability. In: 2019 artificial intelligence for transforming business and society (AITB). IEEE. https://doi.org/10.1109/aitb48515.2019.8947433
40. Soyusiawaty D, Zakaria Y (2018) Book data content similarity detector with cosine similarity (case study on digilib.uad.ac.id). In: 2018 12th international conference on telecommunication systems, services, and applications (TSSA), pp 1–6. https://doi.org/10.1109/TSSA.2018.8708758
41. Thada V, Jaglan D (2013) Comparison of Jaccard, dice, cosine similarity coefficient to find best fitness value for web retrieved documents using genetic algorithm. Int J Innov Eng Technol 2:202–205
42. Thakur N, Mehrotra D, Bansal A, Bala M (2019) Comparative analysis of ranking functions for retrieving information from medical repository. Malays J Comput Sci 32(1):18–30. https://doi.org/10.22452/mjcs.vol32no1.2
43. Truica CO, Radulescu F, Boicea A (2016) Comparing different term weighting schemas for topic modeling. In: 2016 18th international symposium on symbolic and numeric algorithms for scientific computing (SYNASC), pp 307–310. https://doi.org/10.1109/SYNASC.2016.055

44. Valdez D, Pickett AC, Goodson P (2018) Topic modeling: latent semantic analysis for the social sciences. Soc Sci Q 99(5):1665–1679. https://doi.org/10.1111/ssqu.12528
45. Wang C, Song Y, Li H, Sun Y, Zhang M, Han J (2017) Distant meta-path similarities for text-based heterogeneous information networks. In: Proceedings of the 2017 ACM on conference on information and knowledge management, CIKM '17. Association for Computing Machinery, New York, NY, USA, pp 1629–1638. https://doi.org/10.1145/3132847.3133029
46. White RW, Roth RA (2009) Exploratory search: beyond the query-response paradigm. In: Marchionini G (ed) Synthesis lectures on information concepts, retrieval, and services, vol 1. Morgan & Claypool Publishers. https://doi.org/10.2200/s00174ed1v01y200901icr003
47. Zhao J, Sun M, Chiu P, Chen F, Liew B (2021) Know-what and know-who: document searching and exploration using topic-based two-mode networks. In: 2021 IEEE 14th pacific visualization symposium (PacificVis), pp 86–90. https://doi.org/10.1109/PacificVis52677.2021.00019

Information Visualization with ChatGPT

Andreas Stöckl

1 Introduction

With the release of ChatGPT (https://chat.openai.com/), OpenAI has generated tremendous buzz and diverse applications. The chat interface for the latest language model from their house shows remarkable capabilities in various knowledge domains. It offers easy accessibility for a broad group of users due to its simple interface. It is a vivid example of how we can use Artificial Intelligence for cognitive tasks. In this work, we would like to investigate its suitability for information visualization tasks.

Artificial Intelligence in the form of natural language user interfaces has been incorporated into software for visual data analysis, such as Tableau's "Ask Data" [48] and Microsoft's "Power BI Q&A" [29]. These tools allow users to ask questions in plain language and receive instant visual answers without navigating through menus or understanding the data structure. However, using these tools can still be challenging for some users, and integrating more advanced language models like the one behind "ChatGPT" may improve their ease of use.

In recent years, there has been a rise in powerful new language models that are based on the transformer architecture [49]. These models, such as those developed by [7, 33, 34], have been trained on large text datasets [14], utilizing pretraining techniques. As a result, they have achieved remarkable performance on various natural language processing (NLP) benchmarks, as noted by [35].

The objective of this chapter is to investigate the potential benefits of employing these advanced language models in natural language interfaces (NLIs) for data visualization. To accomplish this, we conducted a preliminary user study that compared the performance of ChatGPT with two alternative approaches utilizing Natural Language Processing (NLP) techniques for information visualization.

For comparison, we employ two approaches that we developed in [45]. In one comparative approach, we created a visualization software prototype based on the OpenAI Codex model trained on source code [8].

A. Stöckl (✉)
Department Digital Media, University of Applied Sciences, Roseggerstraße, 15 4600 Wels, Upper Austria, Austria
e-mail: andreas.stoeckl@fh-hagenberg.at

B. Kovalerchuk et al. (eds.), *Artificial Intelligence and Visualization: Advancing Visual Knowledge Discovery*, Studies in Computational Intelligence 1126,
https://doi.org/10.1007/978-3-031-46549-9_17

For a second variant of the prototype, we used traditional techniques for NLI in terms of data visualization. We have incorporated a parser to interpret Natural Language (NL) utterances. This was achieved utilizing the open-source toolkit "Natural Language for Data Visualization" (NL4DV) [30]. It takes as input a dataset and an utterance corresponding to this dataset, returning a JavaScript object notation (JSON) that includes an ordered list of Vega-Lite [37] specifications that are presented as output.

2 Related Work

The reviews by Shen et al. [41] and Srinivasan et al. [44] provide comprehensive coverage of various NLI systems, including their functions and capabilities. The first implementations of NLI trace back to 2001 when it was incorporated into the multimodal interface of an existing information system [9]. This milestone led to the development of a thriving research field that came up with a wide array of approaches [15, 22, 23, 39, 40, 47, 54].

In the realm of visualization recommendations (VisRec), foundational works by Qin et al. [32] and Zhu et al. [55] serve as valuable resources. The classic NLI database subfield is well synthesized by Affolter et al. [2] and Hozcan et al. [31], while visual question answering is overviewed by Mathew et al. [28].

As mentioned in the introduction, we not only utilize the ChatGPT and OpenAI Codex approaches but also implement a variant based on traditional NLP techniques for comparative analysis, using the NL4DV toolkit by Narechania et al. [30]. This implementation makes use of NLTK [25], Stanford CoreNLP [27], Named Entity Recognition (NER) [13], and spaCy [18] to execute NLP tasks like POS tagging, entity recognition, and dependency parsing. Furthermore, it analyses the input query and transmits relevant visualization information, like file attributes and analytical tasks, to a VisRec engine that generates a list of visualization specifications pertinent to the input query.

Descriptive and visualization techniques of information graphics often utilize a grammar of graphics, as detailed by Wilkinson et al. [53], employing newer visualization grammars like Vega [38] and Vega-Lite [37].

ChatGPT, Codex, and other conventional language models employ training methodologies well-suited for NLP but overlook structural information in programming languages, which is crucial for a better understanding of the code. Thus, the CodeT5 model [50], derived from Google's T5 model [36], is presented as an alternative for code generation.

CodeT5 utilizes a multi-stage training process to benefit from the structural information. One aspect of the model learns to complete the code where certain passages or specific identifiers, like variables or function names, are randomly masked. Another learns to identify identifiers in the code inputs. Finally, CodeT5 learns to convert code to text and vice versa. These processes collectively enhance the AI model's comprehension of code.

CodeT5 is trained on 8.35 million code examples paired with comments, and the most extensive version possesses 220 million parameters. This is significantly smaller than ChatGPT and the largest Codex model, which houses 12 billion parameters, used in our prototype. CodeT5's developers claim it achieves top performance on 14 tasks, including text-to-code generation, in Microsoft's CodeXGLUE benchmark [26]. However, no comparison with ChatGPT and OpenAI Codex is available. Unlike ChatGPT and OpenAI Codex, the model is open-source, providing a base for fine-tuning. CodeBERT [12], based on Google's BERT [10], is another model employable for code generation.

3 Interface

Since ChatGPT cannot generate graphics as output, we need a slightly extended setup to make this possible. As an interface, we use Jupyter notebooks, which allows us to execute the Python source code generated by ChatGPT directly as code cells and thus render the graphics. Figure 1 shows a Jupyter notebook with the solution to a task in the form of the source code ChatGPT generated and the graph generated from it with Python. In addition, ChatGPT has generated an accompanying text. As a connection from Jupyter to ChatGPT, the browser plugin "ChatGPT for Jupyter" (https://github.com/jflam/chat-gpt-jupyter-extension) is used for the experiment, which enables the sending of queries from Jupyter cells directly to ChatGPT and to get the answers back in the Jupyter document. The source code generated by ChatGPT is copied into a separate code cell and can be converted directly into the diagram with the Python interpreter.

To test the possibilities of chart generation with the help of OpenAI Codex, we created a web application prototype that allows data visualizations to be created using text inputs and then refined with context-related palettes. These text commands are used to define the basic structure of the visualizations. The user can then use the palettes to set additional specifications, such as those for line width or color. Figure 2 shows the sidebar on the left that allows loading or re-importing data and exporting the results. In the central area, the user can enter text commands via a multiline input field, and the visualization is displayed below. The right column contains graphical user interface (GUI) elements depending on the chart type displayed.

The prototype employs distinct components for data visualization, on top of two alternative solutions for Natural Language Processing (NLP). The NL4DV method leverages Vega-Lite for visualization, while the OpenAI Codex approach utilizes Matplotlib for the same purpose. We delve into the rationale behind selecting various visualization components in the section concerning system architecture. The prototype, illustrated in Fig. 2, is a mixed-initiative system designed for visualization, permitting both Natural Language (NL) input and direct manipulation interactions via a Graphic User Interface (GUI).

For data management, the prototype enables the user to upload .csv files or load pre-existing ones through a selection box. Subsequently, the data fields along with

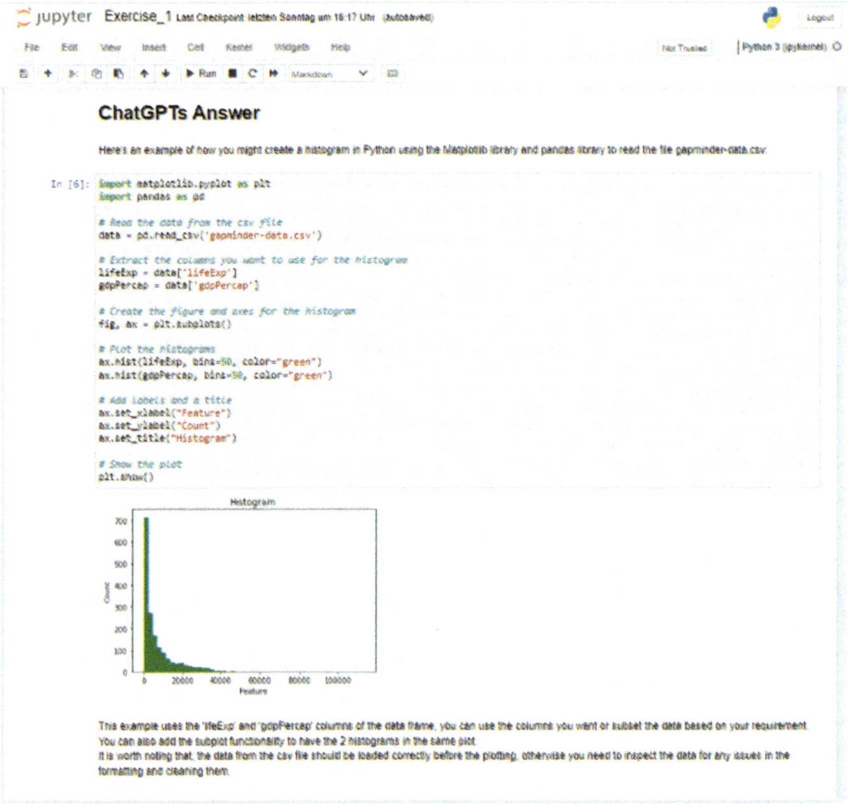

Fig. 1 Jupyter notebook with the solution to a task

their corresponding data types are displayed. The prototype provides export options in the form of Portable Network Graphics (PNG) images and HyperText Markup Language (HTML) exports. The HTML export is facilitated using the mpld3 package, which translates Matplotlib objects into D3js. The comparative prototype equipped with Vega-Lite visualization presents export options directly through the graphic's context menu.

Apart from the opportunity to view a brief tutorial video, the screen's central area presents a multiline text input field for entering instructions in NL. Following this, the produced visualization is showcased. Depending on the prototype variant, this could be a Matplotlib graphic or a Vega-Lite chart.

The right-hand column features several GUI elements along with text commands. The user is allowed to assign a title, label the axes, and set limits. Other elements such as line width, color palette, or bar color are presented according to the created chart type.

In the ChatGPT approach (Fig. 1), the menus of Jupyter are used to upload the data files. In this case, the graphics library used for chart generation depends on the

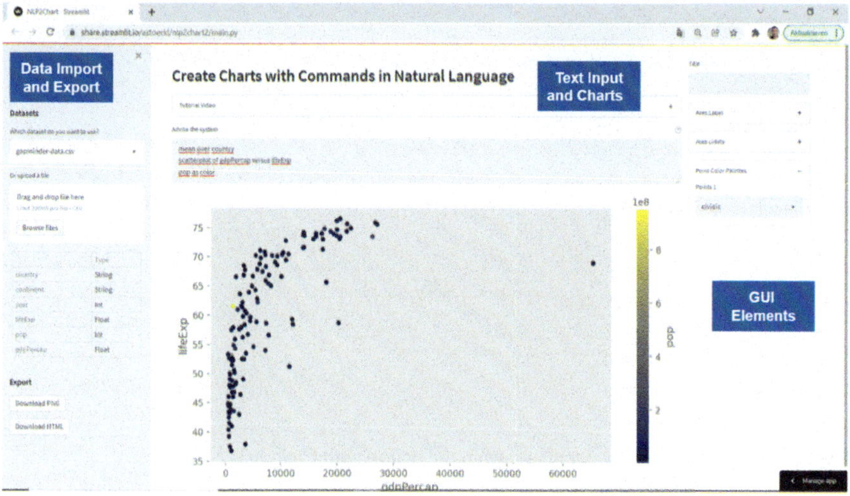

Fig. 2 GUI of the web application

code generated by ChatGPT. In the vast majority of the charts generated by the 11 people in the seven tasks, ChatGPT generates Matplotlib graphics. Only in 4 cases "Seaborn" [51] is used and in another 4 cases, "Plotly" [21].

4 An Example of Dialog with ChatGPT

Now, we'll delve into illustrative usage scenarios to elucidate the potential and constraints of the different methodologies. A more comprehensive analysis is provided in Sect. 5. For demonstration and assessment purposes, we utilize the data from Gapminder.org [16].

First, we upload the corresponding .csv file in the Jupyter notebook. We want to determine the top 15 countries having the highest gross domestic product (GDP) per capita.

The dialog starts with the text input:

> How can I plot the GDP per country with the help of a bar chart in my jupyter notebook. I have the following columns in my csv named gapminder-data: lifeExp, pop, gdpPercap, country, continent, year. The graph should be formatted in the following way: Give "Top GDP Countries" as the title and colour the bars with "Green".

It describes the context, the structure of the data, and how the graph should be formatted. The input generates source code that produces an error (Fig. 3), that we will fix below, because the data was not imported in the source code.

```
: import matplotlib.pyplot as plt

  # group the data by country and calculate the mean GDP per capita
  gdp_by_country = data.groupby('country')['gdpPercap'].mean()

  # sort the countries by GDP in descending order
  gdp_by_country = gdp_by_country.sort_values(ascending=False)

  # plot the data as a bar chart
  gdp_by_country.plot(kind='bar', color='green', title='Top GDP Countries')

  # customize the x-axis labels
  plt.xlabel('Country')
  plt.ylabel('GDP per capita')

  # show the plot
  plt.show()

  ---------------------------------------------------------------------------
  NameError                                 Traceback (most recent call last)
  <ipython-input-3-74224704c030> in <module>
        2
        3 # group the data by country and calculate the mean GDP per capita
  ----> 4 gdp_by_country = data.groupby('country')['gdpPercap'].mean()
        5
        6 # sort the countries by GDP in descending order

  NameError: name 'data' is not defined
```

Fig. 3 Output from the first command from the dialog with ChatGPT

We confront ChatGPT with the error message.

> I have the following error: NameError: name'data' is not defined

Now we generate the correct source code, which creates the chart in Fig. 4. However, this shows all countries, not just the top 15 as required in the task. Therefore, we need to specify this in the next instruction.

> In the plot I want to display the top 15 countries in descending order.

We found the correct solution, as shown in Fig. 5.

5 System Overview of the Web Applications

We've constructed the two versions of the prototypes under comparison with ChatGPT as web applications using Python. In both instances, the "Streamlit" framework [46] is utilized. They can process data in .csv format and allow exporting as PNGs or HTML. However, the two variants diverge in terms of NLP and visualization techniques. As such, our intention is to inspect the outcomes produced by the distinct NL processing and visualization technologies.

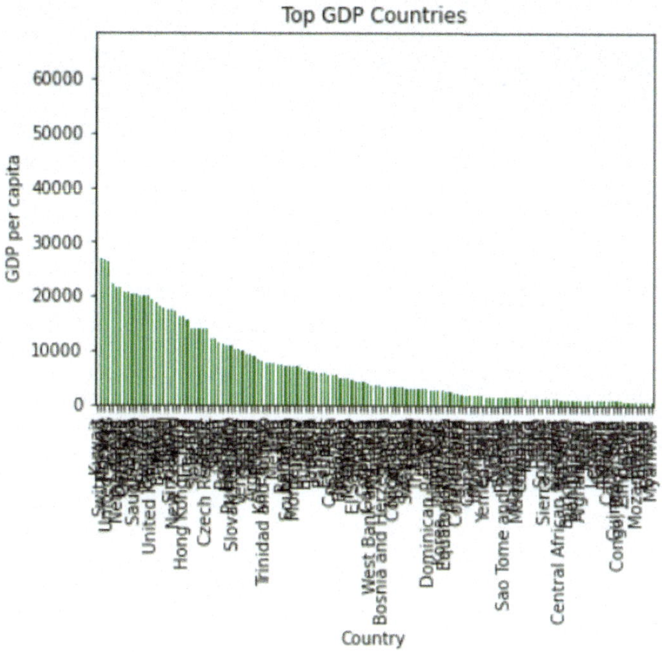

Fig. 4 First version of the resulting diagram in the example

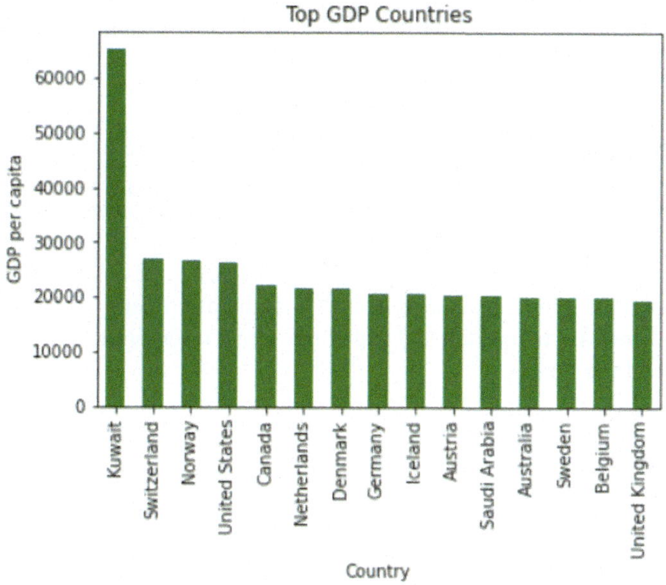

Fig. 5 Presentation of the solution of the task

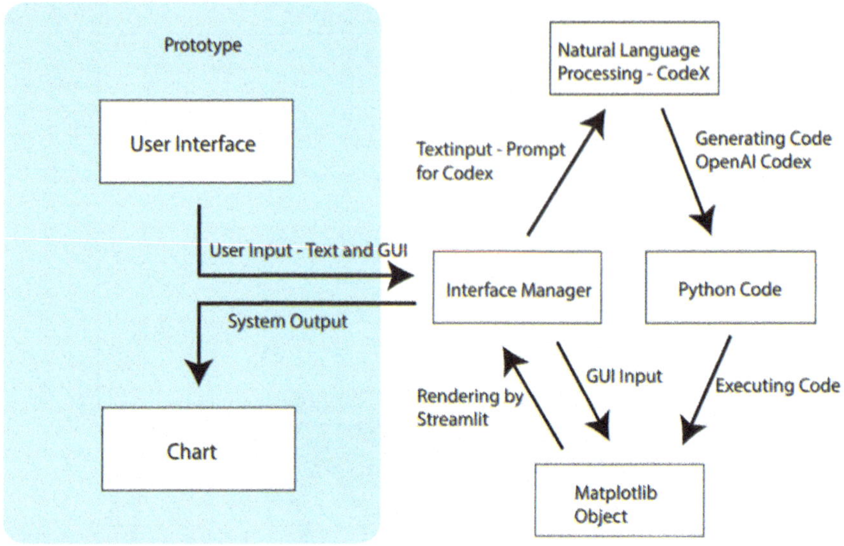

Fig. 6 Architecture of the CodeX-based variant of the prototype

5.1 Architecture of the OpenAI Codex Version

Figure 6 presents a schematic representation of the architecture of the variant utilizing the CodeX language model [8].

5.1.1 Textinput—Prompt for Codex

The construction of the text prompt plays a critical role in determining the quality of the output. Each prompt in the prototype comprises a text stating that the data file chosen in the GUI is set to be loaded in the structure "Load NAMEOFDATAFILE into a dataset", in addition to the text input from the user interface. Triple quotation marks are appended to the start and end of the string, resembling their typical usage in source code comments.

5.1.2 Generating Code with OpenAI Codex

In this variant, the graphics are crafted using Python code, which has been previously created with the aid of the language model. The variant of the generative pre-trained transformer 3 (GPT-3) known as "Codex", which was trained using source code, serves as the language model. We make use of this through OpenAI AI [3], a language model that has undergone pre-training on source code, which we employ in its original, non-fine-tuned state. Moreover, it is not even explicitly defined that Python code is to be generated, which results implicitly from the text prompt. We use the Davinci-codex variant, the most powerful model of the application programming interface (API), allowing input lengths of up to 4.096 tokens.

The Python code that is outputted relies solely on the text input. The text input, along with several other parameters described below, are supplied as input for the model.

- **Temperature**: This is set to 0 in our case. Higher values stimulate the creativity of the language model, ensuring that identical inputs do not always yield the same output. This is suitable for generating text in Natural Language (NL), but not in our scenario where the goal is to generate code as accurately and understandably as possible.
- **top_p and top_k**: These parameters govern the potential to generate varied outputs and are thus set to 1. Since the temperature is already set to 0, these mechanisms no longer have any influence.
- **max_tokens**: This determines the maximum number of tokens generated by the API in a single call and is set to 500. The number should be sufficiently large to generate lengthy code but not excessively so, as that would inflate the cost of API calls.
- **frequency_penalty**: A parameter ranging between –2.0 and 2.0. Positive values impose penalties on new tokens based on their prior frequency in the text, diminishing the likelihood of the model replicating the exact line. This parameter is set to 0.
- **presence_penalty**: A value between –2.0 and 2.0. Positive values impose penalties on new tokens based on their previous appearance in the text, enhancing the chance that the model will introduce new topics. This parameter is also set to 0.

For an in-depth understanding of spamming strategies that utilize parameters such as temperature, top_p, and top_k, please refer to [11, 17, 52].

5.1.3 Code Execution

The produced source code is subsequently executed on a line-by-line basis, barring lines that contain the ".show()" method to prevent immediate graphic displays. Following the execution, the existing Matplotlib object generated by the code execution (if present) is preserved in the application's session.

5.1.4 GUI Input

Specifications to the graph via the GUI elements are made directly as changes to the Matplotlib object and then saved again.

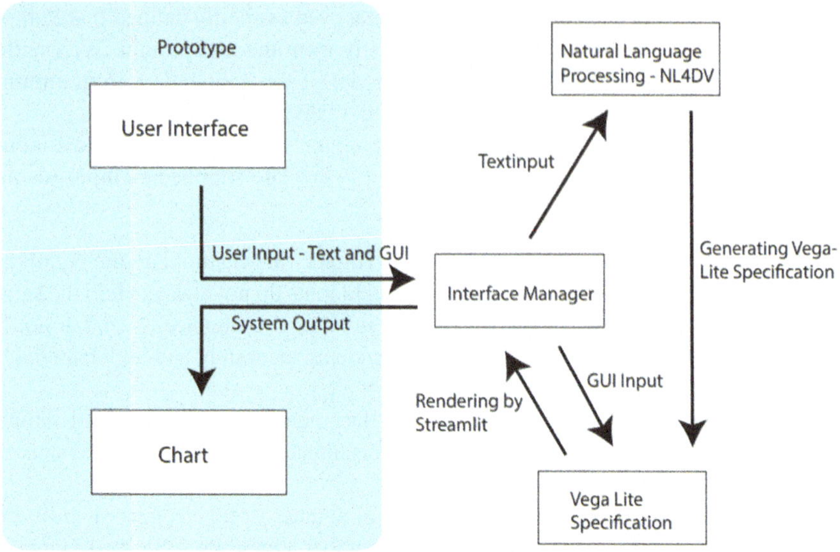

Fig. 7 Architecture of the NL4DV-based variant of the prototype

5.1.5 Rendering by Streamlit—System Output

A Streamlit component for Matplotlib represents the Matplotlib object.

5.2 Architecture of the NL4DV Version

Figure 7 presents a schematic depiction of the architecture of the variant that incorporates the NL4DV module [30].

5.2.1 NLP in the NL4DV Variant

This second variant diverges from the previous one due to its use of a different NLP method. Here, the input text doesn't generate source code; instead, it undergoes Part of Speech (POS) tagging, entity recognition, and dependency parsing. Moreover, heuristics are applied to produce a ranked list of potential charts.

In the prototype, we employ this metric and the highest-ranked chart. The NL4DV version crafts charts as JSONs in line with the Vega-Lite specification [37]. This data is then stored in the web application's session.

5.2.2 GUI Input in the NL4DV Variant

Any modifications to the graphic through the GUI elements are directly implemented as alterations to the saved Vega-Lite object and subsequently stored again.

5.2.3 Rendering by Streamlit—System Output in the NL4DV Variant

A Streamlit component designed for Vega-Lite is responsible for representing the Vega-Lite object.

Hence, the primary distinction between the two variants lies in their NLP processing. The differences in graphics related to the system only result in minor variances in the display and do not influence the system's capabilities.

6 Preliminary User Study

We carried out an initial user study aimed at collecting preliminary feedback on our prototype and the two methodologies, and to evaluate their usability.

6.1 Participants and Setup

We recruited 11 participants (P1-P11, six males, five females) aged from 21 to 33 years through university courses. As per the self-reporting from the participants, seven individuals had never conducted data analysis, and four engaged in data analysis infrequently or only occasionally. When asked about their prior experience with interactive visualization tools, the tools mentioned included Tableau, 3d.js [5], Observable [4], Matplotlib [19], Google Data Studio [42], chart.js https://www.chartjs.org/ and Excel, with 3d.js and Observable being mentioned most often.

The sessions involving the web prototypes were carried out remotely using Microsoft Teams video conferencing software, with the prototypes being hosted by Streamlit [46]. With the participants' consent, we recorded the audio, video, and on-screen activities throughout all the sessions. The duration of these video recordings varied between 40 min and 2 h and 5 min, averaging at 1 h and 20 min.

The session involving ChatGPT was undertaken in a classroom setting, and we utilized the participants' Jupyter documents for evaluation purposes.

There was no maximum number of tries a user can attempt for each task but a maximum of three hours. The participants were not allowed to ask any question, and no discussions between participants were allowed.

6.2 Study Procedure

After a brief introduction to the study's objectives, a short questionnaire on fundamental data and previous knowledge of the subjects was collected using Google Forms. Subsequently, a short introductory video showing the interface and an example was watched by participants for each of the two web variants. Two sessions with the web prototypes were held with nine and two participants and moderated by an experimenter. The in-class session with ChatGPT was held with 11 participants and moderated by an experimenter.

The participants were tasked with a set of seven activities involving the previously introduced Gapminder dataset [16] (see Sect. 4). They were asked to "solve" these tasks, incorporating specific formatting aspects like color, line style, axis label, and title, using the two versions of the prototype. A visual illustration of the kind of diagram was provided alongside each question, and participants were not allowed to ask questions.

1. Construct histograms for the numerical attributes.
2. Plot the numerical attributes per continent using bar charts.
3. Illustrate the GDP per country with a bar chart. The top 15 countries should be presented in descending order.
4. Generate a pie chart depicting the global population distribution by continent.
5. Design a graph that traces the temporal progression of average life expectancy across all countries. Use a line chart for this purpose.
6. Expose the correlation between GDP per capita and life expectancy using a scatterplot. Additionally, employ a color code that signifies population size.
7. Exhibit the relationship between GDP per capita, life expectancy, and population size using a bubble chart.

Following the completion of tasks with the Gapminder dataset, each participant filled out a System Usability Scale (SUS) questionnaire, which included an additional question for comments, for the two web variants [6]. No SUS score was determined for the ChatGPT variant, since the usability strongly depends on the Jupyter documents and these were predefined.

6.3 Results

Participants gave the prototype with the Codex language model an average SUS score of 71.1 (standard deviation (SD) 8.9) and the version with NL4DV an average score of 68.2 (SD 9.7). A score of $\geq$ 68 is considered an indicator of good usability [1].

The comments from the participants revealed a range of diverse opinions concerning the two web versions. For some test participants, the Codex version was easier to use, while the NL4DV version was easier to use for others. This result is also shown in the evaluation of the solutions by individual participants, who had very different solving success rates with the two web variants. The ChatGPT version was easy to use for all participants, and this is reflected in the success rates.

The following remark from the feedback captures this result quite succinctly:

> "Overall, the Vega-Lite version felt like it had more a mind of its own, which led to less blank graphs but also left me feeling less in control".

Moreover, statement—

> "The difficulty is knowing the key words to use"

—is representative of a number of participant responses requesting more documentation and suggestions.

The solution to the tasks was evaluated on the basis of the video material and the Jupyter documents by coding a solved task as 1, an unsolved task as 0, and a partially solved task as 0.5 for each participant. Solutions that generated the correct chart type but did not accurately execute all the formatting requirements were considered as partially solved. Partially solved tasks encompass instances where the histogram lacked the necessary number of bins, or the scatterplot failed to exhibit the correct color coding.

In total, 71 tasks were solved with ChatGPT, 48.5 tasks were solved or partially solved with the Codex version, and 43 with the Vega-Lite version. The spectrum of scores ranged from subjects who solved all tasks correctly with the Codex version to subjects solving only one task. With the NL4DV version, the scatter was not as great, with a maximum of 6 points achieved but no less than 2.5. With the ChatGPT version, the minimum number of solved exercises was 9. This finding further supports the statement mentioned earlier that the deep learning versions offer more control compared to the Codex version, which presents more challenges. The architectural differences and the increased power of the language models contribute to this expected difference in performance. Furthermore, the NL4DV model, utilizing its heuristics, is quicker in achieving the initial result. However, as the test demonstrated, subsequent manipulation of the chart is often not feasible. For instance, determining the number of bins in the histogram or the number of top 15 countries in the bar chart was not possible.

We also assessed the successful completion of the seven tasks and the method used for their execution. As these tasks incorporated some of the most commonly used chart types, it is intriguing to observe the disparities. The 11 subjects attempted to solve each task with all variants, so we obtained 33 solution attempts per task, which resulted in the scores in Table 1.

It is evident throughout that simpler chart types, such as the pie chart (task 4), the histogram (task 1), and basic bar charts (task 2), were well-executed. However, when tasks necessitated multiple data aggregations (such as the line chart in task 5), filter specifications (like the filtering of the top 15 in task 3), and additional data dimensions (like the color code or dot size in the scatterplots in tasks 6 and 7), the rate of successful solutions significantly declined.

In the talsks 1 and 4 the Codex model had an advantage over the NL4DV approach. Here, the higher expressiveness of the model showed the possibility of interpreting

Table 1 Number of solutions per task (Maximum of 33 possible)

Exercise	Total	Codex	NL4DV	ChatGPT
1	28	11	6.5	10.5
2	27.5	6	10.5	11
3	20.5	6	4.5	10
4	28.5	8	10	10.5
5	22.5	6.5	7	9
6	19	5.5	3.5	10
7	16.5	5.5	1	10

things like the number of bins, sorting or filtering, and generating the appropriate Python code. The NL4DV model failed here because its heuristics obviously did not cover these possibilities sufficiently. The Codex model also offered advantages in the scatterplot tasks (tasks 6 and 7) by possibly controlling the dot size or color coding by adapting the text command accordingly. ChatGPT dominated both other models due to its more powerful language model and the simple chat interface.

The NL4DV model generally yielded better results for the basic bar chart (task 2), the pie chart (task 4), and the line chart (task 5). In these cases, the heuristics were more reliable than the generated Python code, which sometimes led to drastically different or even null results with slight alterations. For instance, the input "create a line chart of lifeExp of all countries per year" did not achieve the desired outcome, whereas modifying it to "create a line chart of average lifeExp of all countries per year" generated the expected line chart.

7 Conclusions and Directions for Future Work

The evaluation showed that the participants could solve almost all exercises with the ChatGPT model and its simple chat interface. Dominating the other two approaches.

It is expected that with the progress of fine-tuning, it will be possible to successfully integrate this technology into information visualization products for end users, as Microsoft's "Power BI Q&A" [29] and Tableau's "Ask Data" [48].

The current version of the Codex language model is designed for generating general source code and is not specifically tailored to the task of creating infographics or targeted charts. While fine-tuning is available for the NL (natural language) variants of GPT-3 through the API, it is not yet possible for the Codex model. Once fine-tuning becomes available for Codex, we can expect significant progress in its ability to work with source code and create infographics and targeted charts more effectively.

Another way to try to improve performance is to switch to the open-source models CodeT5 [50], or CodeBERT [12] and fine-tune them, for example, on the Code-

SearchNet dataset [20]. Reference [24] have recently shown the possibilities of fine-tuning programming tasks in Python.

For all model variants, it was shown that it was still tricky for test persons to find the correct text inputs; thus, here, a system that generates suggestions is desirable.

Reference [43] introduced SNOWY, a prototype web application that builds upon the "NL4DV" and "Vega-Lite" frameworks. SNOWY not only generates visualizations from text inputs but also provides suggestions for additional inputs to guide the analysis and refine the graphics. This application facilitates a dialogue with the data, making it beneficial for the deep learning prototype and enhancing usability. By offering suggestions, users are assisted in finding the appropriate text inputs instead of being left to navigate the process on their own.

References

1. System usability scale (2022) https://www.usability.gov/how-to-and-tools/methods/system-usability-scale.html. Accessed 15 Jan 2022
2. Affolter K, Stockinger K, Bernstein A (2019) A comparative survey of recent natural language interfaces for databases. VLDB J 28(5):793–819
3. AI O (2022) Open AI API. https://openai.com/api/. Accessed 07 Jan 2022
4. Bostock M (2022) Explore, analyze, and explain data. https://observablehq.com/. Accessed 15 Jan 2022
5. Bostock M, Ogievetsky V, Heer J (2011) D^3 data-driven documents. IEEE Trans Visual Comput Graphics 17(12):2301–2309
6. Brooke J et al (1996) Sus-a quick and dirty usability scale. Usability Eval Ind 189(194):4–7
7. Brown TB, Mann B, Ryder N, Subbiah M, Kaplan J, Dhariwal P, Neelakantan A, Shyam P, Sastry G, Askell A et al (2020) Language models are few-shot learners. arXiv:2005.14165
8. Chen M, Tworek J, Jun H, Yuan Q, Pinto HPDO, Kaplan J, Edwards H, Burda Y, Joseph N, Brockman G et al (2021) Evaluating large language models trained on code. arXiv:2107.03374
9. Cox K, Grinter RE, Hibino SL, Jagadeesan LJ, Mantilla D (2001) A multi-modal natural language interface to an information visualization environment. Int J Speech Technol 4(3):297–314
10. Devlin J, Chang MW, Lee K, Toutanova K (2018) Bert: Pre-training of deep bidirectional transformers for language understanding. arXiv:1810.04805
11. Fan A, Lewis M, Dauphin Y (2018) Hierarchical neural story generation. arXiv:1805.04833
12. Feng Z, Guo D, Tang D, Duan N, Feng X, Gong M, Shou L, Qin B, Liu T, Jiang D et al (2020) Codebert: a pre-trained model for programming and natural languages. arXiv:2002.08155
13. Finkel JR, Grenager T, Manning CD (2005) Incorporating non-local information into information extraction systems by GIBBS sampling. In: Proceedings of the 43rd annual meeting of the association for computational linguistics (ACL'05), pp 363–370
14. Gao L, Biderman S, Black S, Golding L, Hoppe T, Foster C, Phang J, He H, Thite A, Nabeshima N et al (2020) The pile: an 800gb dataset of diverse text for language modeling. arXiv:2101.00027
15. Gao T, Dontcheva M, Adar E, Liu Z, Karahalios KG (2015) Datatone: managing ambiguity in natural language interfaces for data visualization. In: Proceedings of the 28th annual ACM symposium on user interface software & technology, pp 489–500
16. Gapminder: Gapminder (2022) https://www.gapminder.org/data/. Accessed 07 Jan 2022
17. Holtzman A, Buys J, Du L, Forbes M, Choi Y (2019) The curious case of neural text degeneration. arXiv:1904.09751

18. Honnibal M, Montani I (2017) SpaCy 2: natural language understanding with Bloom embeddings, convolutional neural networks and incremental parsing. To appear
19. Hunter JD (2007) Matplotlib: a 2D graphics environment. Comput Sci Eng 9(03):90–95
20. Husain H, Wu HH, Gazit T, Allamanis M, Brockschmidt M (2019) CodeSearchNet challenge: evaluating the state of semantic code search. arXiv:1909.09436
21. Inc., PT (2015) Collaborative data science. https://plot.ly
22. Kassel JF, Rohs M (2018) Valletto: a multimodal interface for ubiquitous visual analytics. In: Extended abstracts of the 2018 CHI conference on human factors in computing systems, pp 1–6
23. Kim DH, Hoque E, Agrawala M (2020) Answering questions about charts and generating visual explanations. In: Proceedings of the 2020 CHI conference on human factors in computing systems, pp 1–13
24. Li Y, Choi D, Chung J, Kushman N, Schrittwieser J, Leblond R, Eccles T, Keeling J, Gimeno F, Dal Lago A, Hubert T, Choy P, de Masson d'Autume C, Babuschkin I, Chen X, Huang PS, Welbl J, Gowal S, Cherepanov A, Molloy J, Mankowitz D, Sutherland Robson E, Kohli P, de Freitas N, Kavukcuoglu K, Vinyals O (2022) Competition-level code generation with alphacode
25. Loper E, Bird S (2002) Nltk: the natural language toolkit. arXiv:cs/0205028
26. Lu S, Guo D, Ren S, Huang J, Svyatkovskiy A, Blanco A, Clement C, Drain D, Jiang D, Tang D et al (2021) Codexglue: a machine learning benchmark dataset for code understanding and generation. arXiv:2102.04664
27. Manning CD, Surdeanu M, Bauer J, Finkel JR, Bethard S, McClosky D (2014) The Stanford Corenlp natural language processing toolkit. In: Proceedings of 52nd annual meeting of the association for computational linguistics: system demonstrations, pp 55–60
28. Mathew M, Bagal V, Tito R, Karatzas D, Valveny E, Jawahar C (2022) Infographicvqa. In: Proceedings of the IEEE/CVF winter conference on applications of computer vision, pp 1697–1706
29. Microsoft: Microsoft Q & A (2022) https://docs.microsoft.com/en-us/power-bi/create-reports/power-bi-tutorial-q-and-a. Accessed 04 Jan 2022
30. Narechania A, Srinivasan A, Stasko J (2020) NL4DV: a toolkit for generating analytic specifications for data visualization from natural language queries. IEEE Trans Vis Comput Graph (TVCG). https://doi.org/10.1109/TVCG.2020.3030378
31. Őzcan F, Quamar A, Sen J, Lei C, Efthymiou V (2020) State of the art and open challenges in natural language interfaces to data. In: Proceedings of the 2020 ACM SIGMOD international conference on management of data, pp 2629–2636
32. Qin X, Luo Y, Tang N, Li G (2020) Making data visualization more efficient and effective: a survey. VLDB J 29(1):93–117
33. Radford A, Narasimhan K, Salimans T, Sutskever I (2018) Improving language understanding by generative pre-training
34. Radford A, Wu J, Child R, Luan D, Amodei D, Sutskever I et al (2019) Language models are unsupervised multitask learners. OpenAI Blog 1(8):9
35. Rae JW, Borgeaud S, Cai T, Millican K, Hoffmann J, Song F, Aslanides J, Henderson S, Ring R, Young S et al (2021) Scaling language models: methods, analysis & insights from training gopher. arXiv:2112.11446
36. Raffel C, Shazeer N, Roberts A, Lee K, Narang S, Matena M, Zhou Y, Li W, Liu PJ (2019) Exploring the limits of transfer learning with a unified text-to-text transformer. arXiv:1910.10683
37. Satyanarayan A, Moritz D, Wongsuphasawat K, Heer J (2016) Vega-lite: a grammar of interactive graphics. IEEE Trans Visual Comput Graphics 23(1):341–350
38. Satyanarayan A, Russell R, Hoffswell J, Heer J (2015) Reactive vega: a streaming dataflow architecture for declarative interactive visualization. IEEE Trans Visual Comput Graphics 22(1):659–668
39. Setlur V, Battersby SE, Tory M, Gossweiler R, Chang AX (2016) Eviza: a natural language interface for visual analysis. In: Proceedings of the 29th annual symposium on user interface software and technology, pp 365–377

40. Setlur V, Tory M, Djalali A (2019) Inferencing underspecified natural language utterances in visual analysis. In: Proceedings of the 24th international conference on intelligent user interfaces, pp 40–51
41. Shen L, Shen E, Luo Y, Yang X, Hu X, Zhang X, Tai Z, Wang J (2021) Towards natural language interfaces for data visualization: a survey. arXiv:2109.03506
42. Snipes G (2018) Google data studio. J Libr Sch Commun 6(1)
43. Srinivasan A, Setlur V (2021) Snowy: recommending utterances for conversational visual analysis. In: The 34th annual ACM symposium on user interface software and technology, pp 864–880
44. Srinivasan A, Stasko J (2017) Natural language interfaces for data analysis with visualization: considering what has and could be asked. In: Proceedings of the eurographics/IEEE VGTC conference on visualization: short papers, pp 55–59
45. Stöckl A (2022) Natural language interface for data visualization with deep learning based language models. In: 2022 26th international conference information visualisation (IV). IEEE, pp 142–148
46. Streamlit.io: Streamlit cloud (2022) https://streamlit.io/s. Accessed 05 Jan 2022
47. Sun Y, Leigh J, Johnson A, Lee S (2010) Articulate: a semi-automated model for translating natural language queries into meaningful visualizations. In: International symposium on smart graphics. Springer, Berlin, pp 184–195
48. Tableau.com: Tableau's ask data (2022) https://www.tableau.com/solutions/ai-analytics/augmented-analytics. Accessed 04 Jan 2022
49. Vaswani A, Shazeer N, Parmar N, Uszkoreit J, Jones L, Gomez AN, Kaiser Ł, Polosukhin I (2017) Attention is all you need. In: Advances in neural information processing systems, pp 5998–6008
50. Wang Y, Wang W, Joty S, Hoi SC (2021) Codet5: identifier-aware unified pre-trained encoder-decoder models for code understanding and generation. arXiv:2109.00859
51. Waskom ML (2021) seaborn: statistical data visualization. J Open Source Softw 6(60):3021. https://doi.org/10.21105/joss.03021. https://doi.org/10.21105/joss.03021
52. Welleck S, Kulikov I, Kim J, Pang RY, Cho K (2020) Consistency of a recurrent language model with respect to incomplete decoding. arXiv:2002.02492
53. Wilkinson L (2021) The grammar of graphics. In: Handbook of computational statistics. Springer, Berlin, pp 375–414
54. Yu B, Silva CT (2019) Flowsense: a natural language interface for visual data exploration within a dataflow system. IEEE Trans Visual Comput Graphics 26(1):1–11
55. Zhu S, Sun G, Jiang Q, Zha M, Liang R (2020) A survey on automatic infographics and visualization recommendations. Vis Inform 4(3):24–40

Visualization of Similarity Distribution of Lyric Collections

Yuki Nakai and Takayuki Itoh

1 Introduction

Lyric-based music search and analysis are quite useful and interesting because lyrics have been a great factor in pop music songs. Visualization is a useful tool to overview the similarity distribution of the lyrics of song collections and assist the exploratory search of user-desired lyrics. Current collaborative studies on natural language processing and visualization techniques [1, 3, 15] have archived the visualization of the similarity distribution of text documents, and therefore similarity distribution of lyrics can be visualized using such techniques.

We are also working on visualizing the distribution of lyrics for an overview of similarity and exploratory search of lyrics by applying LDA (Latent Dirichlet Allocation). The target users of this study include music listeners, music industry workers, and songwriters. Music listeners can search for songs whose lyrics are similar to their favorite songs while exploring the visualization. Music industry workers can find the trends of lyrics and understand the characteristics of the musicians along with the lyrics. Furthermore, the visualization is useful for songwriters as a reference while surveying the lyrics of past hit songs. The visualization by scatterplots with dimensionality reduction by LDA provides a good overview of the distribution of the lyrics of a song collection. This feature is helpful for professional persons working in the music industry to get an overview of the overall trend of hit songs. Meanwhile, ordinary music listeners often want to explore their favorite songs by finding songs similar to those of a particular artist or searching for songs with lyrics that contain

Y. Nakai · T. Itoh (✉)
Ochanomizu University, Tokyo, Japan
e-mail: itot@is.ocha.ac.jp

Y. Nakai
e-mail: nakai.yuki@is.ocha.ac.jp

B. Kovalerchuk et al. (eds.), *Artificial Intelligence and Visualization: Advancing Visual Knowledge Discovery*, Studies in Computational Intelligence 1126,
https://doi.org/10.1007/978-3-031-46549-9_18

specific content. In such cases, users of this visualization can zoom in on a particular part of similar lyrics starting from an overview of the distribution of lyrics.

Here, the requirements of lyrics visualization are different from the visualization of general documents from the following viewpoints. Firstly, the impression of lyrics is subjective and may be influenced by musical elements other than lyrics, so the criteria for searching for lyrics required by users may vary from person to person. Secondly, lyrics have a higher degree of lexical freedom than formal documents such as newspaper articles and academic papers, and therefore it is often difficult to appropriately calculate the distribution of the lyrics. There have been studies on lyric visualization applying LDA [15], but still, the above issues are open problems.

This chapter introduces our additional studies on lyric visualization using LDA. Firstly, we present the visualization of the distribution of lyrics applying guided LDA instead of the original LDA. Since the guided LDA can interactively consume guided words, users can iteratively visualize the lyric distribution based on the users' viewpoints. Users can observe the differences in individuality and tendency of songs and artists, and the diversity of lyrics, by using the visualization results. This study has been presented as our conference paper [12].

Secondly, we present the visualization of the distribution of lyrics and male and female singers to find the difference of word selections between male and female song writers. In this experiment, we also used LDA and dimensionality reduction to visualize the distribution of lyrics. The songs in the dataset are labeled based on the gender of the singers. Songs by male singers are colored in blue, while songs by female singers are colored in red, in the visualizations. We extract example groups of similar lyrics from the visualization results and discuss how the lyrics in each of the examples are similar.

This chapter introduces case studies with a lyrics dataset of Japanese hit songs (so-called J-POP) since the first motivation of authors for this study is to understand the similarity distribution of the lyrics of J-POP songs.

The remainder of this chapter is as follows. Section 2 introduces related studies, Sect. 3 introduces the theoretical definition of LDA and guided LDA, and Sect. 4 describes the dataset we used for the case studies. Section 5 presents our first study on the comparison between LDA and guided LDA, and Sect. 6 presents our second study on the difference in lyrics between male and female singers. Section 7 summarizes the study and discusses future issues.

2 Related Work

This section first introduces existing work on trend analysis of J-POP song lyrics since the goal of this study is the trend visualization of the lyrics of J-POP songs. Then, the section introduces related studies on the natural language processing and visualization of lyrics not limited to J-POP songs.

The analysis of Japanese pop music lyrics involves a number of issues based on the linguistical uniqueness of the Japanese language, as well as the climate and trends of Japanese society. For this reason, a number of analytical methods for lyrics of Japanese pop music have been published in the Japanese academic community. Kobayashi et al. [9] calculated the usage rates of words used in lyrics of Japanese songs, and discussed the temporal changes of the usage rates. Ohde et al. [14] presented the results of clustering words in lyrics of Japanese songs to analyze the chronological trends and culture of lyrics. This study registered codes for each cluster of the words, and analyzed the frequency of the words and the corresponding codes. Hosoya et al. [5] presented a comparative lyrics analysis of multiple female singer-songwriters applying the random forest method. Kawamura [8] developed a song recommendation method by automatically extracting associative words of search words and calculating features.

Meanwhile, many studies on lyric visualization have been presented for the purpose of overviewing and understanding lyrics. Lyric visualization work is divided into two types: visualization of one song, and visualization of song collections.

As one-song visualization techniques, Oh [13] presented a visual design to tightly align the musical and linguistic features of a song while conveying how the melodic, harmonic, and metric lines of the music fit with the text. Machida et al. [11] represented the content and story of a lyric of a song by multiple icons.

As song-collection visualization techniques, Baur et al. [1] presented an application that allows browsing and exploring a music collection based on its lyrics in a self-organizing map. Burch et al. [3] presented a word cloud design that visualizes the occurrence of words that help to find knowledge about the genre or to compare the content.

LyricsRader presented by Sasaki et al. [15] is one of the most related work to our study presented in this chapter. It uses LDA to archive the embedding of the latent meaning of lyrics and visualize the distribution of lyrics by scatterplots that realize users to interactively explore their interested lyrics from a large-scale song collection. There have been more studies to visualize the lyrics of a large number of songs using LDA [6] and extended LDA models for the analysis of lyrics [16], but to our best knowledge, LyricsRader is one of the first studies on develop a visualization and an explorative user interface for a large number of lyrics. Our study presented in this chapter extends such an existing study of visualization of lyric collections by applying guided LDA and exploring the differences in lyrics between male and female singers.

3 LDA (Latent Dirichlet Allocation)

3.1 Analyzing Lyrics Using LDA

LDA [2] is proposed to discover multiple topics from a set of documents. Figure 1 shows the graphical model of LDA. Shaded vertices in this figure represent the manifest variables, while other vertices correspond to latent variables and unknown param-

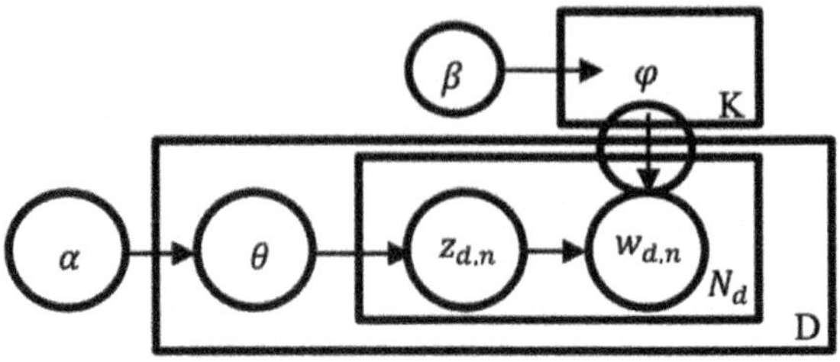

Fig. 1 Graphical model of LDA

eters. Directed edges in this figure represent the dependencies among the variables and parameters. The nested rectangles represent that the variables in the rectangles are repetitively created.

We assume that each lyric is an independent document. Here, D denotes the independent lyrics $X = \{X_1, \ldots, X_n\}$. The lyric X_d consists of N_d words $X_d = \{w_{d,1}, \ldots, w_{d,N_d}\}$. Also, we assume that each word in each lyric has latent topics in this study. K is the number of topics, θ is the topic multinomial distribution parameter for each document, ϕ is the word multinomial distribution parameter for each topic, and α and β are the Dirichlet hyperparameters for θ and ϕ respectively.

The process of document generation from a graphical model is shown below.

1. Select $\theta_d \sim Dir(\alpha)$ for each document X_d
2. Select $\phi_k \sim Dir(\beta)$ for each topic k
3. For each of the N_d words $w_{d,i}$ in document X_d

   - Select topic $z_{d,i} \sim Mult(\theta_d)$
   - Select word $w_{d,i} \sim Mult(\phi_{z_{d,i}})$

Gibbs sampling [4] is applied to LDA to estimate the topic model. Here, the probability that the topic of the i-th word in X_d document is k is given by the following equation:

$$P\left(z_{d,i} = k \mid w_{d,i} = w, z_{-i}, w_{-i}, \alpha, \beta\right)$$

$$\propto \frac{C^{KD}_{kd,-i} + \alpha}{\sum_{k'} C^{KD}_{k'd,-i} + K\alpha} \cdot \frac{C^{WK}_{wk,-i} + \beta}{\sum'_{w} C^{WK}_{w'k,-i} + K\beta}$$

Here, w_{-i} is the set w_{-i} excluded from w and z_{-i} is the set z_i excluded from z. $C^{KD}_{kd,-i}$ is the number of times topic k is assigned to document X_d and $C^{WK}_{w'k,-i}$ is the number of times topic k is assigned to word w. The i-th word is excluded in both cases.

LDA calculates the expected value of the topic multinomial distribution parameter θ for each lyric. The value θ is also explained as the topic mixture ratio of lyrics. We visualize the distribution of songs in a two-dimensional space by applying a dimensionality reduction scheme to the multidimensional values of the probability.

3.2 Analyzing Lyrics Using Guided LDA

Guided LDA [7] extends LDA that sets specific words as reserved words (guide words) to each topic as a preprocessing. This mechanism tends to classify particular words that co-occur with the guide word as representative words of the topic of the guide word. Users can introduce perspectives into the topic classification results by the selection of guided words. Here, guided LDA calculates the probability values of the topics for each lyric as in LDA. In this study, we selected topic themes and guide words based on the following two types.

- season-based topic themes
- event-topic themes

Tables 1 and 2 show the topic themes and topic words employed in this study. We extracted these words by reference to our previous work by Machida et al. [11].

We set the following words shown in Table 3 that frequently appeared as non-important words in the topics while analyzing all lyrics using LDA as stop words.

Table 1 Topic themes and guide words in guided LDA. (1) season-based topic themes

Topic theme	Guide words
Spring	haru (spring), sakura (cherry blossom), sotsugyo (graduation)
Summer	natsu (summer), matsuri (festival), hanabi (fireworks), himawari (sunflower)
Autumn	aki (autumn)
Winter	fuyu (winter), yuki (snow)

Table 2 Topic themes and guide words in guided LDA. (2) event-based topic themes

Topic theme	Guide words
Graduation	sotsugyo (graduation), tomo (friend), wakare (farewell),sayonara (good bye) sakura (cherry blossom), seifuku (school uniform), kadode (departure)
Summer festival	natsu (summer), matsuri (festival), hanabi (fireworks), kori-gashi (sherbert)

Table 3 Stop words

suru (do), aru (be), iru (be), naru (be), nai (not), kono (this), sono (its), ano (that),
sore (it), boku (I), watashi (I), ore (I), anata (you), kimi (you), bokura (we)

We comparatively applied unguided normal LDA as well as guided LDA in our experiments. We applied only Japanese words in our study for lyrics that words of Japanese and other languages (e.g., English) are mixed.

4 Dataset and Implementation

4.1 Lyric Collection Dataset

Our experiment introduced in this chapter applied the songs of the top ten highest-selling CD singles from 1988 to 2007 and the top ten songs on the Billboard Japan annual chart from 2008 to 2020. We constructed the dataset from the above songs, where each song has metadata including the title, artist, lyricist, date, and lyrics. Note that songs whose lyrics do not include any Japanese words are excluded from the dataset. Also, if a song appeared in the above charts in multiple years, the dataset remained only the most recent one. Consequently, we constructed a dataset containing the lyrics of 327 songs.

Natural language processing of Japanese lyrics requires a morphological analysis process. Our study applied MeCab, an open-source software for morphological analysis. The study only applied nouns, verbs, adjectives, adverbs, inspirations, and coordinating verbs, and count their original form as one word. Here, there are several words that are frequently seen in an extremely large number of songs. This study determined such words as "non-important words" and set them as "stop words" to exclude from the analysis.

4.2 Visualization by Scatterplots

We visualize the results of the analysis of the lyrics using LDA and guided LDA as a scatterplot with dimensionality reduction to two dimensions using t-SNE [10]. Here, a single point in the scatterplot represents a single lyric.

5 Comparative Case Study Applying LDA and Guided LDA

This section introduces our first case study applying LDA and guided LDA, and verifies the effectiveness of guide LDA for the analysis of lyrics against the original LDA. Here, we fixed the number of topics as $K = 4$ in the first case study.

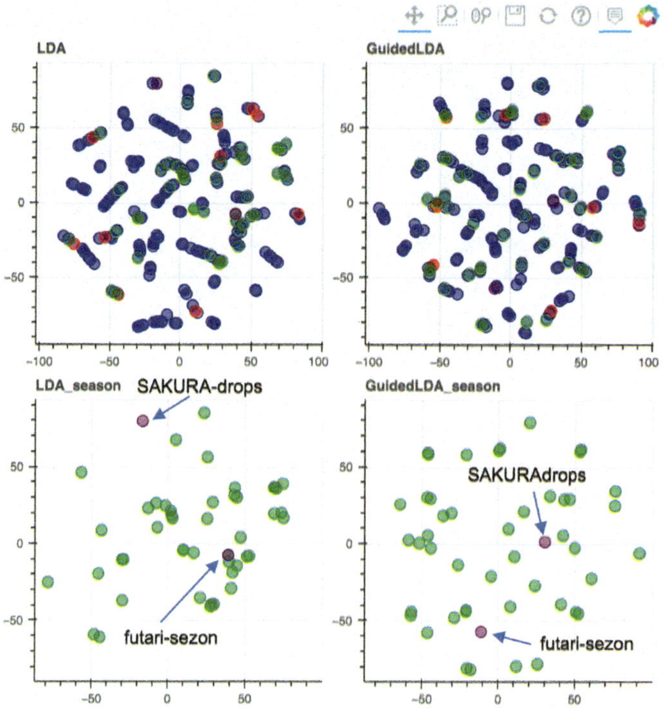

Fig. 2 Visualization example (1). (upper left) Example applying LDA. (upper right) Example applying guided LDA. (lower left) Limited to lyrics including "natsu" (summer) applying LDA. (lower right) Limited to lyrics including "natsu" (summer) applying guided LDA

5.1 Season-Based Topic Themes

This section first introduces the visualization results using the following seasonal words as shown in Figs. 2 and 3.

- spring, cherry, graduation
- summer, firework, sunflower
- autumn
- winter, snow, scarf

Note that we have a graduation ceremony in March and therefore graduation is a spring word in Japan. Also, we regret that fewer autumn songs are included in our dataset and therefore we could not find good seasonal words for autumn.

Colors in the visualization results depict the following songs:

Purple: songs including both "natsu" (summer) and "sakura" (cherry).
Green: songs incluing only "natsu" (summer).
Red: songs including only "sakura" (cherry).
Blue: songs not including "natsu" (summer) or "sakura" (cherry).

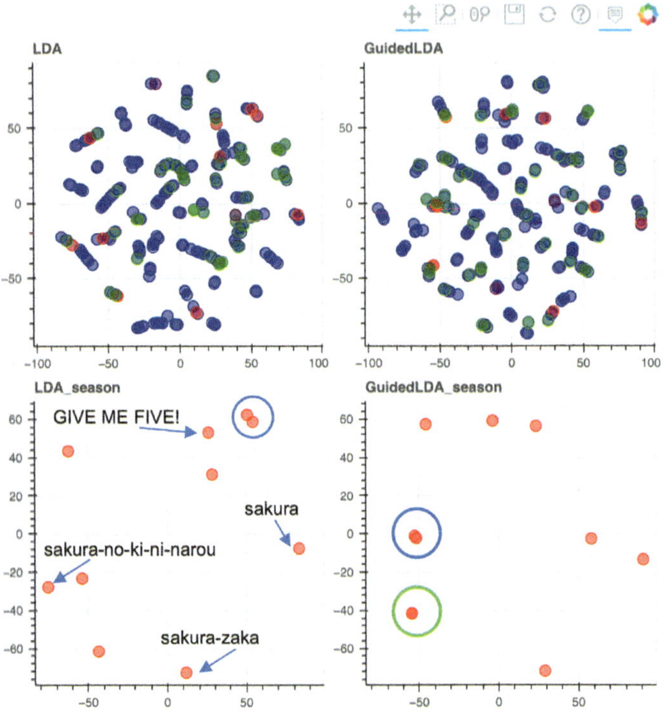

Fig. 3 Visualization example (2). (upper left) Example applying LDA. (upper right) Example applying guided LDA. (lower left) Limited to lyrics including "sakura" (cherry) applying LDA. (lower right) Limited to lyrics including "sakura" (cherry) applying guided LDA

We focused on the two songs displayed in purple in Fig. 2 that include both "natsu" (summer) and "sakura" (cherry). Actually, the two songs include "natsu" (summer) in the lyrics but their stories are not quite related to summer.

The song "SAKURA-drops" indicated in Fig. 2 is a typical song concerned with the above condition. The lyric of this song includes "natsu" (summer) but the story of the lyric is rather related to spring because "sakura" (cherry) appears more frequently in the lyric. The positions of this song in both scatterplots applying LDA and guided LDA are appropriate since they are distant from other songs related to summer.

The song "futari-sezon" indicated in Fig. 2 is another typical song concerned with the above condition. The lyric of this song includes "natsu" (summer) as well as "haru" (spring), "aki" (autumn) and "fuyu" (winter) since the song shows the psychological change of a couple through four seasons. The position of this song in the scatterplot applying LDA is unfortunately close to other songs simply related to summer. On the other hand, the position of this song in the scatterplot applying guide LDA is rather preferable because it is distant from other songs simply related to summer.

Next, we focused on the songs related to spring, closely positioned in the scatterplot applying guided LDA, and actually, their lyrics have common sense, indicated by blue and green circles in Fig. 3(lower-left).

The lyrics of the two songs enclosed by a blue circle include "sakura" (cherry) as well as "seifuku" (uniform) and "kyoshitsu" (classroom) that associated with graduation from high school. The mood of the lyrics of the two songs is actually similar. Both songs are produced by the same lyric writer and performed by the same female idol group. Such two songs are appropriately close in the scatterplot applying guided LDA, while they are unfortunately distant in the scatterplot applying LDA.

The lyrics of the two songs enclosed by a green circle express similar stories that a man reflects his past heartbreak in spring. These songs are performed by the same generation of male singer-songwriters. Again, such two songs are appropriately close in the scatterplot applying guided LDA, while they are unfortunately distant in the scatterplot applying LDA.

As a summary, Figs. 2 and 3 demonstrate that guided LDA brought better visualization results since it closely places the songs that have the common expression of the scenes and stories.

This section also shows the visualization results when we set "natsu" (summer) and "hanabi" (fireworks) as the guide words for one topic, and no guide words for the other topics, as shown in Fig. 4. This figure indicates the lyrics containing the word "natsu" (summer) by red dots, while indicating specific two songs, "hanabi" (fireworks) by Mr. Children and "sayonara-zinrui" (goodbye humanity) by Tama, by green dots, and the other lyrics by blue dots. These two songs include the word "hanabi" (fireworks) in the lyrics, but the lyrics as a whole are not related to summer. The "hanabi" (fireworks) is an independent point in both the scatterplots applying LDA and the guided LDA. However, the distance to the nearby point is farther in the case with the guided LDA than in the case with the LDA. For "sayonara-zinrui" (goodbye humanity), the points are located in the area where summer-related songs such as "rokorosyon" (loco-lotion) by ORANGE RANGE are concentrated in the scatterplot while using LDA. On the other hand, the point corresponding to "sayonara-zinrui" (goodbye humanity) is independent on the scatterplot while using the guided LDA. In other words, the guided LDA provides a better separation of the placement of songs related only to "summer" and songs related only to "fireworks".

5.2 *Event-Based Topic Themes*

The scatterplot shown in Fig. 5 indicates the lyrics that contain the word "sotsugyo" (graduation) and are related to graduation by red dots, the lyrics that do not contain the word "sotsugyo" (graduation) but are related to graduation by purple dots, the lyrics that contain the word "sotsugyo" (graduation) but are not related to graduation by green dots, and the other lyrics by blue dots. In the scatterplot using LDA, the distance between lyrics that contain the word "sotsugyo" (graduation) and are related to graduation (red dots) and lyrics that include the word "sotsugyo" (graduation)

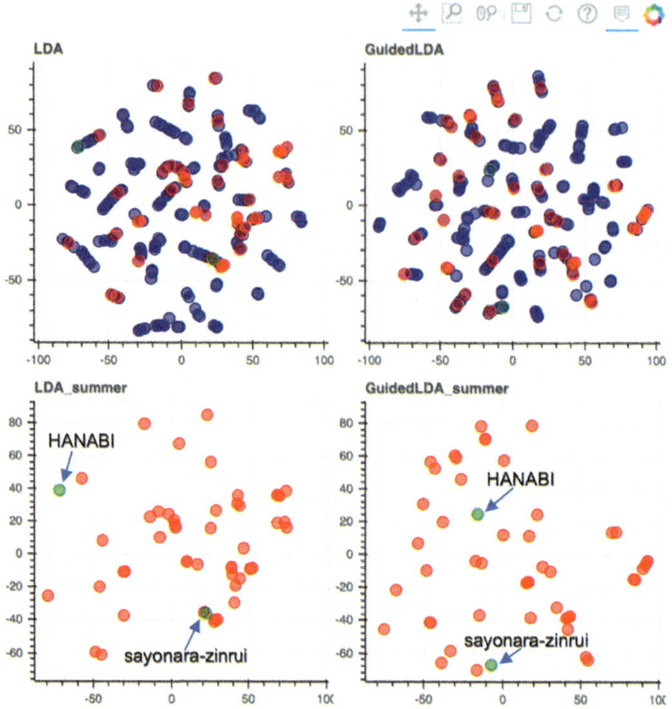

Fig. 4 Visualization example (3). (upper left) Example applying LDA. (upper right) Example applying guided LDA. (lower left) Limited to lyrics including "natsu" (summer), "HANABI" (fireworks) and "sayonara-zinurui" (goodbye humanity) applying LDA. (lower right) Limited to lyrics including "natsu" (summer), "HANABI" (fireworks) and "sayonara-zinrui" (goodbye humanity) applying guided LDA

but are not related to graduation (green dots) are closer. On the other hand, in the scatterplot using guided LDA, lyrics that contain the word "sotsugyo" (graduation) and are related to graduation are farther apart from lyrics that contain the word "sotsugyo" (graduation) but are not related to graduation. In addition, the lyrics related to graduation (red dots and purple dots) are placed close to each other regardless of the presence or absence of the word "sotsugyo" (graduation). Due to the above, guided LDA is better at separating lyrics related to graduation and lyrics not related to graduation.

The scatterplot shown in Figure 6 indicates the lyrics that contain the word "matsuri" (festival) and are related to summer festival by red dots, the lyrics that contain the word "matsuri" (festival) but are not related to graduation by green dots, and the other lyrics by blue dots. In both cases of using LDA and guided LDA, lyrics related to summer festivals and lyrics not related to summer festivals are placed far apart on the scatter plots. Meanwhile, we can see that the distance between the points using guided LDA is further than that using LDA.

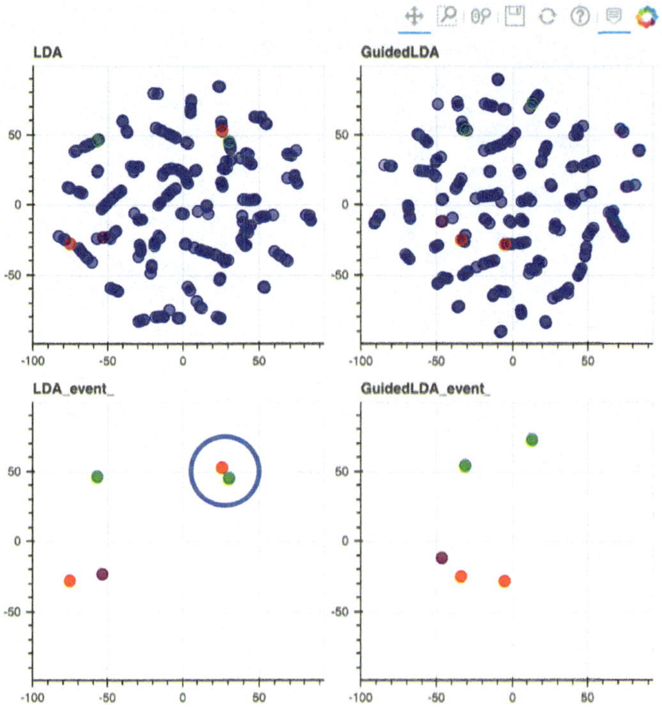

Fig. 5 Visualization example (4). (upper left) Example applying LDA. (upper right) Example applying guided LDA. (lower left) Limited to lyrics including "sotsugyo" (graduation) and related to graduation applying LDA. (lower right) Limited to lyrics including "sotsugyo" (graduation) and related to graduation applying guided LDA

6 Visualization of Difference of Lyrics Between Male and Female Singers

Another interest in our study of the lyrics is how the lyrics of male and songs by female singers differ from each other. To observe the differences in lyrics between male and songs by female singers, we visualized the distribution of lyrics by labeling the songs as male or female.

Here, we labeled the songs in the dataset introduced in Sect. 4 based on the gender of singers before applying LDA. We excluded seven songs whose artists include both male and female singers and labeled 190 songs as songs by male singers, and other 131 songs as songs by female singers.

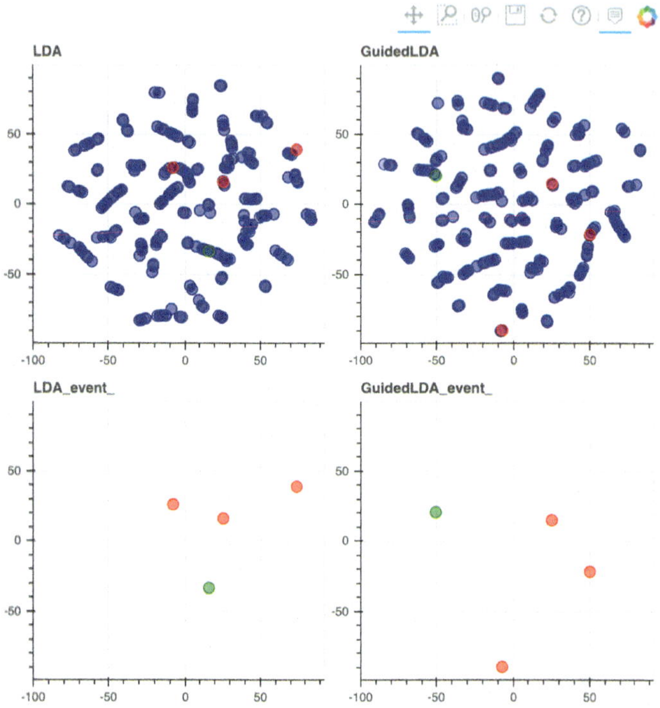

Fig. 6 Visualization example (5). (upper left) Example applying LDA. (upper right) Example applying guided LDA. (lower left) Limited to lyrics including "sotsugyo" (graduation) and related to graduation applying LDA. (lower right) Limited to lyrics including "sotsugyo" (graduation) and related to graduation applying guided LDA

6.1 Topic Analysis of Male and Songs by Female Singers

We observed how male and songs by female singers are categorized by applying LDA to 190 songs by male singers and 131 songs by female singers independently. Here, we fixed the number of topics as $K = 4$ in this case study.

Tables 4 and 5 show the top words of each topic in male and songs by female singers. Here, the mark $*$ in Table 4 indicates the words that appeared as top words only in songs by male singers, while the mark $+$ in Table 5 indicates the words that appeared as top words only in songs by female singers. This result suggests that there are many words that appeared only in the topics of male or songs by female singers exclusively. One of the interesting things is that top words in songs by male singers include male-specific words such as "otoko" (man), while top words in songs by female singers do not include female-specific words.

Table 4 Top words of songs by male singers

Topic M0	ai (love), ikiru (live)*, dare (who), ima (now), mou (already), zutto (throughout), shiru (know), warau (laugh), tsuyoi (string)*, kitto (probably)
Topic M1	yume (dream), kokoro (heart), sora (air), ima (now), kaze (wind), mune (heart)*, sekai (world)*, nani (what), wasureru (forget)*, dare (who)
Topic M2	otoko (man)*, ai (love), natsu (summer)*, koi (love), namida (tear), omae (you)*, yoru (night), aisu (love)*, hoshi (star)*, issho (together)*
Topic M3	ashita (tomorrow)*, China (China)*, mou (already), warau (laugh), hora (hey)*, kyo (today)*, hi (fire)*, ikiru (live)*, hibi (days)*, mawaru (round)*

Table 5 Top words of songs by female singers

Topic F0	ii (good)$^+$, season (season)$^+$, kanji (feeling)$^+$, warau (laugh), koko (here)$^+$, konna (such)$^+$, nani (what), miru (look)$^+$, tension (tension)$^+$
Topic F1	dare (who), hito (human), yume (dream), shiru (know), michi (road)$^+$, doko (where)$^+$, mirai (future)$^+$, kao (face)$^+$, aruku (walk)$^+$, jibun (myself)$^+$
Topic F2	ai (love), zutto (forever), kitto (probably), sora (air), naku (cry)$^+$, namida (tear), dakishimeru (hold)$^+$, mou (already), yoru (night), kaze (wind)
Topic F3	koi (love), ima (now), dare (who), motto (more)$^+$, suki (like)$^+$, kokoro (heart), iu (say)$^+$, ieru (can say)$^+$, kimochi (feeling)$^+$, matsu (wait)$^+$

6.2 *Visualization Examples*

Figure 7 shows the visualization example of the topic analysis and dimensionality reduction. Here, songs by male singers are represented in blue while songs by female singers are in red. This figure indicates specific regions in the scatterplot that we focused on by orange, green and pink circles.

This visualization example shows that there are many groups of closely placed songs. Some of the groups include only songs by male singers while some other groups include only songs by female singers. Also, there are several groups that male and songs by female singers are mixed. The following paragraphs introduce interesting song groups in the visualization example.

Let us focus on a song group indicated by an orange circle in Fig. 7 that includes six male and two songs by female singers. All these songs were love songs. Here, we found that all six songs by male singers include top words categorized in topics M1 or M3. Especially, four of the songs by male singers include "mune" (heart) included only in topic M1 while it did not appear in the top words of songs by female singers. Consequently, this group consists of a similar topic of songs by male singers. Meanwhile, we found that one of the songs by female singers did not include any top words of songs by female singers, which may associate that this song does not sound like a song by a female singer. Another song by another female singer included some top words of songs by female singers, while it also included many words

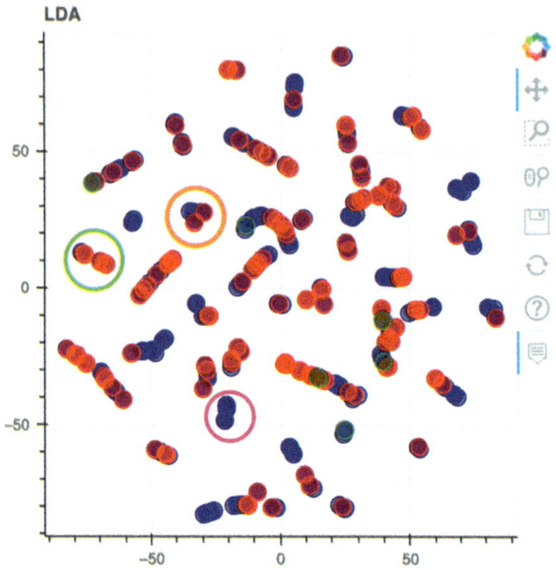

Fig. 7 Visualization of distribution of male/songs by female singers. (1) An orange circle indicates a song group consisting six male and two songs by female singers. (2) A green circle indicates a song group consisting six male and two songs by female singers. (3) A pink circle indecates a song group consisting five songs by male singers

that appeared in the above six songs by male singers, such as "miru" (look), "hito" (human), "ikiru" (live) and "doko" (where), that may associate this song sounds like a make song. Such analysis may assist in understanding the reason for the closeness of topics between male and songs by female singers.

In the second example shown in Fig. 7, a green circle indicates a song group consisting of two male and five songs by female singers. We found that the five songs by female singers included many top words of songs by female singers compared to the two songs by female singers indicated by an orange circle in Fig. 7. All five songs by female singers included the word of the topic F1, and also topics F0 or F3. In other words, all five songs by female singers in this group are quite typical songs by female singers related to multiple topics of songs by female singers. Meanwhile, two songs by male singers in this group were created by the same artist. The two songs included words of the topic F0, F1, and F3. It suggests that the songs in this group are quite related to the topics F0, F1, and F3 and therefore two songs by male singers are mixed in the group.

The third example is a group including only songs by male singers indicated by a pink circle in Fig. 7. Interestingly, each of all five songs by male singers included only one to three top words in the topics M1 or M2, while including some top words in the topics F0 or F1. The result suggests that these five songs by male singers are somewhat gender-neutral because they include a smaller number of male-specific words while including some female-specific words. It is a new discovery for us that there are gender-neutral types of Japanese songs by male singers. Meanwhile, we could not find gender-neutral types of songs by female singers.

These examples suggest that there are a variety of groups of similarly impressed lyrics. This section explained the example including a male-dominant group where songs by female singers did not use female-specific words frequently, a female-dominant group where songs by male singers used female-specific words frequently, and a purely-male group where the songs seem gender-neutral since they did not use either male-dominant or female-dominant words frequently.

7 Future Work

Our future issues are as follows.

The first issue is the handling of lyrics written in multiple languages. Our current implementation omits words in foreign languages and deals with only Japanese words. Since many recent J-POP songs include English phrases, we would like to include the semantics of English phrases in the visualization results. We need to discuss whether to translate them into Japanese lyrics, or whether to treat English words as separate words from Japanese words before extending our implementation.

The second issue is the implementation of user interface functions. The goal of this study is to "obtain visualization results that meet the user's objectives by repeating a series of operations, such as inputting guide words and checking the visualization results," but we have not yet implemented interactive mechanisms to realize this goal. We would like to extend the implementation so that we can interactively display the visualization results of the distribution of lyrics when a guide word is entered.

For the first study applying guided LDA, we would like to develop intelligent methods for selecting guide words. As far as the author experimented, visualization results were not always effective depending on the selection of guide words. We would like to conduct further experiments and discuss what types of guide words are effective to generate better visualization results.

The second study applied two-tone visualization where blue dots depict songs by male singers while red dots depict songs by female singers. Such color representations can be applied to various attributes of the datasets. It is effective to get an overview of the characteristics of the artists/lyricists and the trends of the year or season by assigning the color of each point on the scatterplot corresponding to each song based on meta-information such as artist names, lyricist names, or released year or season. We would like to discover various correlations between lyrics and meta-information by implementing various color representations.

After extending our implementation based on the above issues, we would like to verify the visualization results with a larger number of songs and conduct user evaluation experiments.

8 Conclusion

This chapter proposed a method for quantifying lyric topics using LDA or guided LDA and displaying them as scatterplots by applying dimensionality reduction as a method for visualizing the distribution of lyrics. We introduced our two case studies in this chapter.

The first case study was a comparative study between LDA and guided LDA. Individual viewpoints of users are reflected in the visualizations by selecting guide words while applying guided LDA. This functionality facilitates the interpretation of similarities and differences in lyrics from the personal perspective of the users. Also, the method allows users to create multiple visualization results while selecting guide words. Users can compare the distribution of lyrics from various viewpoints by changing the guide words.

The second case study explored the difference in lyrics between male and songs by female singers by labeling the songs based on the gender of singers before visualizing. There were many groups of similar lyrics in the visualization result and we introduced several examples of interesting groups in Sect. 6. The introduced groups included a male-dominant group including two songs by female singers where female-specific words were not well used, a female-dominant group including two songs by male singers where female-specific words were well used, and a purely-male group where lyrics in this group are somewhat gender-neutral.

References

1. Baur D, Steinmayr B, Butz A (2010) SongWords: exploring music collections through lyrics. In: International symposium on music information retrieval, pp 531–538
2. Blei D, Ng A, Jordan M (2001) Latent dirichlet allocation. In: Advances in neural information processing systems, vol 14
3. Burch M, Fluck T, Freund T, Walzer T, Kloos U, Weiskopf D (2016) Lyrics word clouds. In: 20th international conference information visualisation (IV), pp 51–56
4. Griffiths TL, Steyvers M (2004) Finding scientific topics. Proc Natl Acad Sci 101(suppl 1):5228–5235
5. Hosoya M, Suzuki T (2010) Exploratory analysis of popular songs made by Japanese female singer-songwriters. Jinmonkon 2010, 2010(15):195–202. (in Japanese)
6. Hossain R, Sarker MRKR, Mimo M, Al Marouf A, Pandey B (2019) Recommendation approach of English songs title based on latent dirichlet allocation applied on lyrics. In: 2019 IEEE international conference on electrical, computer and communication technologies (ICECCT), pp 1–4
7. Jagarlamudi J, Daumé III H, Udupa R (2012) Incorporating lexical priors into topic models. In: Proceedings of the 13th conference of the European chapter of the association for computational linguistics, pp 204–213
8. Kawamura Y (2017) A study on music recommendation system considering users' situation based on lyrics information analysis. In: Graduate school research report, science and engineering, vol 47. (in Japanese)
9. Kobayashi Y, Amagasa M, Suzuki T (2015) A chronological variation of lexical indices in the lyrics of popular songs. Jinmonkon 2015(2015):23–30 (in Japanese)

10. Van der Maaten L, Hinton G (2008) Visualizing data using t-SNE. J Mach Learn Res 9(11)
11. Machida W, Itoh T (2011) Lyricon: a visual music selection interface featuring multiple icons. In: 15th international conference on information visualisation (IV2011)
12. Nakai Y, Itoh T (2022) Classification and visualization of lyric collections using guided LDA. In: 26th international conference on information visualisation (IV2022)
13. Oh J (2010) Text visualization of song lyrics. In: Center for Computer Research in Music and Acoustics, Stanford University
14. Ohde A, Matsumoto A, Kaneko T (2013) Lyrics changes of popular songs analyzed by age. Jinmonkon 2013, 2013(4):103–110. (in Japanese)
15. Sasaki S, Yoshii K, Nakano T, Goto M, Morishima S (2014) LyricsRadar: a lyrics retrieval system based on latent topics of lyrics. In: International symposium on music information retrieval, pp 585–590
16. Tsukuda K, Ishida K, Goto M (2017) Lyric jumper: a lyrics-based music exploratory web service by modeling lyrics generative process. In: International symposium on music information retrieval, pp 544–551

Correction to: Artificial Intelligence and Visualization: Advancing Visual Knowledge Discovery

Boris Kovalerchuk, Kawa Nazemi, Răzvan Andonie, Nuno Datia, and Ebad Banissi

Correction to:
B. Kovalerchuk et al. (eds.), *Artificial Intelligence and Visualization: Advancing Visual Knowledge Discovery*, Studies in Computational Intelligence 1126,
https://doi.org/10.1007/978-3-031-46549-9

In the original version of the book, the following belated corrections have been incorporated:

The editor name "Ebad Bannissi" has been changed to "Ebad Banissi" in the Frontmatter.

The book has been updated with the changes.

The updated version of the book can be found at
https://doi.org/10.1007/978-3-031-46549-9

B. Kovalerchuk et al. (eds.), *Artificial Intelligence and Visualization: Advancing Visual Knowledge Discovery*, Studies in Computational Intelligence 1126,
https://doi.org/10.1007/978-3-031-46549-9_19

The manufacturer's authorised representative in the EU is Springer Nature Customer Service Centre GmbH, Europaplatz 3, 69115 Heidelberg, Germany. If you have any concerns regarding our products, please contact ProductSafety@springernature.com

Printed and bound by CPI Group (UK) Ltd, Croydon, CR0 4YY
30/09/2026
02225435-0014